Springer Advanced Texts in Chemistry

Charles R. Cantor, Editor

SPRINGER STUDY EDITION

W.A. Cramer   D.B. Knaff

Energy Transduction in Biological Membranes

A Textbook of Bioenergetics

With 218 Illustrations in 351 Parts

Springer-Verlag
New York Berlin Heidelberg London
Paris Tokyo Hong Kong Barcelona

William A. Cramer
Department of Biological Sciences
Purdue University
West Lafayette, Indiana 47907
USA

David B. Knaff
Department of Chemistry and Biochemistry
Texas Tech University
Lubbock, Texas 79409
USA

Series Editor:
Charles R. Cantor
Human Genome Center
Lawrence Berkeley Laboratory
1 Cyclotron Drive
Berkeley, California 94720
USA

Library of Congress Cataloging-in-Publication Data
Cramer, W. A. (William A.)
  Energy transduction in biological membranes: a textbook of
bioenergetics / W. A. Cramer and D. B. Knaff.
   p. cm—(Springer advanced texts in chemistry)
  ISBN 0-387-97533-0
  1. Membranes (Biology) 2. Bioenergetics. 3. Energy transfer.
 I. Knaff, D. B. (David B.) II. Title. III. Series.
QH601.C73   1989
574.87'5—dc20   89-6107

Printed on acid-free paper.

This Springer-Verlag Study Edition was originally published as a hard cover edition.
© 1991 by Springer-Verlag New York Inc.
All rights reserved. This work may not be translated or copied in whole or in part without the written permission of the publisher (Springer-Verlag, 175 Fifth Avenue, New York, NY 10010, USA), except for brief excerpts in connection with reviews or scholarly analysis. Use in connection with any form of information storage and retrieval, electronic adaptation, computer software, or by similar or dissimilar methodology now known or hereafter developed is forbidden.
The use of general descriptive names, trade names, trademarks, etc., in this publication, even if the former are not especially identified, is not to be taken as a sign that such names, as understood by the Trade Marks and Merchandise Marks Act, may accordingly be used freely by anyone.

Typeset by Caliber Design Planning, Inc.
Printed and bound by Edwards Brothers, Inc., Ann Arbor, Michigan.
Printed in the United States of America.

9 8 7 6 5 4 3 2 1

ISBN 0-387-97533-0 Springer-Verlag New York Berlin Heidelberg (pbk.)
ISBN 3-540-97533-0 Springer-Verlag Berlin Heidelberg New York

ISBN 0-387-96761-3 Springer-Verlag New York Berlin Heidelberg (hbk.)
ISBN 3-540-96761-3 Springer-Verlag Berlin Heidelberg New York

Series Preface

New textbooks at all levels of chemistry appear with great regularity. Some fields like basic biochemistry, organic reaction mechanisms, and chemical thermodynamics are well represented by many excellent texts, and new or revised editions are published sufficiently often to keep up with progress in research. However, some areas of chemistry, especially many of those taught at the graduate level, suffer from a real lack of up-to-date textbooks. The most serious needs occur in fields that are rapidly changing. Textbooks in these subjects usually have to be written by scientists actually involved in the research which is advancing the field. It is not often easy to persuade such individuals to set time aside to help spread the knowledge they have accumulated. Our goal, in this series, is to pinpoint areas of chemistry where recent progress has outpaced what is covered in any available textbooks, and then seek out and persuade experts in these fields to produce relatively concise but instructive introductions to their fields. These should serve the needs of one semester or one quarter graduate courses in chemistry and biochemistry. In some cases, the availability of texts in active research areas should help stimulate the creation of new courses.

CHARLES R. CANTOR

Preface

This book presents the one-semester graduate course in bioenergetics taught in the Department of Biological Sciences at Purdue University. The course consists of 45 lectures (three/week). It has obviously not been possible to cover in the course all of the material presented in this text. Additional material is presented here to allow a choice of topics. In particular, the topic of soluble iron-sulfur proteins in Chapter 4 is expanded in this textbook compared to that which has been taught, using material from the course taught at Texas Tech University. The presentation of active transport (Chapter 9, through section 14) is also from the latter course.

The minimum prerequisites for the course at Purdue are advanced undergraduate biochemistry and one semester of physical chemistry. The students taking this course have come from the interdepartmental biochemistry and plant physiology programs, and the Department of Biological Sciences. Although these students have studied thermodynamics to some extent, most need a review of the concepts. This is one of the purposes of Chapter 1, which has been used to provide a terse review. For this purpose, it relies heavily on many homework problems, assigned in the early part of the course. Given the quantitative nature of many of the concepts and experimental techniques, the philosophy of the course is to use such homework problems throughout as a study guide. This has also been done because many students do not have enough opportunity in other biology and biochemistry courses to apply concepts of physical chemistry.

The discussion in Chapter 1 is directed toward the concepts of the chemical and electrochemical potential, μ_{H^+} and $\tilde{\mu}_{H^+}$, which are the central thermodynamic parameters in a discussion of energy transduction in membranes. Appendix III on calculations of membrane protein hydrophobicity is discussed after Chapter 1 because of its subsequent importance in discussions of membrane structure. Many students have had some introduction to oxidation–reduction potentials (Chapter 2), but have usually not had occasion to consider the concepts in depth, nor to apply them to physiological electron transport reactions that involve mixed one and two electron transfers, pH-dependent and -independent midpoint potentials, and coupling to ATP synthesis. The retention of these concepts is aided by requiring that the students work problems on the application to electron and proton transport systems. Chapter 2 also contains an introduction to the phenomena of long-range electron transfer and to theories and discussion of electron and proton transfer. These first two chapters and the beginning of Chapter 6 on radiant energy transfer contain most of the theoretical background in the book. The experimental basis of the central concept of the proton electrochemical potential is discussed in Chapter 3. In this chapter and in most of the book, the concepts and experimental examples are chosen from chloroplast, mitochondrial, and bacterial membrane systems. The purpose is to present a comparative discussion of the energy transduction problem using data from all of these membrane systems. This approach emphasizes common structural, biochemical, and conceptual motifs, and has been found to be useful in teaching graduate students from varied backgrounds.

Chapters 4–7 discuss the components and pathways for electron transport and proton translocation: Chapter 4 (metalloproteins), Chapter 5 (the quinone connection between electron transport complexes), and Chapter 6 (resonance energy transfer, unique aspects of photosynthesis); chapter 7 (light- and redoxlinked H^+ translocation) is focused on the bacteriorhodopsin, cytochrome oxidase, and cytochrome $b-c_1$ complexes. Chapters 8 and 9 consider the transduction of electrochemical ion gradients with emphasis on transduction of the $\Delta\tilde{\mu}_{H^+}$ to ATP synthesis (Chapter 8) and active transport and protein translocation (Chapter 9). The course finishes with a discussion of the energy requirements for protein translocation (Chapter 9). A one-hour midterm exam has been given to the students at the end of Chapter 4, and a two-hour comprehensive final exam at the end of the course.

The recent advances in molecular and structural biology have had perhaps as great an effect on the field of energy transduction as on any field of biochemistry. The reasons are (1) bioenergetics was previously dominated by spectroscopic data that were crucial, but which often made communication difficult with other areas of biochemistry. The language of molecular biology is universal and its application to chloroplasts, mitochondria, and *E. coli* has facilitated information exchange. (2) The other reason is that it is precisely in the area of energy transducing membranes that studies on the structural biology of membrane proteins have thus far had their greatest successes. The high-resolution analysis of the bacteriorhodopsin molecule by electron diffraction and image reconstruction and the solution of the structure of the bacterial photosynthetic reaction center of *Rps. viridis* by X-ray diffraction analysis have been milestones, making photosynthetic and bioenergetic membrane research the most advanced in the field of membrane structural biology. The significance of these studies was documented by the Nobel Prize in Chemistry for 1988 being awarded to the group at the Max Planck Institute in Martinsried (FRG) who achieved the *Rps. viridis* structure.

The decisions on references for the text made clear to the authors the difference between a textbook and a review article. One can always recognize the contribution of more than one, and often many, research groups, to an experiment or concept. However, because of the nature of a textbook, and also space limitations, we often limited ourselves to one or two references on a subject. These references were chosen through the criteria of students having found them to be accessible and informative, and often include review or overview articles. This method of discussion and referencing emphasizes the logic and flow of concepts, but does not necessarily convey the number of contributors to these subjects nor the historical development of concepts and experiments. It is hoped that the reference(s) supplied in each case will provide an entry point to the complete literature on the particular experiment or concept.

The research of the authors described in the text has been supported by the NIH, NSF, and USDA (W.A.C.), and the NSF, USDA, and Robert A. Welch Foundation (D.B.K.). Grateful thanks to the students who helped to shape the course and text through their critical comments, many colleagues who contributed material for figures and/or provided helpful criticisms of sections of the text, and to postdoctoral and graduate students for much input, particularly from Mick Black, Paul Furbacher, and Bill Widger. The project could not have been finished without the support, patience, and hard work of Hanni Cramer, Janet Hollister, and Sheryl Kelly (Purdue), and Phyllis Huckabee and Judy Leuty (Texas Tech).

West Lafayette, Indiana   William A. Cramer
Lubbock, Texas   David B. Knaff

Contents

Series Preface	v
Preface	vii
Part I Principles of Bioenergetics	1

Chapter 1
Thermodynamic Background   3

1.1	Introduction: The First Law of Thermodynamics	3
1.2	Reaction, Direction, Disorder: The Need for the Second Law	4
1.3	On Entropy and the Second Law of Thermodynamics	6
1.4	Maximum Work	10
1.5	Free Energy	11
1.6	Concentration Dependence of the Gibbs Free Energy	13
1.7	Free Energy Change of a Chemical Reaction	14
1.8	Temperature Dependence of K_{eq}	15
1.9	Other Kinds of Work: Electrical, Chemical Work	16
1.10	Thermodynamics of Ion Gradients	18
1.11	Thermodynamics of $\Delta \bar{\mu}_{H^+}$-Linked Active Transport	19
1.12	Thermodynamics of $\Delta \bar{\mu}_{H^+}$-Linked ATP Synthesis	21
1.13	Nonequilibrium Thermodynamics	22
1.14	"High-Energy" Bonds	23
1.15	Summary	30
	Problems	31

Chapter 2
Oxidation–Reduction; Electron and Proton Transfer   35

2.1	Direction of Redox Reactions	35
2.2	The Scale of Oxidation–Reduction Potentials	36

2.3	Oxidation–Reduction Potential as a Group-Transfer Potential; Comparison of Standard Potentials and pK Values	37
2.4	Calculation of the Potential Change for Linked and Coupled Reactions	38
2.5	Concentration Dependence of the Oxidation-Reduction Potential	39
2.6	Experimental Determination of E and E_m Values	47
2.7	Factors Affecting the Redox Potential	48
2.8	Redox Properties of Quinones and Semiquinones	52
2.9	Midpoint Potentials of Electrons in Photo-Excited States: Application to Photosynthetic Reaction Centers	55
2.10	Electron Transfer Mechanisms	57
2.11	Proton Transfer Reactions	71
2.12	Summary	74
	Problems	75

Chapter 3
Membrane Structure and Storage of Free Energy 78

3.1	Elements of Membrane Structure	78
3.2	Introduction to the Energy Storage Problem	92
3.3	The Chemiosmotic Hypothesis	93
3.4	Measurement of ΔpH and $\Delta\psi$ Across Energy-Transducing Membranes	94
3.5	Relationship Between $\Delta\psi$ and Charge Movement Across the Membrane	103
3.6	Experimental Tests of the Chemiosmotic Hypothesis	103
3.7	A Naturally Occurring Uncoupler: The Uncoupling Protein from Brown Fat Mitochondria	115
3.8	Effect of Uncouplers on Electron Transport Rate	117
3.9	Proton Requirement (H^+/ATP) for Reversible ATP Synthase	118
3.10	Storage of Energy in $\Delta\tilde{\mu}_{Na^+}$	123
3.11	Sufficiency of the Chemiosmotic Framework	124
3.12	Appendix. Ionophores	130
3.13	Summary	135
	Problems	137

Part II Components and Pathways for Electron Transport and H^+ Translocation 139

Chapter 4
Metalloproteins 141

4.1	Heme Proteins, Cytochromes a through d, and o	141
4.2	Occurrence of b Cytochromes	151
4.3	Structure of Cytochrome c	152
4.4	Structure–Function in Mitochondrial Cytochrome c	158
4.5	Residues of Reaction Partners That Are Complementary to Cytochrome c Lysines	162
4.6	Diffusion and Orientability of Cytochrome c	164
4.7	Membrane-Bound c-Type Cytochromes: Cytochromes c_1 and f	165
4.8	Copper Proteins: Plastocyanin	169
4.9	Iron-Sulfur Proteins	171
4.10	Membrane-Bound Iron-Sulfur Proteins	181

Contents

4.11	The Membrane-Bound FeS-Flavoprotein, Succinate: Ubiquinone Oxidoreductase (Complex II)	187
4.12	Summary	189
	Problems	192

Chapter 5
The Quinone Connection 193

5.1	Structures, Stoichiometry, Pools, and Branch Points	193
5.2	Reconstitution of Quinone Function Requires Q_n with $n \geq 3$	198
5.3	The Quinone Pool Is Located Near the Center of the Membrane Bilayer	198
5.4	The Quinone Connection Across the Center of the Membrane	199
5.5	Quinone Lateral Mobility	204
5.6	The Segregation of Electron Transport Components in Thylakoids Requires Lateral Mobility of Quinone	208
5.7	Quinone-Binding Proteins	217
5.8	Quinone Electron Acceptors in Photosynthetic Reaction Centers	223
5.9	Quinone-Binding Proteins in Photosynthetic Reaction Centers	226
5.10	Summary	235
	Problems	237

Chapter 6
Photosynthesis: Photons to Protons 239

6.1	Light Energy Transfer	239
6.2	Use of Energy Transfer as a Spectroscopic Ruler	243
6.3	Light Energy Transfer in Photosynthesis: The Phycobilisome	248
6.4	Structures of Photosynthetic Antenna Pigment-Protein Complexes	252
6.5	Structure of Photosynthetic Reaction Centers	264
6.6	Structure of the Reaction Center Proteins: Transmembrane Charge Separation	266
6.7	Reaction Centers of Plant and Algal PS I and II	277
6.8	Photosynthetic Water Splitting, O_2 Evolution, and Proton Release by PS II	281
6.9	The Cyclic and Noncyclic Electron Transfer Chains	291
6.10	Summary	295
	Problems	297

Chapter 7
Light and Redox-Linked H^+ Translocation: Pumps, Cycles, and Stoichiometry 299

7.1	Introduction	299
7.2	Bacteriorhodopsin, a Well-Characterized Light-Driven H^+ Pump	299
7.3	Cytochrome Oxidase (Mitochondrial Complex IV) as a Proton Pump	311
7.4	The Q Cycle and H^+ Translocation in Complex III and Chloroplast $b_6 f$ Complexes	326
7.5	H^+ Translocation or Deposition Sites in the Mitochondrial, Chromatophore, and Chloroplast Electron Transport Chains; Stoichiometries of H^+ Translocation and ATP Synthesis	346
7.6	Summary	349
	Problems	351

Part III Utilization of Electrochemical Ion Gradients 353

Chapter 8
Transduction of Electrochemical Ion Gradients to ATP Synthesis 355

8.1	Introduction to the Structure and Function of the ATP Synthase	355
8.2	Preparation of H^+-ATPase	360
8.3	Structure of F_0F_1 ATP Synthase	360
8.4	DNA Sequence of *Unc* Operon	368
8.5	Function of the Membrane-Bound Subunits a, b, and c	373
8.6	Mechanism of ATP Synthesis	378
8.7	Thermodynamic and Kinetic Constants for ATP Hydrolysis	383
8.8	Mechanism of Transduction of $\Delta\tilde{\mu}_{H^+}$ to ATP	388
8.9	Other Classes of H^+-Translocating ATPases	390
8.10	Summary	402
Problems		404

Chapter 9
Active Transport 406

9.1	Introduction	406
9.2	Evidence for Protein Carrier-Mediated Transport	407
9.3	Techniques for Studying Transport in Bacteria	408
9.4	Structure of the Cell Envelope of Gram-Negative Bacteria	409
9.5	$\Delta\tilde{\mu}_{H^+}$ Formation in Bacteria	410
9.6	Active Transport of Sugars Coupled to H^+ Cotransport	412
9.7	Kinetic Studies	419
9.8	Structure/Function Considerations	423
9.9	Amino Acid Transport	427
9.10	Sodium-Dependent Transport	428
9.11	Transport Driven by High-Energy Phosphate Intermediates	435
9.12	Periplasmic Transport Systems	440
9.13	Motility	447
9.14	Active Transport in Eukaryotes	449
9.15	Transport or Translocation of Macromolecules	455
9.16	Summary	465

Appendix I
Answers to Problems 466

Appendix II
Physical, Chemical, and Biochemical Constants 468

Appendix III
Prediction of Protein Folding in Membranes 470

References 475

Glossary of Abbreviations 535

Index 537

Preface to Solutions Section 547

Solutions to Homework Problems 549

Errata 577

PART I Principles of Bioenergetics

Chapter 1
Thermodynamic Background

1.1 Introduction: The First Law of Thermodynamics

Thermodynamics describes physical and chemical phenomena in terms of macroscopic properties of matter that are obvious to our senses such as pressure, temperature, and volume. These phenomena are divided into the part being studied, the *system*, and the region around the system that interacts with it called the *surroundings*. The state of a system at a given time is determined by the values of its macroscopic properties which are of two kinds, *intensive* properties such as the pressure, temperature, density, or chemical potential, which are independent of the system size, and *extensive* properties which include mass, volume, and energy. For a single substance, the Gibbs phase rule shows that one extensive property (e.g., mass) and two intensive properties such as pressure and temperature are sufficient to define the state of the system. The energy, E, is a macroscopic property of the system that is useful because (i) it describes the ability of a system to do *work*, (ii) its value is unique for a given state, and (iii) its change, ΔE, between two states is independent of the pathway taken between the states. Because of the latter two properties, energy is called a *state function*. In thermodynamics, energy is divided into the two categories of work and heat, neither of which is a state function because their changes are dependent on the pathway between states. Work is the energy change accomplished by ordered or coherent molecular movement, for example, by the pressure causing a change of volume (mechanical work), matter moving across a concentration gradient (chemical or osmotic work), or electrons moving between two different oxidation potentials (electrochemical work). Energy changes arising from heat result from random molecular motion whose net flow is directed from matter at higher to that at lower temperature.

The *first law of thermodynamics* is well known as the law of conservation of energy: *The total energy of the system and surroundings does not change.* Applied to the system of interest, it also states that addition of heat (ΔQ) and work (ΔW) to the system that have been withdrawn from the surroundings must be reflected in a change of the energy ($\Delta \mathscr{E}$) of the system:

$$\Delta \mathscr{E} = \Delta Q + \Delta W, \tag{1}$$

and for infinitesimal changes

$$d\mathscr{E} = dQ + dW. \tag{1a}$$

The pressure–volume work accomplished by expansion or compression of a gas at pressure, P, through volume change, dV, is

$$dW = -p\,dV, \tag{1b}$$

where dW is positive for work done on the system.

A state function, H, called *enthalpy*, is defined as $H \equiv \mathscr{E} + pV$. This function is useful because, under conditions of constant pressure when the only work done is pressure–volume work (i.e., no chemical, electrical work), the change in enthalpy, ΔH, equals the absorption of heat at constant pressure, ΔQp ($\Delta H = \Delta Q_p$). This can be shown by simple manipulation of the variables in the above statement (1) of the First Law (problem 1).

1.2 Reaction Direction, Disorder: The Need for the Second Law

The first law of thermodynamics arose as a result of the impossibility of constructing a device that can create energy. However, the first law places no limitations on the possibility of transforming energy from one form to another, in particular heat energy to work. The interconvertibility of work and heat is known to be asymmetric in nature. We know from experience that work can always be converted to heat. There are definite limitations, however, to the possibility of converting heat into work. It is not possible to construct a machine that can create work by simply cooling the environment. One statement of the second law, attributed to the 19th century physicists Clausius and Kelvin is: *It is impossible to have a physical or chemical process whose only result is* (i) *to transform into work heat extracted from a source at a single temperature, or* (ii) *to transfer heat from an object at a given temperature to one at a higher temperature.* As a result of the second law, it can be concluded that any heat engine producing work by cycling through different states must include at least two states at different temperatures for the uptake and exchange of heat.

The first law also does not provide a definition of equilibrium or make any prediction about the spontaneity or direction of reactions in the presence of

1.2 Reaction Direction, Disorder: The Need for the Second Law

thermal motion. At temperatures low enough so that heat and molecular motion can be neglected, the equilibrium state of a system is determined by the potential energy ϕ of the reacting quantities decreasing until it is a minimum ($d\phi \leq 0$). In all of these processes energy is conserved (the first law) and, as the potential energy decreases, heat or light may be given off. The first law of thermodynamics and the criterion for equilibrium, that potential energy decreases to a minimum, does not explain many intuitively obvious phenomena and experiments that occur at higher temperatures where molecular motion is significant. For example, it does not explain the phenomenon of evaporation. Because the binding forces between liquid molecules are much stronger in the liquid than in the vapor state, why should the liquid in an open vessel evaporate? The tendency toward evaporation is a result of the statistical propensity of the molecules to fill all available space. One can readily calculate that it is very unlikely in the absence of constraining forces that a large number of molecules would remain confined to a small fraction of the available volume to them.

For example, the probability that one molecule would be found in one half of a box with volume, V, is $(1/2V)/V = 1/2$. The probability that n molecules would all be found in that half is $(1/2)^n$. If n is about 10^{23} for a real box, this event will never be observed. The most likely state is, intuitively, that in which the particles move freely about the box, and at any given time are found to be spread evenly about the box, so that the number of distinct rearrangements of the particles in the box tends toward a maximum. This can be stated mathematically in terms of probabilities. The statement that a population of particles, atoms, or molecules will spontaneously (in the absence of external forces) move toward a state of maximum disorder is the same as that which says mathematically that it moves to a state of maximum probability. This is the statistical view of the second law of thermodynamics.

Exercise 1.1. **Calculate the number of rearrangements, Ω, of identical particles on the two sides, j and k, of a box for four particles, $N = 4(a, b, c, d)$. How many ways, Ω, are there of placing N_j on one side and N_k on the other, with the two sides having the same volume, and each arrangement of the same number of particles in compartment j or k weighed equally?**

$$\text{The answer is: } \Omega = \frac{N!}{N_j! N_k!}, \tag{2}$$

where $N! = N(N-1)(N-2)\ldots(1)$.

For $N_j = 3$, $N_k = 1$, $\Omega = \dfrac{4!}{3! 1!} = 4$, as seen by:

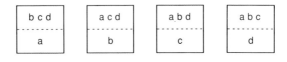

The number of rearrangements with $N_j = N_k = 2$ is $\Omega = \dfrac{4!}{2!2!} = 6$.

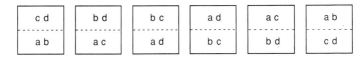

Figure 1.1. Description of two possible arrangements of four identical particles between two sides of a box.

The value of Ω, $\Omega_{N/2, N/2}$ for equal numbers on both sides becomes much greater than the value for unequal numbers as the particle number itself becomes larger. Therefore the relative probability of the state with $N_j = N_k$ becomes increasingly greater with increasing N, relative to the state with N_j very different from N_k, and becomes overwhelmingly more probable when the particle number is macroscopic, i.e., ca. 10^{23}. Diagrams of other spontaneous processes (e.g., shuffling of playing cards) are shown in Chap. 2 of Eisenberg and Crothers (1979).

1.3 On Entropy and the Second Law of Thermodynamics

The entropy (from Gr., *tropē*, change, turn) is a state function whose change in a reaction occurring in the absence of forces and interactions describes the probability that the reaction will go forward (Mahan, 1964; Hill, 1966; Eisenberg and Crothers, 1979). Entropy is a quantity that depends on molecular motion and heat. There are two ways to describe the entropy function, S. First,

$$S = k \ln \Omega, \qquad (3)$$

where $\mathbf{k}$ = Boltzmann's constant = 1.38×10^{-23} J/K, in the statistical formulation due to Boltzmann. The thermodynamic probability function Ω associated with a bulk or macroscopic state is equal to the number of microscopic states of which it is composed, as described in section 1.2. Ω has been converted in Eq. 3 to a logarithmic function. Consequently, the product $\Omega_1 \cdot \Omega_2$ that describes the joint occurrence of two states characterized separately by probabilities or arrangement numbers, Ω_1 and Ω_2, is converted to a sum, $\ln(\Omega_1 \cdot \Omega_2) = \ln \Omega_1 + \ln \Omega_2$. This property of additivity, the same as

1.3 On Entropy and the Second Law of Thermodynamics

that of the state function, energy, confers the properties of a state function to the entropy, S. Then,

$$\Delta S = S(2) - S(1) = k \ln\left(\frac{\Omega_2}{\Omega_1}\right), \tag{3a}$$

In the second formulation of the entropy change due to Carnot, the change dS is proportional to the heat absorbed, dQ_{rev}, for a reversible process. $1/T$ [(absolute temperature)$^{-1}$] is a constant of proportionality and an integrating factor for dQ that makes dQ/T a perfect differential, as required for a state function (Chandrasekhar, 1957). Thus,

$$dS = \frac{dQ_{rev}}{T} \tag{4}$$

and

$$\Delta S = \int_1^2 \frac{dQ_{rev}}{T} = S(2) - S(1). \tag{5}$$

The relationship between the above two expressions (Eqs. 3 and 4) for entropy is illustrated by an example: Consider particles initially confined to one side of a box by a piston (Fig. 1.2A) or a sliding partition (Fig. 1.2B):

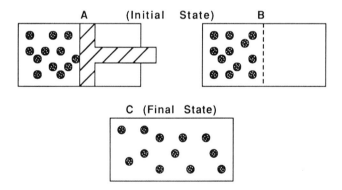

Figure 1.2. Experiment on gas expansion in a chamber carried out in two ways. Initial state for (A) a reversible expansion that can do work by driving a piston, or (B) for an irreversible expansion initiated by raising a partition across the center of the chamber. (C) An identical final state is achieved after experiments (A) and (B). Final volume = twice initial volume. Temperature is kept constant.

The entropy will change to the final state, in which particles fill the box, through two paths, either by pushing the piston or by removing the partition. The final state achieved through either path is shown in Fig. 1.2C. The entropy change can be shown to be $R \ln 2$ for both transitions, independent of path as required for a state function.

Exercise 1.2. Calculate the entropy change for the two transitions described in Fig. 1.2.

In the first experiment (Fig. 1.2A) one can reverse the change by driving the piston backwards by adding heat to the system. The temperature is held constant by placing the system in contact equilibrium with a heat sink. The change in piston position and molecular distribution is a result of nothing else than the absorption of heat. From Eq. 4, the entropy change from initial (Fig. 1.2A) to final (Fig. 1.2C) state would be:

$$dS = \frac{dQ_{rev}}{T},$$

for an incremental change of added heat. From the first law of thermodynamics (energy conservation), $dE(T) = dQ + dW$, i.e., the change in energy is equal to the heat put in plus the work done on the system. E is a function only of T for a dilute gas in which intermolecular interactions are neglected. Since the temperature is constant,

$$d\mathscr{E}(T) = 0,$$

and

$$dQ = -dW$$

from Eq. 1, and

$$dS = \frac{-dW}{T} = \frac{pdV}{T},$$

from Eqs. 1b and 4.

For the dilute gas,

$$pV = RT$$

for 1 mol; then,

$$dS = \frac{RT}{T} \cdot \frac{dV}{V}$$

$$= R\frac{dV}{V},$$

and

$$\Delta S = R \int_{V_1}^{V_2} \frac{dV}{V} = R \ln \frac{V_2}{V_1} = R \ln 2,$$

since

$$V_2 = 2V_1 \quad (\text{Fig. 1.2}).$$

The second method of calculating the entropy change for the change in particle rearrangements shown in Fig. 1.2 is to calculate the number of ways

1.3 On Entropy and the Second Law of Thermodynamics

of distributing the particles in the two compartments after the partition is removed to allow a redistribution of particles into the entire volume of the box. For a large number of particles, on the order of a mole, this redistribution can never be reversed. If the number of particles is N, and there are N_j and N_k in each compartment, then the relative probability of having a given number N_j and N_k on each side of the partition is proportional to the number of different ways $W_{j,k}$, in which this may occur (Eq. 2):

$$W_{j,k} = \frac{N!}{N_j! N_k!}.$$

For the problem in Fig. 1.2A, the initial state has all N particles on one side before removal of the partition. Therefore,

$$W_{j,k} = W_{N,0} = \frac{N!}{N! 0!} = W_1$$

$$\equiv 1.$$

(the symbol $\equiv$ means, "equivalent by definition"). After the partition is removed, there are $N/2$ particles on each side if N is a large number. Then

$$W_{j,k} = W_{N/2, N/2} = \frac{N!}{\frac{N!}{2} \frac{N!}{2}} = W_2.$$

From the Boltzmann expression for S,

$$\Delta S = k \ln \frac{W_2}{W_1}$$

$$= k \ln \frac{N!}{\frac{N!}{2} \frac{N!}{2}},$$

since $W_1 = 1$. The solution uses an approximation for large values of N,

$$\ln N! \simeq N \ln N - N. \tag{6}$$

Then,

$$\Delta S = k \left[N \ln N - N - \frac{2N}{2} \ln \frac{N}{2} + N \right]$$

$$= kN \ln \frac{N}{N/2} = kN \ln 2;$$

the gas constant $R \equiv N\mathbf{k}$ if N is the number of particles in a mole (Avogadro's number, 6.02×10^{23}), and,

$$\Delta S = R \ln 2$$

for 1 mol of particles.

Example 2 considered an entropy change that was the same when calculated according to the view of (i) Carnot or (ii) Boltzmann. These views described a reversible change involving the input of heat and the performance of work, or an irreversible change involving a redistribution of particles with $dQ = 0$ and $dS > 0$. The finding that ΔS was independent of the two paths chosen in this problem supports the statement that entropy is a state function.

In general, reactions have irreversible components so that

$$dS \geq \frac{dQ_{rev}}{T} \qquad (7)$$

for a reversible or irreversible transition, and

$$dS \geq 0 \qquad (8)$$

for the case $dQ = 0$ (e.g., the case of Fig. 1.2B). These two expressions summarize the second law of thermodynamics mathematically, for which an alternative statement to that on p. 4 would be: *There is a state function, S, called entropy. In an irreversible process the entropy of the system and surroundings will increase, whereas it will remain constant in a reversible process. The total entropy of a system and surroundings never decreases.* In the absence of external forces ($dQ_{rev} = 0$), reaction directions are determined by changes toward more probable particle distributions corresponding to positive changes of ΔS.

1.4 Maximum Work

Maximum work is done by a system in reversible transitions where the entropy changes are solely due to heat exchange and not particle rearrangements that increase entropy without heat exchange. This follows from the first law and the fact that entropy is a state function:

$$\Delta \mathscr{E} = \Delta Q + \Delta W,$$

where ΔW is positive for the work done on the system. For reversible and irreversible changes,

$$\Delta \mathscr{E}_{rev} = \Delta Q_{rev} + \Delta W_{rev},$$

and

$$\Delta \mathscr{E}_{irrev} = \Delta Q_{irrev} + \Delta W_{irrev}.$$

Since energy is a state functon, $\Delta \mathscr{E}_{rev} = \Delta \mathscr{E}_{irrev}$ for a transition between the same states. Then

$$\Delta Q_{rev} + \Delta W_{rev} = \Delta Q_{irrev} + \Delta W_{irrev}.$$

$$\Delta Q_{rev} - \Delta Q_{irrev} = \Delta W_{irrev} - \Delta W_{rev}.$$

1.5 Free Energy

Because entropy is a state function

$$\Delta S_{rev} = \Delta S_{irrev},$$

where

$$\Delta S_{rev} = \frac{\Delta Q_{rev}}{T},$$

and

$$\Delta S_{irrev} > \frac{\Delta Q_{irrev}}{T}.$$

It follows that $\Delta Q_{rev} > \Delta Q_{irrev}$, if T is constant (isothermal). Therefore, at constant T, $\Delta W_{irrev} > \Delta W_{rev}$, and more (positive) work must be done (on the system) to obtain the same result in an irreversible transition. For work done by the system that is negative, it follows that

$$|\Delta W_{rev}| > |\Delta W_{irrev}| \qquad (9)$$

and more work is done by a system in a reversible, compared to an irreversible, transition (Problem 2). Discussion of the properties of theoretical (e.g., Carnot) heat engines for generating work can be found in many texts on thermodynamics and physical chemistry.

1.5 Free Energy

The inequality $\Delta S \geq \Delta Q/T$ specifies the direction of reactions in which only the heat change and temperature are specified. Inequalities specifying the direction of chemical reactions under other conditions are useful: From the first law, $dE = dQ + dW$, where dW can include pressure–volume, chemical, and electrical work (i.e., dW_{p-V}, dW_{chem}, $dW_{elect.}$).

If $dW = dW_{pV} = -p\,dV$ for pressure–volume work, then,

$$d\mathscr{E} = dQ - p\,dV;$$
$$dQ = d\mathscr{E} + p\,dV$$
$$= d(\mathscr{E} + PV)_p, \qquad \text{at constant } p,$$

where $\mathscr{E} + PV \equiv$ enthalpy $\equiv H$, the heat change at constant pressure. Then, $dQp = dH$.

From the second law, $dS \geq dQ/T$,

$$TdS \geq dH, \qquad \text{by substitution,}$$

and

$$dH - TdS \leq 0. \qquad \text{Rewriting,}$$
$$d(H - TS)_{p,T} \leq 0.$$

The Gibbs Free Energy, G, is defined as

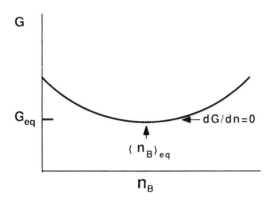

Figure 1.3. Dependence of free energy, G, on the molar concentration, n_B, of product B when the total concentration of reactants and products, $n_B + n_A$, is constant. Equilibrium occurs when G is at a minimum, determined by $dG/dn_i = 0$.

$$G \equiv H - TS; \qquad dG_{p,T} = dH - TdS \qquad (10)$$

The second law of thermodynamics is then reformulated to describe reactions at constant temperature and pressure:

$$dG_{p,T} \leq 0, \quad \text{with only } p\text{-}V \text{ work.} \qquad (11)$$

The reaction is at equilibrium and reversible if

$$dG_{p,T} = 0. \qquad (11a)$$

This is shown in a plot of the free energy of a hypothetical reaction, $A \to B$, as a function of the ratio of the number of moles of A, n_A, to those of B, n_B (Fig. 1.3).

At equilibrium, the free energy, G, is at a minimum and the slope of G is zero ($dG = 0$) for small displacements of G in either direction. At all points on this graph away from equilibrium, the second law says that the tendency of G will be to decrease to the minimum value shown on the graph. That is,

$$dG < 0.$$

This expression is exactly analogous to that for the approach to equilibrium in a mechanical problem described by the potential energy function, ϕ. Free energy changes are thought of most easily when separated into ΔH and $T\Delta S$ terms. At low temperatures the ΔH term dominates, and at high temperatures the sign of the ΔG change and the direction of the reaction is mainly determined by the $T\Delta S$ term. Evaporation of a liquid or the folding and unfolding of a protein molecule are typical of problems considered in chemistry or biochemistry. In these cases the ΔH term arises from the making and breaking of bonds, and the ΔS term from changes in the number of spatial configurations of the molecules, or range of velocities available to the molecular populations. One needs the $T\Delta S$ term to explain why liquids evaporate because the ΔH term alone, bond energy in the liquid state, would tend to prevent it. However, water molecules will tend to occupy the space above the liquid where there is more volume available per molecule than in the liquid (i.e., ΔS is positive for liquid $\to$ vapor transition).

1.6 Concentration Dependence of the Gibbs Free Energy

To isolate the dependence of the free energy, G, on pressure, p, and concentration, c, one can write $G \equiv \mathscr{E} + pV - TS$, using $H \equiv E + pV$. Then, $dG = Vdp - SdT$ for a reversible transition involving only p–V (expansion work) when pressure and temperature are not held constant. If T = constant, then $dG = Vdp$. Integration of this expression, with the assumption that the dependence of G on gas pressure will be the same as its dependence on solute concentration for dilute solutions, yields

$$G = G° + nRT \ln c, \tag{12}$$

with $G°$ the standard free energy corresponding to a concentration, $c = 1$ molar, so that the logarithm of the concentration term is zero.

Exercise 1.3. Derive formula (12) for the concentration dependence of free energy. From the perfect gas law, $pV = nRT$, and the relation $dG = Vdp$,

$$V = \frac{nRT}{p},$$

and

$$dG = nRT \frac{dp}{p};$$

then,

$$\int_{G°}^{G} dG = nRT \int_{p°}^{p} \frac{dp}{p}, \tag{12a}$$

with $G°$ and $p°$ constants of integration. Integrating,

$$G - G° = nRT \ln(p/p_0); \quad \text{rearranging,}$$

$$G = G° + nRT \ln(p/p_0),$$

and for dilute solutions, concentrations c and c_0 would be substituted for p and p_0. Then, one sees from this expression that the free energy change, dG, which is used to calculate the work made available from physiological metabolite and ion gradients does, in fact, depend only on the change or gradient in c and not on its absolute value. Thus,

$$dG = nRT \frac{dc}{c},$$

and

$G = G° + nRT \ln(c/c_0)$ is the final result for the dependence of G on c.

For the standard state defined as 1 molar, standard conditions of temperature (25°C) and pressure (1 atm), $c_0 = 1$ M, $G°$ is the standard free energy, and

$$G = G° + nRT \ln c.$$

1.7 Free Energy Change of a Chemical Reaction

Considering the reaction

$$n_A A + n_B B \to n_C C + n_D D$$

where the n's refer to the number of moles of each compound A, B, C, and D participating in the reaction, and combining the free energy expressions for each compound, the free energy change, ΔG, for the reaction is:

$$\Delta G = \Delta G^\circ + RT \ln \frac{\text{(products)}}{\text{(reactants)}}$$
$$= \Delta G^\circ + RT \ln \frac{(c_C)^{n_C}(c_D)^{n_D}}{(c_A)^{n_A}(c_B)^{n_B}}, \quad (13)$$

where the cs in parentheses are the effective concentrations, or activities, of compounds A, B, C, and D. At equilibrium, $\Delta G = 0$, and therefore

$$\Delta G^\circ = -RT \ln \left\{ \frac{(c_C)^{n_C}(c_D)^{n_D}}{(c_A)^{n_A}(c_B)^{n_B}} \right\}_{\text{equil.}}$$

The quantity in the bracket is defined as the equilibrium constant, so

$$\Delta G^\circ = -RT \ln K_{eq} = -2.3 RT \log_{10} K_{eq}; \quad \Delta G^\circ < 0 \text{ when } K_{eq} > 1,$$
$$\Delta G^\circ > 0 \text{ when } K_{eq} < 1. \quad (14)$$

Exercise 1.4. An example that may be of interest in thinking about the relation,

$$\Delta G^\circ = -2.3 RT \log_{10} K_{eq},$$

is the application to acid–base reactions, $AH \to A^- + H^+$, for which the $pK = -\log_{10} K_{eq}$. Then, $\Delta G^\circ = 2.3 RT \cdot (pK)$, and ΔG° decreases by $2.3 RT = 1.36$ kcal/mol (25°C) for each unit decrease in pK of the acidic group.

1.7.1 Comments on ΔG°

ΔG° is the free energy change for conversion of a standard amount of reactant [1 molar for solutes, 1 atmosphere of pressure [1.013×10^5 Pascal (Pa)] for gases, at 25°C] to the same standard amount of product. This complete conversion may be thought of as involving first a conversion of the reactant to an equilibrium mixture (with a negative ΔG because the reaction goes toward equilibrium) and then conversion of the equilibrium mixture to the products (with a positive ΔG since the reaction is going away from equilibrium). If $K_{eq} > 1$, then the negative ΔG step outweights that with positive ΔG. The 1 M standard state for c_0 and G° is either extrapolated from measurements on more dilute solutions (certainly, substrates for enzymatic reactions are never used at 1 M concentration) or calculated from K_{eq}, but in any case does not play any role in calculations of ΔG and $\Delta G°$ because it has the value of 1 in the concentration quotient.

It is a common misconception that the direction of a reaction can be determined from the value of $\Delta G°$. It is true that $\Delta G° < 0$ if $K_{eq} > 1$, and $\Delta G° > 0$ if $K_{eq} < 1$. But, the actual concentrations or activities involved in a reaction are not necessarily those operating at equilibrium. Therefore the sign of $\Delta G°$ does not necessarily determine whether or not a reaction goes forward or not. The same holds for $\Delta E°$, the change of standard potential, in oxidation–reduction reactions. It is the sign of ΔG and ΔE, the energy and electrical potential terms that include the operating concentrations in the reaction, that predict its direction. However, the sign of $\Delta G°$ and $\Delta E°$ can indicate the direction at the midpoint of the reaction where the concentration of products, p, equals that of the reactants r [i.e., $(p)/(r) = 1$, $\ln[(p)/(r)] = \ln 1 = 0$, and $\Delta G = \Delta G°$ in Eq. (13)]. $\Delta G°$ may also provide this information if the reaction is near, rather than exactly at, its midpoint. But, if the reaction is far from the midpoint, i.e., $(p)/(r) \gg 1$, or $(p)/(r) \ll 1$, then the sign of $\Delta G°$ may tell one little or nothing because the concentration term, $RT\ln[(p)/(r)]$, will contribute importantly to the total ΔG. One useful property of $\Delta G°$ is therefore that the free energy change at the midpoint of the reaction, ΔG_m, is often equated to $\Delta G°$. In real solutions this is an approximation because of non-ideal concentration effects and ionic strength interactions.

1.8 Temperature Dependence of K_{eq}

$$\text{If } \Delta G° = -RT\ln K_{eq}, \quad \text{then}$$

$$\ln K_{eq} = -\Delta G°/RT$$

$$= -(\Delta H° - T\Delta S°)/RT$$

$$= -\frac{\Delta H°}{R}\left(\frac{1}{T}\right) + \frac{\Delta S°}{R}$$

Differentiating with respect to T alone, one obtains the Van't Hoff equation

$$\frac{d}{dT}(\ln K_{eq}) = +\frac{\Delta H°}{RT^2}, \tag{15}$$

because

$$\frac{d}{dT}\left(\frac{1}{T}\right) = -\frac{1}{T^2},$$

and the assumption is made that $\Delta S°$ and $\Delta H°$ are independent of temperature, so that

$$\frac{d}{dT}(\Delta S°) = 0, \quad \frac{d}{dT}(\Delta H°) = 0.$$

This equation says that K_{eq} will increase with temperature [i.e., $d(\ln K_{eq})/dT$ is positive] if $\Delta H°$ is positive so that heat is absorbed as the temperature is increased (endothermic reaction). Conversely, if $\Delta H°$ is negative, the reaction is exothermic, and heat is given off to the surrounding environment during the reaction, then $d(\ln K_{eq})/dT$ will also be negative. Thus, the reaction will go forward to a greater extent with increasing temperature if $\Delta H°$ is positive, and to a lesser extent if $\Delta H°$ is negative. This property of chemical reactions is essential for their stability. If one had the opposite sign relationship in the Van't Hoff equation, then all reactions in nature would be thermally unstable and explosive. That is, if reactions that evolve heat would go forward to a larger extent as heat is evolved, then one has the following scenario: A reaction evolving heat goes forward to a small extent at an initial temperature; the temperature of the reaction increases as heat is evolved; it goes forward to a larger extent; more heat is evolved; it continues to go forward, more heat ... boom! Thus, the Van't Hoff equation expresses the negative feedback required in all stable reactions. This is one example of the principle of Le Chatalier operating in all of nature: *An external influence, disturbing the equilibrium of a body, induces processes in the body that weaken the effects of this influence.*

1.9 Other Kinds of Work: Electrical, Chemical Work

For a reversible transition at constant p, T, involving only pressure–volume (p–V) work,

$$dG_{p,T} = 0. \tag{11a}$$

If other kinds of work, W_{other}, are involved (e.g., chemical, electrical work), then

$$dG_{p,T} = +dW_{other} = (dW_{other})_{max} \tag{16}$$

for the reversible transition.

For electrical work, $\Delta G = \Delta W_e =$ (moles of charge moved through voltage, ΔE) $\cdot \Delta E = (n_a zF) \cdot \Delta E$ where ΔE is the electrical potential difference, in volts, across which a compound with charge z^+ is transferred, F is the Faraday constant (96,487 coulombs/mol or 23.06 kcal mol^{-1} V^{-1}), and n_i is the number of moles of compound i transported across ΔE. When the potential exists across a membrane, this membrane potential is usually written as $\Delta \psi$. Then, for electrical work, the free energy level, $G_{(elec)}$, corresponding to a potential E or ψ, from which one can derive the Nernst equation (Chap. 2, section 2.5; Chap. 3, section 3.4.2), is

$$G_{(elec)} = n_i zF \cdot E \quad \text{or} \quad G_{(elec)} = n_i zF \cdot \psi; \quad \text{and} \tag{17a}$$

$$\begin{aligned}\Delta G_{(elec)} &= n_i zF \cdot \Delta E, \quad \text{or}\\ &= n_i zF \cdot \Delta \psi,\end{aligned} \tag{17b}$$

for work done transporting n_i mol through an electrical gradient. It is under-

1.9 Other Kinds of Work: Electrical, Chemical Work

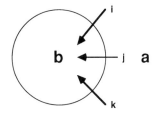

Figure 1.4. Transport of solutes i, j, k from compartment a into b.

stood that pressure and temperature are constant, and the subscripts have not been included.

The differential free energy change corresponding to a small change, dn_i, in n_i at an electrical potential of E or ψ is:

$$dG_{i(\text{elec})} = dn_i zF \cdot E, \quad \text{or} \quad dG_{i(\text{elec})} = dn_i zF \cdot \psi \tag{18}$$

1.9.1 Chemical Work and Chemical Potential

The chemical potential, μ_i, of compound i is the free energy per mole,

$$\left[\frac{\partial G}{\partial n_i}\right]_{T,p,n_j \neq n_i} \equiv \mu_i$$

The symbol ∂ stands for partial derivative, the derivative with respect to n_i with T, p, when the other $n_j \neq n_i$ are held constant.

The μ value is an important quantity for transport problems because it is the change of free energy of a system per mole of component transported in or out of the system. Consider a two-component system (outside solution, a, cytoplasm, b). Then, the free energy $G = G^a + G^b$. If dn_i, dn_j, dn_k moles of components $i, j,$ and k are transferred from the solution outside, a, to the inside of the cell, b (Fig. 1.4), then

$$dn_i^b = dn_i = \text{moles of } i \text{ gained by phase } b, \tag{18a}$$

$$dn_j^b = dn_j = \text{moles of } j \text{ gained by phase } b,$$

$$dn_i^a = -dn_i = \text{moles of } i \text{ lost by phase } a, \tag{18b}$$

and

$$dn_j^a = -dn_j = \text{moles of } j \text{ lost by phase } a.$$

The change in free energy of the system, $dG_{i(\text{chem})}$, for the chemical work, dW_c, done on it through movement of dn_i only is:

$$dG_{i(\text{chem})} = (dG^a + dG^b) = dW_c \tag{18c}$$

$$= \frac{\partial G^a}{\partial n_i^a} \cdot dn_i^a + \frac{\partial G^b}{\partial n_i^b} \cdot dn_i^b \tag{18d}$$

$$= \left(\frac{\partial G^b}{\partial n_i^b} - \frac{\partial G^a}{\partial n_i^a}\right) \cdot dn_i, \tag{18e}$$

which from Eq. 18a and Eq. 18b above,

$$= (\mu_i^b - \mu_i^a) \cdot dn_i,$$

from the definition given above for chemical potential. Thus,

$$dG_{i(\text{chem})} = \sum \mu_i \cdot dn_i, \quad (19)$$

where the chemical potential of component i is summed ($\sum$) over the compartments a and b.

In general, the free energy change arising from movement of dn_i at a position where the chemical potential in μ_i is:

$$dG_i^a = \mu_i^a \, dn_i^a.$$

For net transport, $\mu_i^a \neq \mu_i^b$. The reacton will go forward if $\mu_i^a > \mu_i^b$, because then $dG < 0$. At equilibrium, $\mu_i^a = \mu_i^b$, because $dG = 0$.

The combined free energy change for electrical (Eq. 18) and chemical work (Eq. 19) arising from movement of dn_i at a position where the electrical potential is ψ and the chemical potential, μ_i, is

$$dG_i = dn_i \cdot zF\psi + \mu_i \cdot dn_i \quad (20)$$

Dropping the subscript i, the sum of the electrical (e) and chemical (c) terms is then dG_{ec}.

$$dG_{ec} = \mu \cdot dn + zF\psi \cdot dn = \tilde{\mu} \cdot dn, \quad (21)$$

where the electrochemical potential, $\tilde{\mu} = \mu + zF\psi = dG/dn$, or

$$\tilde{\mu} = \mu^0 + RT \ln(c/c_0) + zF\psi, \quad (22)$$

which is the complete expression for the electrochemical potential.

Just as the difference in chemical potential is a measure of the escaping tendency of uncharged molecules due to a concentration gradient, the electrochemical potential describes this tendency of charged molecules due to both a concentration gradient and an electrical potential. When applied to H$^+$ gradients across energy transducing membranes, Eq. 22 (or Eq. 24) is the basic equation of the chemiosmotic hypothesis for energy coupling (Chap. 3).

1.10 Thermodynamics of Ion Gradients

For protons, $z = 1$, and

$$\tilde{\mu}_{H^+} = \tilde{\mu}_o + RT \ln(H^+) + F\psi,$$

or

$$\tilde{\mu}_{H^+} = \tilde{\mu}_o + 2.3RT \cdot \log_{10}(H^+) + F\psi.$$
$$= \tilde{\mu}_o - 2.3RT \cdot \text{pH} + F\psi \quad (23)$$

since pH $= -\log H^+$. We are interested in $\Delta\tilde{\mu}_{H^+}$ across a membrane with the initial (reactant) state, "out" (or "in"), and the final product state "in" (or "out"). Then, for the $\Delta\tilde{\mu}_{H^+}$ formed, for example, by H^+ movement from the chloroplast stroma to the lumen, $\Delta\psi = \psi_{in} - \psi_{out}$, $\Delta pH = pH_{in} - pH_{out}$, $(\tilde{\mu}_o)_{in} = (\tilde{\mu}_o)_{out}$, and $\Delta\tilde{\mu}_{H^+} = (\tilde{\mu}_{H^+})_{in} - (\tilde{\mu}_{H^+})_{out}$.

Remembering that $(-\log H^+) = pH$,

$$\Delta\tilde{\mu}_{H^+} = F \cdot \Delta\psi - 2.3RT \cdot \Delta pH \tag{24}$$

This is the fundamental equation of the chemiosmotic hypothesis.

The "proton-motive force," Δp, often referred to in the bioenergetic literature, is defined as

$$\Delta p = \frac{\Delta\tilde{\mu}_{H^+}}{F} = \Delta\psi - \frac{2.3RT}{F} \cdot \Delta pH,$$

in volts or millivolts, or

$$\Delta p = \Delta\psi - 59 \cdot \Delta pH \text{ at } 25°C, \tag{24a}$$

in millivolts.

Similarly, for an electrochemical gradient involving a chemical gradient of sodium (Na^+) ions (Chaps. 3 and 9),

$$\Delta\tilde{\mu}_{Na^+} = F \cdot \Delta\psi + 2.3RT \cdot \log_{10}\frac{(Na^+)_{final}}{(Na^+)_{init.}}, \tag{25}$$

with $\Delta\psi = (\psi_{final} - \psi_{init.})$.

1.11 Thermodynamics of $\Delta\tilde{\mu}_{H^+}$-Linked Active Transport

The uptake of solute S (charge z) from the initial state, "out," to the final state, "in," is accompanied by the uptake (*symport*) of n protons (Fig. 1.5A), that is driven by the proton electrochemical potential, $\Delta\tilde{\mu}_{H^+}$. If all of the free energy available in the $\Delta\tilde{\mu}_{H^+}$ is stored in the electrochemical potential, $\Delta\tilde{\mu}_S$, of substrate accumulation by a symport mechanism,

$$\Delta G_{total} = 0.$$
$$\Delta G_{total} = n \cdot \Delta\tilde{\mu}_{H^+} + \Delta\tilde{\mu}_S = 0. \tag{26}$$

Solving for $\Delta\tilde{\mu}_{H^+}$,

$$\Delta\tilde{\mu}_{H^+} = 2.3RT\log_{10}\frac{(H_i^+)}{(H_o^+)} + F \cdot \Delta\psi$$

$$= F \cdot \Delta\psi - 2.3RT \cdot \Delta pH.$$

In analogy to Eqs. 22–25,

$$\Delta\tilde{\mu}_S = 2.3RT\log_{10}\frac{S_i^{+z}}{S_o^{+z}} + zF \cdot \Delta\psi, \tag{26a}$$

with parentheses in the quotient for effective concentrations or activities omitted for simplicity.

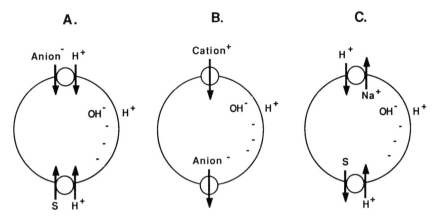

Figure 1.5. Diagrammatic representation of symport (A), uniport (B), and antiport (C) reactions. (Based on Rosen and Kashket, 1978.)

from Eq. 26,

$$2.3RT \cdot \log \frac{S_i^{+z}}{S_o^{+z}} + zF \cdot \Delta\psi = 2.3RT \cdot n\Delta\text{pH} - nF \cdot \Delta\psi;$$

dividing by $2.3RT \equiv Z$,

$$\log_{10} \frac{S_i^{+z}}{S_o^{+z}} = n \cdot \Delta\text{pH} - (n + z) \cdot \frac{\Delta\psi}{Z}, \tag{27}$$

with $2.3RT/F = 59$ mV at 25°C, and $\Delta\psi$, expressed in millivolts, is the expression for accumulation of solute S with charge z by a symport mechanism. The solute uses n protons ($n \cdot H^+$) per solute molecule, with the transport driven by a transmembrane pH gradient (ΔpH) and membrane potential ($\Delta\psi$).

Special cases: For $n = 0$, The mechanism is called uniport (Fig. 1.5B), is driven only by $\Delta\psi$ because the protons have been removed from the problem, and then

$$\log_{10} \frac{S_i^{+z}}{S_o^{+z}} = -z \cdot \frac{\Delta\psi}{Z}. \tag{28}$$

Thus, to achieve an accumulation ratio of 100 for a singly charged species, a $\Delta\psi$ of -118 mV would be required at 25°C.

For an antiport as described in Fig. 1.5C above, the initial state for solute movement is i and the final state is o. Then,

$$\Delta\tilde{\mu}_S = 2.3RT \cdot \log_{10} \frac{S_o^{+z}}{S_i^{+z}} - zF \cdot \Delta\psi,$$

with signs reversed from Eq. 26a because of reversal of initial and final states, and $\Delta\psi \equiv \psi_{\text{in}} - \psi_{\text{out}}$, as before, according to convention.

Using $\Delta\tilde{\mu}_{H^+}$ as in the symport case above, with $n \cdot H^+$ utilized per solute molecule transported, and combining $\Delta\tilde{\mu}_{H^+}$ and $\Delta\tilde{\mu}_S$,

$$\log_{10}\frac{S_i^{+z}}{S_o^{+z}} = (n-z) \cdot \frac{\Delta\psi}{Z} - n \cdot \Delta\text{pH}. \tag{29}$$

If $n = z$ as it appears to be in some antiports, then the charge movement would be neutral, $\Delta\psi$ removed from the problem, and

$$\log_{10}\frac{S_i^{+z}}{S_o^{+z}} = -n \cdot \Delta\text{pH};$$

for example,

$$\frac{S_i^{+z}}{S_o^{+z}} = 10^{-1},$$

if $\Delta\text{pH} = +1$ and $n = 1$.

1.12 Thermodynamics of $\Delta\tilde{\mu}_{H^+}$-Linked ATP Synthesis

As expressed in the chemiosmotic hypothesis (Chap. 3), the synthesis of ATP is linked to the utilization of the proton electrochemical gradient across the membrane, from the side of relatively positive, "p," proton potential, to the side that is relatively negative, "n":

$$\text{ADP} + P_i + n\text{H}_p^+ \rightarrow \text{ATP} + \text{H}_2\text{O} + n\text{H}_n^+.$$

If all of the free energy stored in the proton electrochemical potential is utilized ($\Delta\tilde{\mu}_{H^+} < 0$) for ATP synthesis, the total free energy change is zero. Thus,

$$\Delta G_{tot} = n \cdot \Delta\tilde{\mu}_{H^+} + \Delta G_{ATP} = 0; \tag{30}$$

substituting for ΔG_{ATP},

$$-n \cdot \Delta\tilde{\mu}_{H^+} = \Delta G^\circ + 2.3RT \cdot \log_{10}\frac{(\text{ATP})}{(\text{ADP})(P_i)}; \tag{31}$$

then,

$$\log_{10}\frac{(\text{ATP})}{(\text{ADP})(P_i)} = -\frac{1}{2.3RT}(\Delta G^\circ + n \cdot \Delta\tilde{\mu}_{H^+}). \tag{32}$$

For H^+ movement linked to ATP synthesis in chloroplasts and mitochondria (Chaps. 3, 7, and 8), the initial and final states for the H^+ gradient linked to ATP synthesis are inside (lumen) $\rightarrow$ outside (stroma) for chloroplasts, and outside (intermembrane space) $\rightarrow$ inside (matrix) for mitochondria.

1.13 Nonequilibrium Thermodynamics

The conditions for true equilibrium are never truly met in biological cells or organelles because of the coupling between reactions. Thus, the product of one reaction becomes the reactant of a subsequent reaction. In the case of two coupled reactions, (a) and (b), where the $-\Delta G$ produced by (a) is fully utilized by (b) and completely coupled to it, the total ΔG will be close to zero, satisfying the conditions for equilibrium (see Eqs. 26 and 30). In reality, two systems are never perfectly coupled. Real systems "slip" to some extent and are therefore somewhat irreversible. Nonequilibrium thermodynamics provides a theoretical framework for dealing with these problems (Caplan, 1971; Rottenberg, 1979, 1986). The rates, J_a and J_b, of the two coupled reactions, (a) and (b), close to equilibrium are assumed to be proportional to the "driving forces," which in the case of energized membranes are the negative free energy gradients, $-\Delta G_a$ and $-\Delta G_b$. Thus,

$$J_a = -(L_{aa}\Delta G_a + L_{ab}\Delta G_b)$$
$$J_b = -(L_{ba}\Delta G_a + L_{bb}\Delta G_b). \tag{33}$$

which are the so-called phenomenological equations linking forces and flows. The coupling coefficients are the L values, with L_{ii} a "straight" coefficient relating each flux to its conjugate force and L_{ij} the "cross" coefficient providing the coupling between reactions. On a microscopic level, where the system is ideally reversible, or in an ideal biochemical system with a reversible enzymatic system, one has the reciprocity relationship

$$L_{ab} = L_{ba}. \tag{33a}$$

Applying the phenomenological equations to coupled reactions of electron transport $(J_e, \Delta G_e)$ and phosphorylation $(J_p, \Delta G_p)$,

$$J_e = -(L_e\Delta G_e + L_{ep}\Delta G_p) \tag{34a}$$
$$J_p = -(L_{pe}\Delta G_e + L_p\Delta G_p). \tag{34b}$$

"Level flow" and "static head" are defined as the conditions $\Delta G_p = 0$ (no backpressure from ΔG_p), and $J_p = 0$ (large back-pressure from ΔG_p). For the former case,

$$J_p/J_e = L_{pe}/L_e,$$

and for the latter,

$$-\Delta G_e/\Delta G_p = L_p/L_{pe}.$$

Linearity and reciprocity have been demonstrated in oxidative phosphorylation (Lemasters and Billica, 1981). For such a two-flow system, the degree of coupling, q, is defined as

$$q \equiv L_{ep}/\sqrt{L_e L_p}, \tag{35}$$

with $q = +1$ and 0 for completely coupled and uncoupled systems. The phenomenological stoichiometry, Z, is defined as

$$Z \equiv \sqrt{L_p/L_e}, \tag{36}$$

so that the conditions for level flow and static head can be rephrased as

$$\text{Level flow:} \quad J_p/J_e = qZ, \tag{37}$$

and

$$\text{Static head:} \quad -\Delta G_e/\Delta G_p = Z/q. \tag{38}$$

The energy ratio at static head will be greater, the greater the degree of coupling, q. Z is equal to the ATP/electron transport flux ratio and the stoichiometry of ATP synthesis when the degree of coupling $q = 1$. Z is usually assumed to equal 3 for an NAD-linked substrate in mitochondria.

For conditions intermediate between level and static flow, the following relationship can be derived from Eqs. 34a,b and 35:

$$J_p/J_e = Z[q + Z(\Delta G_p/\Delta G_e)]/[1 + qZ(\Delta G_p/\Delta G_e)] \tag{39}$$

1.14 "High-Energy" Bonds

"High-energy" bonds are unstable bonds that are readily hydrolyzed and are defined as having a $\Delta G°$ at pH 7 more negative than that of typical phosphate esters, i.e., $\Delta G°' \leq -7$ kcal/mol (Table 1.1). The majority of the high-energy compounds involve phosphate or sulfur compounds (Wald, 1969), and high-energy compounds other than the phosphate anhydride compounds shown

Table 1.1. Standard free energy of hydrolysis of some phosphate anhydride and ester compounds

Compound	$\Delta G°$ (kcal/mol)[a]
Phosphoenolpyruvate	−14.8
1,3-Diphosphoglyceric acid	−11.8
Creatine phosphate	−10.0
Acetyl phosphate	−10.0
Phosphoarginine (pH 8.0)	−8.0
ATP → ADP + P_i ($+Mg^{++}$)	−7.7
ATP → ADP (pH 8.0)	−8.4
ATP → ADP (pH 9.5)	−10.4
Glucose-1-phosphate	−5.0
Pyrophosphate	−4.0
Fructose-6-phosphate	−3.8
Glucose-6-phosphate	−3.3
Glycerol-1-phosphate	−2.2

[a] pH = 7.0, except where noted; temperature, 25°C.
From Jencks, 1976; Bridger and Henderson, 1983.

in Table 1.1 include thiol acyl esters, sulfonium compounds, acyl imidazole, and acyl amino acids. The central position of ATP in the hierarchy of phosphate group potentials (Lehninger, 1971) can be seen in Table 1.1, so the ADP/ATP couple can mediate phosphate flow from phosphorylated compounds with very negative $\Delta G^{\circ\prime}$ (phosphoenolpyruvate, 1,3-diphosphoglycerate, creatine phosphate, for example), and can act as both a phosphoryl donor and acceptor. ATP does not act as the immediate source of energy and phosphoryl groups in many biosynthetic reactions. CTP, GTP, and UTP, which have free energies of hydrolysis very similar to that of ATP, are preferentially used for different biosynthetic pathways. They are the general precursors, respectively, for biosynthesis of lipid, protein and cellulose, and polysaccharides.

The high-energy compounds are energy-rich because of (i) the positive ΔS° arising from the increased number of states available in the phosphate product due to resonance (Oesper, 1950), and (ii) the negative ΔH° from electrostatic repulsion (Hill and Morales, 1951) caused by the negatively charged terminal phosphate groups (Fig. 1.6A and B). The contribution of entropy or resonance stabilization to the large free energy of hydrolysis is a consequence of the atomic properties of phosphorus, i.e., large bond lengths with weaker binding energies (Wald, 1969). This leads to a partial double bond character of the orthophosphate P–O bonds. The average P–O bond length in orthophosphate is 1.54 Å compared to 1.73 Å expected for a single P–O bond (Pauling, 1960). The greater electron delocalization is synonymous with resonance stabilization and the positive contribution to the ΔS° of reaction (Table 1.2).

The large negative ΔH° of the ATP molecule arises from the three or four negative charges that it has in the absence of Mg^{2+} (Fig. 1.6A), or the one or two that exist in its presence that result in net charge repulsion between the terminal phosphate groups (Table 1.2). The net charge is also controlled by the pK for $HATP^{3-} \to H^+ + ATP^{4-}$ which is in the physiological range of 6.5–7.0 (Alberty, 1969). The ΔG° of hydrolysis is thus dependent on pH (Fig. 1.6C; also, Klotz, 1967), as well as the presence of the cation Mg^{2+} that can bridge across the β–γ oxygens of ATP (Fig. 1.6A). The fact that the pK values of orthophosphate, ADP, and ATP are all below 7.0 means that ATP synthesis in energy transducing membranes can be measured at alkaline pH (pH $\simeq$ 8) by a pH change:

$$HPO_4^{2-} + ADP^- + H^+ \xrightarrow{Mg^{2+}} ATP^{2-} + H_2O.$$

Another major contribution (Factor 3, Table 1.2) to the large positive ΔS° and negative ΔG° of hydrolysis of phosphate anhydride compounds may arise from differential solvation by H_2O of products and reactants (George et al., 1970; de Meis et al., 1985). A large ΔS° of hydrolysis has been found to accompany hydrolysis of charged orthophosphate, pyrophosphate, and polyphosphate compounds, all of which have appreciable solvation energies. For example, the solvation energies (kcal/mol, 25°C) of $H_2PO_4^-$, HPO_4^{2-}, and PO_4^{3-} are 76, 299, and 637. This suggests that the large $+\Delta S^\circ$ of hydrolysis arises from a smaller amount of solvated or ordered water in the products

1.14 "High-Energy" Bonds

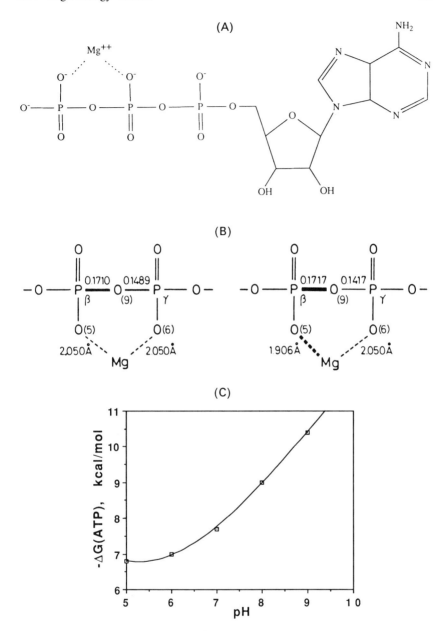

Figure 1.6. (A) Structure of ATP-Mg^{2+} at alkaline pH showing net negative charge on phosphate groups. (B) Calculated effect of symmetric (left) and asymmetric (right) locations of Mg^{2+} bound to O(5) and O(6) on the bonds between P_β and O(9) and P_γ-O(9) of a methyltriphosphate compound of ATP. NMR spectra show that Mg^{2+} binds more tightly to the β- than to the γ-phosphoryl group (asymmetric case, right), thereby weakening the terminal P_γ–O_9 bond as shown by a decrease in the overlap population (Terada et al., 1984). (C) pH dependence of free energy of hydrolysis of ATP.

Table 1.2. Physical-chemical factors contributing to large negative $\Delta G°$ of hydrolysis of phosphate anhydride compounds

Factor	$-\Delta H°$ or $+T\Delta S°$
1. Electrostatic charge repulsion	$-\Delta H°$
2. Large number of resonant forms of orthophosphate	$+T\Delta S°$
3. Smaller amount of solvated or ordered water in less highly charged products	$+T\Delta S°$

because they carry a smaller average charge than the reactants. This factor would also predict that the magnitude of the $\Delta G°$ associated with ATP synthesis would be smaller in a nonpolar environment.

1.14.1 Experimental Determination of the $\Delta G°$ for ATP Hydrolysis

It has been shown using ^{32}P- and ^{18}O-labeled phosphate that enzymatic hydrolytic cleavage of ATP to ADP involves breakage of the O_9–P_γ bond (Fig. 1.6B), the terminal phosphate gaining an oxygen atom from water, the β–γ oxygen remaining with the ADP (Bridger and Henderson, 1983), and a phosphoryl freed that can be transferred to an acceptor.

It has been estimated that under alkaline conditions in the presence of 10^{-2} M Mg^{2+}, where $\Delta G° = -9$ kcal/mol, $\Delta H° = -4$ kcal/mol and $T\Delta S° = 5$ kcal/mol (Alberty, 1969). A $\Delta G°$ of -9 kcal/mol corresponds to a $K_{eq} > 10^6$, which is too large to be measured accurately through the single ATP hydrolysis reaction. Therefore, the K_{eq} for ATP hydrolysis must be measured in two or more coupled reactions such as (a) the glutamine synthetase and glutamine hydrolysis reactions (Rosing and Slater, 1972), (b) the acetate kinase and phosphate acetyltransferase reactions (Guynn and Veech, 1973), or other reactions coupled by ATP:

(i) Glutamic acid + NH_4^+ + ATP $\xrightarrow{K_I}$ Glutamine + ADP + P_i

(ii) Glutamine + H_2O $\xrightarrow{K_{II}}$ Glutamic acid + NH_4^+

The sum of reactions (i) and (ii) is:

$$ATP + H_2O \rightarrow ADP + P_i;$$

with $K_\Sigma = K_I K_{II}$,

$$\Delta G_\Sigma° = \Delta G_I° + \Delta G_{II}° = -RT \ln K_\Sigma.$$

Similarly,

(iii) acetate$^-$ + ATP^{4-} $\rightleftarrows$ acetyl phosphate^{2-} + ADP^{3-}

(iv) acetyl phosphate^{2-} + CoA $\rightleftarrows$ acetyl CoA + P_i^{2-}

(v) Sum: ATP^{4-} + acetate$^-$ + CoA $\rightleftarrows$ acetyl-CoA + ADP^{3-} + Pi^{2-}

1.14 "High-Energy" Bonds

The equilibrium constant for (i) was measured and recalculated to be 700 at 37°C, pH 7 after correction for concentrations using activity coefficients and Mg^{2+} chelation. K_{II} for the glutaminase reaction was determined by microcalorimetry (Benzinger et al., 1959) to be 229 at 37°C, yielding an equilibrium constant for the ATP hydrolysis reaction at pH 7 and 37°C of 1.6×10^5. Because of the charge repulsion of the terminal phosphate groups, the magnitude of the $\Delta G°$ of ATP hydrolysis is about 1 kcal/mol larger at pH 8.0 compared to 7.0 (Table 1.1; Fig. 1.6C). Equation (v) is the combination of the ATP and acetyl-CoA hydrolysis reactions. Determination of the $\Delta G°$ for acetyl-CoA hydrolysis led to $\Delta G°$ (pH = 7) values for ATP hydrolysis at 38°C, of -7.6 and -8.5 kcal/mol in the presence (10^{-3} M) and absence of Mg^{2+} (Guynn and Veech, 1973). The effect of low pH and higher $[Mg^{2+}]$ in decreasing the $|\Delta G°|$ again demonstrates the repulsive effect of the negative charge on the phosphate groups and its contribution to the $\Delta H°$ of reaction.

1.14.2 Energetics of Enzyme-Bound ATP ↔ ADP Interconversion

The above energetic considerations apply to ADP and ATP and the appropriate enzymes in solution. The ΔG requirement for ATP synthesis at the active site of the mitochondrial ATP synthase enzyme is, however, quite small ($K_{eq} \simeq 1$) and the ATP synthetic reaction readily reversible because the enzyme binds ATP much more tightly than ADP and P_i. The large ΔG requirement for ATP synthesis is used to force a structural change in the enzyme allowing release of the ATP (Chap. 8, section 8.6). There is substantial precedent for such equilibrium constant changes in kinase enzymes involved in ATP ↔ ADP interconversion, such as 3-phosphoglycerate kinase and pyruvate kinase (Nageswara Rao et al., 1978, 1979).

1.14.3 Measurement of Internal Adenine Nucleotide Concentrations by ^{31}P Nuclear Magnetic Resonance (NMR)

The noninvasive technique of ^{31}P-NMR spectroscopy can be used to measure extra- and intracellular and intraorganelle concentrations of nucleotide tri- and diphosphates, as well as other phosphorylated compounds, and inorganic phosphate (P_i), simultaneously with the ΔpH across the membrane. The pH gradient can be measured because the chemical shift of the orthophosphate resonance is pH-dependent in the region of the phosphate pK, and the P_i peaks from the external and internal media can be resolved. The two peaks collapse into one ($\Delta pH < 0.1$) in the presence of an uncoupler of phosphorylation. The increase in ATP in energized mitochondria at the expense of ADP is illustrated in Fig. 1.7 which shows the ^{31}P-NMR spectrum of mitochondria before and after oxygenation.

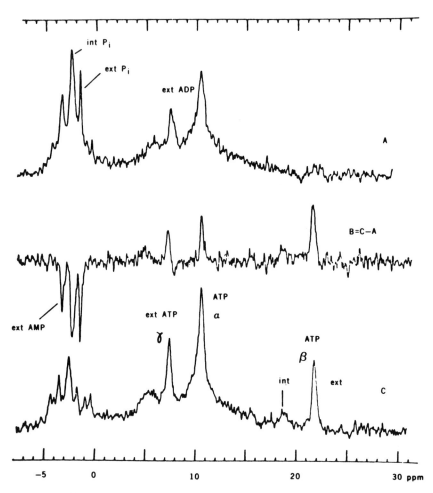

Figure 1.7. ^{31}P-NMR spectrum measured at 145.7 MHz of mitochondria before (A) and after (C) oxygenation. The γ- and β-phosphate resonances of ATP were observed at approximately 5.1 ppm and 18.7 ppm (internal) in the presence of Mg^{2+}, and at 5.3 and 21.5 ppm (external) in ATP free of divalent metal ions. (B) Difference between (C) and (A). (From Shulman et al., 1979. Reprinted with permission from *Science*, vol. 205, pp 160–166, 13 July 1979. Copyright 1979 by the AAAS.)

1.14.4 Caged ATP

A biochemically inert ("caged") 2-nitrobenzyl derivative of ATP that is light sensitive, P^3-1-(2-nitro)phenylethyladenosine-5'-triphosphate (Fig. 1.8), has provided a tool for studying kinetics on a millisecond time scale of ATP-dependent biochemical reactions through photolysis of the caged ATP. Biologically active ATP is selectively released from its cage by irradiation with a short (e.g., 30 ns from a laser) nondestructive (wavelength approx. 360 nm) light pulse (Kaplan et al., 1978; McCray et al., 1980). The ATP production is

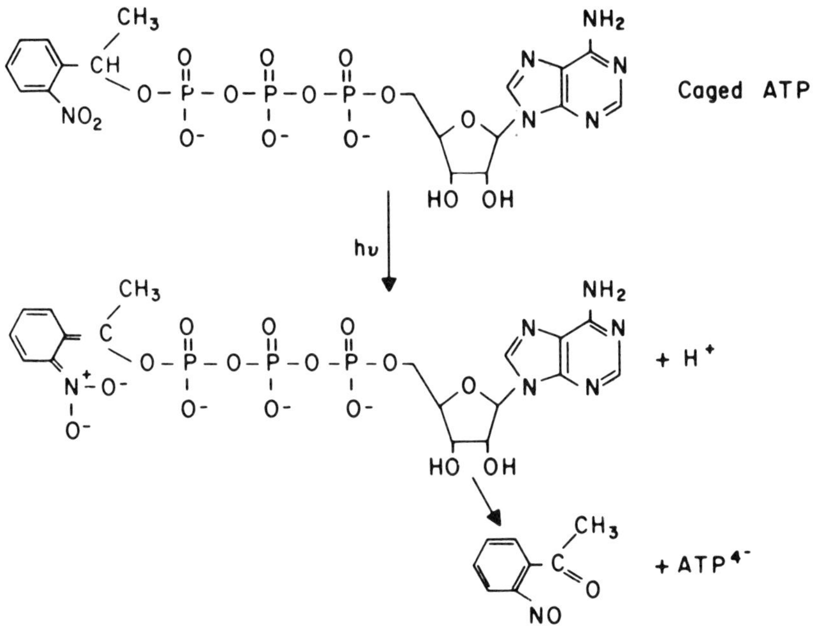

Figure 1.8. Structure of caged ATP and its photolytic reaction products. The extinction coefficient of the caged ATP is 660 M^{-1} cm^{-1} at 347 nm, a wavelength long enough that the light pulse does not cause protein damage. (Reprinted with permission from Pierce et al., *Biochemistry*, 22, pp 5254–5261. Copyright 1983 American Chemical Society.)

pH dependent and proportional to the laser energy up to 2.5 mM caged ATP, from which a 20% yield, 500 μM ATP, can be obtained. The use of caged ATP is advantageous in the study of kinetics of ATP-dependent reactions, because it eliminates the loss of time resolution due to diffusion of ATP to the active site.

1.14.5 Speculation on the Central Role of ATP in Evolution

It is an interesting question as to why the purine-containing nucleoside adenosine, rather than guanosine, uridine, or cytidine is the central intermediate in energy flow and is also centrally involved in the important coenzymes, NADH, FADH$_2$, and acetyl-CoA. The answer can be given that oxidative phosphorylation produces ATP, but again why ATP? The central role of adenosine may be related to the fact that it can be synthesized, in an environment mimicking the primordial environment, from five molecules of HCN, a compound present in the primordial atmosphere (Wald, 1969). In the evolution of energy storage in the phosphate group potential, ADP/ATP may have been preceded by the simpler and less versatile pyrophosphate PP$_i$ (Baltscheffsky et al., 1986).

1.15 Summary

Thermodynamics describes physical and chemical phenomena in terms of the bulk or macroscopic properties of matter. There are three commonly stated laws of thermodynamics, of which the first two are discussed here. The *first law of thermodynamics* is a statement of the law of *conservation* of energy: the total energy of the system and surroundings does not change. The *second law of thermodynamics* describes the limitations of converting heat into work: *it is impossible to have a process whose only result is to tranform into work heat extracted from a source at a single temperature, or to transfer heat from an object at a given temperature to one at a higher temperature.* An alternative statement is that there is a state function called *entropy. In an irreversible process the entropy of the system and surroundings will increase, whereas it will remain constant in a reversible process. The total entropy of system and surroundigs never decreases.* There are two mathematical descriptions of the entropy, S: (1) the "statistical" formulation due to Boltzmann, $S = k \ln \Omega$; (2) the "thermodynamic" formulation: $dS \geq dQ_{rev}/T$. The entropy, like the energy, is a state function. For problems that can be described by both formulations, the entropy change between final and initial states is identical, as required for a state function. When $dQ_{rev} = 0$ in an isolated system, $dS \geq 0$; $dS = 0$ for a reversible reaction. Away from equilibrium, the reaction direction is determined by $dS > 0$. When the reaction includes changes of mechanical (pressure–volume) work ($dW = pdV$) as well as heat, the reaction direction is specified at constant pressure and temperature by an inequality for the *Gibbs free energy* state function, G: $\Delta G \leq 0$; $\Delta G = 0$ at equilibrium. The reaction direction is determined both by the standard or midpoint free energy change and by the concentration of products and reactants. The relevant state function for transport problems involving chemical work is the molar Gibbs free energy or chemical potential, μ, for which the approach to equilibrium is defined by $\Delta \mu \leq 0$. For electrochemical work carried out by energy-transducing membranes involving ion gradients and gradients of membrane potential, the appropriate state function is the *electrochemical potential*, $\tilde{\mu}$: $\Delta \tilde{\mu} \leq 0$. When $\tilde{\mu}$ is written in term of the difference in *trans*-membrane potential, $\Delta \psi$, and the *trans*-membrane gradient of the hydrogen ion concentration, ΔpH, then the change in *proton electrochemical potential*, $\Delta \tilde{\mu}_{H^+} = F \cdot \Delta \psi - 2.3RT \cdot \Delta$pH. This is the fundamental equation of the *chemiosmotic hypothesis* for energy transduction. An equation of exactly the same form can be written for gradients of other ions such as Na^+. The equation illustrates that electrical and chemical gradients using an ion flux can be used interchangeably for energy transduction.

High-energy bonds are unstable bonds, usually involving phosphate or sulfur anhydrides, that are readily hydrolyzed, and have a $\Delta G°$ at pH $7 \leq -7$ kcal/mol. The bond lability arises from (1) a negative $\Delta H°$ caused by electrostatic repulsion of the negatively charged terminal phosphate groups, (2) a positive $\Delta S°$ caused by the large number of orthophosphate resonance forms, and (3) a lower degree of solvation of less highly charged products. The $\Delta G°$ for ATP synthesis from enzyme-bound ADP and phosphate can be small compared to the $\Delta G°$ in solution, because the ATP is bound more tightly to the enzyme.

Problems

1. Show that the enthalpy change, ΔH, of a system is equal to the heat change, Q_p, under the common laboratory situation of constant pressure. Use Eq. 1 for the first law of thermodynamics, and the assumption that the only work done arises from expansion–contraction (pressure–volume) work.
2. Consider a situation where the energy change, ΔE, is + 20 kilojoules (kJ). Calculate the amount of work done by the system for (a) reversible and (b) irreversible transitions from the same initial to the same final state (ΔS the same) in which the heat change is (a) + 40 kJ and (b) + 30 kJ, respectively. For which transition is more work done by the system?
3. Show that the free energy change for a reversible reaction at constant temperature and variable pressure in the absence of non-P–V work is $dG = Vdp$.
4. Calculate the entropy change for a reversible expansion of 1 mol of an ideal gas from 1 L to 10 L at 27°C.
5. (a) Calculate the entropy change accompanying the melting of 9 g of ice to water at 0°C. The heat of melting is 5.98 kJ/mol. (b) By what factor would the same molar quantity of an ideal gas have to expand in order to achieve the same increase of entropy?
6. Derive $\Delta G° = -RT \ln K_{eq}$ by carrying out the steps omitted in the derivation in section 1.7.
7. The $\Delta G°$ for binding of O_2 to ferromyoglobin is -7.5 kcal/mol. What is the equilibrium constant K_{eq}? $\Delta H°$ was found to be -16.4 kcal/mol from a measurement of the rate of change of K_{eq} with temperature. What is the value of $\Delta S°$ for the reaction at 25°C?
8. For the reaction $H_2O(l) \rightarrow H_2O(g)$, $\Delta H° = 40.1$ kJ/mol, and $\Delta S° = 107.4$ J/mol-deg. Calculate the temperature at which water will spontaneously boil, i.e., equilibrium lies on the side of $H_2O(g)$.
9. Calculate the standard free energy change for reduction of ferricytochrome c by $(NH_3)_5 Ru(II)$ (Chap. 2, section 10.5) if the $\Delta H°$ and $S°$ for the reaction at 25°C = -11.9 kcal/mol and -25.8 e.u., respectively.
10. Consider the thermal denaturation of chymotrypsin. At what temperature is it half-denatured at equilibrium if the standard enthalpy and entropy change are 418 kJ/mol and 1.32 kJ/°K-mol?
11. The standard free energy for hydrolysis of glucose-6-phosphate is -13.8 kJ/mol. Blood glucose is formed through the catalysis of this reaction by glucose-6-phosphatase in the liver. What is the free energy change of this

reaction at 37°C if the steady-state concentrations (mM) of glucose-6-phosphate, phosphate, and glucose in the liver and blood, respectively, are 1, 5, and 5? Remember: $2.3RT = 5.68$ kJ/mol (1.36 kcal/mol) at 25°C.

12. (a) The values of the standard enthalpy change and free energy change for transfer of benzene (in benzene) to benzene in water at 291 K, are respectively, 0 and 4.64 kcal/mol. Calculate the entropy change for this transfer. What is the sign of the entropy change and what is the reason for this sign? (b) Similarly, calculate and briefly explain the nature of the entropy change for the thermal denaturation of β-lactoglobulin at 25°C, pH = 3, and 5 M urea, where the values of the enthalpy and free energy changes are -21 and $+0.6$ kcal/mol.

13. In the case of protein folding, observed values of $\Delta H°$ and $\Delta S°$ for formation of α-helices from randomly coiled homopolymers for different amino acids, are:

	$\Delta H_{obs}°$ (cal)	$\Delta S°$ (cal mol^{-1} deg^{-1})
Gly	-500	-3
Ala	-200	-0.5
Leu	$+100$	$+1$
Val	$+200$	$+1.5$
Glu	$-1,100$	-3.5
Lys	-900	-3

Which of the above amino acids, initially at a standard concentration in the random coil conformation, will spontaneously form > 50% α-helix at room temperature (25°C)?

14. The fraction, α, of oligonucleotide molecules in a hairpin, as compared to a coil, state at temperature T can be determined spectrophotometrically from the absorbance at 260 nm which is greater in the coil conformation. Calculate $\Delta G°$ as a function of temperature for the hairpin-coil transition of an 11-mer oligonucleotide with the melting profile for the fraction of molecules, α, in the hairpin state at equilibrium shown in Fig. 1.9.

15. Osmotic work due to semipermeability of membranes is one example of chemical work that occurs in biological systems. The osmotic pressure, $P \simeq RTc$, can be calculated by equating the chemical potential for pure solvent on one side of the membrane with that for solvent in the presence of impermeable solute at concentration, c, on the other. Calculate the osmotic pressure at room temperature in a liposome containing impermeable sucrose at 0.1 M relative to an external solution of pure H_2O.

16. (a) What is the arithmetic change in $\Delta G°$ that accompanies a 10-fold increase in K_{eq} at 27°C? (b) What is the effect of a decrease of 1.36 kcal/mol in $\Delta G°$ on the equilibrium constant of a reaction at 25°C?

17. Given the reaction r ⇌ p proceeding at 0°C with a $\Delta G°$ of -1.25 kcal/mol for the reaction r → p, what is the direction of this reaction when it operates between concentrations of r and p that are: (a) 5 mM and 5 mM, (b) 9.1 and 0.91 mM, (c) 0.91 and 9.1 mM, (d) 0.09 and 9.91 mM?

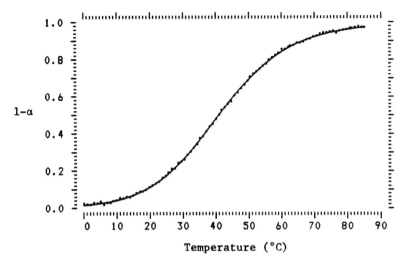

Figure 1.9. Melting profile of an 11-mer, $T_m = 40.3°C$ ($\alpha = 0.5$) (Nadeau, 1985).

18. The formula for transfer of hydrocarbon solutes from water to the interior of an sodium dodecyl sulfate (SDS) micelle (0.1 M NaCl, ~85 SDS ions/micelle at 25°C) is $\mu_{mic}° - \mu_w° = -1934 - 771n_c$, in cal/mol (Tanford, 1980). Calculate the chemical potential change for transfer of ethane and pentane.

19. Calculate (a) the free energy required to neutralize, by protonation at pH 7, 50% of the glutamic acid in a membrane protein to be inserted in a membrane, if $\Delta G = -1.36 (pK - pH)$ and the $pK = \log K_a = 3.7$; and (b) the standard free energy of stabilization, $\Delta G°$, of a 20-residue membrane-spanning α-helix, if the $\Delta H°$ originates from hydrogen bond formation (assume -5 kcal/mol per bond), there are 16 hydrogen bonds in the 20-residue polypeptide, and $T\Delta S° = +25$ kcal/mol relative to the unfolded state (Engelman et al., 1986).

20. The standard free energy and enthalpy of hydrolysis of the "high-energy" compound, ATP, are approximately -8 and -4 kcal/mol. What is the magnitude and sign of the standard entropy change at 25°C for this reaction?

$$ATP + H_2O \rightarrow ADP + P_i$$

What are the reasons for the magnitude and sign of the changes of $\Delta H°$ and $\Delta S°$?

21. If $\Delta H° = -4$ kcal/mol for ATP hydrolysis, calculate $\Delta G°$ for this reaction at 25°C if the equilibrium constant is 2×10^5 at 37°C.

22. If one would try to measure the $\Delta G°$ of ATP hydrolysis directly without using two or more coupled reactions, it would be difficult because the K_{eq} is so large. How much ATP is present at equilibrium starting with only ATP (1 mM) if K_{eq} at pH 8 = 10^6?

23. Consider a proton symport mechanism operating for transport of a divalent ($z = -2$) anion into *E. coli* cells or right-side-in vesicles with a negative internal membrane potential. Assume $2H^+$ ($n = 2$) transported per anion. (a) What is the accumulation ratio for $\Delta\psi = -59$ mV and $\Delta pH = 1$ (inside = 8, outside = 7)? (b) What is the accumulation ratio for transport of an uncharged solute under the same conditions?

24. Consider transport of Cl^- into inside-out membrane vesicles, $\Delta\psi$ positive inside. (a) Derive an expression for $\log S_i/S_o$ according to a uniport mechanism. (b) By how much does the accumulation ratio change when $\Delta\psi$ is increased from $+59$ to $+177$ mV?

25. Historically, the $\Delta G^{\circ\prime}$ of ATP was measured using two coupled reactions to be -7.7 and -6.9 kcal/mol using (a) the hexokinase reaction (glucose + ATP → glucose-6-phosphate + ADP, with $\Delta G^{\circ\prime} = -4.4$ kcal/mol), and (b) the galactokinase reaction (gal + ATP → gal-1-phosphate + ADP, with $\Delta G^{\circ\prime} = -1.9$ kcal/mol). Determine the reaction that should be coupled to the kinase reactions (a) and (b) as well as its expected $\Delta G^{\circ\prime}$.

26. Using the value for the equilibrium constant of ATP hydrolysis at pH 7 and 37°C determined from the glutamine synthetase and glutaminase reactions, calculate the ΔG° for ATP hydrolysis at 25°C and pH 8.0.

27. The ideal conditions of level flow ($\Delta G_p = 0$) cannot be utilized in order to measure the stoichiometry of ATP synthesis, because the concentrations of ATP, ADP, and phosphate required for $\Delta G_p = 0$ are not experimentally attainable. Calculate the value of J_p/J_e for conditions between level and static flow with $\Delta G_p = 42$ kJ/mol, $\Delta G_e = -210$ kJ/mol from an NAD-linked substrate, $Z = 3$, and $q = 0.97$ (Lemasters, 1984).

28. Show that the free energy change for a reversible reaction involving p–V work and electrical work for movement of n equivalents of an ion of charge ($+z$) through a potential ΔE is: $\Delta G = +nzF\Delta E$ at constant pressure and temperature. What is the free energy change for movement of 1 mol of electrons (charge $= -1$) through a potential of 1 V?

Chapter 2
Oxidation–Reduction; Electron and Proton Transfer

2.1 Direction of Redox Reactions

It was shown in Chap. 1 that a free energy change $\Delta G_{p,T}$ can be associated with a change of electrical potential. In the present case, we consider the application to oxidation–reduction potentials. Thus,

$$\Delta G = +nzF \cdot \Delta E = -RF \cdot \Delta E, \tag{1}$$

with $z = -1$ for electrons, for the free energy change in a reversible reaction associated with electrical work driven by a change in oxidation–reduction potential, ΔE, and

$$\Delta G° = -nF \cdot \Delta E°, \tag{1a}$$

for the changes in standard potential and free energy.
Since

$$\Delta G° = -RT \ln K_{eq}, \tag{2}$$

$$\Delta E° = \frac{RT}{nF} \ln K_{eq}. \tag{3}$$

Because reactions tend to go forward if $\Delta G < 0$, and $\Delta G = -nF\Delta E$, the criterion for a forward direction of an oxidation–reduction reaction is

$$\Delta E > 0. \tag{4}$$

From (3),

$$K_{eq} > 1$$

when $\Delta E° > 0$.

Electron transfer tends to proceed, by these criteria, in the direction of more positive E and $E°$. Alternatively, a strong reductant results from a redox couple with a negative oxidation–reduction potential, and a relatively strong oxidant from a couple with a positive potential.

It is important to note that the oxidation potentials E and $E°$ are not state functions. A consequence is that these functions are not necessarily additive. That is, for a series of sequential redox reactions $A \rightarrow B \rightarrow C \rightarrow D$, $\Delta E_{AD} \neq \Delta E_{AB} + \Delta E_{BC} + \Delta E_{CD}$, although the free energy as a state function is additive i.e., $\Delta G_{AD} = \Delta G_{AB} + \Delta G_{BC} + \Delta G_{CD}$ (see section 2.5.1).

2.2 The Scale of Oxidation–Reduction Potentials

Absolute values of the oxidation–reduction potential, like the absolute value of the free energy, have no meaning in nature. The absolute values are set by the reference or standard that is chosen, which in this case is the hydrogen

Table 2.1. Oxidation–reduction potentials of some important redox couples

	Standard or midpoint potential (mV)	
Half-reaction	$E°$ (pH = 0)	$E°'$ (E_{m7})
A_o (ox) + e^- → A_o (red)[a]	—	−900
Pheophytin (ox) + e^- → Pheophytin (red)	—	−600
Ferredoxin (ox) + e^- → Fd (red)	—	−432
$2H^+ + 2e^- \rightarrow H_2$(g) [Reference]	0.0	−414 (25°C)
O_2(g) + e^- → O_2^-	—	−330[b]
$NAD^+ + H^+ + 2e^- \rightarrow NADH$[c]	−113	−324
FMN (ox) + $2e^- + 2H^+$ → $FMNH_2$	+209	−205
O_2(l) + e^- → O_2^-	—	−160[a]
cyt b + e^- → cyt b (red)[d]	—	approx. −50
menadione + $2e^- + 2H^+$ → menadiol[e]	+422	+8
ubiquinone + $2e^- + 2H^+$ → ubiquinol	—	+60
cyt c + e^- → cyt c (red)	—	+250
plastocyanin + e^- → plastocyanin (red)	—	+370
cyt a + e^- → cyt a (red)	—	+250–400
$Fe(CN)_6^{3-} + e^- \rightarrow Fe(CN)_6^{4-}$	—	+450 (0.5 M NaCl)
$O_2 + 4H^+ + 4e^- \rightarrow 2H_2O$	+1,230	+815
Tyrosine/tyrosine (aq)	—	+930
$P680^+ + e^- \rightarrow P680$	—	+1,170

[a] Primary electron acceptor of photosystem I of oxygenic photosynthesis (section 6.7).
[b] cf., Wood (1987).
[c] The NAD^+/NADH couple is unusual in having an $H^+:e^-$ ratio = 0.5.
[d] From cytochrome b_6–f complex.
[e] This compound has been used to treat pathological disorders in the cytochrome b–c_1 complex by serving as a redox bypass of this region of the respiratory chain (Chap. 7, section 4).

electrode at pH = 0 (concentration of H^+ = 1 M) under standard conditions of temperature and pressure.

$$2H^+ + 2e^- \rightleftarrows H_2(g); \qquad E° = 0.0 \text{ V}$$

An electrochemical reaction written in this form is called a half-cell reaction because it describes the redox events occurring at one of the two electrodes that could be used to complete a current-carrying electrolytic cell. The free energy changes associated with a half-cell reaction are:

$$\Delta G° = -nFE°, \qquad (5a)$$

and

$$\Delta G = -nFE \qquad (5b)$$

One can see from the hydrogen half-cell reaction that its equilibrium constant and $E°$ are both pH dependent. Its standard potential (in mV) at pH = 7, $E°'$, is -414 mV at 25°C (see example 3 below).

Using the hydrogen electrode as the standard, a short list of half-cell redox reactions of biochemical interest is shown in Table 2.1.

2.3 Oxidation–Reduction Potential as a Group-Transfer Potential; Comparison of Standard Potentials and pK Values

The standard potential should be considered as a measure of the tendency to donate or accept electrons. The more negative the potential, the greater the tendency to donate (reducing ability), the more positive, the greater the affinity for the electron (oxidizing ability). Thus, in Table 2.1, reduced ferredoxin and NAD(P)H are strong reductants, and ferricyanide [$Fe(CN)_6^{3-}$] and molecular O_2 are strong oxidants. A strong reductant is a good electron donor, and a strong oxidant a good electron acceptor. There is an important analogy with a strong and weak acid as a good proton donor and acceptor, respectively, and this analogy can be extended to high-energy bonds and group transfer potentials (Table 2.2). The standard potential, $E°$, of an oxidation–reduction reaction is analogous to the pK of an acid–base reaction. Just as the pK is the pH at which an acid is half-protonated, the $E°$ is the redox potential under standard conditions at which an electron donor-acceptor is 50% reduced. For this reason, the notation E_m is used as an approximation for $E°$ when ligand binding (see section 2.5) or concentration effects do not enter the problem. When the midpoint potential is affected by proton or ligand binding, the $E°$ is defined as the midpoint potential under standard conditions (1 M protons or other binding ligand).

Table 2.2. Oxidation–reduction potential considered as a group transfer potential

	Proton-transfer potential (acid–base)	Electron-transfer potential (redox potential)	Group-transfer potential (high-energy bond)
Equation	$A^- + H^+ \to AH$	$A^+ + e^- \to A$	$A \sim PO_4 \to A + PO_4$
Measure of transfer potential	$pK_a = \dfrac{\Delta G°}{2.3RT}$,	$E° = -\dfrac{\Delta G°}{nF}$,	$\Delta G°$ per mole phosphate transferred

Based on Klotz (1967, 1986).

From the Henderson–Hasselbalch equation, it is known that a pH buffer exerts its effect over a range of ± 1 pH unit around its pK; similarly, a one-electron redox buffer tends to stabilize the redox potential over a range of ± 59 mV (25°C) around its $E°$ or E_m.

2.4 Calculation of the Potential Change for Linked and Coupled Reactions

1. Consider $A \to B \to C$, so

$$A + n_1 e^- \to B, \quad E° = E_{AB}°$$

$$B + n_2 e^- \to C, \quad E° = E_{BC}°,$$

with the net reaction

$$A + (n_1 + n_2)e^- \to C, \quad E° = E_{AC}°$$

(Eisenberg and Crothers, 1979).

Note that for the purposes of balancing the redox equations and calculating the net free energy change, the convention in biochemistry is to write the redox equation with the oxidized species on the left-hand side of the equation.

The total standard free energy change, ΔG_{AC}, is

$$\Delta G_{AC}° = \Delta G_{AB}° + \Delta G_{BC}°. \tag{6}$$

From Eq. 5a, b,

$$-(n_1 + n_2)FE_{AC}° = -n_1 FE_{AB}° - n_2 FE_{BC}°$$

$$E_{AC}° = \frac{n_1 E_{AB}° + n_2 E_{BC}°}{(n_1 + n_2)}, \tag{7}$$

and

$$E_{AC} = \frac{n_1 E_{AB} + n_2 E_{BC}}{(n_1 + n_2)} \tag{8}$$

(see section 2.5.1).

2.4 Calculation of the Potential Change for Linked and Coupled Reactions

If $n_1 = n_2$,
$$E_{AC}° = \frac{E_{AB}° + E_{BC}°}{2}, \tag{8a}$$

and
$$E_{AC} = \frac{E_{AB} + E_{BC}}{2}. \tag{8b}$$

2. Consider the reduction of B by A, given the half-cell potentials.

$$A(ox) + n_1 e^- \to A(red), \quad E° = E_A°; \quad G° = G_A°$$
$$B(ox) + n_2 e^- \to B(red), \quad E° = E_B°; \quad G° = G_B°,$$

with the charge-balanced reaction,

$$\left(\frac{n_2}{n_1}\right) A(red) + B(ox) \to \left(\frac{n_2}{n_1}\right) A(ox) + B(red),$$

for transfer of n_2 electrons from A to B. The standard free energy change, $\Delta G_{AB}°$, for transfer of the n_2 electrons from A to B is

$$\Delta G_{AB}° = G_B° - \frac{n_2}{n_1} G_A°. \tag{9}$$

From Eqs. 5a, b,

$$-n_2 F \cdot \Delta E_{AB}° = -n_2 F E_B° - \frac{n_2}{n_1}(-n_1 F E_A°),$$

and then,
$$\Delta E_{AB}° = E_B° - E_A°; \tag{10a}$$

similarly,
$$\Delta E_{AB} = E_B - E_A \tag{10b}$$

so that the change of standard potential in a complete redox reaction is the difference of the standard potentials for the two half-cell reactions. For the reaction to proceed in the forward direction, $\Delta E > 0$ and $E_B° > E_A°$, i.e., in the directon of more positive $E°$ and higher electron affinity.

2.5 Concentration Dependence of the Oxidation–Reduction Potential

From Chap. 1, $G = G° + RT \ln(c/c_o)$,

$$\Delta G = \Delta G° + RT \ln \frac{(products)}{(reactants)},$$

$$\Delta G = -nF\Delta E,$$

and

$$\Delta G° = -nF\Delta E°$$

for a reaction, $[\text{ox}] + ne^- \to [\text{red}]$, involving transfer of n electrons.
Then

$$\Delta E = -\frac{\Delta G°}{nF} - \frac{RT}{nF}\ln\frac{(\text{red})}{(\text{ox})}$$

$$= \Delta E° - \frac{RT}{nF}\ln\frac{(\text{red})}{(\text{ox})},$$

$$= \Delta E° - 2.3\frac{RT}{nF}\log_{10}\frac{(\text{red})}{(\text{ox})}, \tag{11a}$$

and for a half-cell reaction, one would write,

$$E = E° - 2.3\frac{RT}{nF}\log_{10}\frac{(\text{red})}{(\text{ox})} \tag{11b}$$

Note the change of sign in front of the concentration term compared to Eq. 1.13. Expressions (11a) and (11b) state the Nernst equation for redox reactions.

Since $2.3\frac{RT}{F} = 59.1$ mV at 25°C,

$$\Delta E(\text{mV}) = \Delta E° - \frac{59}{n}\log_{10}\frac{(\text{red})}{(\text{ox})}, \tag{12}$$

and for a half-cell reaction, one would write

$$E = E° - \frac{59}{n}\log\frac{(\text{red})}{(\text{ox})}. \tag{13}$$

The fact that the $E°$ values can be considered as midpoint potentials is illustrated in Eqs. 11 and 12, because $E = E°$ and $\Delta E = \Delta E°$ when the concentration of reductant equals that of oxidant. In the absence of ligand binding, one would write:

$$\Delta E = \Delta E_m - \frac{59}{n}\log_{10}\frac{(\text{red})}{(\text{ox})} \tag{12a}$$

$$E = E_m - \frac{59}{n}\log_{10}\frac{(\text{red})}{(\text{ox})} \tag{13a}$$

As discussed for ΔG (Chap. 1, section 6), the direction of an oxidation–reduction reaction is determined by the concentrations of products and reactants as well as the ΔE_m, i.e., by the ΔE.

2.5 Concentration Dependence of the Oxidation–Reduction Potential

Exercise 2.1. Graph the level of reduction, cyt $c(\text{red})/[\text{cyt } c(\text{red}) + \text{cyt } c(\text{ox})] = \text{cyt } c(\text{red})/\text{cyt } c(\text{totl})$, as a function of the external redox potential, E.

For a half-cell reaction involving transfer of one e^- ($n = 1$):

$$\text{cyt } c(\text{ox}) + e^- \rightarrow \text{cyt } c(\text{red})$$

$$E = E_m - 59 \log \frac{c(\text{red})}{c(\text{ox})}$$

Note, for every 10-fold increase in $c(\text{red})/c(\text{ox})$, E decreases by 59 mV. (Think about the analogy with acid-base problems).

The solution would be:

For $c(\text{ox}) + e^- \rightarrow c(\text{red})$,

$$E = E_m - 2.3 \frac{RT}{nF} \log_{10} \frac{c(\text{red})}{c(\text{ox})}, \quad \text{or}$$

or

$$E = E_m - 59 \log_{10} \frac{c(\text{red})}{c(\text{ox})},$$

for $n = 1$, at 25°C, in mV. The parentheses defining effective concentration or activity in the concentration quotient have again been removed for simplicity. Two graphical aids are that $c(\text{red}) = c(\text{ox})$ when $E = E_m$ and $c(\text{red})/c(\text{ox})$ changes by a factor of 10 for every change of 59 mV in the potential. So, $c(\text{red})/c(\text{total}) = 0.5$ when $E = E°$, 0.91 when $E = E_m - 59$ (mV), and 0.09 when $E = E_m + 59$ (mV). Then the graphical solution is:

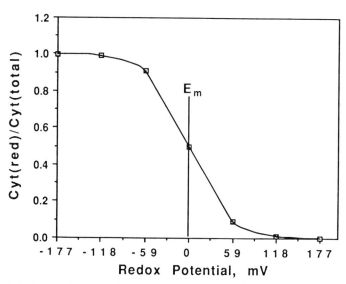

Figure 2.1. Redox titration of an $n = 1$ component such as cytochrome c. Note that the plot of $\log c(\text{red})/c(\text{ox})$ will be a straight line.

Exercise 2.2. Determine the potential change for two physiological coupled one e⁻ half-cell reactions, the reduction of cytochrome c by cytochrome b:
The two half-cell reactions are

$$\text{cyt } b(\text{ox}) + e^- \rightarrow \text{cyt } b(\text{red})$$

$$\text{cyt } c(\text{ox}) + e^- \rightarrow \text{cyt } c(\text{red}), \quad \text{and the overall reaction is:}$$

$$\text{cyt } b(\text{red}) + \text{cyt } c(\text{ox}) \rightarrow \text{cyt } b(\text{ox}) + \text{cyt } c(\text{red})$$

From Eq. 12a, the change in the potential of the reaction is:

$$\Delta E = \Delta E_m - 59 \log \frac{b(\text{ox}) \cdot c(\text{red})}{b(\text{red}) \cdot c(\text{ox})},$$

with

$$\Delta E_m = E_m(\text{cyt } c) - E_m(\text{cyt } b).$$

Note, for every 10-fold increase (decrease) in $\frac{b(\text{ox})}{b(\text{red})} \cdot \frac{c(\text{red})}{c(\text{ox})}$, the ratio of products to reactants, the ΔE decreases (increases) by 59 mV.

Exercise 2.3. Determine the half-cell potential for a reaction involving 2e⁻ ($n = 2$) and H⁺ transfer, e.g., the hydrogen electrode reaction or reduction of quinone to quinol.
The half cell-reactions are:

$$2H^+ + 2e^- \rightarrow H_2(g), \quad \text{the } H_2 \text{ electrode reaction,}$$

or

$$\text{Quinone} + 2e^- + 2H^+ \rightarrow \text{Quinol (QH}_2).$$

Considering the quinone–quinol reaction, the potential at a particular pH value is

$$E = E° - \frac{59}{2} \log \frac{(QH_2)}{(Q)(H^+)^2},$$

or

$$E = E° + \frac{59}{2} \log(H^+)^2 - \frac{59}{2} \log \frac{(QH_2)}{(Q)};$$

then,

$$E = (E° - 59 \text{ pH}) - \frac{59}{2} \log \frac{(QH_2)}{(Q)},$$

and

$$E = E_{mh} - \frac{59}{2} \log \frac{(QH_2)}{(Q)}, \tag{14}$$

2.5 Concentration Dependence of the Oxidation–Reduction Potential

where $(E_{mh} = E° - 59\,pH)$ is the midpoint potential at a particular pH value. Note: (a) The standard potential, $E°$, of the hydrogen electrode is defined as 0.0 V at pH 0, 1 atm pressure, and 20°C. The standard potential, $E°$, of the 2-electron benzoquinone–hydroquinone couple at pH = 0 and 25°C is +681 mV. (b) The standard potential for reactions such as the hydrogen electrode reaction $(2H^+ + 2e^- \rightleftarrows H_2)$, with equal numbers of electrons and protons transferred, decreases by 59 mV for each unit increase in pH value, so $\Delta E_{mh}/\Delta pH = -59$ mV. Equation 14 is valid whenever the number of electrons and protons in the redox reaction is the same. One exception will be the $NAD(P)^+/NAD(P)H$ couple, for which $NAD^+ + 2e^- + H^+ \rightleftarrows NADH$; $E_{m7} = -0.32$ V, and $\Delta E_{mh}/\Delta pH = -59/2$ mV. (c) For $n = 2$ reactions such as the hydrogen electrode or quinone reduction, the value of E decreases by $59/2$ mV for every 10-fold increase in the quotient QH_2/Q.

Exercise 2.4. pH-Dependent midpoint potential, E_m, of *Pseudomonas aeruginosa* cytochrome *c*:

This reaction is pH dependent because of a heme propionic acid that becomes protonated, with $pK_{red} > pK_{ox}$ ($K_{red} < K_{ox}$) for proton dissociation from ferro- and ferricytochrome *c*, respectively. It is almost always true, as in the present example, and exercise 3 for quinone reduction, that $pK_{red} > pK_{ox}$, because the reduced compound with the additional electronic charge is expected to be more readily protonated. One can then derive the dependence of E_{mh} on the H^+ concentration, and the proton dissociation constants, K_r and K_o, for the reduced and oxidized cytochrome. The total amount, S_r or S_o, of reduced and oxidized compound c is:

$$(S_r) = (c_r) + (H^+ \cdot c_r),$$

and

$$(S_o) = (c_o) + (H^+ \cdot c_o),$$

for the protonated and unprotonated forms, with dissociation constants defined by:

$$\frac{(c_r)(H^+)}{(c_r \cdot H^+)} = K_r, \quad \text{and} \quad \frac{(c_o)(H^+)}{(c_o \cdot H^+)} = K_o.$$

For the reduction of the cytochrome, $S(ox) + e^- \rightarrow S(red)$,

$$E = E_{mh} - \frac{2.3RT}{F} \log_{10} \frac{(S_r)}{(S_o)}.$$

If $pK_r > pK_o$, there are three different regions for the redox reactions:

(i) the basic pH region, $pH > pH_r$, reactant and product unprotonated,

$$c_o + e^- \rightarrow c_r; \quad E_{mh} = E_{mb},$$

the constant limiting E_m at basic pH.

(ii) intermediate pH, $pK_r > pH > pK_o$,

$$c_o + e^- + H^+ \to c_r \cdot H^+;$$

E_{mh} as in exercise 3,

$$\Delta E_{mh}/\Delta pH = -59 \text{ mV}.$$

(iii) acidic pH, $pK_r > pK_o > pH$.

$$c_o \cdot H^+ + e^- \to c_r \cdot H^+; \qquad E_m = E_{ma},$$

constant limiting E_m at acidic pH.

A plot of the midpoint potential versus pH is shown in Fig. 2.2, where the limiting E_m values, E_{mb} and E_{ma}, are indicated.

The effect of preferential ligand (e.g., H^+) binding to the reduced form ($pK_r \gg pK_o$) on the midpoint potential of the redox couple can be seen in Fig. 2.2. Stabilization of the reduced form means that the compound is more readily reduced, i.e., it is a better oxidant and has a more positive E_{mh}. It can be seen in Fig. 2.2 that the effect of increasing ligand concentration, or lower pH, is to increase the E_{mh}.

The discussion of the effect of H^+ concentration on E_m applies to any other ligand. Thus, for an arbitrary ligand, L, in the region of intermediate concentration for the case $K_r \ll K_o$:

$$c_o + L + e^- \to c_r \cdot L,$$

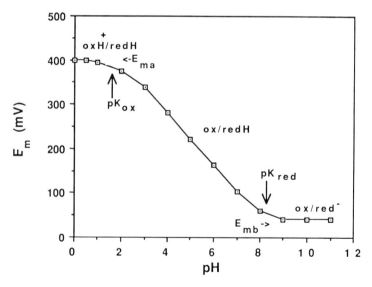

Figure 2.2. Dependence of midpoint potential on pH for $pK_r \gg pK_o$. (Based on Dutton, 1978.)

2.5 Concentration Dependence of the Oxidation–Reduction Potential

and

$$E = E_{mL} - 59 \log_{10} \frac{(c_r \cdot L)}{(c_r)(L)}$$

$$= E° + 59 \log_{10}(L) - 59 \log_{10} \frac{(c_r \cdot L)}{(c_o)}; \tag{15}$$

the midpoint potential including the influence of L is:

$$E_{mL} = E° + 59 \log_{10}(L), \tag{16}$$

where $E°$ is the standard potential in the presence of the ligand at a concentration of 1 M.

A graph of the dependence of midpoint potential on the concentration of ligand, L, for $K_r \ll K_o$ would have the same appearance as Fig. 2.2 for H^+ binding. The limiting midpoint potential in Fig. 2.2 at high pH would correspond to the midpoint in the absence of ligand binding. The preferential binding of L to the reduced form, as shown by Eq. 16, would then cause an increase of 59 mV for each 10-fold increase in the concentration of L for the case $K_r \ll L \ll K_o$.

Exercise 2.5. Cytochrome reduction by quinol:
The special feature of this reaction is that it involves the reduction of a one-electron, pH-independent redox component by a compound whose oxidation-reduction involves two electrons and is pH dependent.

$$QH_2 + 2\,cyt(ox) \to Q + 2\,cyt(red) + 2H^+.$$

The two half-reactions are:

$$QH_2 \to +2e^- + 2H^+; \quad E_1° = E°(Q/QH_2)$$

$$cyt(ox) + e^- \to cyt(red); \quad E_2° = E_m[c(ox)/c(red)].$$

From Eqs. 10 and 12,

$$\Delta E = \Delta E° - \frac{59}{2} \log \frac{(Q) \cdot (c_r)^2 \cdot (H^+)^2}{(QH_2) \cdot (c_o)^2},$$

with $\Delta E° = E_2° - E_1°$.

Then

$$\Delta E = (\Delta E° + 59\,\mathrm{pH}) - \frac{59}{2} \log \frac{(Q)}{(QH_2)} \cdot \left(\frac{c_r}{c_o}\right)^2.$$

If one expands the argument of the logarithm, one can see that the term arising from the cytochrome redox reaction describes a one-electron oxidation–reduction:

i.e., $\quad \dfrac{59}{2} \log_{10} \dfrac{(Q)}{(QH_2)} \cdot \left(\dfrac{c_r}{c_o}\right)^2 = \dfrac{59}{2} \log_{10} \dfrac{(Q)}{(QH_2)} + 59 \log \dfrac{(c_r)}{(c_o)}.$

2.5.1 Dependence of ΔE on Reaction Pathway: Reduction of O_2 to H_2O

ΔE, unlike ΔG, depends on the path of the reaction (section 2.4). As an example, consider the four different pathways for reduction of O_2 to H_2O involving the intermediates, superoxide (O_2^-) and hydrogen peroxide (H_2O_2) (Fig. 2.3). One path would be the four-electron reduction:

$$O_2 + 4e^- + 4H^+ \rightarrow 2H_2O, \quad \text{with } E_{m7} = 0.82 \text{ V}.$$

The relationships between the E_{m7} values for the different reactions can be determined from Eq. 8. The value for one of them [e.g., $E_{m7}(O_2/O_2^-)$] can be solved from the E_{m7} and n values for the O_2/H_2O (0.82 V, $n = 4$) and O_2^-/H_2O (1.20 V, $n = 3$) couples.

From Eq. 8,

$$E_{AB} = \frac{(n_1 + n_2)E_{AC} - n_2 E_{BC}}{n_1}$$

Then,

$$E_{AB} = \frac{(1 + 3)(0.82) - (3)(1.20)}{(1)} = -0.32 \text{ V}$$

2.5.2 Medical Relevance of Oxygen Free Radicals

Oxygen radicals have been implicated as causative agents in aging and disease (Halliwell, 1987). O_2^- is produced in activated phagocytic cells, is highly reactive in a hydrophobic environment, can cross membranes as $HO_2^{\cdot}$ (pK = 7.4), and is catalytically removed by superoxide dismutase, $2O_2^- + 2H^+ \rightarrow H_2O_2 + O_2$. The reactive hydroxyl radical, $OH^{\cdot}$, or a related species, can be formed by ferrous (e.g., Fe^{2+}-cytochrome c) or cuprous ion reduction of H_2O_2 to $OH^{\cdot}$ and OH^-. Site-specific $OH^{\cdot}$ formation can cause strand breaks in DNA, deplete glutathione levels, liberate metals into tissue, stimulate protease

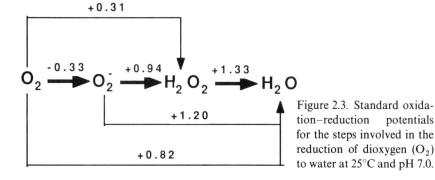

Figure 2.3. Standard oxidation–reduction potentials for the steps involved in the reduction of dioxygen (O_2) to water at 25°C and pH 7.0.

activity, and cause lipid peroxidation. The phagocytes of the animal host defense system are able to eliminate invading microorganisms by generating the strong oxidants H_2O_2, OH^--like free radicals, and oxidized halogens such as HOCl. All of these oxidants arise from a common precursor, $O_2^{\overset{\cdot}{-}}$. The $O_2^{\overset{\cdot}{-}}$ is produced in very large quantities by the phagocytes, after the cells have been activated by contact with their targets, through a membrane-bound oxidase enzyme that catalyzes a one electron reduction of O_2 by NADPH (Babior, 1987). Much of the information on this oxidase has come from patients with chronic granulomatous disease, a group of inherited disorders in which oxidase activity is reduced or absent. These patients suffer from severe bacterial infections of the deep tissues.

2.6 Experimental Determination of E and E_m Values

A common method for measurement of redox potentials of colored redox proteins is to use a spectrophotometric cuvette in which electrodes can be inserted to measure simultaneously the oxidation–reduction potential and the optical density change (Fig. 2.4). The spectrophotometric assay is made at wavelengths where there is a local maximum in the absorbance difference of reduced and oxidized states. For the redox protein cytochrome c, such a wavelength might be 550 nm, the peak of the reduced cytochrome α-band,

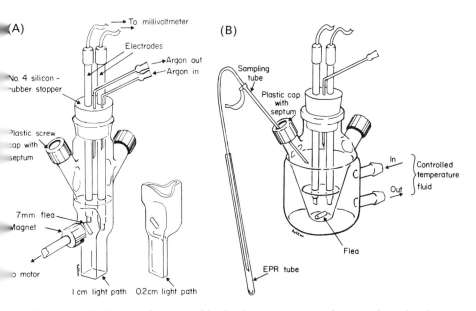

Figure 2.4. (A) Cuvette for anaerobic simultaneous spectrophotometric and redox measurements of stirred samples. (B) Vessel for determination of redox potential in samples that can be transferred to an EPR tube (Dutton, 1978).

Table 2.3. Some commonly used redox buffers

Redox buffers[a]	E_{m7} (pH = 7) (mV)
Neutral red	−320
Anthraquinone-2-sulfonate	−220
2-Hydroxy-1,4-naphthoquinone	−140
Indigo-disulfonate	−120
2,5-Dihydroxy-1,4-benzoquinone	−60
Indigo-tetrasulfonate	−40
1,4-Naphthoquinone	+65
1,2-Naphthoquinone	+135
2,6-Dichlorophenol-indophenol	+220
1,4-Benzoquinone	+260
Ferricyanide (one e^- buffer) – a commonly used oxidant	+450

[a] One- and two-electron ($n = 1$ and $n = 2$) redox buffers will buffer the potential over a range of ±59 mV and ±29.5 mV around the E_m, corresponding to 90% oxidation and 90% reduction of the buffer. Other necessary properties of the redox mediators are that they must (i) react reversibly with both the electrode and the protein to be titrated, and (ii) not change state upon oxidation or reduction, interfere with titration of the optical density changes, nor introduce artificial pathways of electron transport.

and a wavelength at which the oxidized form has a small extinction coefficient and a flat spectrum (Fig. 4.3).

Measurement of redox potential of $A(ox) \pm e^- \rightarrow A(red)$: A platinum electrode is used to measure the voltage generated by electrons flowing to reduce A(ox) relative to the voltage of a reference cell, e.g., $AgCl + e^- \rightarrow Ag + Cl^-$. A problem arises if compound A is large and does not readily diffuse to the platinum electrode, e.g., if A is a protein or a component bound to membrane. In this case one uses mediator compounds that diffuse between the platinum electrode and compound A. The mediator compounds are redox buffers in the same way that proton donors and acceptors are pH buffers. Just as a pH titration would be conducted by adding acid and base, a redox titration is carried out reversibly by adding oxidant and reductant. The latter can be done chemically (e.g., with sodium dithionite, $S_2O_4^{2-}$, whose active species is SO_2^-) or electrically by supplying current to an electrode. A list of some commonly used redox mediators is shown in Table 2.3.

2.7 Factors Affecting the Redox Potential

As stated above, the midpoint potential E_m is a measure of the electron affinity of a compound. The more positive the E_m, the higher the affinity for the electron, and the more stable the reduced form. In considering the effect of different environmental influences on the E_m, it is always a question of the effect on the electron affinity of the redox center or, phrasing it another way, the effect on the relative stability of the oxidized versus the reduced form.

2.7 Factors Affecting the Redox Potential

Some important parameters having an effect on the E_m of redox compounds in solution are ionic strength, ligand binding, solvent polarity, ligand chemistry and geometry, and local charges or fields.

2.7.1 Ionic strength

The solution E_m of small metalloorganic compounds is affected by ionic strength because of the different charge of reduced and oxidized forms. For the ferricyanide $[Fe(CN)_6^{3-}]$/ferrocyanide $[Fe(CN)_6^{4-}]$ couple, $E_m = +0.36$ V at zero ionic strength and $+0.45$ V at high (> 0.5 M) ionic strength. It can be derived from Debye–Hückel theory that the free energy change, ΔG, due to an ion of charge z interacting with an ionic solution, is $\Delta G = -\kappa z^2$, where κ is a constant proportional to the square root of the ionic strength (Eisenberg and Crothers, 1979). Thus, the more highly charged form, reduced ferrocyanide in this case, is more stable in the presence of higher ionic strength, so that its E_m would increase with increasing ionic strength.

2.7.2 Ligand Binding

As discussed above (exercise 2.4), preferential ligand binding to the reduced or oxidized forms of a redox compound will stabilize these species and, respectively, increase or decrease the E_m. The ligand can be a proton (Fig. 2.2), or another ligand such as CO bound to ferrous heme, or phosphate or chloride binding to cytochrome c (Fig. 2.5).

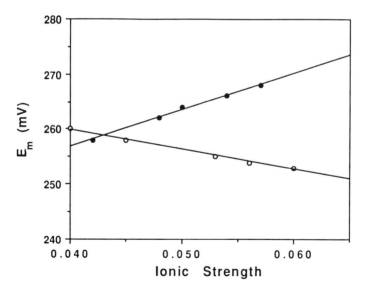

Figure 2.5. Dependence of cytochrome c E_m on ionic strength in the absence of binding ions (o) and in the presence of Cl$^-$ and Na$^+$ (●). (Based on Schejter and Margolit, 1970.)

An example of ionic strength and ligand effects is the effect of ion binding on the net charge and midpoint potential of cytochrome c:

It can be inferred from the decrease of E_m with increasing ionic strength, in a medium where there is no special ion binding (open circles in Fig. 2.5), that the net charge of oxidized cytochrome c must be greater than that of the reduced cytochrome, and is thus stabilized to a greater extent at higher ionic strength. The net charge of the reduced form of horse cytochrome c at pH 7 is $+7$ [$+9$ from the basic amino acid residues (pI of horse $c = +10.4$), -2 from the heme propionic acids], and $+8$ in the oxidized state after loss of one electron. However, in the presence of ions that bind to cytochrome c, particularly chloride and phosphate, the slope of the E_m versus ionic strength function is positive (Fig. 2.5), implying that the reduced form now has a greater net charge than the oxidized form. This is explained by the ability of ferri-(oxidized) cytochrome c to bind 2 Cl^-, $H_2PO_4^-$, or $MgADP^-$ ions (Margoliash et al., 1970; Schejter and Margalit, 1970), with little or no binding by the reduced form. In addition, the reduced, but not the oxidized, form has a tendency to bind one cation (e.g., Na^+). Thus, the net charge of the reduced form would be $+8$ with one bound sodium, and that of the oxidized cytochrome would be $+6$ with two bound Cl^- ions, so that the slope of the E_m versus ionic strength plot in the presence of Cl^- and Na^+ should be steeper than that in the presence of the nonbinding medium (Fig. 2.5).

Summary of changes in net charge of cytochrome c due to ion binding

No ion binding		Ion binding
c(ox)	$+8$	$+6\,(+2Cl^-)$
c(red)	$+7$	$+8\,(+Na^+)$

2.7.3 Effect of Axial Heme Ligands and Side Chains on Metallo-Porphyrin E_m

The more basic the axial ligand (Fig. 2.6), the better it is as an electron donor, the greater its attraction for the more positively charged Fe^{3+} form of the iron atom, and therefore the greater its tendency to stabilize the Fe^{3+} oxidized form, i.e., decrease the E_m. The effect of such ligands can be estimated through their pK_a. The greater the basicity of the ligand, the larger its effect in decreasing the midpoint potential (Table 2.4).

Thus, cytochromes with two axial histidine ligands usually have a more negative E_m than those with one methionine ligand along with the histidine, because the imidazole nitrogen is a better electron donor than the methionine sulfur. Thus, the E_{m7} values of the *bis*-histidine, histidine-methionine, and *bis*-methionine complexes of the iron porphyrin, mesoheme, are -220, -110,

2.7 Factors Affecting the Redox Potential

Figure 2.6. Iron-porphyrin with two pyridines as the axial ligands.

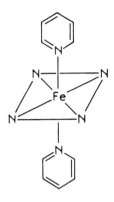

Table 2.4. Effect of basicity of extra ligands on the potential of iron-protoporphyrin

Extra ligand	pK_a	E_{m7} (mV)
Nicotine	3.2	200
Pyridine	5.2	160
α-Picoline	6.2	115
Histidine	6.0	>80
Pilocarpine	7.0	>50

From Falk, 1964.

and +20 mV, respectively (Moore and Williams, 1977). The relative orientation of the two imidazole rings has a small effect on the heme E_m value for bis-histidine coordination (Babcock et al., 1985). The E_m for an orthogonal orientation can be 50 mV more positive than that corresponding to a parallel alignment (Walker et al., 1986).

2.7.4 Effect of Membrane Potential

The equivalent effect of a change in redox potential, ΔE, or of membrane potential, $\Delta \psi$, on the free energy of an electron (Chap. 1, Eq. 18) implies that the midpoint redox potential of a redox center in a membrane is affected by the presence of a membrane potential. The change in E_m of the redox center will be opposite in sign to that of the membrane potential at the center. That is, for two centers in equilibrium with $\Delta \psi = 0$ on opposite sides of a membrane bilayer, $\tilde{\mu}_{in} = \tilde{\mu}_{out}$, and $E_{mi} = E_{mo}$: If the membrane is energized, so $\Delta \psi_{in} = -100$ mV, the E_m values must change to maintain electrochemical equilibrium, and $E_{mi} = E_{mo} + 100$ mV (Walz, 1979; for applications, cf., Hinkle and Mitchell, 1970; Takamiya and Dutton, 1977).

2.8 Redox Properties of Quinones and Semiquinones

Besides the two-electron oxidation–reduction of quinones (Exercise 2.3), one-electron reactions involving the semiquinone can occur, generally with different midpoint potentials. The two electron quinone reaction is

$$Q + 2e^- + 2H^+ \rightleftarrows QH_2, \tag{17a}$$

with midpoint potential $\equiv E_{mh}$.

The reactions involving the semiquinone are:

$$Q + e^- + H^+ \rightleftarrows QH^\cdot, \tag{17b}$$

with midpoint E_{m1} at the given pH,

and

$$QH^\cdot + e^- + H^+ \rightleftarrows QH_2, \tag{17c}$$

with midpoint potential $\equiv E_{m2}$.

The other reaction linking the fully reduced (quinol), oxidized (quinone), and half-reduced (semiquinone) components is that of "dismutation":

$$2QH^\cdot \rightleftarrows Q + QH_2,$$

with the semiquinone formation constant using equilibrium concentrations,

$$K_s = \{(QH^\cdot)^2/(Q)(QH_2)\}, \tag{17d}$$

or

$$\{(Q^{\cdot-})^2/(Q)(QH_2)\}$$

at alkaline pH,

a measure of the stability of the semiquinone. The pK values of $(QH_2 \rightarrow QH^- + H^+)$ and $(QH^\cdot \rightarrow Q^{\cdot-} + H^+)$ are approximately 11 and 5–6, respectively, for duroquinol and durosemiquinone. The structure of the latter along with that of the simplest quinone, p-benzoquinone, and its associated quinol (hydroquinone) is shown in Fig. 2.7.

In general, $E_{mh} \neq E_{m1} \neq E_{m2}$ for the different quinone redox reactions because the affinity of the quinone (Q) for one electron reduction to the semiquinone (QH$^\cdot$) is different from its affinity for cooperative reduction to the quinol (QH$_2$) by two electrons, and also different from the affinity of the semiquinone for one electron reduction to the quinol. If the E_m of the (Q/QH$^\cdot$) reaction decreases, then that of the (QH$^\cdot$/QH$_2$) reaction must increase to balance the free energy change for the net (Q/QH$_2$) reaction. This follows from the fact that the path through the semiquinone is a second route, through two one-electron reactions, from a common initial state (quinone) to the same final state (quinol). In addition, the more unstable the QH· (lower value of K_s) the stronger QH· is as a reductant, and the more negative is the E_m of Q/QH·. Thus, there should be some definite relations between E_{mh}, E_{m1}, E_{m2}, and K_s.

2.8 Redox Properties of Quinones and Semiquinones

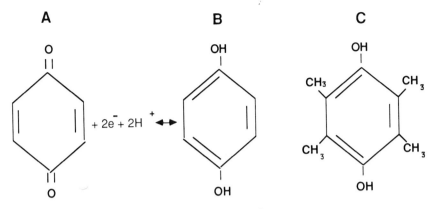

Figure 2.7. Structure of (A) *p*-benzoquinone and (B) its associated quinol, hydroquinone, with $E° = +681$ mV at 25°C, and (C) the tetramethyl-substituted quinol (duroquinol), with an $E° = +480$ mV at 25°C of the Q/QH_2 couple.

First, consider the relationship between E_{m1}, E_{m2}, and K_s: Let QH_2, $QH^{\cdot}$, and Q mix and freely adjust their concentrations in order to equilibrate through reactions 17b–d above. Let E_1 and E_2 be the working potentials of reactions 17b and 17c:

$$E_1 = E_{m1} - 59\log_{10}\frac{(QH^{\cdot})}{(Q)(H^+)}$$

$$E_2 = E_{m2} - 59\log_{10}\frac{(QH_2)}{(QH^{\cdot})(H^+)}$$

At equilibrium, $\Delta G = 0$ for coupled oxidation–reduction between reactions 17c and 17b, that is

$$[QH^{\cdot} + (QH^{\cdot} + e^- + H^+)] \rightleftarrows [QH_2 + (Q + e^- + H^+)],$$

with net reaction,

$$2QH^{\cdot} \rightleftarrows Q + QH_2.$$

Then, it can be seen that the equilibration between reactions 17b and 17c occurs through the dismutase reaction (17d), and at equilibrium,

$$E_1 = E_2,$$

which is defined as E_{12}.

Thus, at equilibrium between the two semiquinone reactions,

$$E_{12} = E_{m1} - 59\log_{10}\frac{(QH^{\cdot})}{(Q)(H^+)}, \tag{17e}$$

and

$$E_{12} = E_{m2} - 59\log_{10}\frac{(QH_2)}{(QH^{\cdot})(H^+)}; \tag{17f}$$

then,

$$E_{m1} - 59\log_{10}\frac{(QH^\cdot)}{(Q)(H^+)} = E_{m2} - 59\log_{10}\frac{(QH_2)}{(QH^\cdot)(H^+)},$$

and

$$E_{m1} - E_{m2} = 59\log_{10}\left\{\frac{(QH^\cdot)^2}{(Q)(QH_2)}\right\} = 59\log_{10}K_s, \tag{18}$$

where K_s is the semiquinone formation constant defined in Eq. 17d (Clark, 1960).

If $K_s > 1$, indicating that the semiquinone species is stable, $E_{m1} > E_{m2}$, and the couple $Q/QH^\cdot$ operates at a more oxidizing (more positive) potential than the couple $QH^\cdot/QH_2$. However, if K_s is less than one, and the semiquinone is unstable, then $E_{m1} < E_{m2}$, and the $Q/QH^\cdot$ couple provides a stronger reductant than the $QH^\cdot/QH_2$ couple. K_s has been estimated to be 10^{-10} in a nonpolar membrane environment (Mitchell, 1976), where the concentration of semiquinone would therefore be too small to be meaningful. In the case of semiquinone species thought to interact with proteins, the K_s has been found to be 10^{-1}–10^{-2} in mitochondria and chromatophores (Chap. 5), which corresponds to $E_{m1} - E_{m2} = -60$ to -120 mV and a significant concentration of semiquinone.

2.8.1 Relationship of Midpoint Potential of the Quinone/Quinol $n = 2$ Reaction to the Semiquinone $n = 1$ Reactions

To determine the relationship of the midpoint potential of Eq. 17a above to that of Eqs. 17b and c, write

$$Q + e^- + H^+ \rightleftarrows QH^\cdot, \tag{17b}$$

with free energy change $\Delta G = \Delta G_1$;

$$QH^\cdot + e^- + H^+ \rightleftarrows QH_2, \tag{17c}$$

with $\Delta G = \Delta G_2$. Equation 17a above is the sum of reactions 17b and 17c; i.e.,

$$Q + 2e^- + 2H^+ \rightleftarrows QH_2,$$

with the free energy change

$$\Delta G(Q/QH_2) = \Delta G_1 + \Delta G_2.$$

Then,

$$\Delta G(Q/QH_2) = -nFE(Q/QH_2) = -2FE(Q/QH_2) = \Delta G_1 + \Delta G_2.$$

Substituting for ΔG_1 and ΔG_2,

2.9 Midpoint Potentials of Electrons in Photo-Excited States

$$\Delta G(Q/QH_2) = -2FE(Q/QH_2)$$

$$= -F\left(E_{m1} - 59\log_{10}\frac{QH^{\cdot}}{Q}\right) - F\left(E_{m2} - 59\log\frac{QH_2}{QH^{\cdot}}\right)$$

$$= -F(E_{m1} + E_{m2}) + 59F\log\frac{QH_2}{Q}.$$

Then,

$$E(Q/QH_2) = \frac{E_{m1} + E_{m2}}{2} - \frac{59}{2}\log\frac{QH_2}{Q}, \quad (19)$$

and

$$E_m(Q/QH_2) = (E_{m1} + E_{m2})/2, \quad (20)$$

the average of the two semiquinone reactions.

If $K_s < 1$, then from Eq. 18, $E_{m1} < E_m < E_{m2}$, (21)

so that the strength of the reductant in the semiquinone/quinone couple > that of quinol/quinone > that of quinol/semiquinone. Similarly,

if $K_s > 1$, $E_{m1} > E_m > E_{m2}$. (22)

From Eqs. 18 and 22 it can be seen that the splitting of E_{m1} and E_{m2} around E_{mh} is symmetric, so that the two cases described in Eqn. 21 and 22 can be arranged in an E_m level diagram (Fig. 2.8).

The above formal treatment of quinones can also be applied to flavins and flavoproteins which are also two-electron compounds with a defined half-reduced (flavin semiquinone) state (Chap. 4.11).

2.9 Midpoint Potentials of Electrons in Photo-Excited States: Application to Photosynthetic Reaction Centers

The energy, W, in a photon of light is

$$W = h\nu = hc/\lambda \quad (23)$$

where h = Planck's constant = 6.63×10^{-34} J-s, ν and λ are the frequency and

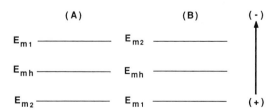

Figure 2.8. Relative E_m values for $K_s < 1$ (A) and $K_s > 1$ (B); $|K_s|$ the same in (A), (B).

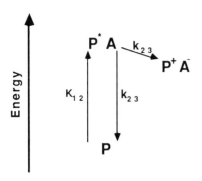

Figure 2.9. Simple energy level diagram for excitation of pigment molecule by absorbed light $h\nu$ with rate constant k_{12}, and deexcitation by fluorescence (rate, k_{21}) or electron transfer from the excited state to acceptor A (rate, $k_{23} \simeq 10^{12}$ s^{-1}).

Table 2.5. Comparison of energy ($h\nu$) of the ground-excited state (P*) transition in photosynthetic reaction centers with calculated $\Delta\mu_m$ value; E_m of P*

Reaction center complex	Energies (eV)		E_m values (V)	
	$h\nu$ (λ max of P)	$\Delta\mu_m$ (eV)[a]	$E_m(P^+/P)$[b]	$E_m(P/P^*)$[c]
P680	1.82	1.80	+1.0	−0.80
P700	1.77	1.75	0.50	−1.25
P870	1.42	1.39	+0.45	−0.95

[a] $\Delta\mu_m$ calculated from the average of absorption and emission energies.
[b] Determined by redox titration of ground state of reaction center.
[c] $E_m(P/P^*) = E_m(P^+/P) - \Delta\mu_m/e$.
From Blankenship and Prince, 1985.

wavelength (4–7 × 10^{-7} m for the visible region of the spectrum) of light, and c is the speed of light, 3 × 10^8 m·s^{-1}. Consider the free energy change in the excitation of a light-absorbing pigment molecule, P, from the ground state to the first excited singlet state from which fluorescence can occur (Fig. 2.9). This energy is equal to the energy, W, of the absorbed photon (Eq. 23) if the excitation does not cause a change in (i) volume or (ii) the number of molecular substates associated with the excited state relative to the ground state (Parson, 1978), and (iii) if P* does not relax through its vibrational-rotational substates before the electron transfer can take place. Some such relaxation, occurring on a time scale of ∼ 10^{-12} s, will generally occur. Then the midpoint chemical potential change, $\Delta\mu_m$, for the transition from the ground state to the excited state, should be equated to the energy of the photon associated with the fluorescence emission peak from the first excited singlet state, or alternatively the average of the absorption and emission energies (Seely, 1978).

The midpoint potential differences of the ground and excited states of the reaction center chlorophyll and bacteriochlorophyll complexes utilized in plant and bacterial photosynthesis are compared (Table 2.5), with the photon energies calculated from the wavelength of the absorbance peak of the reaction center (Chaps. 5 and 6). The E_m values of the excited state, P*, can be calculated from the $\Delta\mu_m$ value and the measured E_m of the ground (P$^+$/P) state (Seely,

1978; Blankenship and Prince, 1985). That is, $E_m(P/P^*) = E_m(P^+/P) - \Delta\mu_m/e$. The operating chemical potential of the excited state, which depends on its population, is dissipated somewhat because the electron transfer rate (rate constant, k_{23}, to the primary acceptor A in Fig. 2.9) is very fast, approximately 10^{12} s^{-1} (Parson, 1978).

2.10 Electron Transfer Mechanisms

The electron transfer reaction from a donor, D, to acceptor, A, can be depicted as:

$$D + A \rightleftarrows D - A \overset{k_{et}}{\rightleftarrows} D^+ - A^- \rightleftarrows D^+ + A^-,$$

where k_{et} is the rate constant for forward electron transfer from donor to acceptor in the precursor complex. One can focus on the rate constant k_{et} if only intramolecular transfer is considered. k_{et} depends on (i) the $\Delta G°$ of the reaction, (ii) the distance and orientation between donor and acceptor sites, and (iii) the nature of the intervening medium.

2.10.1 Dependence of Electron Transfer Rate on $\Delta G°$

Electron transfer reactions involving redox centers in solution are classified as outer sphere (reactants do not share a common ligand) or inner sphere. The present discussion is limited to the outer sphere case for which the interaction of the participating electronic orbitals of the two reactants is relatively weak. The energetics of the electron transfer is discussed in the context of the electron moving in the potential energy well of the reactant nuclei (Fig. 2.10). This formalism is similar to that used to discuss the theory of fluorescence (Chap. 6.1). It is assumed that the internuclear distances of reductants, products, and associated solvent do not change during the time of transfer of the electron from reactant to product. This is formally described by requiring that the transfer take place at the intersection of the potential energy surfaces of reactants and products (solid arrow in Fig. 2.10), where the nuclear distances and energies are the same. The system thus must be activated by an energy, $\Delta G^\ddagger$, to this point (Fig. 2.11). It is also quantum mechanically possible for the transition from the initial to final state to occur by nuclear tunneling (wavy arrow in Fig. 2.10).

The effective rate constant, k, for the electron transfer reaction is:

$$k = \kappa(r) \cdot v \cdot e^{-\Delta G^\ddagger/RT}, \tag{24}$$

where κ (kappa) is the transmission coefficient for electron transfer which is dependent on the distance, r, between donor and acceptor, R is the gas constant, and T the absolute temperature. $\Delta G^\ddagger$ is the free energy of activation related to the rearrangement of reactants necessary to accomplish the electron

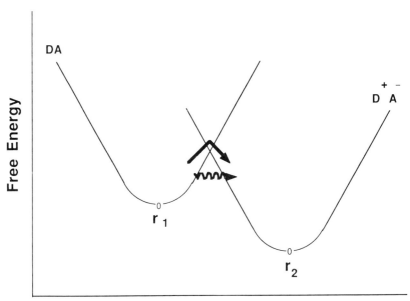

Figure 2.10. One-dimensional potential energy diagram for electron transfer between a donor (D) and acceptor (A) which have formed an ecounter complex (DA) by diffusion. The nuclear and solvent coordinates are plotted on an arbitrary scale on the abscissa, with the equilibrium positions, r_1 and r_2, different because of configuration changes, and separated in energy by $\Delta G°$. Two mechanisms of reaching r_2 from r_1 are (i) to surmount (solid arrows) the activation barrier ($\Delta G^\ddagger$) defined by the intersection between reactant and product potential energy curves. The electron transfer would then occur when the nuclear coordinates of reactant and product are identical. (ii) The jagged arrows show a pathway of electron transfer via nuclear tunneling, or crossing of the reactant DA to the product D^+A^- curve. The reorganizational energy, λ (not shown; see Fig. 2.11) is that required to move all of the nuclei from r_1 to r_2 without electron transfer having taken place.

transfer. For intramolecular reactions, the collision frequency ν would have a maximum value corresponding to vibrational frequencies, $\sim 10^{13}$ s^{-1} and for bimolecular reactions, the maximum value of ν would correspond to the diffusion-limited collision frequency in the liquid phase, $\sim 10^{11}$ M^{-1} s^{-1}. An approximate expression for $\Delta G^\ddagger$ can be derived in terms of the $\Delta G°$ for the reaction and the reorganization energy, λ, of the activated complex (Marcus, 1965; DeVault, 1984; Marcus and Sutin, 1985):

$$\Delta G^\ddagger = (\lambda + \Delta G°)^2/4\lambda \tag{25}$$

neglecting any work required to bring the reactants together. When $\Delta G° = 0$, $\Delta G^\ddagger(0) = \lambda/4$. The reorganization energy, λ, is that required to change the separation of the nuclei from r_1 to r_2 in the absence of electron transfer (Fig.

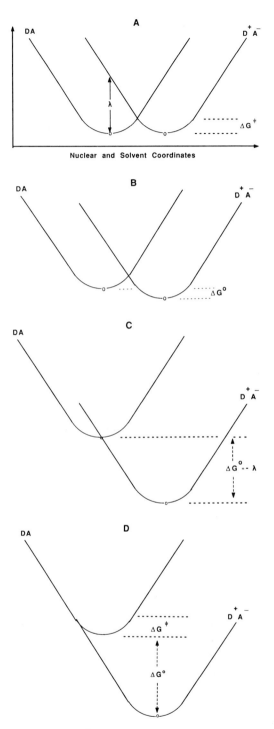

Figure 2.11. Dependence of $\Delta G^{\ddagger}$ on the value of ΔG° defined by relative position of reactant (DA) and product (D^+A^-) potential energy curves. (A) $\Delta G^{\circ} = 0$, $\Delta G^{\ddagger} = \lambda/4$; (B) $\Delta G^{\circ} < 0$. (C) $\Delta G^{\ddagger} = 0$ when $\Delta G^{\circ} = -\lambda$. (D) When $\Delta G^{\circ} > |-\lambda|$, $\Delta G^{\ddagger}$ is again > 0, the "inversion" region.

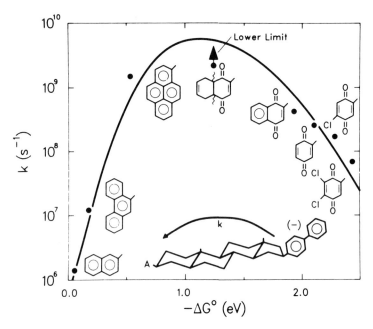

Figure 2.12. Intramolecular electron transfer rate constants for electron transfer between biphenyl radical anions and various organic acceptors, A, as a function of the $\Delta G°$. The "Marcus inversion" of the electron transport rate can be seen for $|\Delta G_o| > 1.0$ eV; solvent, 2-methyltetrahydrofuran, at 23°C; cf, Closs and Miller, 1988. (Reprinted with permission from Miller et al., *J. Am. Chem. Soc.*, 106, pp 3047–3049. Copyright 1984 American Chemical Society.)

2.10). λ contains two components: An internal component (λ_i) arising from changes of bond lengths that can be calculated from molecular vibrational coordinates, and an external component (λ_o), the solvent reorganization energy.

Equation 25 shows that $\Delta G^{\ddagger}$ decreases as $\Delta G°$ is made increasingly negative, or the reaction increasingly exothermic, and becomes zero in this approximation when $-\Delta G° = \lambda$. $\Delta G^{\ddagger}$ increases upon a further increase in the magnitude of $\Delta G°$. These changes of $\Delta G°$ correspond to a lowering of the nuclear potential energy function of the products relative to that of the reactants (Fig. 2.11).

Experimental confirmation of the increase in rate constant up to a maximum value with increasing $|\Delta G°|$, and then a decrease in electron transfer rate constant with increasingly negative values of $\Delta G°$, was obtained in studies of intramolecular long distance electron transfer between biphenyl radical anions and various organic electron acceptors (Miller et al., 1984b). The rate constant reached a maximum value of $\sim 2 \times 10^9$ s^{-1} for $\Delta G° \simeq -1$ eV (23 kcal/mol), as qualitatively predicted by Figs. 2.11A–C. The rate constant then decreased by one or two orders of magnitude when $\Delta G°$ increased in magnitude from -1 to -2 eV (Fig. 2.12), as described by Figs. 2.11C, D.

2.10.2 Forward Rate Constant

The expression for the forward rate constant, k_{et} or k_{12}, of the electron transfer reaction

$$D - A \underset{k_{21}}{\overset{k_{12}}{\rightleftarrows}} D^+ + A^-$$

can be expressed in terms of the self-exchange rate constants, k_{11} and k_{22}, which can be obtained for solution reactions (Marcus, 1965; Marcus and Sutin, 1985). k_{11} and k_{22} are defined as

$$D^*(red) + D^+(ox) \overset{k_{11}}{\rightleftarrows} D^{*+}(ox) + D(red),$$

and

$$A^*(red) + A^+(ox) \overset{k_{22}}{\rightleftarrows} A^{*+}(ox) + A(red).$$

The equilibrium constant, $K_{eq} = k_{12}/k_{21}$, for the electron transfer reaction from D to A. From Eq. 25,

$$\Delta G_{12}^{\ddagger} = [\lambda_{12}^2 + 2\lambda_{12}\Delta G^{\circ} + (\Delta G^{\circ})^2]/4\lambda_{12},$$

$$= \frac{\lambda_{12}}{4} + \frac{\Delta G^{\circ}}{2} + \frac{(\Delta G^{\circ})^2}{4\lambda_{12}} \qquad (26)$$

Again, from Eq. 25, values of $\Delta G^{\ddagger}$ for the self-exchange reactions for which $\Delta G^{\circ} = 0$ (Fig. 2.11C) are:

$$\Delta G_{11}^{\ddagger} = \lambda_{11}/4, \quad \text{and} \quad \Delta G_{22}^{\ddagger} = \lambda_{22}/4. \qquad (27)$$

From the nature of outer sphere reactions in which the reactants perturb each other to a minimum extent, it was assumed that the reorganizational energy, λ_{12}, contains equal contributions from the two reactants, i.e., that the reorganizational energy, λ_{12}, = the arithmetic mean of λ_{11} and λ_{22} for the individual self-exchange reactions (Marcus, 1965):

$$\lambda_{12} = (\lambda_{11} + \lambda_{22})/2. \qquad (28)$$

From Eqs. 26 and 28,

$$\Delta G_{12}^{\ddagger} = \frac{\lambda_{11} + \lambda_{22}}{8} + \frac{\Delta G^{\circ}}{2} + \frac{(\Delta G^{\circ})^2}{4\lambda_{12}},$$

and from (27),

$$= \frac{\Delta G_{11}^{\ddagger} + \Delta G_{22}^{\ddagger} + \Delta G^{\circ}}{2} + \frac{(\Delta G^{\circ})^2}{4\lambda_{12}} \qquad (29)$$

Thus, the activation energy is equal to one-half of the sum of the activation energies of the self-exchange reactions and the standard free energy change for the transfer reaction, plus another term which is small if the reorganizational energy $\lambda_{12} \gg \Delta G^{\circ}$. Neglecting the last term in Eq. 26, and using Eq. 24 to continue the derivation,

$$k_{12} = \kappa_{12} v e^{-\Delta G_{12}^{\ddagger}/RT}$$
$$= \kappa_{12} v e^{-(\Delta G_{11}^{\ddagger} + \Delta G_{22}^{\ddagger} + \Delta G^{\circ})/2RT}.$$

Since
$$k_{11} = \kappa_{11} \cdot v \cdot e^{-\Delta G_{11}^{\ddagger}/RT},$$
$$k_{22} = \kappa_{22} \cdot v \cdot e^{-\Delta G_{22}^{\ddagger}/RT},$$

and
$$K_{eq} = e^{-\Delta G^{\circ}/RT},$$

then
$$k_{12} = \frac{\kappa_{12}}{\sqrt{\kappa_{11}\kappa_{12}}} \cdot \sqrt{k_{11} k_{22} K_{eq}},$$

assuming that the frequency factors v are identical, and remembering that
$$e^{-\Delta G_{11}^{\ddagger}/2RT} = \sqrt{e^{-\Delta G_{11}^{\ddagger}/RT}}, \quad \text{etc.}$$

When
$$\kappa_{12}/\sqrt{\kappa_{11}\kappa_{22}} = 1,$$

then
$$k_{12} = \sqrt{k_{11} k_{22} K_{eq}}, \tag{30}$$

the "cross-relation."

If the above assumptions for ΔG° and the transmission coefficient are not made, then the more general expression is $k_{12} = \sqrt{k_{11} k_{22} K_{eq} f}$ with the factor f often close to unity.

The theoretical prediction for k_{12} values from the cross-relation and measured values for inorganic electron transfer reactions generally agree within a factor of 10 with exceptions occurring when: (i) redox couples are not substitution-inert, (ii) the ΔG° of the reaction is very large, and/or (iii) work involved in precursor and product formation cannot be neglected ((Marcus and Sutin, 1985; Holwerda et al., 1978).

2.10.3 Application of the Cross-Relation to Redox Reactions

If
$$k_{12} = \sqrt{k_{11} k_{22} K_{eq}},$$

then
$$\log_{10} k_{12} = \log_{10} \sqrt{k_{11} k_{22}} + \frac{\log_{10} K_{eq}}{2};$$

since $\Delta E^{\circ} = 59 \log_{10} K_{eq}$,
$$\log_{10} k_{12} = C + \Delta E^{\circ}/118, \tag{31}$$

2.10 Electron Transfer Mechanisms

where the constant

$$C = \log_{10} \sqrt{k_{11} k_{22}}.$$

$\Delta E°$ is the difference, in mV, of the standard potentials of the reacting redox couples. If $\log k_{12}$ is plotted versus the $\Delta E°$ of the reaction, the slope (mV^{-1}) will be 1/118. The dependence of the rate constant, k_{12}, on K_{eq} and $\Delta E°$ reiterates the point that the rate will be greater, the larger the K_{eq} (the more negative the $\Delta G°$, and the more positive the $\Delta E°$).

2.10.4 Electron Transfer Reactions Between Proteins in Solution

The cross-relation (Eqs. 30 and 31) has been applied to a set of electron transfer reactions between several small soluble copper (Cu) and (Fe) iron redox proteins whose redox properties are summarized in Table 2.6.

The rate constant data for the reactions shown in Table 2.6 could be fit fairly well by Eq. 30 if the following best-fit values for the self-exchange rate constants were used:

Redox protein	k_{ii} (M^{-1} s^{-1})
c_{553} (B. filiformis)	2.8×10^8
c_{551} (Ps. aeruginosa)	4.6×10^7
Ps. Azurin	9.9×10^5
Alc. azurins	2.6×10^5
Plastocyanins	6.6×10^2
Cytochrome c	1.5×10^2

Within this group of redox proteins there was little kinetic selectivity. One exception was the reduction of acidic negatively charged plastocyanin by the positively charged basic horse heart cytochrome c (last entry, Table 2.6).

Reduction of Cytochromes c by Flavins and Flavoproteins

The predicted exponential relation (Eq. 11) between k_{12} and $\Delta E°$ is qualitatively borne out in the general trend for the k_{12} measured for reduction of different c-type cytochromes by the semiquinone of the flavin-protein, flavodoxin (Fig. 2.13). However, not all of the oxidants fit the relation. Three samples showing relatively low reactivity are mitochondrial cytochromes (H, I, J) for which the heme is known from structural data to be recessed from the protein surface. Thus, for flavins and flavoproteins that react with cytochrome c near the exposed heme edge (Chap. 4), the distance of the heme edge to the protein surface is a factor that can cause deviations from the predicted reaction rate.

Quinol Reactions

The cross-relation (31) describes the $\Delta E°$ dependence of the reduction rate of cytochrome c by a series of substituted benzoquinols. The $\Delta E°$ values are

Table 2.6. Comparison of observed and measured rate constants of soluble electron transport proteins

Oxidant	E_m(mV)	Reductant	E_m(mV)	$k(M^{-1}\,s^{-1})$ (obs)	$k(M^{-1}\,s^{-1})$ (calc)	$\dfrac{k(\text{calc})}{k(\text{obs})}$
1. Azurin (Ps.)	304	Cytochrome c_{553}	335	$1.43\,(10^7)$	$5.3\,(10^6)$	0.4
2. Azurin (Alc. f.)	266	Cytochrome c_{553}	335	$5.23\,(10^5)$	$8.5\,(10^5)$	1.6
3. Azurin (Alc. spp.)	260	Cytochrome c_{553}	335	$2.11\,(10^6)$	$2.3\,(10^6)$	1.1
4. Plastocyanin (bean)	350	Cytochrome c_{553}	335	$4.00\,(10^5)$	$5.0\,(10^5)$	1.3
5. Plastocyanin (Sc.)	350	Cytochrome c_{553}	335	$7.36\,(10^5)$	$5.6\,(10^5)$	0.8
6. Cytochrome c	260	Cytochrome c_{553}	335	$3.96\,(10^4)$	$5.7\,(10^4)$	1.4
7. Azurin (Alc. spp.)	260	Cytochrome c_{551}	286	$2.96\,(10^6)$	$2.5\,(10^6)$	0.8
8. Azurin (Ps.)	304	Cytochrome c_{551}	286	$6.10\,(10^6)$	$6.0\,(10^6)$	1.0
9. Azurin (Alc. f.)	266	Cytochrome c_{551}	286	$2.10\,(10^6)$	$1.9\,(10^6)$	0.9
10. Cytochrome c	260	Cytochrome c_{551}	286	$6.7\,(10^4)$	$8.2\,(10^4)$	1.2
11. Cytochrome c	260	Azurin (Ps.)	304	$1.6\,(10^3)$	$5.2\,(10^3)$	3.2
12. Plastocyanin (parsley)		Cytochrome c_{551}	286	$7.5\,(10^5)$	$7.8\,(10^5)$	1.0
13. Plastocyanin (parsley)		Cytochrome c (horse heart)	260	$1.0\,(10^6)$	$2.2\,(10^3)$	0.002

From Wherland and Pecht, 1978.

2.10 Electron Transfer Mechanisms

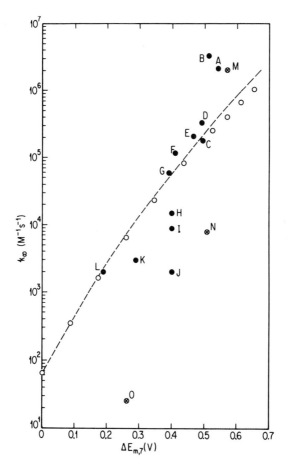

Figure 2.13. Semilog plot of k_{12} extrapolated to infinite ionic strength versus ΔE_m for oxidation of flavodoxin semiquinone by different cytochromes c. Marcus exponential relation with $v = 5 \times 10^9$ M^{-1} s^{-1} (o) and $\Delta G^{\ddagger}$ ($\Delta E_m = 0$) = 10.7 kcal/mol. c, cytochrome from A, *Rhodospirillum tenue*; B, *Euglena* sp. *c*-552; C, *Rhodomicrobium vannielli*; D, *Rb. capsulata* c_2; E, *Rh. rubrum* c_2; F, *Ps. aeruginosa c*-551; G, *P. denitrificans* c_2; H, *Candida krusei c*; I, tuna c; J, horse c; K, *C. thiosulfatophilum c*-555; L, *Ectothiorhodospira halophila c*-551; M, ferricyanide; N, cobalt phenanthroline; O, ferric EDTA. (Reprinted with permission from Tollin et al., *Biochemistry*, 23, pp 6345–6349. Copyright 1984 American Chemical Society.)

calculated for transfer from the anionic quinol to cytochrome c (Rich and Bendall, 1980):

$$QH^- + c(ox) \rightarrow QH^{\cdot} + c(red)$$

When the data are plotted in this way the slope of $\log k_{12}$ versus $\Delta E°$ from Eq. (31) is $(118 \text{ mV})^{-1}$ (Fig. 2.14).

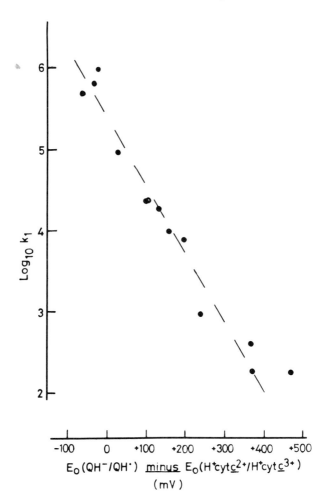

Figure 2.14. Marcus plot of the relationship between the reduction rate of cytochrome c and $\Delta E°$ (QH'/QH$^-$) for a series of substituted benzoquinols reducing cytochrome c (Rich and Bendall, 1980).

2.10.5 Dependence of k_{et} on Distance between Donor and Acceptor

The theory of Hopfield (1974) predicted a dependence of electron transfer rate, k_{et}, on distance between donor and acceptor, which follows from the theory of electron tunneling (DeVault, 1984). When the κ factor of Eq. 24 incorporates this distance dependence as $\exp[-\alpha(r - r_o)]$, then

$$k_{et} \text{ is proportional to } \exp[-\alpha(r - r_o)], \tag{32}$$

where the constant $\alpha = 1.4$ Å^{-1}. r and $(r - r_o)$ are the center–center and edge–edge distances, respectively, if r_o is the sum of the radii of donor and acceptor.

2.10 Electron Transfer Mechanisms

Substituting for κ, a complete expression for k_{et} would be:

$$k_{et} = \nu \exp[-\alpha(r - r_o)] \exp[-\Delta G^{\ddagger}/RT], \quad (33)$$

and using Eq. 25 for $\Delta G^{\ddagger}$,

$$k_{et} = \nu \exp[-\alpha(r - r_o)] \exp[-(\lambda + \Delta G^{\circ})^2/4\lambda RT]. \quad (34)$$

A value of $\alpha = 1.2$ Å^{-1} was obtained empirically from studies of electron transfer reactions in rigid media between aromatic molecules (Miller et al., 1984a). The value of ν was found to be 10^{13} s^{-1} and the electron transfer across a 10 Å thick insulating barrier to take place in 10^{-10} s after generation of aromatic radical anions by short pulse radiolysis (Miller et al., 1984b).

When the electronic wave functions are taken into account, the coefficient of the exponential factor in Eq. 34 at room temperature becomes (Ulstrup and Jortner, 1975; Jortner, 1976):

$$\frac{2\pi}{h} |T_{RP}|^2 \frac{1}{\sqrt{4\pi \lambda TR/N}},$$

where h is Planck's constant, T_{RP} is the matrix element describing electronic overlap between the reactant and product states, R is the gas constant, and N is Avogadro's number.

Long-Distance Electron Transfer in Proteins

Long-distance electron donor transfer is defined as that occurring over a distance several times the limiting π–π overlap distance (~ 3 Å) of donor and acceptor. There are several examples of electron transfer reactions involving redox proteins in which donor–acceptor distances are fixed, known, and in the range 12–20 Å (Table 2.7). One method of measuring the distance dependence was to attach the inorganic pentamine-ruthenium (II/III), $(NH_3)_5Ru^{2+/3+}$ [$a_5Ru^{2+/3+}$], which is a reductant of intermediate strength ($E_m = 85$ mV), to specific histidine or tyrosine residues. The pentamine-ruthenium can be photochemically reduced and then in turn act as an intramolecular reductant for specific amino acid residues (His or Tyr) in cytochrome c, azurin, and myoglobin (Scott et al., 1985). The thermodynamic parameters of heme c were not significantly perturbed in "ruthenated" cytochrome c (Mayo et al., 1986). Electron transfer from the Ru(II) to the oxidized endogenous redox group could be initiated photochemically by pulse radiolysis of the excited state (*) of added rubidium bipyridyl, $Ru(bpy)_3^{2+}$, in EDTA solution, which then rapidly reduced the initially oxidized and bound pentamine ruthenium, a_5Ru. For the case of cytochrome c (Mayo et al., 1986):

$$a_5Ru(His-33)^{3+} \text{ cyt } c \text{ (heme}^{3+}) \xrightarrow{h\nu[Ru(bpy)_3^{2+*}]} a_5Ru(His-33)^{2+} \text{ cyt } c \text{ (heme}^{3+})$$

$$\xrightarrow{\text{slow}, k_1} a_5Ru(His-33)^{3+} \text{ cyt } c \text{ (heme}^{2+})$$

Table 2.7. Some examples of long fixed distance electron transfer through known residues of redox proteins

Donor	Protein	Acceptor	Distance (Å)	Rate (s^{-1})	Reference
$Ru^{II} \cdot His\,33$	Cyt c	Fe(III)	12	30–50	Nocera et al. (1984)
					Mayo et al. (1986)
$Ru^{II} \cdot His\,83$	Azurin	Cu(II)-S(Cys 112)[a]	12	1.9	Kostic et al. (1983)
Cr · Tyr-83 (oxidant)	Plastocyanin	Cu(I)-S(Cys 84)[a]	10–12	10^6	Brunschwig et al. (1985)
Zn-porphyrin	Hemoglobin	Fe(III)	20	100	Peterson-Kennedy et al. (1984)

[a] Copper liganding is CuN_2SS^* from 2 His, Met, and Cys.
From Scott et al., 1985; Mayo et al., 1986.

These intraprotein electron transfer rates are much smaller than those measured for biphenyl radical anions–organic acceptors (Fig. 2.12), but it is likely that different mechanisms apply. In a_5 Ru(His-33)-cyt c, electron transfer from Ru^{2+} to Fe^{3+}-heme probably occurs through the two amino acids, Asn-31 and Leu-32, that are known from the structure of cytochrome c (Chap. 4), to occupy the intervening space between the attached Ru and the heme (Fig. 2.15). These data are important in demonstrating electron transfer through a pathway including amino acid residues. The possibility of a special role for conjugated aromatic amino acids in facilitating electron transfer through cytochrome c was suggested by studies in which the Phe-87 of the yeast iso-1-cytochrome c, that is thought to be in the pathway of electron transfer, was replaced by Tyr, Gly, and Ser residues through site-directed mutagenesis. Electron transfer from the reduced cytochrome c to Zn-substituted cytochrome c peroxidase was 10^4 faster for the wild-type Phe-87 and the Tyr-87 variant, than for the Gly and Ser mutants (Liang et al., 1987). However, a caveat to the conclusion that the aromatic residues have a special conductive role is the possibility that the mutations cause conformational changes that also affect electron transfer efficiency. The Ser mutant has been shown to be conformationally altered at residues near and remote from the mutation site, and to cause a large increase in solvent accessibility to the heme group (Louie et al., 1988). The latter changes might not occur with the Tyr mutant that is closer in size to the original Phe residue.

Preliminary data on the distance dependence of the electron transport rate, k_{et}, in proteins and protein complexes indicates that $k_{et} = k_o e[-\alpha(r - r_o)]$ (Eq. 32). The value of α in blue copper proteins, 1.4 $Å^{-1}$ (Gray, 1986), is similar to that discussed above, but α may have a smaller value, 0.7–0.9 $Å^{-1}$, in heme proteins (Mayo et al., 1986).

2.10 Electron Transfer Mechanisms

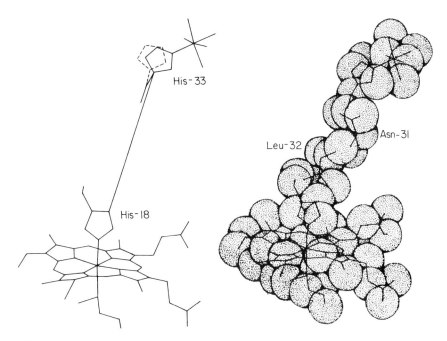

Figure 2.15. Computer-generated view of (left) the a_5Ru(His-33) donor (Tryp-33 of tuna cyt c replaced by His dashed/solid, without/with a_5Ru) and heme acceptor in cytochrome c (Chap. 4, section 4.3) and (right) the intervening Asn-31 and Leu-32 residues. The dots represent the Van der Waals surfaces of the displayed atoms (Mayo et al., 1986).

2.10.6 Nuclear Tunneling

The activation energy term in Eqs. 33 and 34 may not apply to reactions at cryogenic temperatures. The mechanism of nuclear tunneling through the potential energy barrier, as depicted in the path described by the wavy arrow in Fig. 2.10, would provide the only mechanism at low temperatures (< 100 K) where the reactants do not have sufficient energy to reach the activation barrier. At sufficiently low temperatures, the reaction becomes temperature-independent since it occurs at the *zero*-point vibrational energy level of the reactants. Temperature-dependent and -independent behavior of an electron transfer reaction at high (>100–150 K) and low (<100 K) temperatures, respectively, implies that the reaction at low temperature may involve tunneling of nuclei through the activation energy barrier.

Operation of a tunneling mechanism in photosynthetic electron transport was inferred from the temperature independence of the rate of light-induced cytochrome oxidation in the purple photosynthetic bacterium *Chromatium vinosum* (DeVault and Chance, 1966). The half-time for cytochrome oxidation at room temperature was 2 μs, and the activation energy over a physiological temperature range was 3.3 kcal/mol. As the temperature was lowered to

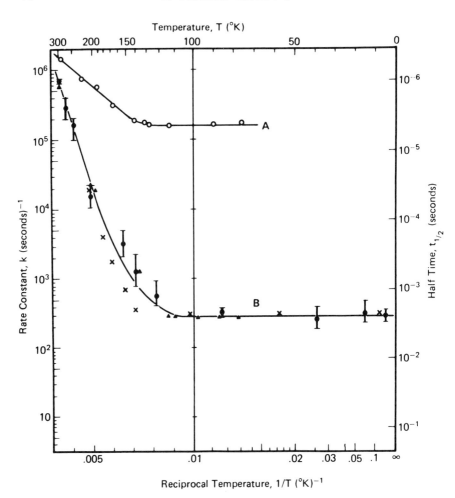

Figure 2.16. Temperature dependence of the photooxidation of cytochrome c in the photosynthetic bacteria (A) *Rb. sphaeroides* (Kihara and McCray, 1973) and (B) *Chromatium vinosum* (DeVault and Chance, 1966), strain D, whole cells; subchromatophores (▲) (Dutton et al., 1971) and (×) (Hales, 1976). (Based on DeVault, 1984.)

cryogenic levels, the half-time at 120 K decreased to 2 ms. For several different bacteria, the half-time hardly changed as the temperature was lowered to the level of liquid helium, $< 10\ °K$ (Fig. 2.16).

Exercise 2.6. Calculate the rate constant for reduction of the reaction center of the photosynthetic bacterium, *Rps. viridis*.

The rate constant for the reduction of the reaction center bacteriochlorophyll special pair by its donor, cytochrome c-558, can be calculated (Marcus and Sutin, 1985) from Eq. 24:

$$k_{12} = \nu \kappa_{12}(r) e^{-\Delta G^{\ddagger}/RT}; \quad \text{if } \nu = 10^{13} \text{ s}^{-1},$$

then

$$\nu \kappa_{12}(r) = 10^{13} \exp[-\alpha(r - r_o)] = 10^{13} \exp[1.2(r - r_o)],$$

with $\alpha = 1.2 \text{ Å}^{-1}$, for the reduction proceeding between fixed sites. The center–center separation of the chlorophyll and closest heme ring is ~ 21 Å, and the edge–edge distance, $(r - r_o)$, $\simeq 10$ Å. $\Delta G^{\ddagger}$ was calculated from Eq. 25, the known ΔE_m of 170 mV between the reaction center bacteriochlorophyll and the cytochrome c, the resulting $\Delta G°$ of -16 kJ/mol, and an estimate of 57 kJ/mol or 0.6 eV for λ (from the average of the λ's for the cytochrome c and BCh1 self-exchanges), to be 7.1 kJ/mol. With $RT = 2.5$ kJ/mol at 25°C,

$$k_{12} = 10^{13} \, e^{-12} e^{-7.1/2.5} = 10^{13} \, e^{-15}$$
$$= 10^{6.5} \text{ s}^{-1} = 3 \times 10^6 \text{ s}^{-1},$$

as compared to the measured value of 4×10^6 s^{-1} (Holten et al., 1978).

2.11 Proton Transfer Reactions

There is a dearth of experimental data on the mechanism of intramembrane proton transfer reactions within or between proteins. This is partly because the spectral changes accompanying the protonation–deprotonation reactions are much harder to define than for oxidation–reduction. However, methods now exist to study these reactions on a msec time scale in bacteriorhodopsin (Chap. 7), and laser pulse techniques have been used to measure the kinetics of H$^+$ transfer reactions from laser-excited dye molecules to other indicator dyes serving as H$^+$ acceptors (Huppert et al., 1981).

2.11.1 Studies of Proton-transfer Kinetics in Model Systems with a Light-Induced Proton Pulse

Light-absorbing dye molecules that are stronger acids in their first excited singlet state (∗) than in their ground (o) state (i.e., pK^* < p$K°$) can be used as fast ($\lesssim 1$ ns) H$^+$ injectors. The dye acts as an effective H$^+$ emitter if the H$^+$ transfer occurs more rapidly than the fluorescence lifetime, typically a few nanoseconds. The fluorescence emission peak of the neutral form is at a shorter wavelength than the anion when pK^* < p$K°$ (Fig. 2.17). The pK values are related to the standard free energy of the protonation–deprotonation reaction. The change in pK can be very large, often 5–6 pH units. For example, it is >5 pH units for the proton emitter 8-OH-pyrene-trisulfonate whose emission maxima are 445 and 510 nm for the neutral and anionic forms,

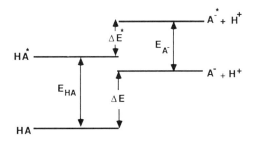

Figure 2.17. Electronic energy levels showing ground (o) and excited (∗) state for an acidic dye (HA) and its conjugate base (A⁻); $E_{HA} = hv_{AH}$ and $E_{A^-} = hv_{A^-}$. Relative $\Delta G°$ values for H^+ dissociation from ground and excited state, are proportional to ΔE and ΔE^* (Adapted from Huppert et al., 1981.)

respectively. The general relation between the pK of the ground and excited state, whose energies are related as shown in Fig. 2.17, is

$$(pK^* - pK°) = N(hv_{A^-} - hv_{AH})/2.3RT, \qquad (35)$$

where h is Planck's constant, N is Avogadro's number, and v_A and v_{AH} are the frequencies of the anionic and neutral form at the peak of the respective fluorescence emission spectra.

These laser techniques which allow transient acidification (10–50 μM H^+) within a few nanoseconds have been used to study the properties of H^+ transfer between reactants absorbed to the same lipid micelle or phospholipid membrane. Studies on a model H^+ emitter (2-naphthol) and H^+ acceptor (bromcresol green) absorbed to detergent micelles showed that the H^+ electrochemical potential on the micelle surface was in equilibrium with that in the bulk solution, in spite of the fast ($k = 2 \times 10^{11}$ M^{-1} s^{-1}) H^+ transfer reactions occurring on the surface (Nachliel and Gutman, 1984).

2.11.2 Proton Translocation Through Hydrogen Bond Networks in Proteins

The membrane bilayer-spanning structure of proteins responsible for light- (bacteriorhodopsin; see Chap. 7, section 7.2) and respiratory- (cytochrome oxidase; see Chap. 7, section 7.3) or ATP-driven proton pump activities across the 40 Å hydrophobic bilayer of energy-transducing membranes (Chap. 7) implies that these H^+ translocation networks use hydrogen bonded chains made of the amino acid side chains of these proteins (Nagle and Morowitz, 1978; Nagle and Tristram-Nagle, 1983), and/or transient *trans*-protein water channels (Varó and Keszthelyi, 1985). The amino acid side chains that would be able to form a hydrogen bonded chain (Fig. 2.18) are probably those with pK values low enough that they will not bind protons too tightly. These might be aspartic and glutamic acids, histidine, serine, threonine, and tyrosine.

The H^+ Network Through Bacteriorhodopsin

The integral membrane protein bacteriorhodopsin, which functions as a light-driven H^+ pump, is known from electron diffraction and topographical analy-

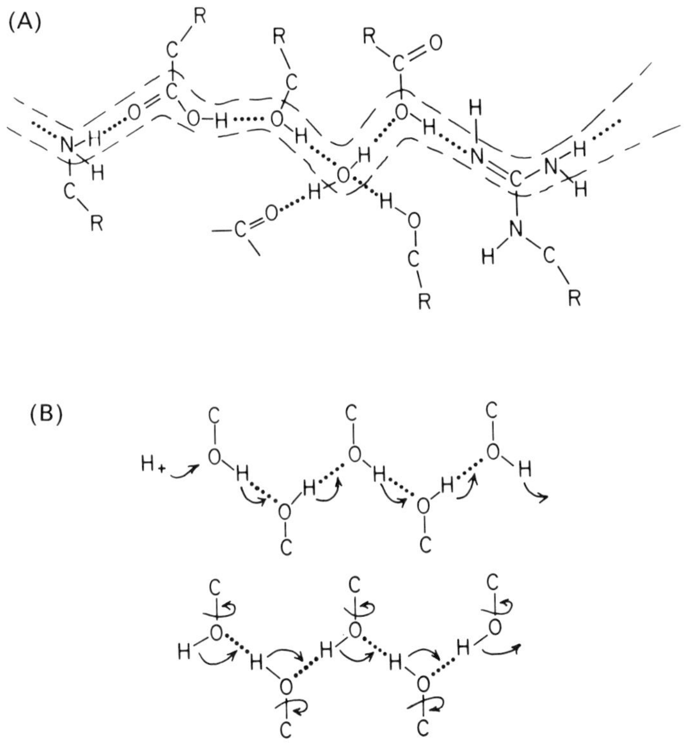

2.11 Proton Transfer Reactions

Figure 2.18. (A) Hydrogen-bonded chain from different amino acid chains of membrane proteins and bound H_2O molecules. R groups are the remainder of side chains leading to peptide bond. Path of proton channel is indicated by dashed lines (Nagle and Tristram-Nagle, 1983). (B) Two-step or hop-turn mechanism of H^+ conduction (Nagle and Tristram-Nagle, 1983) in which the proton first hops to an adjacent group leaving a defect (I), and the hole-defect can then rotate (II) to assume the configuration in (I), thereby translocating one H^+.

sis to be folded in the membrane as seven membrane-spanning α-helices. The retinal that triggers light-driven H^+ translocation across the membrane upon absorption of a photon is linked by a Schiff base to lysine-216 in membrane-spanning peptide VII. The complete mechanism and pathway of H^+ translocation through the protein and across the membrane is not known, although the involvement of particular Asp and possibly Tyr residues has been shown by a combination of site-directed mutagenesis and spectroscopic analysis. (Chap. 7, section 7.2.)

Velocity of H^+ Translocation in the Presence of a Membrane Potential

The mean velocity, v, iof movement of an ion with electrophoretic mobility, m, in the presence of a potential, $\Delta\psi$, in a membrane of thickness, d, is

$$v = \frac{m \cdot \Delta\psi}{d}, \qquad (36)$$

where $\Delta\psi$ is the membrane potential and $\Delta\psi/d$ is the electric field across the membrane. Since the dimensions of $\Delta\psi$, d, and v are volts, cm, and cm/s, respectively, the dimensions of m are cm^2/V-s. If the m values for H^+ hopping and bond rotation (Fig. 2.18) in a protein or membrane were similar to the values for ice, 10^{-4}–10^{-3} cm^2/V-s (Kunst and Warman, 1980), the mean velocity of H^+ movement would be approximately 10 cm/s if $\Delta\psi = 0.03$ volt, and $d = 40$ Å. The transit time for the H^+ translocation across a 50 Å membrane would be on the order of tens of nanoseconds (Problem 45), much faster than the characteristic rate-limiting steps of physiological electron transport chains or the proton pump rates that occur on a millisecond time scale. It then seems likely that the rate limiting step of H^+ translocation across energy-transducing membranes involves protein structural changes.

2.12 Summary

The discussion of free energy relations in Chap. 1 is extended to oxidation–reduction reactions:

$$\Delta G = -nF\Delta E; \qquad \Delta G^\circ = -nF\Delta E^\circ,$$

where ΔE and ΔE° are the changes in net and standard oxidation–reduction (redox) potential for a reaction. E° is a measure of the affinity of a redox compound for an electron in the same way as a pK describes the affinity of an acid–base buffer for a proton. The relationship between the ΔE° for a reaction and its equilibrium constant is $\Delta E^\circ = RT/nF \cdot \ln K_{eq}$. The criteria for spontaneous or forward reactions are $\Delta E > 0$ and ΔE° (or ΔE_m) > 0, so that electron transfer proceeds in the direction of more positive E and E° or E_m. Unlike the free energy, G, the oxidation–reduction potential, E, is not a state function. E° or E_m values are affected by the local electrostatic environment, which is influenced by factors such as ionic strength, ligand binding, and in the case of heme compounds, the basicity of axial heme ligands such as histidine and methionine. The range of E_{m7} values for electron transfer reactions in photosynthetic and respiratory electron transport systems that have been assayed extends from -0.9 V, for the primary electron acceptor of photosystem I of the oxygenic photosynthetic electron transport chain, to $+1.2$ V for the P680 reaction center chlorophyll on the oxidizing side of photosystem II.

The E_m is pH-dependent if the proton affinity is different for oxidized and reduced states, with $\Delta E_m/\Delta pH = -59$ mV at 25°C when the pK of the reduced state is much greater than that of the oxidized form. Such pH-dependent E_m values are important in the study of the mechanisms of H^+ translocation across the membrane bilayer. The E_m values of the quinone/semiquinone [$E_{m1}(Q/QH^{\cdot})$], semiquinone/quinol [$E_{m2}(QH^{\cdot}/QH_2)$], and quinone/quinol [$E_m(Q/QH_2)$] redox couples of these H^+ carriers in photosynthetic and respiratory electron transport chains, are related by the formulae: $E_{m1} - E_{m2} = 59 \log_{10} K_s$, where K_s is the semiquinone dismutation constant, and $E_m = (E_{m1} + E_{m2})/2$.

Mechanisms of Electron and Proton Transfer

The rate constant, $k = \kappa(r) v \exp[-\Delta G^{\ddagger}/RT]$ for electron transfer between electron donor and acceptor, is dependent on the $\Delta G°$ and distance (r). The activation free energy, $\Delta G^{\ddagger} \simeq (\lambda + \Delta G°)^2/4\lambda$, with λ a reorganizational energy associated with the change from the reactant to the product complex. These formulae predict that the k_{et} will increase as the $-\Delta G°$ for the electron transfer increases, but this effect will reach a limit when the reactant and product potential energy curves "cross-over" at large values of $-\Delta G°$. Solution electron transfer rates between species 1 and 2 can often be described by the formula: $k_{12} = \sqrt{k_{11}k_{22}K_{eq}}$, where k_{11} and k_{22} are self-exchange rate constants. The rate of long-distance (10–20 Å) electron transfer in proteins can be described by the expression $k = v \exp[-\alpha(r - r_0)] \exp[-(\lambda + \Delta G°)^2/4\lambda RT]$, where the distance coefficient, α, has been found by experiment to have values ranging from 0.8–1.4 Å^{-1}. Electron transfer at room temperature can occur over distances of 10–20 Å on a time scale of micro and milliseconds, respectively. The observation of light-induced millisecond electron transfer in photosynthetic bacteria at cryogenic temperatures ($T = 5$–100 K) implies that the electron transfer mechanism under these conditions involves tunneling of reactant nuclei through the activation energy barrier.

The time course of proton transfer reactions can be studied in model systems using as an H^+ emitter the excited-state of a dye molecule that has a pK much below that of the ground state. Such studies showed that protons emitted on the surface of a lipid micelle are in equilibrium with the bulk solution. The mechanism of H^+ translocation in a proton pumping protein such as bacteriorhodopsin has been shown from mutagenesis experiments to involve an H-bonded network of amino acids (such as Asp, Glu, His, Ser, Thr, or Tyr) and/or water channels.

Problems

29. Using values of R and F in appropriate units, evaluate the coefficient $2.3RT/nF$ in millivolts at 0°C, 25°C, and 37°C for $n = 1$ and $n = 2$.
30. (a) Derive an expression for the relationship between

    $$\Delta E° \quad \text{and} \quad K_{eq},$$

    given

    $$\Delta G° = -RT \ln K_{eq}.$$

31. The half-reactions for the oxidation–reduction of cytochromes b and c,

along with the standard potentials are:
$$b(ox) + e^- \to b(red), \quad E° = +0.05 \text{ V at } 25°C$$
$$c(ox) + e^- \to c(red), \quad E° = +0.25 \text{ V at } 25°C$$
Consider the reduction of cytochrome c by cytochrome b, i.e., reactants are $b(red)$, $c(ox)$, and products are $b(ox)$ and $c(red)$: What is the redox potential change, ΔE, for the reaction when the steady-state ratio of reduced to oxidized cytochrome is: $b(red):b(ox) = 10:1$, $c(red):c(ox) = 1:10$)? What is the free energy change, ΔG, for the reaction when (a) one electron is tranferred between cytochrome b and c, and (b) when a total of two electrons are transferred from b to c using the same one electron reaction?

32. Mitochondria can use the free energy made available from the transfer of 2 electrons in the redox reaction described in the previous problem to drive the coupled synthesis of ATP, $ADP + P_i \to ATP + H_2O$ at 25°C. The $\Delta G°$ for this ATP synthesis is $+8$ kcal/mol. If the reaction operates with concentrations of 10^{-5} M and 10^{-3} M for ADP and phosphate, calculate the maximum concentration of ATP at which it can operate.

33. An oxidation–reduction titration of cytochrome c is carried out by measuring an absorbance change, ΔA, at 550 nm, from which is derived the concentration of the reduced component. (a) Plot the fractional concentration of reduced cytochrome, $c(red)/c(tot.)$ versus the experimentally manipulated redox potential, E. (b) Plot the ratio of reduced to oxidized cytochrome, $c(red)/c(ox)$, versus E on semilog paper.
$$c(ox) + e^- \to c(red), \quad E_m = +0.26 \text{ V at } 25°.$$

34. Construct a graph like that in the previous problem for a redox component with the same E_m as cytochrome c, but with $n = 2$.

35. Construct a graph like that in the previous problem, and in addition take into account pH effects and plot at pH $= 9$ for component "X":
$$X(ox) + 2e^- + 2H^+ \to X(red), \quad E_{m7} = +0.26 \text{ V}.$$

36. (a) Graph the pH dependence of the E_m of P. aeruginosa cytochrome c if the H^+ dissociation constants K_r and K_o are 10^{-9} M and 10^{-3} M, and $E_{ma} = +500$ mV. (b) What value would you estimate for E_{mb}? (c) What is the effect on E_{ma} of increasing the concentration 100-fold of a ligand that would bind very tightly and exclusively to the oxidized form at acidic pH?

37. Consider the redox scheme for the reduction of O_2 to H_2O through different pathways: Show that the midpoint potentials at pH 7 for the different redox couples (Fig. 2.3) are, or are not, consistent with the free energy change between initial and final states being independent of path. The potentials are $+0.31$ V, -0.33 V, $+0.94$ V, $+1.33$ V, $+1.20$ V, and $+0.82$ V.

38. Calculate the E_{m7} values of the semiquinone couples that interact with

the chloroplast herbicide binding protein if $K_s = 4 \times 10^{-2}$ and the E_{m7} of the Q/QH_2 couple is $+90$ mV.

39. Calculate the most reduced value possible for the E_m value of the excited state of the P865 reaction center from the facultative aerobic green bacterium C. aurantiacus, if the E_m of the P^+/P ground state is $+0.36$ V. n.b., 1 eV $= 1.6 \times 10^{-19}$ j.

40. Calculate the second-order rate constant according to the Marcus theory approximation for the reduction of B. filiformis cytochrome c_{553} by Ps. aeruginosa azurin.

41. Consider the reduction, described by the "cross relation" approximation, of the heme and copper proteins, Ps. aeruginosa cytochrome c_{551}, Ps. aeruginosa azurin, chloroplast plastocyanin, and horse heart cytochrome c, with self-exchange rate constants 4.6×10^7, 9.9×10^5, 6.6×10^2, and 1.5×10^2 M^{-1} s^{-1}, respectively, by Bumilleriopsis cytochrome c_{553} with the self-exchange rate constant, 2.8×10^8. Which of the above electron transfer reactions occurring in solution at 25°C will have the largest forward rate constant, k_{12}, and which the smallest, and what will be the value of this rate constant for the fastest and slowest reactions? Other data are in Table 2.6.

42. From the E_m and $\log k_{12}$ values for various $(QH^{\cdot}/QH^-)$ couples, plot $\log k_{12}$ for the rate of cytochrome c reduction by the different anionic quinols (QH^-) versus the ΔE_m for the reaction. Assume that the E_m for cytochrome $c = +254$ mV. How does the data fit the prediction made by eqn. (31)? [E_m value (mV), $\log_{10} k_{12}$]: Hydroquinone ($+482, 2.95$); 2-methylbenzoquinone ($+412, 3.97$); 2,3-dimethylhydroquinone ($+390, 4.29$); 2,6-dimethylhydroquinone ($+350, 4.38$); trimethylhydroquinone ($+279, 4.93$); tetramethylhydroquinone ($+221, 5.81$); 2-methyl-5-isopropylbenzoquinone ($+353, 4.36$); 2,5-dichlorohydroquinone ($+623, 2.28$); 2,6-dichlorohydroquinone ($+621, 2.67$); tetrachlorohydroquinone ($+726, 2.78$); ubiquinol-1 ($+191, 5.68$); plastoquinol-1 ($+239, 5.98$).

43. Consider the reduction of the copper protein plastocyanin azurin, or stellacyanin in the Cu(II) oxidized state by $Cr(phen)_3^{3+/2+}$. Assume that the reorganization energies λ_{11} and λ_{22} of $Cr(phen)_3^{3+/2+}$ and the Cu(II)/Cu(I) proteins are each about $+23$ kcal/mol, that $\Delta G°$ for reduction of the copper protein is also ~ -23 kcal/mol, the rate constant for reduction $\sim 10^6$ s^{-1}, and the characteristic vibration frequency $v = 10^{13}$ s^{-1}. (a) What is the activation energy of the reaction? (b) Write an expression for the rate constant for electron transfer as a function of edge–edge distance of separation. (c) Calculate the edge–edge distance of the electron transfer reaction if the α in Eqs. 32–34 $= 0.8$ Å^{-1}.

44. Calculate the pK change of the first excited singlet state of the proton emitting dye 8-hydroxypyrenetrisulfonate.

45. Estimate the transit time for a proton to travel in an ice-like environment across a 40 Å membrane bilayer through the ATP synthase protein if the imposed membrane potential is 80 mV.

Chapter 3
Membrane Structure and Storage of Free Energy

3.1 Elements of Membrane Structure

Chloroplasts, mitochondria, and Gram-negative bacteria all share the property of being bounded by a pair of membranes (Keegstra et al., 1984). The two membranes differ in passive permeability properties, the outer membrane being a porosity shield that is more permeable to small molecules. The porosity of the outer membrane is conferred by specific porin proteins, with mitochondrial and bacterial porins allowing passage of solutes of molecular weight < 10,000 and 700, respectively. The mitochondrial pore-forming protein forms at least part of the hexokinase receptor complex in tumor mitochondria, which may relate to the large increase in binding of hexokinase observed in rapidly growing cancer cells, whereby ATP generated by oxidative phosphorylation may be efficiently trapped as glucose-6-phosphate (Nakashima et al., 1986). The bacterial and mitochondrial, and possibly the chloroplast, membrane systems also contain adhesion sites between the two membranes. In the bacteria these sites are known to mediate transfer of newly synthesized lipid and lipopolysaccharide from inner to outer membrane, whereas the mitochondrial sites allow import of proteins across both membranes. The mitochondrial, bacterial, and chloroplast membrane systems all require energy for this import.

There are some structural differences between the three membrane systems: (i) The inner membranes of mitochondria (Figs. 3.1C and 3.3B) and bacteria contain the energy transduction proteins, whereas in chloroplasts these proteins are located in a third membrane system, the thylakoid membranes encapsulated within the two outer membranes (Figs. 3.1A and 3.3A). Most of the thylakoid membrane mass is organized in stacked interconnected (grana) discs. (ii) Extensive infolding of the mitochondrial inner membrane into cristae creates a large surface area. (iii) Chloroplasts and mitochondria differ from

3.1 Elements of Membrane Structure

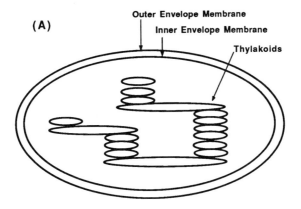

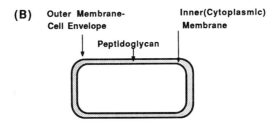

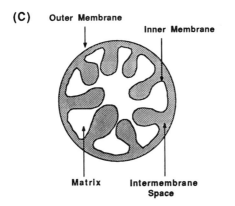

Figure 3.1. Schematic structural organization of (A) the chloroplast, (B) a Gram-negative bacterium such as an *E. coli* cell, and (C) a mitochondrion. The mitochondrion is an oval structure with approximately the same dimensions as a log phase *E. coli* cell, 1–2 µm × 0.5–1.0 µm. The dimensions of a mitochondrion depend on the tissue source and pathological state, and those of a bacterial cell depend on growth phase. The long dimension of the chloroplast is approximately 5 µm (based on Keegstra et al., 1984).

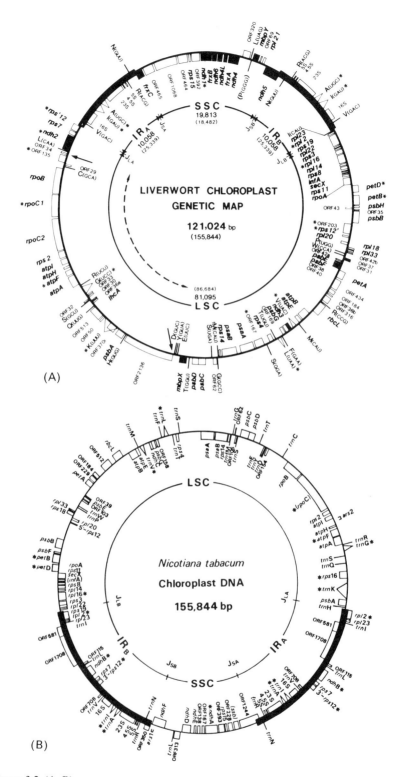

Figure 3.2 (A, B)

3.1 Elements of Membrane Structure

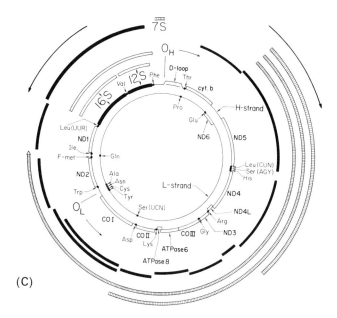

(C)

Figure 3.2. Genetic map of (A) liverwort (119 different genes; Ohyama et al., 1986; Ohyama et al., 1988) and (B) tobacco (122 different genes; Sugiura et al., 1986) chloroplast genomes. Approximately 60 genes are involved in transcription–translation, ~20 for electron transport and ATP synthesis, and there are ~30 open reading frames (ORFs). (C) Genetic and transcription maps of the HeLa cell mitochondrial genome. (Attardi et al., 1986.)

bacteria in their need and ability to export CO_2 fixation products and tricarboxylic acid (TCA) cycle intermediates, respectively, into the cytoplasm, and also to import ADP and phosphate, and export ATP, using phosphate transport and ADP/ATP carriers.

The structural similarities between these membranes, and some of their intrinsic proteins, imply that they share a common evolutionary background, that the electron transport chains of photosynthetic bacteria and cyanobacteria may be used for both respiration and photosynthesis, and that the mechanisms of free energy storage are fundamentally very similar or identical. Mitochondria and chloroplasts probably evolved by endosymbiosis between a protoeukaryote containing a nucleus and eubacteria containing apparatus for photosynthesis (cyanobacteria or purple photosynthetic bacteria) and/or oxidative phosphorylation. Most of the oligomeric protein complexes of these organelles are coded for by both organelle and nuclear DNA, and synthesized on organelle and cytoplasmic ribosomes. Organelle evolution may have occurred by progressive deletion of organelle DNA genes and acquisition of the lost functions by the nucleus (Wallace, 1982). Thus, organelle genomes (Fig. 3.2A–C) are much smaller than those of bacteria. The coding capacity ($3-10 \times 10^8$ daltons) of plant mitochondrial DNA (mtDNA) is greater than that of animals

($\sim 10^7$ daltons), with chloroplast DNA being of intermediate size (120–160 kilo base-pairs; $\leq 10^8$ daltons). The human mtDNA is a closed circular molecule of 16.6 kilobases (kb) present in approximately 10^4 copies/cell. The small size of mammalian mtDNA is illustrated further by the fact that it codes for only 13 mitochondrial proteins. Using the fact that the mtDNA is maternally inherited, the technique of restriction endonuclease analysis of human and animal mtDNA has become an important tool in the study of human evolution (Lewin, 1987). The sequences of the *frx* genes in the liverwort chloroplast genome (Fig. 3.2A) resemble those of [4Fe-4S] bacterial ferredoxin as well as an Fe-protein of N_2-fixing bacteria, implying an evolutionary connection with nitrogen-fixing cyanobacteria. The chloroplast *mbp*X and *mbp*Y gene sequences have significant identity with the *mal*K–*mal*F and hisP–hisQ genes of *E. coli* involved in maltose and histidine transport, respectively. There are also seven open reading frames that exhibit high sequence homology with seven genes coding for subunits of the human mitochondrial NADH-dehydrogenase (Ohyama et al., 1988), indicating that (i) chloroplasts and mitochondria have diverged from a common origin, and (ii) an NADH-plastoquinone oxidoreductase may be involved in the chloroplast electron transport chain.

The number of mitochondria in animal cells, corresponding to 20% of the cell volume, is greater than usually found in plant cells. Mitochondria of plant cells have a higher respiratory rate than those of animal cells (Table 3.1), but it is about an order of magnitude smaller than the specific rate of chloroplast photosynthetic electron transport.

Energy transduction or oxidative phosphorylation in mammalian mitochondria is carried out in the mitochondrial inner membrane by a number

Table 3.1. Comparison of respiratory rates of mitochondria from various tissues and cells

	Respiratory rates[a]		
	Substrate		
Organism	Succinate	α-Ketoglutarate	NADH
Guinea pig liver[b]	90	30	—
Beef heart	120	90	—
Potato tubers	510	260	330
Mung bean hypocotyls	450	—	510
Spinach leaves	240	—	320
Neurospora crassa	260	140	450
Saccharomyces cerevisiae	200	140	530

[a] Units: nmol O_2/min/mg protein; multiply by 4 to obtain electron transfer rate.
[b] Liver cells contain 1,000–2,000 mitochondria, $\sim 1/5$ of the cell volume.
From Douce (1985).

3.1 Elements of Membrane Structure

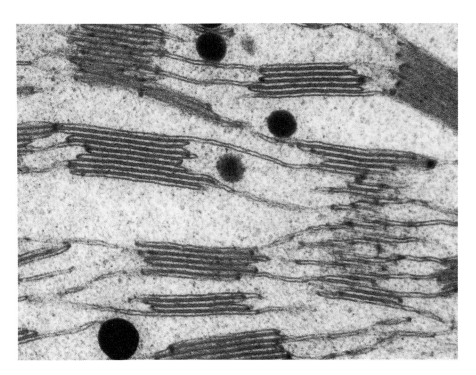

Figure 3.3. (A) Thin-section electron micrograph through a portion of a spinach chloroplast showing interconnected appressed grana and unstacked stroma thylakoid membranes. The flattened sac-like structure of individual thylakoids can be noted. The tangentially sectioned granum on the right shows continuity of the inner space, the angle between grana and stroma membranes, and the even spacing of the latter. × 80,000 (Staehelin, 1986).

(B)i

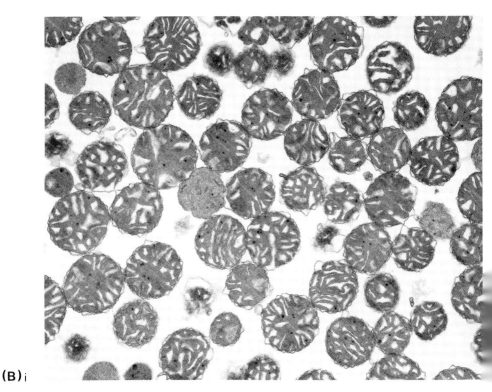

Figure 3.3. (B) Thin-section electron micrographs of rat-liver mitochondria showing inner (cristae) and outer membranes. (i) Untreated control, showing the morphological effect of (ii) the lipid-soluble uncoupler, dinitrophenol (DNP, 10 μM), and (iii) the cholesterol- and triglyceride-lowering drug (antihyperlipidemic) clofibrate (1 mM) (Woods et al., 1977). The details of the mitochondrial morphology are also tissue-specific. Morphological changes including changes in the ratio of inner to outer membrane are observed in a number of pathological situations. Mitochondrial myopathies, in which defects in muscle mitochondria are the most prominent features, often involve structural abnormalities in the mitochondria. (*cf*., Carafoli and Roman, 1980.)

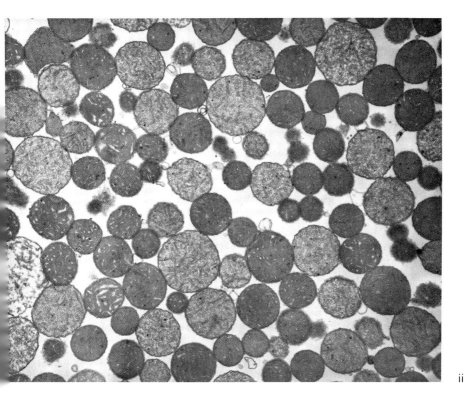

ii

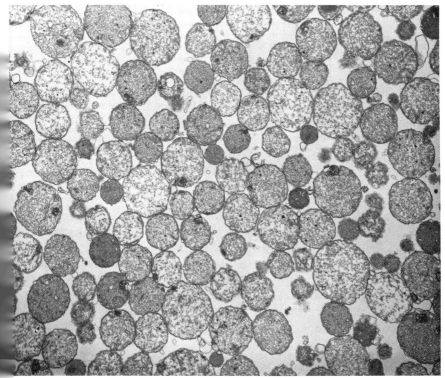

iii

Figure 3.3B *(continued)*

of membrane-bound respiratory enzyme complexes, the dehydrogenases (complexes I and II, the NADH and succinate dehydrogenases), the cytochrome $b-c_1$ complex (III), cytochrome oxidase (IV), and the ATP synthase. The intrinsic membrane protein complexes of respiration and photosynthesis are linked by extrinsic membrane proteins, a c-type cytochrome in mitochondria and chromatophores, and plastocyanin in chloroplasts, mediating electron transport to the oxidase or reaction center, and a pool of diffusible lipophilic quinone (Chap. 5). The activities of the separable respiratory complexes I–IV, and the composition and stoichiometry of the redox centers of complexes III and IV in mammalian mitochondria ae shown (Table 3.2).

The large pool of lipophilic quinone and a high potential extrinsically bound c type (Fe) cytochrome or (Cu) plastocyanin are ubiquitous features of photosynthetic energy transducing membranes. The cytochrome $b-c_1$ complex is present in all known mitochondria from fungal to human, chloroplasts, and photosynthetic and denitrifying bacteria, but not in *E. coli*. Higher plant chloroplasts contain three membranous electron transport complexes, the cytochrome b_6-f complex analogous to the $b-c_1$ complex, and the reaction center complexes of photosystems I and II. Some properties of these complexes are summarized (Tables 3.3 and 3.4).

The inner membranes of mitochondria and *E. coli*, as well as chloroplast thylakoid membranes, are typically 75% protein and 25% lipid (Table 3.5) by weight, which might be expected to lead to close packing of the protein. However, because much mitochondrial inner membrane protein (e.g., F_1 ATPase) can be removed by nonionic detergent treatment without disrupting the bilayer, and a large amount of the protein of complexes III and IV is known from electron microscope studies to protrude extensively from the bilayer, much of the protein mass is situated outside the bilayer and the intrinsic protein is relatively freely dispersed in the membrane (Fig. 3.4).

The lipid acyl chains of the thylakoid and mitochondrial lipids are highly unsaturated (Table 3.6). The high level of unsaturation of the lipid fatty acyl chains, approximately five double bonds on the average per lipid molecule (Table 3.6A), indicates that the membrane bilayer should have a low viscosity. The lipophilic quinone and the intrinsic electron transport, ATPase, and chloroplast light harvesting complexes should then be highly mobile in the plane of the membrane. The major polar lipids of thylakoid membranes are monogalactosyldiacylglycerol (MGDG) and digalactosyldiacylglycerol (DGDG), which typically comprise approximately 45% and 25%, respectively, of the polar lipids of the thylakoid membrane (Table 3.5). The major lipids of the inner mitochondrial membrane, like that of Gram-negative bacteria, are phospholipids. The presence of sulfolipids in thylakoids and phospholipids in mitochondria and bacterial membranes is a major source of their negative surface-charge density (section 3.4.2).

Linear representations of the respiratory chain and the green plant and bacterial photosynthetic electron transport chains are shown in Fig. 3.5A–C.

3.1 Elements of Membrane Structure

Table 3.2. Proteins and protein complexes of the bovine heart mitochondrial membrane

Component	Concentration (nmol/mg protein)	Activities of complexes (μmol e$^-$/min/mg)	Redox groups	E_m (mV)
Complex I[a]	0.1	NADH $\to$ CoQ[b], 150	FMN, [2Fe-2S], [4Fe-4S]	-370 to -20 (see Chap. 4)
Complex II	0.2	Succinate $\to$ Q$_2$[b], 50	FAD, [2Fe-2S], [4Fe-4S], [3Fe-XS]	-270 to $+140$ (see Chap. 4)
Complex III	0.25–0.5	Q$_2$H$_2$ $\to$ cyt c(Fe^{3+}), 300–600	b_{566}, b_{562}, c_1, [2Fe-2S]	-90, $+50$, $+220$, $+290$
Complex IV	0.6–1	Cyt c(Fe^{2+}) $\to$ oxidase, 25–50	a, a_3, 2Cu	$+300$–400
Cytochrome c	0.8–1.0	—	c heme	$+260$
ATP synthase	0.5	—	—	—
ADP/ATP translocase	4.0	—	—	—
NADH–NADP$^+$ transhydrogenase	0.05	—	—	—
Ubiquinone	6–8	—	—	—
Phospholipid	500	—	—	—

[a] Approximate stoichiometries in mammalian mitochondria:
Complex I: Complex III: Cytochrome c: Complex IV: UQ = 1:3:8:8:64.
Complex IV: ATPase: ADP-ATP translocase = 1:1:5. (cf. Wainio, 1970; Tazgoloff, 1982.)
[b] CoQ is ubiquinone-10, Q$_2$ is ubiquinone-2 (Chap. 5).
From Capaldi (1982).

Table 3.3. Properties of isolated chloroplast electron transport complexes

Complex	Activity	No. polypeptides
Photosystem II	$2H_2O \rightarrow O_2 + 4e^- + 4H^+$ (600 μmol O_2/mg Chl-hr)[a]	4–9
Cytochrome b_6–f	$QH_2 \rightarrow$ plastocyanin (100 s^{-1}, with isolated complex)	4–5
Photosystem I	plastocyanin (red) $\rightarrow NADP^+$ (200 μmol e^-/mg Chl-hr)	~7

[a] pH optimum shifted from 8 to 6 in isolated PSII; multiply by 4 to obtain electron transfer activity.

Table 3.4. Molecular weight and E_m of chloroplast and algal electron transport proteins

Proteins	M_r ($\times 10^3$)	E_{m7} (mV)
PSII:		
P680 reaction center core		+1,200 ($P680^+$/P680)
		−600 (Pheo/$Pheo^-$)
D1 polypeptide (binds Q_B, Fe)	33	+50 (Q_B/Q_B^-)
D_2 polypeptide (binds Q_A, Fe)	33	−20 (Q_A/Q_A^-)
Cytochrome b_{559} (2 hemes)	9, 4	+400 mV; lower potential ascorbate-reducible component often present
Accessory Chl-binding polypeptides	43, 47	—
1–2 accessory polypeptides	22, 24	—
Extrinsic PSII polypeptides[a]	17, 23, 33	—
b_6–f:		
subunit IV	15	—
Rieske [2Fe-2S]	20	+290
Cytochrome b_6 (2 hemes)	23	−50
Cytochrome f	35	+350
Plastocyanin	10.5	+370
PSI:		
Core: P700 protein (in complex)	83, 82	+500
Ferredoxin binding	22	−900 for primary acceptor, A_o (Table 6.8)
2[4Fe-4S]	9	
Plastocyanin binding	19	
Accessory polypeptides:		
Ferredoxin-$NADP^+$ reductase	33–36	−360
Ferredoxin	11	−430

[a] Stoichiometry: 1 33-kD:1–2 23-kD:1–2 17-kD:220 Chl:4Mn:2 PQ:2 b-559 heme.

3.1 Elements of Membrane Structure

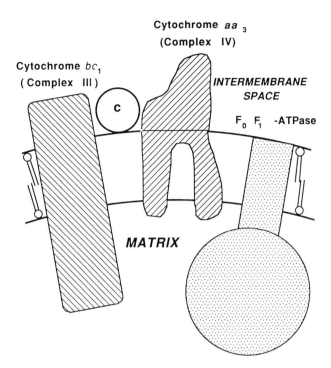

Figure 3.4. Distribution of intrinsic and extrinsic protein mass in the mitochondrial inner membrane. These three complexes comprise 35% of mitochondrial inner membrane protein. Fifteen to thirty percent of the mass of these complexes is confined to the bilayer. (After Hackenbrock, 1981.) Data for complex III comes from analysis of two-dimensional crystals by electron diffraction and image reconstruction at ~ 25 Å resolution from which a dimeric complex was found to be embedded in the membrane with dimensions of $75 \times 85 \times 140$ Å (long dimension perpendicular to the membrane plane), 50% of the mass extending 70 Å from the matrix side and $\sim 20\%$ extending 30 Å on the cytoplasmic side (Leonard et al., 1981).

Table 3.5. Lipid composition of (A) thylakoid (Gounaris et al., 1983) and (B) mitochondrial inner membranes (Colbeau et al., 1971)

	Lipid composition (mol fraction × 100)					
A. Spinach chloroplast thylakoids (2.4 mol lipid/mol Chl)	MGDG[a]	DGDG	PG	SQDG	PC	OPL
	47	24	14	7.5	3.0	4.5
	(mol fraction × 100)					
B. Rat liver mitochondrial inner membrane	PC	PE	CL	SM	PI	LPC
	40.5	39	17	2	1.5	0.5

[a] MGDG, monogalactosyldiacylglycerol; DGDG, digalactosyldiacylglycerol; PG, phosphatidylglycerol; SQDG, sulfoquinonosyldiacylglycerol; PC, phosphatidylcholine; OPL, other phospholipids; PC, phosphatidylcholine; PE, phosphatidylethanolamine; CL, cardiolipin; SM, sphingomyelin; PI, phosphatidylinositol; LPC, lysophosphatidylcholine.

Table 3.6. Fatty acids of (A) thylakoids (Gounaris et al., 1983) and (B) mitochondrial inner membranes (Colbeau et al., 1971).

	Fatty acid composition (mol fraction × 100)									
	16:0	16:1	16:3	18:0	18:1	18:2	18:3	20:3	20:4	22:6
A. Spinach chloroplast thylakoid	9.0	2.0	1.0	0.5	2.0	3.5	72.0	—	—	—
B. Rat liver mitochondrial inner membrane[a]	22.0	3.0	—	20.0	10.5	13.0	—	1.0	17.0	4.0

[a] Calculated from phosphatidylcholine, phosphatidylethanolamine, and cardiolipin.

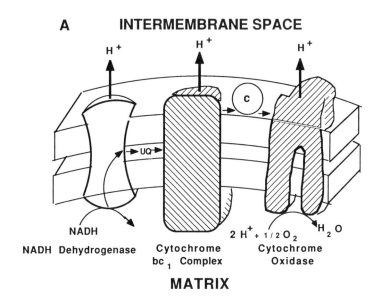

Figure 3.5. Electron transport chain of (A) mammalian mitochondria showing electron flow through, and H^+ translocation across the NADH dehydrogenase, cytochrome b–c_1, and cytochrome oxidase complexes in the respiratory chain. (B) The branched respiratory chain of *E. coli*. Flavoprotein (F) dehydrogenases include D-lactate, NADH, and succinate dehydrogenases, and pyruvate oxidase; Q is UQ-8; d and o are high- and low-affinity oxidases used in stationary (low ambient O_2) and logarithmic phases, respectively (Anraku and Gennis, 1987). (C) Photosynthetic noncyclic electron transport chain of green plants and algae, showing electron flow, H^+ deposition associated with H_2O-splitting in photosystem II, and H^+ translocation by the cytochrome b_6–f complex using electrons donated by plastoquinol.

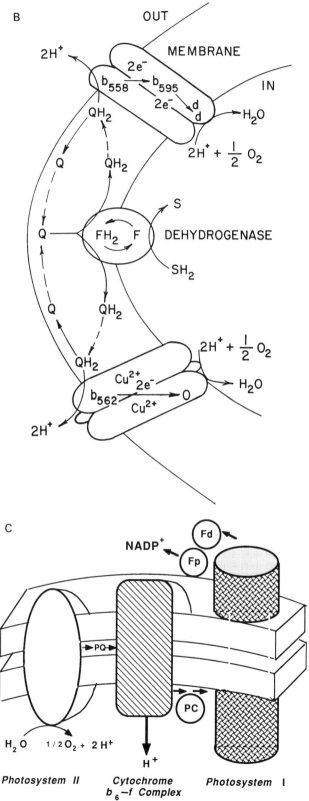

3.2 Introduction to the Energy Storage Problem

Three general mechanisms have been proposed for storage of the free energy made available from electron transport: (i) chemical high-energy bonds, (ii) bulk phase or localized electrochemical gradients, and (iii) conformational changes in the ATP synthase (Chap. 8).

(i) The chemical model assumes that high-energy bonds ("$\sim$") are generated through electron transfer and that these high-energy bonds are transferred through a series of compounds until they are finally transferred to ATP (ADP $\sim$ P). This model was based on analogy with the substrate level phosphorylation in glycolysis and the Krebs cycle, and perhaps most specifically with the glyceraldehyde-3-phosphate dehydrogenase reaction in which the substrate is converted to 1,3-diphosphoglyceric acid (Fig. 3.6), with the following features (Racker, 1965): (i) The oxidation of glyceraldehyde-3-phosphate by NAD^+ and formation of a nonphosphorylated thioacyl high-energy intermediate precedes phosphorylation; (ii) two electrons are transferred per phosphorylation; (iii) the $\Delta G°$ of the 1,3-diphosphoglyceric acid $\simeq$ -12 kcal/mol (Chap. 1).

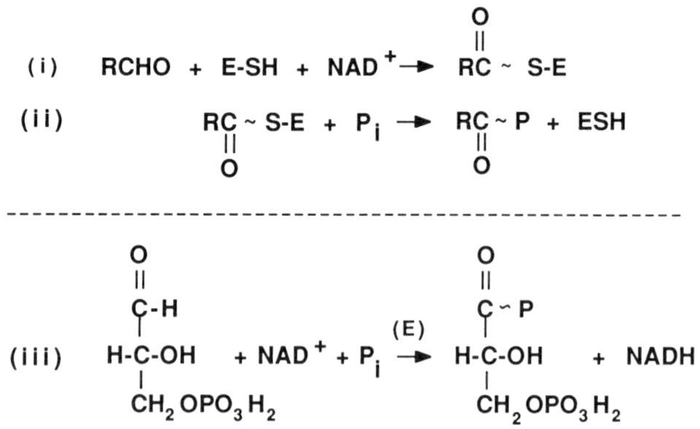

$$\text{(i)} \quad RCHO + E\text{-}SH + NAD^+ \rightarrow RC(=O) \sim S\text{-}E$$

$$\text{(ii)} \quad RC(=O) \sim S\text{-}E + P_i \rightarrow RC(=O) \sim P + ESH$$

$$\text{(iii)} \quad \begin{array}{c} O \\ \parallel \\ C\text{-}H \\ | \\ H\text{-}C\text{-}OH \\ | \\ CH_2 OPO_3 H_2 \end{array} + NAD^+ + P_i \xrightarrow{(E)} \begin{array}{c} O \\ \parallel \\ C \sim P \\ | \\ H\text{-}C\text{-}OH \\ | \\ CH_2 OPO_3 H_2 \end{array} + NADH$$

Figure 3.6. The glyceraldehyde-3-phosphate dehydrogenase reaction. (Based on Racker, 1965.) The overall reaction (iii) proceeds via steps (i) and (ii) as described in the text.

Through consideration of reactions such as these, a mechanism for oxidation phosphorylation involving a series of high-energy chemical intermediates coupled to the transfer of electrons from hypothetical cytochrome A(red) to cytochrome B(ox) was proposed (Chance and Williams, 1956):

$$\text{cyt A(red)} + \text{cyt B(ox)} + I \rightarrow \text{cyt A(ox)} - I + \text{cyt B(red)}$$

$$\text{cyt A(ox)} \sim I + X \rightarrow \text{cyt A(ox)} + X \sim I$$

$$X \sim I + P \rightarrow X \sim P + I$$

$$X \sim P + ADP \underset{(\text{ATPase})}{\rightleftarrows} ATP + X,$$

where I $\sim$ X and X $\sim$ P are hypothetical nonphosphorylated and phosphorylated high-energy intermediates, respectively. The chemical model was not experimentally successful because the phosphorylated and nonphosphorylated high-energy intermediates could not be isolated. The model is mentioned partly for historical reaons and partly because phosphorylated intermediates do function in some ATPase (Chap. 8) and active transport proteins (Chap. 9).

3.3 The Chemiosmotic Hypothesis

The experimental impasse regarding high-energy chemical intermediates in the pathway of electron transport-linked ATP synthesis led to the proposal that "the elusive character of the energy-rich intermediates of the orthodox chemical coupling hypothesis would be explained by the fact that these intermediates do not exist" (Mitchell, 1961). An alternative hypothesis for energy coupling was offered in which ATP synthesis could be driven by a proton electrochemical potential, $\Delta\tilde{\mu}_{H^+}$ (Chap. 1, section 10). The underlying viewpoint of this "chemiosmotic" hypothesis, derived from studies on enzyme-catalyzed group translocation, was that "if the processes that we call metabolism and transport represent events in a sequence, then not only can metabolism be the cause of transport, but also transport can be the cause of metabolism" (Mitchell, 1961).

The ideas that chemical potential and work can be stored in an ion gradient coupled to electron transport are also found in earlier work of Lundegårdh (1945), who had noted in a paper on "absorption and exudation of inorganic ions by the roots," the possibility of an oxidation–reduction reaction being anisotropically organized on a membrane so that H^+ would be produced on one side and consumed on the other.

The tenets of the chemiosmotic hypothesis are:

1. Energy transducing membranes are vesicular, sealed, and impermeable to protons except for the pathways involved in redox-mediated or protein-catalyzed H^+ translocation.
2. Energy is stored in a pH gradient or membrane potential which are energetically equivalent, with the electrochemical potential, $\Delta\tilde{\mu}_{H^+} = F \cdot \Delta\psi - 2.3RT \cdot \Delta pH$ (Chap. I, Eqs. 22–24).
3. The $\Delta\tilde{\mu}_{H^+}$ is formed by vectorially alternating H^+ and e^- carriers in the electron transport chain (Fig. 3.7). The extrusion of protons occurs in the H^+–e^- carrier loop when electrons are transferred from the H^+ carrier to the e^- carrier. This original formulation of the hypothesis predicted that the H^+/e^- ratio for proton translocation must always equal 1. It did not include *trans*-membrane proteinaceous H^+ pumps or a "Q cycle" as alternative mechanisms for H^+ translocation (Chap. 7) to account for observed H^+/e^- ratios > 1.
4. An H^+ flux is coupled to the ATP synthase/ATPase catalyzed by the large

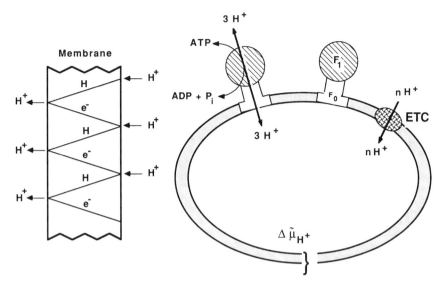

Figure 3.7 (Left). Electron transport chain vectorially organized for net H$^+$ translocation according to Mitchell (1966).

Figure 3.8 (Right). Reversible F$_0$F$_1$ ATPase and electron transport chain (ETC) in H$^+$-impermeable membrane generating a $\Delta\tilde{\mu}_{H^+}$-positive inside. The anisotropy of this orientation is characteristic of chloroplast thylakoids, chromatophores, and submitochondrial particles (SMP).

    multisubunit F$_0$F$_1$ ATPase. Each reaction is aniostropic with respect to this flux. The synthesis reaction is coupled to an H$^+$ flux driven by the $\Delta\tilde{\mu}_{H^+}$ from the $\tilde{\mu}_{H^+}$-positive side of the membrane (p-side). The reverse reaction of ATP hydrolysis results in H$^+$ translocation in the opposite direction (Fig. 3.8).

5. Uncouplers of energy transduction were predicted to be lipid-soluble weak acids (e.g., dinitrophenol) or bases (e.g., methylamine) that can catalyze the equilibration of H$^+$ or OH$^-$ across the membrane (sections 3.6.7, 3.12 below).

The related concept of energy storage in electrochemical proton gradients localized within the membrane is discussed below (see section 3.11).

3.4 Measurement of ΔpH and $\Delta\psi$ Across Energy-Transducing Membranes

The measurement of ΔpH or $\Delta\psi$ is not trivial. Different probes are used for each measurement and there are experimental problems associated with each probe. The general problems are (i) permeation of the internal probes; the

3.4 Measurement of ΔpH and Δψ Across Energy-Transducing Membranes

bacterial outer membrane, in particular, can form a barrier to these probes. (ii) Quantitative use of these probes requires determination of intraorganelle or intracellular volume. (iii) Vesicle heterogeneity will give rise to a distribution of values of ΔpH and Δψ.

3.4.1 ΔpH

The pH gradient can be measured through the distribution of (i) a radiolabeled weak acid such as 5,5-dimethyl-2,4-oxazolidinedione (DMO, Fig. 3.9A), acetate, or butyrate that can accumulate in the unprotonated charged form in a neutral or alkaline compartment such as the intraorganelle space of mitochondria. (ii) A labeled weak base (e.g., methylamine or hexylamine) can be trapped as a protonated charged form in the acidic compartment of energized thylakoids, submitochondrial particles, or chromatophores (Fig. 3.9B).

The assumptions of the method are:

1. The unionized or neutral form of the weak acid, A, or base, B, equilibrates freely across the membrane; e.g., $[B]_{in} = [B]_{out}$.
2. The pK_a of the weak acid or base is the same inside and outside the organelle; i.e., $pK_{a(in)} = pK_{a(out)}$.

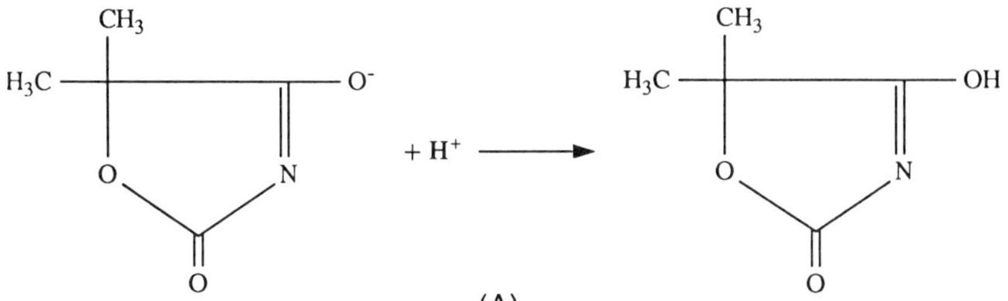

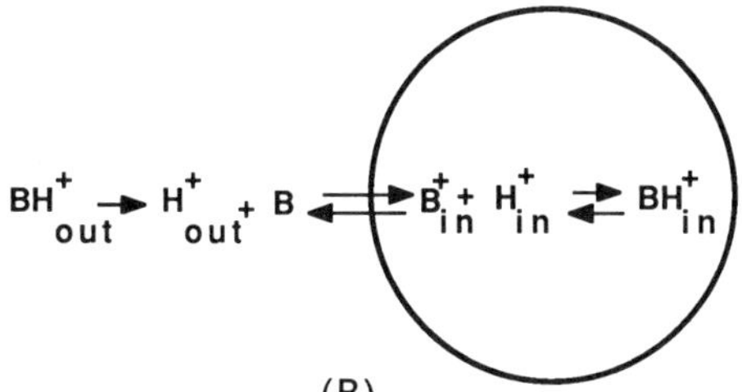

Figure 3.9. (A) Structure of DMO, $pK = 6.1$. (B) Distribution of a weak base across an energized thylakoid, chromatophore, or SMP membrane with a low internal pH.

If the total concentration, (B_T), of a weak acid or base $= (B + BH^+)$,
Then,

$$\frac{(B)_{T,in}}{(B)_{T,out}} = \frac{(B)_{in} + (BH^+)_{in}}{(B)_{out} + (BH^+)_{out}} \tag{1a}$$

Since,

$$(H^+)_{in} = K_a \frac{(BH^+)_{in}}{(B)_{in}},$$

and

$$(H^+)_{out} = K_a \frac{(BH^+)_{out}}{(B)_{out}}; \tag{1b}$$

then,

$$\frac{(H^+)_{in}}{(H^+)_{out}} = \frac{(BH^+)_{in}}{(BH^+)_{out}},$$

since

$$(B)_{in} = (B)_{out}. \tag{1c}$$

Substituting (1b) into (1a),

$$\frac{(B)_{T,in}}{(B)_{T,out}} = \frac{K_a + (H^+)_{in}}{K_a + (H^+)_{out}}$$

$$= \frac{(H^+)_{in}}{(H^+)_{out}}, \tag{1d}$$

if

$$K_a \ll (H^+)_{in}, (H^+)_{out}$$

[as in Eq. (1e) below]. This result describes the pH-dependent partition across the membrane of the weak acid or base, which can be used to measure $(H^+)_{in}$ if $(H^+)_{out}$ is known.

Use of pH-Indicating Dyes

The pH gradient across the thylakoid membrane, SMP or chromatophores, can also be measured by the quenching, Q, of fluorescence of amine dyes with a high pK such as 9-aminoacridine (Fig. 3.10A). As discussed above for DMO, the dye technique is based on the assumptions that (i) the neutral form of the basic dye, B, freely equilibrates across the membrane and that in the lumen it can be protonated in the energized membrane to the cationic form (BH^+) leading to a higher concentration in the lumen. (ii) Due to the higher concentration, the dye fluorescence is quenched to a lower level, F_e, in the energized membrane relative to the level F_d in the deenergized membrane.

3.4 Measurement of ΔpH and Δψ Across Energy-Transducing Membranes

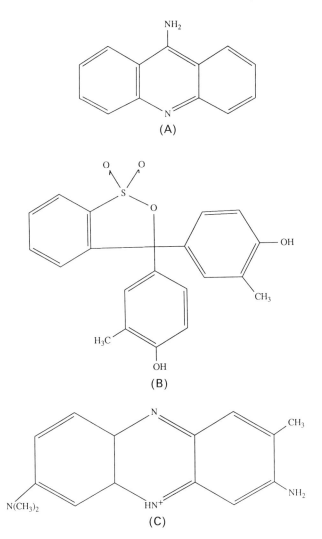

Figure 3.10. Structures of probes for ΔpH. (A) 9-Aminoacridine; [pK of ring = 10.0; a fluorescent probe (λ_{exc}, 420 nm, λ_{em}, 530 nm)]. (B) Cresol red (pK 8.3). (C) Neutral red cation (pK 6.6).

The degree of quenching, $Q = 1 - F_e/F_d$, is then proportional to B_i/B_{init} (ratio of the internal dye concentration, B_i, in the energized membrane to the initial internal concentration, B_{init}) and the volume fraction, v ($v \equiv V_{in}/V_{tot}$), of the internal space. The pH gradient (ΔpH) across the membrane (pH$_i$ ≤ pH$_o$) is related from Eq. 1d, for $K_a \ll (H^+)_{in}$ and $K_a \ll (H^+)_{out}$, to Q and v (Schuldiner et al., 1972; Haraux and de Kouchkovsky, 1980):

$$Q = (B_i/B_{init})(v);$$

then,

$$B_i = (B_{init})\left(\frac{Q}{v}\right);$$

and the external dye concentration,

$$B_{out} = B_{init} - (B_i) \cdot (v) = B_{init}(1 - Q);$$

the ratio

$$B_i/B_{out} = Q/(1 - Q)(v),$$

and from Eq. 1d, relating the ratio of internal to external dye concentration to that of H^+,

$$\Delta pH \simeq \log_{10}\left(\frac{Q}{1 - Q}\right)\left(\frac{1}{v}\right). \tag{1e}$$

The pH-indicating dyes cresol red (Fig. 3.10B) and phenol red are used as external pH indicators through pH-dependent changes in absorption spectra. Neutral red (Fig. 3.10C) has been used to indicate millisecond pH transients in the internal lumenal space of the thylakoid when nonpermeant buffers are used to eliminate the external pH response (Junge et al., 1986).

3.4.2 Measurement of $\Delta\psi$

The uptake and accumulation of lipid-soluble cations and anions (Fig. 3.11) can be used to assay the transmembrane electrical potential $\Delta\psi$ in membrane systems that have, respectively, a negative and positive transmembrane potential, although these cations do not readily penetrate untreated outer membranes of intact Gram-negative bacteria such as *E. coli*. The use of lipid-

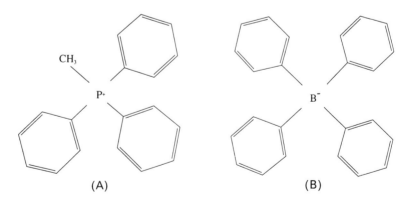

Figure 3.11. Chemical structures of lipid-soluble ions. (A) Methyltriphenylphosphonium. (B) Tetraphenylborate.

3.4 Measurement of ΔpH and Δψ Across Energy-Transducing Membranes

soluble cations to measure $\Delta\psi$ has been checked with microelectrodes with *E. coli* inner membrane vesicles isolated from large cells where the two methods yielded similar values for $\Delta\psi$ (Felle et al., 1980). [However, these two methods have yielded discrepant results in mitochondria where the use of microelectrodes indicated that the value of $\Delta\psi$ was much smaller (-10 to -20 mV) than indicated by the ion distribution method (Tedeschi, 1980).]

The theoretical basis of the method using lipid-soluble cations (LSC^+) to measure $\Delta\psi$ is as follows:

At equilibrium:
$$\tilde{\mu}_{in} = \tilde{\mu}_{out}.$$

Then,
$$RT\ln(LSC^+)_{in} + F\psi_{in} = RT\ln(LSC^+)_{out} + F\psi_{out},$$

if $\tilde{\mu}$ is the same inside and outside the membrane, and $z = 1$; then,

$$\Delta\psi \equiv \psi_{in} - \psi_{out} = \frac{RT}{F}\ln\frac{(LSC^+)_{out}}{(LSC^+)_{in}}. \tag{2}$$

A value of $\Delta\psi = -59$ mV, negative inside, would be sensed by a 10-fold ratio of $(LSC^+)_{in}/(LSC^+)_{out}$. An experimental problem is that excess uptake of LSC^+ can decrease the $\Delta\psi$. The lipid-soluble cation technique, like other techniques for sensing $\Delta\psi$, can be tested by generating potassium diffusion potentials of known magnitude from a potassium gradient across the vesicle and addition of the ionophore valinomycin (Appendix, 3.12).

A potassium diffusion potential can be generated across closed membrane vesicles by loading or preparing the vesicles with a known concentration of K^+ inside and then diluting the vesicles into a medium containing a known K^+ concentration in the presence of the ionophore valinomycin (see Appendix 3.12). When valinomycin is added, it will carry K^+ from one side of the membrane to the other, down the concentration gradient. However, very little K^+ will flow before the membrane potential, negative inside, increases to a level where it prevents further K^+ efflux. At this point, there is an equilibrium between the $\Delta\psi$ and the potassium concentration gradient. At equilibrium,

$$\Delta\tilde{\mu}_{K^+} = 2.3RT\log\frac{K^+_{in}}{K^+_{out}} + F\Delta\psi = 0.$$

Then,
$$\Delta\psi = -\frac{2.3RT}{F}\log\frac{K^+_{in}}{K^+_{out}} \tag{3}$$

The value of the K^+ diffusion potential generated by membrane vesicles made in the presence of 100 mM K^+ and then diluted 100-fold in the presence of valinomycin into a K^+-free medium would be -118 mV at 25°C. Note that the basis of this calculation is that the K^+ concentrations inside and outside the vesicle do not change significantly upon addition of valinomycin (see section 3.5).

The Carotenoid Band Shift as an Indicator of the Transmembrane $\Delta\psi$ in Chloroplasts and Chromatophores

The transmembrane electric field is strong enough, approximately 2×10^5 V/cm (arising from a $\Delta\psi$ of 100 mV across the insulating layer of the membrane which is approximately 50 Å thick), to change the energy levels of pigment (Fig. 3.13 A) or artificial dye (Fig. 3.13 B) molecules with large permanent or inducible dipole moments in their excited states. The electric field can cause a shift in the excited state energy level relative to the ground state, a Stark effect, resulting in a change of the absorption spectrum (Fig. 3.12). The carotenoids (e.g. β-carotene, Fig. 3.13A) are examples of polarizable molecules without a permanent dipole moment. The shift, $\Delta\lambda_m$, of the peak wavelength, λ_m, is approximately (Wraight et al., 1978):

$$\Delta\lambda_m \simeq \frac{\lambda_m^2}{2hc} \cdot [(\Delta\alpha)(E_0 + E)] \cdot E, \qquad (4)$$

in cm, where $\Delta\alpha$ is the difference in polarizability between excited and ground

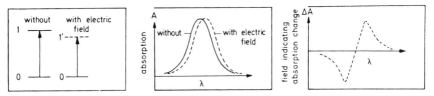

Figure 3.12. The mechanism of electrochromism. (Left) Shift of excited state energy level in electric field, resulting in (center) a shift of the spectrum, a red shift if the excited state energy level is decreased by the field. (Right) Difference spectrum, presence minus absence of field (Witt, 1979).

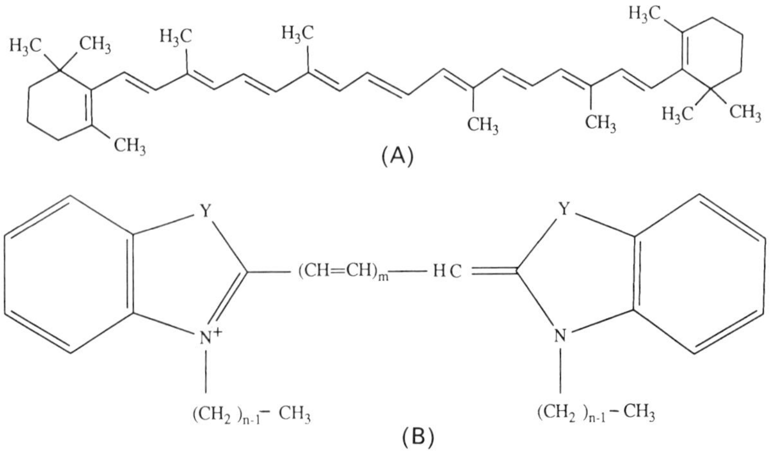

Figure 3.13. (A) Structure of β-carotene. (B) Structures of the cationic potential-sensing cyanine dyes (right), where $Y = O$, S, or $C(CH_3)_2$, n varies from 2 to 18, and $m = 1-3$. (After Sims et al., 1974.)

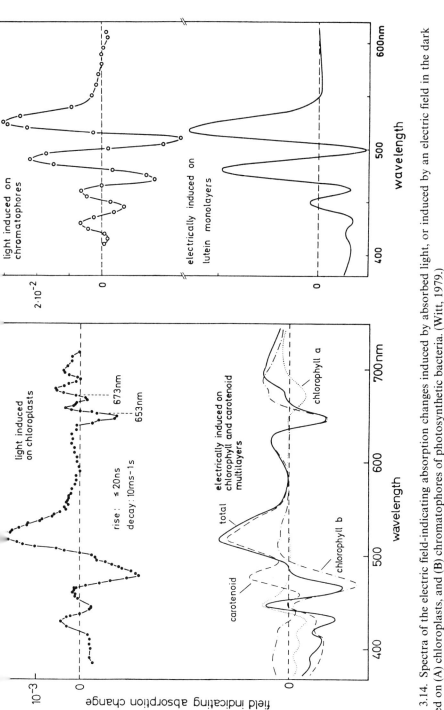

Figure 3.14. Spectra of the electric field-indicating absorption changes induced by absorbed light, or induced by an electric field in the dark imposed on (A) chloroplasts, and (B) chromatophores of photosynthetic bacteria. (Witt, 1979.)

states, $\sim 10^{-21}$ cm^3, E is the electrical field strength across the membrane bilayer, $\sim 2 \times 10^5$ V/cm [for calculation (Problem 50); convert to esu volt units: 1 V (esu) = 300 V (Standard International Units)]. $E_0 \gg E$ is a permanent field ($\sim 2 \times 10^6$ V/cm) hypothesized to exist locally between the pigments to explain the linear dependence of $\Delta\lambda_m$ on field strength. The letters h and c represent Planck's constant and the speed of light, respectively.

Typical electric-field indicating carotenoid absorption changes are shown in Fig. 3.14A and B for chloroplasts and chromatophores. The quantitative relationship between the amplitude of the carotenoid absorbance change and the actual value of $\Delta\psi$ is usually determined by calibration of the carotenoid absorbance change against potassium diffusion potentials of known magnitude.

Surface Potential and Transmembrane Potential

In addition to net charge movement from one bulk phase to the other across the membrane, the transmembrane $\Delta\psi$ is also determined by the difference between the surface potentials on the opposing membrane surfaces due to

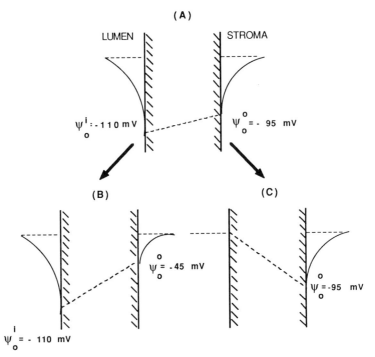

Figure 3.15. (A) Relation between thylakoid surface potential (ψ_o) and *trans*-membrane $\Delta\psi$. Effect on the external surface potential, ψ_o^o, internal surface potential, ψ_o^i, and $\Delta\psi$, of (B) an increase in external ionic strength (+5 mM MgSO$_4$) or a decrease in pH, and (C) light-induced H$^+$ uptake into the lumen, on negative surface charge and potential. The negative surface charge densities are -0.034 and -0.025 coulomb/m^2 on lumen and stroma sides, respectively, in an ionic environment equivalent to 5 mM KCl and pH 7.0 (based on Barber, 1982).

differences in the negative surface charge density. For example, the inner surface of the chloroplast thylakoid membrane has a larger negative surface charge density than the outer surface resulting in a negative $\Delta\psi \equiv (\psi_i - \psi_o)$, under conditions of low ionic strength in the dark (Fig. 3.15A). The effect of increasing the ionic strength or decreasing the pH on one side (e.g., external), will be to screen the surface charge and decrease the magnitude of the surface potential on that side of the membrane. For the external side, this would increase the size of the negative $\Delta\psi$ (Fig. 3.15B). The light-induced H^+ uptake to the lumen that is characteristic of thylakoid membranes would tend to compensate the negative internal surface charge and to decrease the magnitude of a negative $\Delta\psi$ (Fig. 3.15C). The membrane surface charge also affects the conformation of phospholipid head groups and may cause changes in the orientation of the dipole moment of the phosphocholine group which can be assayed by ^{2}H-NMR (Seelig et al., 1987). The resulting changes in dipole potential may be sufficiently large that they can trigger conformational changes in membrane proteins and facilitate protein insertion into membranes (Chap. 9.15).

3.5 Relationship Between $\Delta\psi$ and Charge Movement Across the Membrane

The number of protons or monovalent ions that must be translocated across the membrane per unit area in order to generate a $\Delta\psi$ of approximately 100 mV is quite small because of the membrane capacitance.

The $\Delta\psi$ is equal to Q/C, where C is the specific membrane capacitance, which for biological membranes is approximately equal to 1 μfarad/cm^2, and Q is the charge per unit area.

Then, if $\Delta\psi = 0.1$ V, and C = 10^{-6} farads/cm^2,

$$Q = 10^{-7} \text{ coulombs/cm}^2.$$

Because the charge on one proton is 1.6×10^{-19} coulombs, the number of protons translocated per unit area would be 6×10^{11}/cm^2. The number of protons translocated per square micron is then 6×10^3, or 0.6 protons per 10^4 Å^2 of membrane surface. Thus, translocation of only 20 protons across the membrane of vesicles 300 Å in diameter would generate a membrane potential of 100 mV.

3.6 Experimental Tests of the Chemiosmotic Hypothesis

3.6.1 Proton Movement

Light-, respiratory-, or ATP-driven H^+ or Na^+ movement across a closed organelle or bacterial membrane is always associated with energization of that membrane (e.g., photosynthetic or purple membrane system; Fig. 3.16A–B;

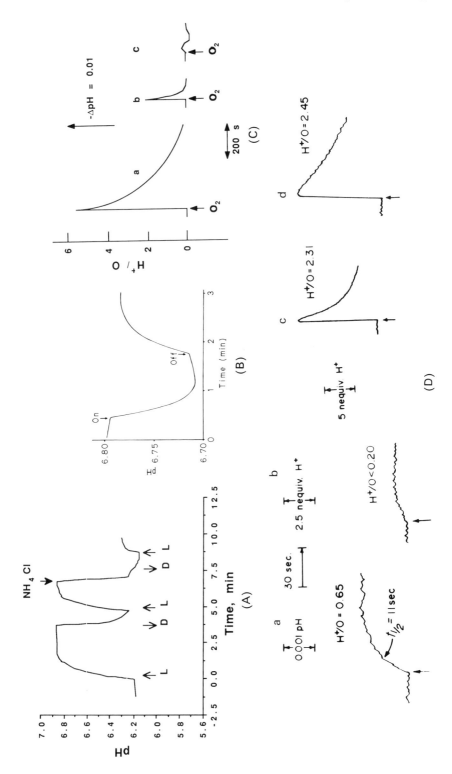

3.6 Experimental Tests of the Chemiosmotic Hypothesis

◁ Figure 3.16. pH changes measured upon energization of (A) illuminated chloroplasts with the artificial e⁻ carrier, pyocyanine, initial pH 6.2, showing a light-induced alkalinization that is dissipated by uncoupler, NH_4Cl; actinic light (L), on (↑), off (D, ↓). (Based on Neumann and Jagendorf, 1964.) (B) Purple membranes from *Halobacterium halobium* illuminated at 560 nm. (Oesterhelt and Hess, 1973.) (C) Anaerobic mitochondria containing 2 mM of the NADH-linked substrate, β-hydroxybutyrate [curve a, no addition; the actual value of the H^+/O ratio for an NADH-linked substrate is believed to be approximately twice that measured in the original experiment, depicted here for historical reasons; curve b, +50 μM dinitrophenol; curve c, plus detergent to disrupt the electrical integrity of the membrane]. (Based on Mitchell and Moyle, 1965.) (D) H^+ efflux induced by an O_2 pulse (arrow) in *E. coli* cells, (a) untreated, in the presence of (b) uncoupler to short-circuit the H^+ flow, and in the presence of (c) thiocyanate (SCN^-) as a lipid-soluble counter anion, and (d) colicin E1, a channel-forming protein that can depolarize *E. coli* cells (Gould and Cramer, 1977).

respiratory membranes, Fig. 3.16C, D). The light-driven H^+ movement is inwardly directed and thus causes an alkalinization of the medium in chloroplasts and chromatophores of photosynthetic bacteria (Fig. 3.16A). This also occurs through respiratory activity in inside-out submitochondrial particles (SMP), formed as shown in Fig. 3.17. The H^+ movement associated with energization by respiratory activity or light, of mitochondria (Fig. 3.16C) photosynthetic and nonphotosynthetic bacteria, or inside-out thylakoid membrane preparations, is outwardly directed.

3.6.2 Experimental Tests: ATP Synthesis with an Artificial $\Delta\tilde{\mu}_{H^+}$

ATP Synthesis in Chloroplasts Caused by an Acid–Base Transition

In a dramatic experiment that emphasized the energetic capability of a *trans*-membrane ΔpH for ATP synthesis, chloroplasts were incubated in the dark

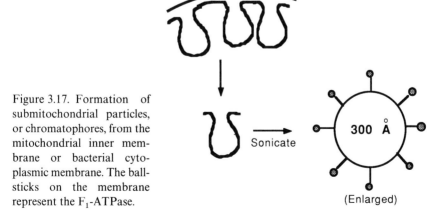

Figure 3.17. Formation of submitochondrial particles, or chromatophores, from the mitochondrial inner membrane or bacterial cytoplasmic membrane. The ball-sticks on the membrane represent the F_1-ATPase.

(Enlarged)

Table 3.7. ATP synthesis in chloroplasts caused by an acid–base transition

Reaction mixture (+ 10 mM succinate, 30 μM DCMU)	pH, initial	ATP (nmol/mg Chl)
Complete	3.8	140
	7.0	12
$-P_i$	3.8	12
$-ADP$	3.8	4
$-Mg^{2+}$	3.8	60
$-$Chloroplasts	3.8	4

From Jagendorf and Uribe (1966).

in the presence of a permeant acid, and then transferred to a basic solution containing ADP and P_i. The chloroplasts synthesized large amounts of ATP in the dark acid–base transition (Jagendorf and Uribe, 1966). The detailed protocol was:

(i) Chloroplasts were buffered at pH 4.0 in the dark in the presence of an organic acid such as succinate that can penetrate the membrane in its neutral form, and with an electron transport inhibitor to ensure the absence of electron transport activity.

(ii) The acidified chloroplast mixture was transferred to a solution at pH 8.0 containing ADP, $^{32}P_i$, and enough buffer to neutralize the acid carried over.

(iii) The reaction was stopped with trichloroacetic acid, and the ATP level measured (Table 3.7).

The efficiency of this dark acid–base ATP synthesis was dependent on the initial acid stage pH, the nature of the permeant acid, and had an optimum basic stage pH (+ADP) of approximately 8.0.

This experiment was important conceptually and historically because it demonstrated synthesis of ATP in a situation where a high-energy chemical intermediate generated stoichiometrically by electron transport could not function as a precursor to ATP synthesis, because: (i) ATP was synthesized in the dark in the presence of an electron transport inhibitor; (ii) synthesis of ca. 150 nmol ATP/mg Chl corresponds to ~ 1 ATP per 6–7 chlorophyll molecules, approximately 100 times the content of the chloroplast electron transport chain components (stoichiometry of reaction centers to cytochrome complexes = 1 : 600 Chl). Neither the presence of the electron transport inhibitor DCMU, or other electron transport inhibitors, nor the redox state of the chloroplasts, affected the amount of ATP synthesized during the dark acid–base transition (Miles and Jagendorf, 1970).

The interpretation of these data is that the free energy needed for ATP synthesis, ΔG_{ATP}, is stored in the proton electrochemical potential, $\Delta \tilde{\mu}_{H^+} = F \cdot \Delta \psi - 2.3RT \cdot \Delta pH$ (Chap. 1), which equaled $-2.3RT \cdot \Delta pH$ in the acid–base transition experiment. Since the protons in the initial state are outside (stroma), and in the final state inside (lumen), $\Delta pH = pH_{in} - pH_{out} = 4 - 8 = -4$, and $\Delta \tilde{\mu}_{H^+} = -1.36 \, (-4.0) = +5.44$ kcal/mol. Under physiological con-

3.6 Experimental Tests of the Chemiosmotic Hypothesis

ditions, a $\Delta\tilde{\mu}_{H^+}$ of similar magnitude would be generated by H^+ translocation linked to photosynthetic and respiratory electron transport (Fig. 3.18A). Free energy will be made available for ATP synthesis when this proton gradient is discharged from the inside (lumen) to the outside (stroma) through the ATP synthase enzyme complex (Fig. 3.18B). In the absence of compensatory ion movements that would prevent build-up of a $\Delta\psi$ arising from the H^+ movement, the proton flux and resulting ATP synthesis is limited. The presence of a compensatory ion movement such as K^+ influx in the presence of valinomycin can prevent the formation of this membrane potential (Fig. 3.18C). The existence of a negative internal $\Delta\psi$, as in mitochondria or *E. coli* cells, can facilitate vectorial H^+ flux through the ATP synthase (Fig. 3.18D), as expected for an energetically favorable $F\Delta\psi$ component of the $\Delta\tilde{\mu}_{H^+}$.

The amount of free energy made available for ATP synthesis from the $\Delta\tilde{\mu}_{H^+}$ generated by electron transport is dependent on the number of protons, n_{H^+}, translocated through the potential, $\Delta\tilde{\mu}_{H^+}$, per ATP synthesized. That is,

$$\Delta G_{ATP} = n_{H^+} \cdot \Delta\tilde{\mu}_{H^+}$$
$$= n_{H^+} \cdot [-2.3RT \cdot \Delta pH] \text{ in the acid-base experiment.}$$

For the chloroplast acid–base ATP synthesis, the initial state of the protons is inside (lumen) and the final state outside (stroma). Therefore, for the proton movement associated with ATP synthesis,

$$\Delta pH = pH_{out} - pH_{in}$$
$$= +4.0, \text{ so that the free energy made available for ATP}$$

synthesis at 25°C is

$$\Delta G_{ATP} = -5.44 \cdot n_{H^+}, \tag{5}$$

which

$$= -16.3 \text{ kcal/mol if } n_{H^+} = 3 \text{ (see Table 3.10).}$$

Experimental Tests: ATP Synthesis in Submitochondrial Particles Caused by a ΔpH and $\Delta\psi$

When submitochondrial particles are subjected to an artificially imposed $\Delta\tilde{\mu}_{H^+}$ generated by an acid–base transition and a diffusion potential in the presence of ADP and $^{32}P_i$, [^{32}P]ATP is synthesized (Fig. 3.19). The $\Delta\tilde{\mu}_{H^+}$ including the diffusion potential is generated by diluting potassium-deficient vesicles at pH 5.0 into a medium at pH 7.5 containing 100 mM K^+ in the presence of valinomycin. As shown in Fig. 3.19, The presence of valinomycin is essential. The facilitated inward flow of K^+ has two roles: (i) it prevents the generation of a negative internal $\Delta\psi$ by efflux of a small number of uncompensated H^+ from the internal membrane space that would limit the outwardly directed H^+ flux used for ATP synthesis. In the present case, H^+ flux out of the small (~ 300 Å diam.) SMP vesicles would quickly generate a $\Delta\psi$ that would limit the flux (Problems 48 and 49). (ii) The out $\rightarrow$ in K^+ gradient provides a $\Delta\tilde{\mu}_{K^+}$ to drive the H^+ efflux.

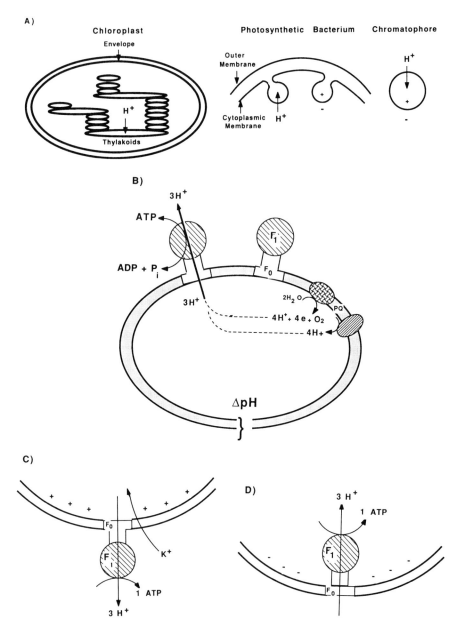

Figure 3.18. Schematic diagrams illustrating (A) formation of the ΔpH or $\Delta\tilde{\mu}_{H^+}$ across the thylakoid and bacterial membrane, (B) its discharge across the thylakoid ATP synthase resulting in ATP synthesis (based on Junge and Jackson, 1982), and (C) its discharge across the ATP synthase of thylakoids, submitochondrial particles, or chromatophores that is larger in the presence of a compensatory movement of K^+, i.e., in the presence of a high external concentration of K^+ and valinomycin (Appendix, section 3.12); (D) discharge of proton gradient as in (B), but in untreated mitochondrial or bacterial membranes; H^+ flow is toward negative side of $\Delta\psi$.

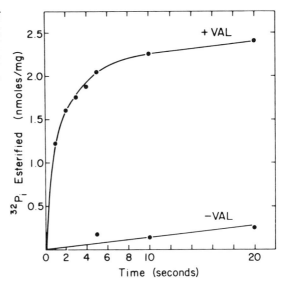

Figure 3.19. Time course of $^{32}P_i$ esterification following a $\Delta\tilde{\mu}_{H^+}$ transition from pH 5.0 ($-K^+$) to 7.5 ($+100$ mM K^+) (Thayer and Hinkle, 1975).

3.6.3 Topography of ATP Synthesis and H$^+$ Flux

The direction of proton flux coupled to ATP synthesis or hydrolysis in mitochondrial and bacterial membranes has a sidedness opposite to that of thylakoids (Fig. 3.18D). The H$^+$ flux associated with ATP synthesis in mitochondria and bacterial cells is driven mainly by the membrane potential term ($F \cdot \Delta\psi$) of the electrochemical potential. The negative internal $\Delta\psi$ is created by H$^+$ extrusion linked to electron transport in mitochondria and bacteria, which in turn drives the uptake of H$^+$ from the periplasmic or intermembrane space through the membrane-bound segment of the ATPase (Chap. 8). The ATP synthetic reaction for different energy-transducing membranes, including the direction of the H$^+$ flux, can be written as:

$$ADP + P_i + nH_i^+ \to ATP + H_2O + nH_o^+$$

(chloroplasts, chromatophores, SMP)

$$ADP + P_i + nH_o^+ \to ATP + H_2O + nH_i^+ \quad \text{(mitochondria, bacteria)}, \quad (6)$$

where n, designated as n_{H^+} in subsequent discussion, is the number of H$^+$ translocated across the membrane per mole of ATP synthesized or hydrolyzed.

3.6.4 Experimental Tests: ATP Synthesis in Thylakoids Caused by an Externally Imposed Electrical Field

High transmembrane fields can be generated by the imposition of external electric fields as shown schematically in Fig. 3.20. When an external field of 1,000–2,000 V/cm is imposed across the lipid vesicles (Fig. 3.20), the voltage

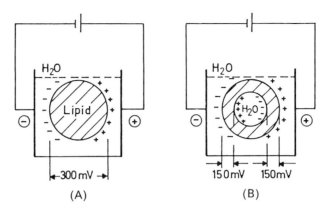

Figure 3.20. Principle of the external electric field method. Two electrodes in aqueous solution are separated by 1 mm and 300 V are applied, so that the field strength is 3,000 V/cm. The voltage across a 1 μm (10^{-4} cm) distance is then 300 mV (A), distributed as 150 mV across the two high resistance sides (B). The opposite polarity across the lipid annulus induced by the electrode plates on the two sides of the vesicle is shown (Witt, 1979).

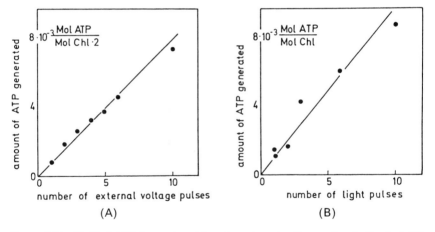

Figure 3.21. Yield of ATP formation driven by external voltage pulses in the dark (A) and light pulses (B). Duration of both pulses = 30 ms (Witt, 1979). Characteristic time for release of ATP from the ATPase, 50 μs (Hamamoto et al., 1982).

across the vesicle is approximately -100 mV across the left half of the vesicle which has the same polarity as would a light-induced electrical field (positive inside), and ca. $+100$ mV across the right half. The externally imposed potential of the correct sign ($+\Delta\psi$ in the case of thylakoids) can drive ATP synthesis at half the rate (Fig. 3.21A) measured with pulses of light (Fig. 3.21B). The ability of the artificially generated $\Delta\psi$ to drive ATP synthesis in thylakoids, even though the contribution of $\Delta\psi$ in thylakoids is very small in the physiological steady-state, is a proof of the energetic equivalence of $\Delta\psi$ and ΔpH.

3.6 Experimental Tests of the Chemiosmotic Hypothesis 111

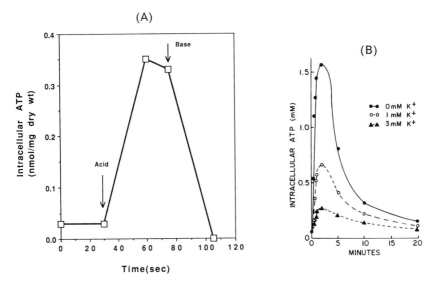

Figure 3.22. ATP synthesis in *E. coli* cells driven (A) by a pulse of acid (HCl) that lowered the extracellular pH from 7.6 to 4.7 and a subsequent addition of Tris base that raised it to 7.6 (adapted from Grinius et al., 1975), and (B) by a K$^+$-diffusion potential (Wilson et al., 1976b).

3.6.5 Experimental Tests: ATP Synthesis and Active Transport in *E. coli* Cells Generated by a ΔpH or a Δψ

ATP synthesis can be driven by (i) the ΔpH created by a pulse of acid added to *E. coli* cells treated with EDTA to permeabilize the outer membrane (Fig. 3.22A), or (ii) by the Δψ, using the diffusion potential that is generated when valinomycin is added to cells containing a high internal potassium concentration (Fig. 3.22B). Similarly, proline transport in inner membrane vesicles can be driven by a potassium diffusion potential, when the vesicles are made in the presence of K$^+$ and then diluted in the presence of valinomycin into a low K$^+$ concentration (Hirata et al., 1973; Appendix, 3.12). One can calculate that the amount of K$^+$ efflux that is needed to generate a Δψ of -100 mV is less than 0.1% of the *E. coli* internal K$^+$ concentration of ~ 0.2 M (Problem 47).

3.6.6 Experimental Tests. Reconstitution of Purple Membrane Vesicles Catalyzing Proton Uptake and Adenosine Triphosphate Formation

The ability of illuminated purple membranes of the purple halophilic bacterium *Halobacter halobium* to exchange protons with the external medium (Fig. 3.16B) can be reconstituted in synthetic phospholipid membrane vesicles

Table 3.8. Photophosphorylation by reconstituted soybean lipid vesicles containing bacteriorhodopsin[a]

Additions	Glucose-6-P (nmol)[b]	Glucose-6-P (nmol/mg bR)
Complete system (illuminated)	15.5	594
(dark)	0.6	23
Minus bR	0	0
Plus uncoupler (1799)	0	0
Minus coupling factor	2.9	110

[a] Vesicles contain a hydrophobic protein fraction from beef-heart mitochondria along with bacteriorhodopsin (bR) and coupling factor.
[b] ATP synthesis measured with glucose and hexokinase.
From Racker and Stoeckenius (1974).

by incorporating the purified bacteriorhodopsin pigment–protein into artificial membrane vesicles (Racker and Stoeckenius, 1974). Protons are taken up by the reconstituted vesicles, a direction of H^+ movement opposite to that of the native purple bacteria. When the mitochondrial ATP synthase was also incorporated into the artificial vesicles, photophosphorylation (20 mol ATP:mol bR) could be reconstituted (Table 3.8). The ability of these reconstituted vesicles to support light-dependent ATP and uncoupler-sensitive ATP synthesis in the presence of incorporated mitochondrial ATPase complex was, historically, important evidence that influenced a change in viewpoint of many in the bioenergetics field regarding the nature of the high-energy intermediate in membranes.

3.6.7 Experimental Tests: Mechanism of Action of Uncouplers of Phosphorylation

It was proposed in the chemiosmotic hypothesis that lipid-soluble uncouplers of phosphorylation (Table 3.9), which are usually weak acids, act by increasing the permeability of the coupling membrane to protons. This hypothesis has been confirmed (Fig. 3.23).

Uncouplers were found to increase the proton conductance of planar bilayer membranes by 2–3 orders of magnitude, with the maximum effect occurring near the pK_a of the uncoupler (Fig. 3.23A), suggesting that both forms of the uncoupler (e.g., FCCP) are translocated across the membrane in a shuttle-like mechanism that has been proposed to involve a monomeric form (Fig. 3.24A), or a dimer of the protonated and unprotonated forms of the coupler (Fig. 3.24B). The rate limiting step for proton transport was estimated to be ca. 10^3 s^{-1} ($\Delta\psi \to 0$) for transport of the unprotonated A^- form of the uncoupler across the membrane (Benz and McLaughlin, 1983). Another model for the action of dinitrophenol, based on the ability of DNP to accelerate H^+ exchange between octanol and H_2O, suggested that it

Table 3.9. Chemical structures, pK values, and uncoupling activities of representative uncouplers

Uncoupler (MW)	Structure	pK_a	Uncoupling activity — Respiration	Uncoupling activity — ATPase
SF 6847 (282.39)		6.83	10 nM	3 nM
S-13 (383.23)		6.4 (in 10% ethanol)	20 nM	7 nM
S-6 (392.67)		6.3	150 nM	100 nM
FCCP (254.17)		6.2 (in 10% ethanol)	70 nM	15 nM
CCCP (204.62)		5.95	110 nM	35 nM
TTFB (323.92)		5.5 (in 50% ethanol)	30 nM	
2,4-Dinitrophenol (184.11)		4.1	24 μM	8 μM

FCCP, carbonyl cyanide p-trifluoromethoxyphenylhydrazone; CCCP, carbonyl cyanide m-chlorophenylhydrazone; SF 6847, 3,5-di(tert-butyl)-4-hydroxybenzylidenemalononitrile; S-13, 2',5-dichloro-3-(tert-butyl)-4'-nitrosalicylanilide; S-6, 4',5-dichloro-3-(p-chlorophenyl)salicylanilide; TTFB, 4,5,6,7-tetrachloro-trifluoromethylbenzimidazole (Terada, 1981).

might shuttle protons locally between protein acid-base groups (Terada et al., 1983).

The protonophoric activity of the uncouplers on artificial liposome membranes is mostly determined by two factors: the pK_a, weaker acids having greater potency (Table 3.9), and the partition coefficient between the liposome and aqueous buffer phases (Miyoshi and Fujita, 1987). The most potent uncoupler listed in Table 3.9 is SF-6847. Its ability to collapse the ΔpH of chloroplast thylakoid membranes is shown in Fig. 3.23B. The extra potency of this compound appears to be correlated with the flatness of its structure, which may allow a more facile shuttle action, since rotation of the malonitrile group about the ring was found to decrease the efficiency of SF-6847 derivatives (Terada et al., 1984).

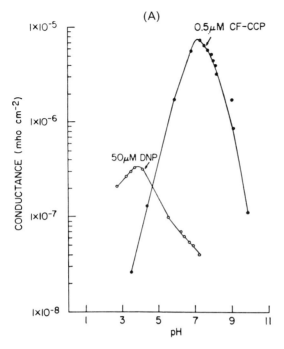

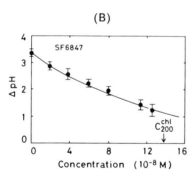

Figure 3.23. (A) Uncoupler-dependent H^+ conductance and its pH dependence in planar lipid bilayers (Hopfer et al., 1968; cf. Liberman and Topaly, 1968). (B) Effect of the uncoupler SF-6847 on the ΔpH of illuminated thylakoids. The arrow marks the concentration of uncoupler that causes a two-fold increase in electron transport rate (Miyoshi and Fujita, 1987).

Is the Active Ion H^+ or OH^-?

Much of the data involving H^+ fluxes could equally well be interpreted in terms of an oppositely directed OH^- current. Two experiments directed to this question are discussed later (Chap. 8, section 8.3.4, and Chap. 9, section 9.6.3). In addition, the existence of H carriers such as quinone in the electron transport pathways implies that H^+ and not OH^- is transported and translocated.

3.7 A Naturally Occurring Uncoupler

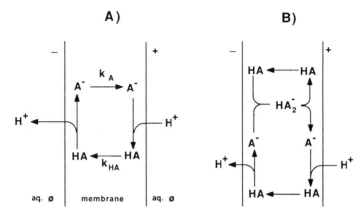

Figure 3.24. Models for H^+ conduction across bilayer membranes mediated by uncouplers. (A) Monomolecular model including four parameters that account for the ability of the weak acid FCCP to transport protons across phospholipid bilayer membranes and uncouple oxidative phosphorylation: For phosphatidylethanolamine bilayers, surface pK between 6.0 and 6.4; adsorption coefficient onto membrane-solution interface, $\beta_A = 3 \times 10^{-3}$ cm; rate constant for movement of HA across membrane, $k_{HA} = 10^4$ s^{-1}; k_A, dependent on $\Delta\psi$ and membrane dielectric constant = 700 s^{-1} when $\Delta\psi \to 0$ and $\varepsilon = 27$. (based on Benz and McLaughlin, 1983.) (B) Bimolecular model first suggested by Finkelstein (1970) where HA and A$^-$ are the neutral and anionic forms, and the return pathway is driven by the anionic HA_2^- which is a complex of A$^-$ and HA.

3.7 A Naturally Occurring Uncoupler: The Uncoupling Protein from Brown Fat Mitochondria

Animals adapted to a cold environment possess a mitochondrial-based non-shivering thermogenesis mechanism located in the brown adipose, or fatty, tissue that allows them to generate heat (Nicholls and Rial, 1984). The origin of the heat is the free energy of electron transport that would be transduced to ATP in well-coupled mitochondria. This free energy is released as heat in brown adipose tissue mitochondria because of the presence of an uncoupling protein, the most abundant protein in these mitochondria, which returns H^+ extruded by the respiratory chain back into the mitochondria, thus bypassing the ATP synthase (Klingenberg, 1985). From hydropathy plots (Appendix III), the folding of the uncoupling protein in the membrane has been predicted to occur as shown in Fig. 3.25. The protein shows many structural similarities to the mitochondrial adenine nucleotide translocator discussed below (section 3.9.1).

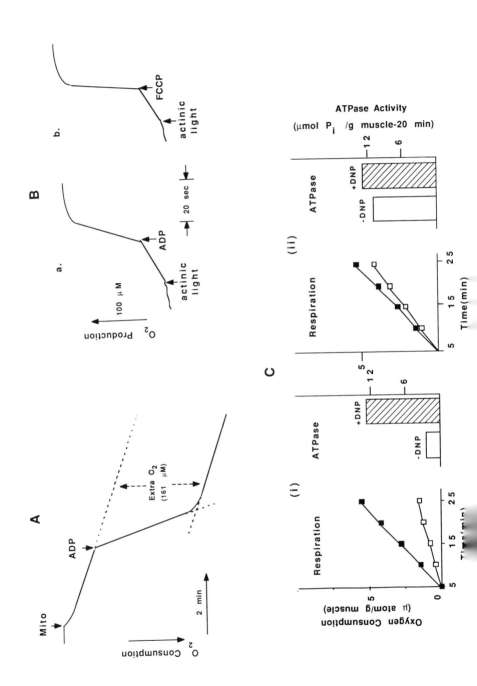

3.8 Effect of Uncouplers on Electron Transport Rate

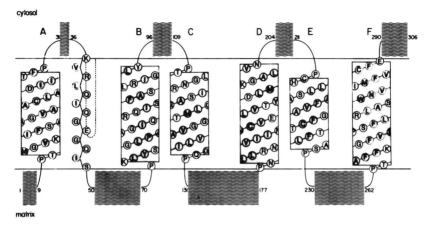

Figure 3.25. Proposed membrane folding pattern of the uncoupling protein from brown fat mitochondria. The protein contains 306 amino acids and has a molecular weight of 33,215 (Aquila et al., 1985).

3.8 Effect of Uncouplers on Electron Transport Rate

Mitochondria and chloroplasts that are in a state of high $\Delta\tilde{\mu}_{H^+}$ due to respiratory or photosynthetic electron transport activity are said to be in "state 4" (Chance and Williams, 1956) or, in the terminology of nonequilibrium thermodynamics, static head (Chap. 1, section 1.13). Under this condition, the rate of electron transport in "tightly coupled" mitochondria (Fig. 3.26A), or chloroplasts (Fig. 3.26B), is accelerated by addition of ADP in the presence of P_i (state 3), or the addition of uncoupler (state 3u). Addition in state 3 of "energy-transfer inhibitors" such as DCCD or oligomycin, which bind to the ATP synthase complex and prevent utilization of $\Delta\tilde{\mu}_{H^+}$, cause inhibition of electron transport that can be released by an uncoupler. The interpretation of the increase in electron transport rates by ADP or uncouplers, relative to those

---

Figure 3.26. (A) Effect of ADP (320 μM) in the presence of Pi on the rate of respiratory electron transport using succinate as the substrate for rat liver mitochondria. ADP/O = 2.0. (Adapted from Lemasters, 1984.) (B) Effect of ADP + P_i (a) and uncoupler (b) on the rate of photosynthetic electron transport. (↑) and (↓), light on and off; chlorophyll concentration of thylakoid membranes, 10 μg/ml; ADP, 0.5 mM; P_i, 10 mM; uncoupler, FCCP, 1 μM. Light intensity, 10^3 J/m²·s. (C) (i) Normal and (ii) loose coupling in muscle mitochondria, the latter measured in mitochondria of a woman suffering from hypermetabolism and generalized muscle weakness. The mitochondria have normal ADP:O ratios. However, they have nearly maximum respiratory rates in the absence of ADP, which together with a high level of ATPase activity [see figure, in the absence of dinitrophenol (DNP)] lead to low efficiencies of ATP synthesis. (Adapted from Luft et al., 1962.)

at static head, is that the product of the electron transport reaction, $\Delta\tilde{\mu}_{H^+}$, is decreased through utilization by the ATP synthesis reaction and dissipated by addition of the protonophoric uncouplers. At static head, the large $\Delta\tilde{\mu}_{H^+}$ tends to drive reversed electron transport reactions.

The stimulation of respiratory (or photosynthetic) electron transport by (ADP + P_i) can be used to calculate the stoichiometry of ATP synthesis to oxygen consumed or produced (ATP/O, ATP/$2e^-$, or P/O ratio) in respiration and photosynthetic electron transport. A major unresolved question in the calculation of these ATP/O ratios is whether the smaller rate of basal oxygen consumption or production measured in the absence of ADP should be subtracted from the rate measured in its presence in order to obtain the true rate of coupled electron transport.

The phenomenon of "loose coupling" (i.e., lack of stimulation of electron transport by ADP or uncoupler) in organelles from normal tissue that are capable of physiological rates of ATP synthesis may be explained, particularly at high $\Delta\tilde{\mu}_{H^+}$, by a high leak rate of H^+. This has been ascribed to a nonlinear relationship between $\Delta\tilde{\mu}_{H^+}$ and (i) respiration and (ii) the H^+ leak rate (Krishnamoorthy and Hinkle, 1984), possibly due to intrinsic uncoupling of the electron transport chain enzymes (Zoratti et al., 1986). "Loose coupling" in mitochondria can also be pathological, as in the case of the "Luft disease" (Fig. 3.26C, ii).

3.9 Proton Requirement (H^+/ATP) for Reversible ATP Synthase

The value of n_{H^+} in Eqs. 5, 6 has been determined from the value of $\Delta\tilde{\mu}_{H^+}$ and ΔG_{ATP} measured at static head in chloroplasts or mitochondria (Table 3.10). The ΔpH can be determined from the distribution of a labeled probe (section

Table 3.10. H^+:ATP ratio in different energy-transducing membrane systems

Membrane source	H^+:ATP	Reference
Chloroplasts	3	McCarty (1978); Hangarter and Good (1982)
Chloroplasts	1.9 ± 0.55	Lemaire et al. (1985)
Chromatophores	3.5 ± 1.3	Clark et al. (1983)
SMP	3	Sorgato et al. (1982); Berry and Hinkle (1983); Scholes and Hinkle (1984)
E. coli	2–4	Vink et al. (1984)
	2.5–4 (Aerobic)	Kashket (1982)
	3 (Anaerobic)	Kashket (1983)
	2	Perlin et al. (1986)
	2	Driessen et al. (1987)

3.9 Proton Requirement (H^+/ATP) for Reversible ATP Synthase

3.4.1 above), or together with the internal levels of ATP, ADP, and P_i, by ^{31}P-NMR (Chap. 1). A value of $n_{H^+} = 3$ in thylakoid membranes was derived also from measurement of the $\Delta\tilde{\mu}_{H^+}$ that was just sufficient for net ATP synthesis from either $\Delta\Psi$ or ΔpH components of the $\Delta\tilde{\mu}_{H^+}$, and the ΔG_{ATP} that was measured at the energetic threshold for ATP synthesis (Hangarter and Good, 1982). Values of $n_{H^+} = 3$ have also been determined for the ATP synthetic reaction in mitochondria, SMP, and chromatophores. Although there are data indicating a similar value in *E. coli*, there are also measurements showing that the value of n_{H^+} in bacterial cells may be smaller and also dependent on the metabolic state and level of energization of the membrane. For example, $n_{H^+} = 2$ in *E. coli* with succinate as the carbon source and saturating O_2, but $n_{H^+} = 2$–4 under conditions of O_2 limitation (Vink et al., 1984). A summary of the H^+/ATP ratios measured for ATP synthesis, or for H^+ translocated per ATP hydrolyzed, obtained in different membrane systems is shown in Table 3.10.

The complete circuit for flow of H^+ and substrates involved in mitochondrial ATP synthesis and utilization of ATP in the cytoplasm (Fig. 3.27)

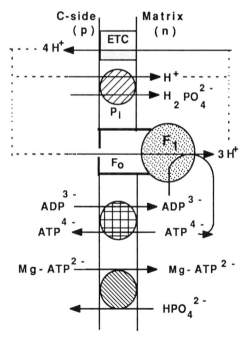

Figure 3.27. H^+ translocation, from negative (n) to positive (p) $\tilde{\mu}_{H^+}$, phosphate transport, electrogenic adenine nucleotide (ATP/ADP) exchange and relatively slow ATP-Mg/phosphate exchange across the mitochondrial inner membrane associated with synthesis of one ATP molecule from ADP and orthophosphate on the F_1 ATPase.

involves electrogenic exchange of matrix ATP for cytoplasmic ADP through the adenine nucleotide translocator (Klingenberg, 1985), with net translocation of one charge, a slower ATP-Mg/phosphate exchange (Aprille, 1988) that can influence the matrix adenine nucleotide pool, and electroneutral transport of phosphate through the inner membrane. An electroneutral phosphate (P_i^-) proton symport involving the uptake (H^+ symport) of one proton per phosphate transported, or exchange of one OH^- (Fig. 3.27), is carried out by an M_r 33,000 or a 33 + 35 kDa dimeric transport protein, that is primarily responsible for phosphate supply in oxidative phosphorylation (Kaplan et al., 1986). The requirement of net uptake of one H^+ for phosphate transport, together with three H^+/ATP used in the synthetic reaction (Table 3.10), increases the total number of H^+ translocated per ATP synthesized or hydrolyzed in mitochondria to four. Twenty-five percent of this energy requirement (one of four protons) is utilized for transport. The total H^+ requirement/ATP in the case of thylakoid membranes is three, since phosphate transport in the chloroplast is located in the outer envelope. The H^+ uptake (or OH^- release) associated with phosphate transport was largely responsible for the relatively low $H^+:O$ stoichiometry found in the first determinations of the O_2-induced H^+ extrusion from mitochondria (Mitchell and Moyle, 1965). More accurate $H^+:O$ values could be obtained when phosphate transport was inhibited with N-ethylmaleimide (Brand et al., 1976).

3.9.1 Function and Structure of Adenine Nucleotide Translocator (Klingenberg, 1985; Vignais et al., 1985)

Unlike transport systems in bacteria, most mitochondrial and chloroplast transport reactions involve an antiport or counterexchange. The exchange transport of ADP and ATP (Fig. 3.27) is the fastest transport system in eukaryotic cells relying on respiration, and the adenine nucleotide translocator is also the most abundant membrane protein in all mitochondria (~ 10–15% of the total membrane protein) except those from brown adipose tissue where the uncoupling protein is dominant. The mechanism for driving the preferential efflux of ATP across the mitochondrial inner membrane is electrical: In the presence of an arppreciable $\Delta\psi$, the energetics will be favorable for extrusion of ATP (ATP^{4-} or $MgATP^{2-}$) with an extra negative charge relative to ADP (ADP^{3-} or $MgADP^-$). The relationship between the external (e) and internal (i) ATP and ADP concentrations, and $\Delta\psi$, was found to be (Klingenberg and Rottenberg, 1977):

$$\log\left\{\frac{(ATP/ADP)_e}{(ATP/ADP)_i}\right\} = 0.85F \cdot \Delta\psi + \text{constant} \qquad (7)$$

The higher value of the $(ATP/ADP)_e$ ratio is consistent with the finding that the maximum value of ΔG_p measured outside mitochondria is greater than that measured outside submitochondrial particles, approximately 15 vs. 12

3.9 Proton Requirement (H$^+$/ATP) for Reversible ATP Synthase

kcal/mol. At high rates of respiration and large values of the $\Delta\psi$, as much as 75% of the maximal respiratory rate is controlled by the adenine nucleotide translocator (Wanders et al., 1984; Wanders and Westerhoff, 1988). The outer mitochondrial membrane has no role in ATP–ADP exchange or P_i transport since it is permeable to small metabolites.

The ATP/ADP carrier protein which has a high lysine content is thought to carry a net positive charge of three. Thus, in the absence of Mg^{2+} the substrate–carrier complex would be: $C^{3+} + ADP^{3-} \rightarrow C - ADP$; $C^{3+} + ATP^{4-} \rightarrow (C - ATP)^-$ (Klingenberg, 1985), so that the ADP complex is neutral and the ATP complex would carry one ($-$) charge, providing the electrical basis for the exchange. Specific inhibitors of the adenine nucleotide carrier protein carry at least three negative charges. The inhibitors carboxyatractyloside (CAT), atractyloside (ATR), and bongkrekic acid (BKA) have been useful in purifying the carrier protein as a dimer of M_r 30,000 subunits. Furthermore, because BKA binds to the protein only on the matrix (m) side of the membrane, and CAT and ATR only to the cytosolic (c) side, the carrier can be locked into "c"- or "m"-states. The carrier mechanism may utilize a reorientating single-site $\Delta\psi$-gated pore mechanism, whereby ATP^{4-} can enter in the "m"-state, and be released to the cytosol when the carrier is in the "c"-state (Fig. 3.28A). Sequence and topography studies suggest a folding pattern for the monomeric protein in the membrane (Fig. 3.28B) similar to that of the uncoupling protein (Fig. 3.25) with which it shares significant sequence and structure homology. The different conformation of the "c"- and "m"-states defined by the CAT and BKA complexes was documented by a difference between these two states in the pattern of lysine residues accessible from both sides of the membrane to the reagent, pyridoxal-5-phosphate (Fig. 3.28B).

Primary Biliary Cirrhosis and Dilated Myocardiomyopathy: Diseases Generating Antibodies Against the Adenine Nucleotide Translocator

Primary biliary cirrhosis is a progressive disease destructive of the liver. It is an autoimmune disease serologically characterized by the presence of antimitochondrial antibodies. Sera from 13 patients with the cirrhosis were examined for cross-reactivity to antibodies against the adenine nucleotide translocator isolated from heart, kidney, and liver mitochondria. All of the patients contained antibody against the liver enzyme, while 10 of the 13 sera did not cross-react against the carrier from heart, and none cross-reacted with that isolated from kidney (Shultheiss et al., 1983, 1984). Six of these sera inhibited adenine nucleotide transport in liver mitochondria, whereas none of the sera inhibited transport in heart and kidney mitochondria. None of the sera from 20 normal patients, four patients with antimitochondrial pseudolupus syndrome, and three with syphilis, contained antibody against the nucleotide translocator. Thus, the adenine nucleotide translocator is an autoantigen in this disease. Organ-specific and functionally active autoantibodies that inhibit

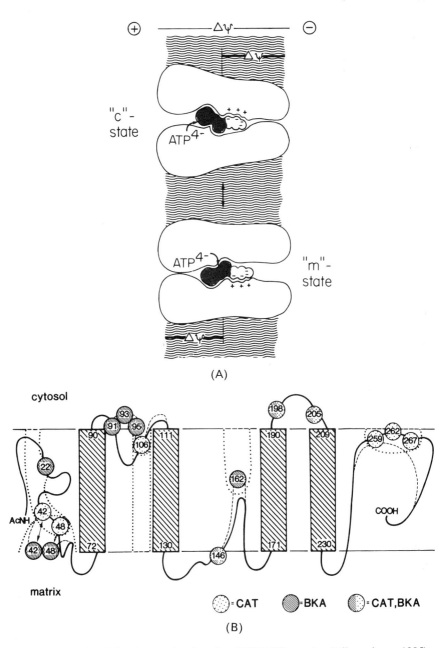

Figure 3.28. (A) Molecular mechanism for ADP/ATP carrier (Klingenberg, 1985); (B) Proposed transmembrane folding pattern of ADP/ATP carrier, and conformational changes of some lysine residues in the transition between "m" and "c"-states. Full circles represent lysines accessible and labeled by pyridoxal-5-phosphate in the CAT complex ("c"-state). Rectangles indicate the accessible lysines of the BKA complex ("m"-state) (Bogner et al., 1986). Inaccessible lysines are marked by a dashed circle.

the ADP/ATP exchange rate in heart mitochondria have also been found in a statistically significant (17/18) group of patients afflicted with dilated cardiomyopathy (Shulthiess and Bolte, 1985).

3.10 Storage of Energy in $\Delta\tilde{\mu}_{Na^+}$

Na^+ and K^+ are well known to be asymmetrically distributed across eukaryotic plasma membranes through the function of the Na^+, K^+-ATPase (Chap. 8). Na^+ can also substitute for H^+ as the energy-coupling ion in alkalophilic and alkali-tolerant bacteria. Thus, as noted in section 3.3, the general equations for the thermodynamics of ion gradients (Chap. 1, section 3.10) can be directly applied to Na^+ gradients:

$$\Delta\tilde{\mu}_{Na^+} = F \cdot \Delta\psi - 2.3RT \cdot \log_{10} \frac{[Na^+]_{final}}{[Na^+]_{init.}} \quad (8)$$

The reason for the evolutionary switch to a Na^+-based energetic system in alkalophiles that grow optimally at external pH values near 10 is probably that the ΔpH gradient in such organisms is very unfavorable for ATP synthesis linked to net H^+ uptake or any transport linked to H^+ symport. The alkali-tolerant marine bacterium, *Vibrio alginolyticus* extrudes Na^+ in an electrogenic process dependent upon respiration (Fig. 3.29) that is resistant to protonophoric uncouplers. The $\Delta\tilde{\mu}_{Na^+}$ generator is located in the NADH dehydrogenase of *V. alginolyticus*. Another mechanism of Na^+ export from the cell, decarboxylation of oxaloacetate to pyruvate demonstrated in the bacterium, *Klebsiella aerogenes*, is discussed in Chap. 9 in connection with Na^+-dependent active transport. Motility in *V. alginolyticus* as well as ATP synthesis, can be supported by an artificial Na^+ gradient, but not a ΔpH (Dibrov et al., 1986). Thus, the $\Delta\tilde{\mu}_{Na^+}$ can support the three main types of work generated in a bacterial cell: chemical (ATP synthesis), osmotic (active transport), and mechanical (motility). An extensive discussion of Na^+-based systems is provided by Skulachev (1988).

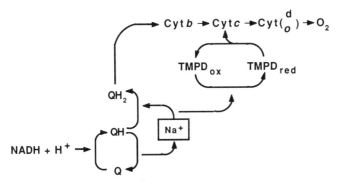

Figure 3.29. Respiration-coupled Na^+ pump in *V. alginolyticus*. (Adapted from Tokuda and Unemoto, 1984.)

3.11 Sufficiency of the Chemiosmotic Framework

The chemiosmotic hypothesis was controversial when it was proposed in the early 1960s. Its central emphasis on free energy storage in electrochemical ion gradients was shown to be correct in the ensuing years. The important details of protein structure and function in the mechanisms of electron transport, and H^+ and solute translocation, are currently areas of intense investigation due to the explosion of information on membrane protein sequence and structure. An important point in the chemiosmotic model that has been raised by several laboratories (Dilley, 1971; Dilley et al., 1987; Ferguson, 1985; de Kouchkovsky et al., 1984; Westerhoff et al., 1984; Williams, 1961, 1962, 1978) is whether the H^+ flux or chemical potential can be transferred directly from its generating source (e.g., electron transport chain) to its site of utilization (ATP synthase or transport protein). This model for direct or local H^+ transfer contrasts with the concept of the chemiosmotic model in which protons are transferred from the generating source to the delocalized aqueous phase before transfer to the site of utilization (sink). The difference in the thermodynamics of the localized and delocalized models is that in the former the intermediate proton spaces into which protons are pumped need not have the same level of $\tilde{\mu}_{H^+}$. H^+ that are delocalized after pumping into the bulk lumen, periplasmic, or intermembrane water-like space would be at the same $\tilde{\mu}_{H^+}$ level. The possibility of *lateral* H^+ transfer over distances on the order of 10^2 Å, has a precedent in transmembrane proton pumps, such as those of bacteriorhodopsin and cytochrome oxidase (Chap. 7).

The experimental data suggesting a localized model for the energy-linked H^+ flux can be classified (Ferguson, 1985) as structural, kinetic, and thermodynamic (Table 3.11).

These data indicate the complexity of detailed analysis of the proton- and ion-linked coupling mechanisms in organelles and bacteria. Almost all experiments agree that there is a correlation between net energy transduction and $\Delta\tilde{\mu}_{H^+}$ in H^+-linked systems over the whole range of values of these parameters. A fundamental experimental problem is that measurement of transmembrane ΔpH and $\Delta\psi$ is not a trivial problem (see section 3.4). Careful measurements require determination of intravesicular volume, and each probe of ΔpH and $\Delta\psi$ has unique experimental problems associated with its use. In general, the most lipophilic probes give the highest values of ΔpH and $\Delta\psi$. Many of the experimental questions about details of the chemiosmotic mechanism concern the lack of linearity and the uniqueness of the relation between ΔG_{ATP} or ΔG_{solute} and $\Delta\tilde{\mu}_{H^+}$. All experiments that indicate a complex relationship between ΔG_{ATP} or ΔG_s and $\Delta\tilde{\mu}_{H^+}$ should perhaps include a second method for estimating ΔpH and/or $\Delta\psi$. However, the large number of experiments indicating a lack of correlation or of a unique correlation between thermodynamic flux (e.g. ATP synthesis) and energy input ($\Delta\tilde{\mu}_{H^+}$) implies that the delocalized chemiosmotic model may not be completely or sufficiently detailed in its description of the mechanism. One modification that would not affect the principles would be to consider the heterogeneity of (i) the coupling sites and/or (ii) the membrane vesicle populations.

3.11 Sufficiency of the Chemiosmotic Framework

Table 3.11. Summary of experiments that are inconsistent with the chemiosmotic model of energy coupling, or more consistent with a "localized" model

Experiment	Comments on experiment
A. *Structural*	
1. Nonvesicular mitochondrial membrane fragments coupled for energy transduction (Storey et al., 1980)	Chemiosmotic model requires vesicular membrane; (Fig. 3.1); nonvesicular nature of this preparation may need better characterization, such as definition of lack of solute-impermeable space and of a transmembrane $\Delta\psi$ (Ferguson, 1985).

2. Most of the protons taken up by illuminated energized thylakoid membranes (Fig. 3.16A) are bound to the membrane and not translocated to the inner lumen space. The number of protons taken up by thylakoids at pH 7 is approximately 1 μeq H^+/mg Chl, and the internal space of the chloroplast thylakoids is approximately 10–100 μl/mg Chl, depending on external osmotic strength. Then, the internal pH would be 10^{-6} mol $H^+/10^{-5}$ L, or 10^{-6} mol $H^+/10^{-4}$ L, if all the protons taken up were translocated to the internal lumen, generating an internal pH = 1–2. The lumen pH, however, is approximately 4–5 in fully energized thylakoid membranes, implying that 99% of the protons taken up from the medium are bound to the membrane and not free in the lumen, and that membrane protonation as well as the ΔpH, should be relevant to the mechanism of energy storage (Dilley et al., 1987).

3. (i) Identity of sequestered ($-RNH_2$) protonable groups in thylakoids, determined using acetylation by acetic anhydride, occurring on extrinsic PS II proteins, LHC protein, cyt *f*, and CF_o subunit *c*; these H^+ binding groups (~ 30 nmol/mg Chl = 1 H^+/35 Chl) are thought to define at least part of the localized H^+-transfer pathway. (ii) Localized or delocalized coupling can be measured by the absence or presence, respectively, of a delay in ATP synthesis as a function of flash interval (see below, Fig. 3.30). Localized coupling defined in this way predominates in low salt or high Ca^{2+} (Chiang and Dilley, 1987).	Large ($\sim 1,000$ Å) spatial separation between (i) most of photosystem II units that are in appressed membranes of granar stacks, and (ii) ATP synthase in nonappressed membranes (Staehelin, 1986) implies that the pathway of localized H^+ transfer through special protein or lipid networks would be longer than any known pathway. For a photosynthetic unit of 400–600 Chl, average dimension of 250 Å and 1 H^+/35 Chl, the spacing between individual sequestered H^+ would be one per 10 Å for the case of linear alignment of the sequestered H^+ buffering groups. H^+ uptake from the lumen and H^+ flow across the ATP synthase have been found to occur simultaneously, arguing that the H^+ flow does not involve intramembrane domains (Junge, 1987).

Table 3.11 (continued)

Experiment	Comments on experiment

B. *Kinetics*

4. Delay in chloroplast ATP synthesis as a function of H^+ buffer capacity in lumen: ATP synthesis measured as a function of flash length showed a small lag of 5 ms. This increased to ~ 50 ms in the presence of valinomycin-KCl after which the rate was independent of valinomycin (Fig. 3.30), showing that a membrane potential drives ATP synthesis during the first 50 ms, but a localized or delocalized H^+ gradient afterward. The H^+ buffering capacity of the lumen can be increased by addition of exogenous permeant buffer, which according to the chemiosmotic hypothesis, should increase the time required for a critical pH change in the lumen. However, a concentration of bicarbonate or Tris buffer that should delay acidification of the lumen by 350 or 1,500 ms, respectively, caused no increase in the delay (Fig. 3.30).

5. Substitution of ^2H_2O for H_2O, and thereby the less mobile $^2H^+$ deuteron for the proton, caused an increase in the control ratio of electron transport in chloroplasts, while the transmembrane ΔpH and the phosphorylation efficiency were decreased. The data were interpreted in terms of a lateral resistance to protons and a lateral pH difference on the surface of the membrane between H^+ sources and sinks. This resistance would be increased in ^2H_2O because of the lower mobility of the heavier $^2H^+$ (de Kouchkovsky et al., 1982).	(i) True pH in pure ^2H_2O = pH reading on lab meter + 0.4. (ii) based on simplest kinetic energy considerations, ratio of velocities of ^1H and ^2H should be $\sqrt{2}$, or possibly a factor of 4 in ice (Kunst and Warman, 1980); (iii) effects of isotope substitution could be complicated, involving conformational changes of many proteins.

C. *Thermodynamic.* $\Delta G_{ATP} = n_{H^+} \cdot \Delta \tilde{\mu}_{H^+}$, for changes between the bulk phases, according to the chemiosmotic model.

6. Substantial ΔG_{ATP} and photophosphorylation in cells of *H. halobium* in the absence of measurable bulk phase $\Delta \tilde{\mu}_{H^+}$ (Helgerson et al., 1983; Michel and Oesterhelt, 1980).	Difficult to understand according to localized as well as delocalized model, since proton-pumping bacteriorhodopsin molecules are arranged on the cell membrane in patches that exclude the ATP synthase.

3.11 Sufficiency of the Chemiosmotic Framework

Table 3.11 (*continued*)

Experiment	Comments on experiment
7. Substantial ΔG_{ATP} or phosphorylation in mitochondria, submitochondrial particles, or chromatophores when $\Delta \tilde{\mu}_{H^+}$ is greatly decreased (Baccarini-Melandri et al., 1977; Wilson and Forman, 1982).	ΔG_{ATP} may sometimes be overestimated because of adenylate kinase activity generating ATP (Ferguson, 1985). For (6) and (7), if $\Delta \tilde{\mu}_{H^+}$ is too small to account for ATP formation, even with altered H^+ stoichiometry, then the energy available from localized energy coupling is too small if the membrane is laterally homogeneous (Nagle and Dilley, 1986).
8. Large or unaffected $\Delta \tilde{\mu}_{H^+}$ in mitochondria and chloroplasts when ΔG_{ATP} or ATP synthesis is inhibited, and electron transport stimulated by anaesthetic-type uncouplers (chloroform and halothane) in mitochondria (Rottenberg, 1983), or gramicidin or palmitic acid in chloroplasts (Pick et al., 1987). Chronic ethanol feeding of rats caused an 8% decrease of ΔG_{ATP} in state 4 while $\Delta \tilde{\mu}_{H^+}$ remained unchanged (Rottenberg et al., 1985). (The case is the opposite of C.6 above in this table.)	Anaesthetics caused increased lipid fluidity, which may allow increased respiratory rate and alter membrane environment of ATP synthase. Also, because of experimental difficulty in measurement of $\Delta \psi$ and ΔpH, it is perhaps important in these contradictory cases to always measure $\Delta \psi$ and ΔpH by at least two methods.
9. Other examples of nonunique relations between ATP synthesis and $\Delta \tilde{\mu}_{H^+}$ modified by uncouplers, inhibitors, or light intensity in the photosynthetic bacterium *Rb. capsulatus*; apparent threshold value of $\Delta \tilde{\mu}_{H^+}$ for ATP synthesis (Baccarini-Melandri et al., 1977). ATP synthesis less sensitive to $\Delta \tilde{\mu}_{H^+}$ reduction by ionophores than by light intensity in *Rb. capsulatus* and also in thylakoids (Sigalat et al., 1985).	It is perhaps better to measure ΔpH and $\Delta \psi$ by two methods, as on (8); in addition, uncouplers and inhibitors may bind nonspecifically to critical membrane proteins such as ATPase and merely alter their function.

Table 3.11 (*continued*)

Experiment	Comments on experiment
10. Active transport analog to cases (7), (8) for ATP synthesis: Light-driven alanine transport in the photosynthetic bacterium *R. sphaeroides* was almost independent of the magnitude of $\Delta\psi$, with a threshold value of -80 to -100 mV. Active transport and ATP synthesis showed linear dependence on electron transport rate at $\sim$ constant value of $\Delta\psi$ or $\Delta\tilde{\mu}_{H^+}$ (Elferink et al., 1983), suggesting that electron transport might directly provide free energy needed for alanine transport. Such a utilization of ΔG_{et} or an increased H^+ stoichiometry in a symport mechanism would also explain high accumulation levels of lactose relative to the $\Delta\tilde{\mu}_{H^+}$ (Konings and Booth, 1981). Chemical mechanism of dithiol (—SH) $\leftrightarrow$ disulfide (—S—S—) interchange has been proposed for regulation of substrate-binding affinity by electron transport (Konings and Robillard, 1982). Direct use of ΔG_{et} in photophosphorylation by chloroplasts was suggested by the ability of low concentrations of uncouplers to stimulate ATP synthesis while decreasing the $\Delta\tilde{\mu}_{H^+}$ (Giersch, 1983), although the latter effect could only be obtained in intact and not broken chloroplasts (Haraux, 1985).	Activation of transport system, ATP synthase could have nonlinear dependence on $\Delta\psi$ and still be compatible with delocalized model. The (—SH) $\leftrightarrow$ (—S—S—) interchange model could have a regulatory role in either model for energy transduction.
11. Linear relationship between inhibition H^+-ATPase pump by photoaffinity-labeled ATP and inhibition of reverse electron transport is inconsistent with delocalized H^+-coupling between ATPase and redox enzymes (Herweijer et al., 1985).	

3.11 Sufficiency of the Chemiosmotic Framework

Table 3.11 (*continued*)

Experiment	Comments on experiment
12. Alkalophilic bacteria: *Bacillus alkalophilus* can grow well at pH values of 11, where the ΔpH contribution ($\Delta\tilde{\mu}/F = +120$ mV) to $\Delta\tilde{\mu}_{H^+}$ is thermodynamically very unfavorable for H^+-linked ATP synthesis. In addition, ATP synthesis can be driven by both respiration and a diffusion potential at pH 7, but only respiration at pH 9 (Guffanti and Krulwich, 1984).	To partly cope with the unfavorable $\Delta\tilde{\mu}_{H^+}$, these bacteria use a sodium gradient, $\Delta\tilde{\mu}_{Na^+}$, to drive active transport, but they must somehow use the $\Delta\tilde{\mu}_{H^+}$ for ATP synthesis since they have an H^+-ATPase (Hicks and Krulwich, 1986).

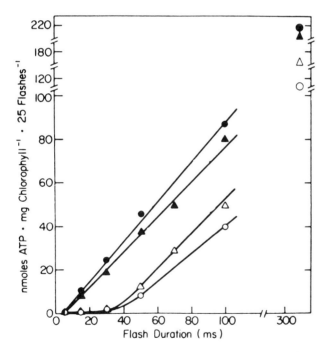

Figure 3.30. The effect of HCO_3^- (10 mM), with or without valinomycin (0.1 μM)-KCl, on ATP formation from short light flashes. The amount of buffer in the lumenal space was determined from uptake of [^{14}C] bicarbonate, uptake of ^3H_2O, and correction for trapped suspending medium using [^{14}C]inulin (MW ≃ 5,000). (△) and (▲) HCO_3^- with and without valinomycin-KCl; (○) and (●), control with and without valinomycin-KCl (Ort et al., 1976; cf., Horner and Moudrianakis, 1983).

Two detailed localized models that have tried to take the data of Table 3.11 into account are (i) the "mosaic coupling" model (Westerhoff et al., 1984) and the "microchemiosmotic" hypothesis (Haraux, 1985). The mosaic model would add the following postulate to those stated above for the chemiosmotic hypothesis: "Single redox or light-driven H^+ pumps do not share with other primary H^+ pumps the space (or domain) into which they pump protons, but they do share these domains with other individual H^+-coupled ATP synthase or solute transport proteins." This model assumes a functional organization of proton sources and sinks in small functional units, with the electric field not the same in all units because of heterogeneity in (a) spacing between units and (b) dielectric constant due to lipid and protein heterogeneity. The microchemiosmotic model does not require a one:one relation between sources and sinks, but a lateral as well as transmembrane H^+ flux. This surface H^+ flux between H^+ sources and sinks would occur through different proteins whose effective H^+ resistance, and $\tilde{\mu}_{H^+}$ level would be heterogeneous over the surface of the membrane. The $\tilde{\mu}_{H^+}$ would tend to be high near a primary H^+ pump (e.g. electron transport chain) and low in the region of a sink (e.g., ATP synthase, H^+-symport protein) or a nonspecific leak (Haraux, 1985). Membrane heterogeneity could also arise from local variations in lipid content, lipid fluidity, the H-bond network in proteins (section 2.11), and bound water (Nagle and Dilley, 1986).

A quantitative feature of both localized models is an equivalent electrical circuit model of the energy-transducing membrane in which sources, sinks, *trans*-membrane H^+ pathways, and lateral pathways for H^+ flux have separate lumped resistance values which allow calculation of the observed relationships between thermodynamic fluxes and forces (Westerhoff et al., 1984; de Kouchkovsky et al., 1984). However, in order that a significant resistance exists between intramembrane H^+ source and the bulk phase, it appears that the source must be shielded by lipid and/or protein from the aqueous phase because, when an H^+ generator has aqueous contact with the bulk phase, the $\Delta\tilde{\mu}_{H^+}$ will equilibrate between these phases within nanoseconds (Nachliel and Gutman, 1984).

3.12 Appendix. Ionophores

Ionophores are small (MW $\simeq$ 1,000) lipid-soluble molecules (Läuger, 1972; Pressman, 1976; Nicholls, 1982) that greatly decrease the electrical resistance ($\sim 10^8$ ohm/cm^2 in 1 M KCl in the absence of ionophore) of lipid bilayer membranes (Fig. 3.31). Ionophores are classified as (i) mobile carriers (Fig. 3.32A, left), [e.g., valinomycin (Fig. 3.32B), nigericin, nonactin (Fig. 3.32C), and FCCP (Table 3.9)] that carry 10^3–10^4 ions/molecule-s. and whose activity is frozen out at temperatures below the membrane lipid phase transition; (ii) channels (Fig. 3.32A, right) (e.g., gramicidin) that form single conducting units

3.12 Appendix. Ionophores: Generation and Detection of $\Delta\psi$ and ΔpH   131

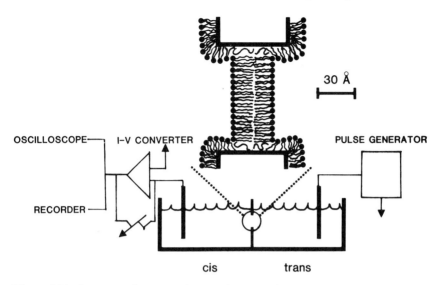

Figure 3.31. Apparatus for measuring conductance through a planar bilayer membrane. (Contributed by J.W. Shiver.)

in planar membrane bilayers that can be detected by their discrete appearance, high conductance ($\sim 10^7$ ions/channel-s) and relative insensitivity to temperature in membranes with a phase transition.

Valinomycin is one of a group of antibiotics produced by *Streptomyces* that exert inhibitory activity through their ionophoretic capacity. Valinomycin contains three repeating units of (L-lactate)-(L-valine)-(D-hydroxyisovalerate)-(D-valine) (Fig. 3.32B). The dehydrated K^+ ion is coordinated precisely and specifically ($\sim 10^4$ greater affinity than Na^+) to carbonyl groups in the hydrophilic interior of the macrocycle, whose exterior is nonpolar and thereby soluble in the membrane bilayer. Because valinomycin is neutral, it carries the single (+) charge of the bound K^+ ion. As it diffuses across the membrane, it functions as a K^+ uniport (Fig. 3.33A), similar in this sense to gramicidin, which can function as an H^+ uniport (Fig. 3.33B). The ionophores nigericin and monensin lose a proton as they bind K^+ or Na^+, respectively, and function as K^+–H^+ or Na^+–H^+ antiports or exchangers (Fig. 33C, D). Gramicidin A is a hydrophobic linear polypeptide antibiotic consisting of 16 residues, 15 amino acids, and a COOH-terminal ethanolamine: HCO-L-Val-Gly-L-Ala-D-Leu-L-Ala-D-Val-L-Val-D-Val-L-Trp-D-Leu-L-Trp-D-Leu-L-Trp-D-Leu-L-Trp-NHCH$_2$CH$_2$OH. Gramicidin A forms a *trans*-membrane ion channel that permits passive diffusion of monovalent cations with diameters up to 5 Å, and can be crystallized as a 26 Å long tube consisting of two antiparallel β strands (Wallace and Ravikumar, 1988). The generally accepted structure for the channel consists of two left-handed $\beta^{6.3}$-helical dimers that

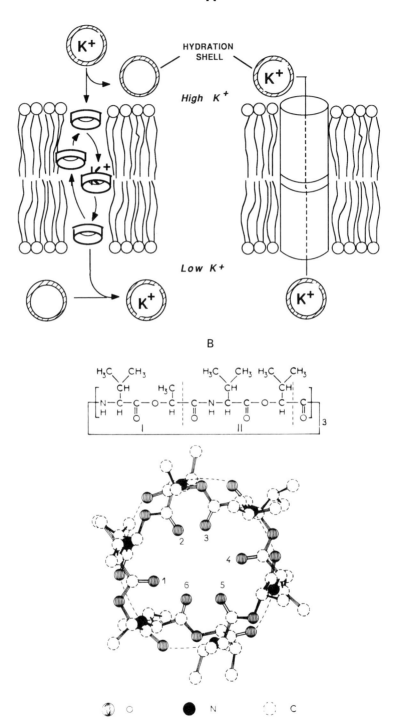

Figure 3.32 (A, B)

3.12 Appendix. Ionophores: Generation and Detection of $\Delta\psi$ and ΔpH

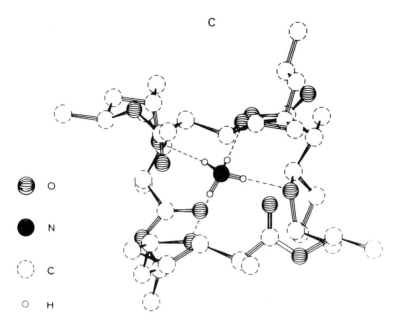

Figure 3.32. (A) Schematic view of the K^+ ionophore valinomycin acting as a mobile carrier on the dehydrated ion (left) and q a channel such as gramicidin (right). (Adapted from Nicholls, 1982.) (B) the valinomycin "bracelet" (MW = 1,110; ionic selectivity, $Rb^+ > K^+ > Cs^+ > Ag^+ > Tl^+ > NH_4^+ > Li^+$), the K^+ valinomycin complex has a charge of +1; the ionophore nigericin (MW = 724; selectivity, $K^+ > Rb^+ > Na^+ > Cs^+ > Li^+$) carries a single ($-$) charge and the K^+-nigericin or H^+-nigericin complex is neutral (not shown). (C) Nonactin channel containing an NH_4^+ ion. (Drawings for panels B and C from A. Pullman.)

A. Valinomycin
   Cation specificity: $Rb^+ > K^+ \gg Na^+ > Li^+$

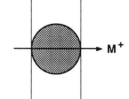

B. Gramicidin
   $H^+ > Rb^+ > K^+ > Na^+ > Li^+$

C. Nigericin ($K^+ \leftrightarrow H^+$)
   $K^+ > Rb^+ > Na^+ > Li^+$

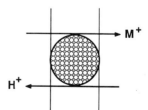

D. Monensin ($Na^+ \leftrightarrow H^+$)
   $Na^+, Li^+ > K^+, Rb^+$

Figure 3.33. Ionophore-mediated uniport activity of valinomycin (A) and gramicidin (B), and antiport activity of nigericin (C) and monensin (D).

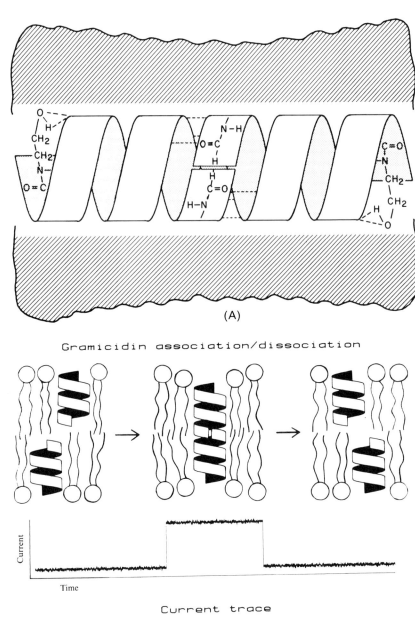

Figure 3.34. (A) Representation of the gramicidin channel as a left-handed NH_2-terminal to NH_2-terminal $\beta^{6.3}$ helical dimer with the channel entrances at the COOH-termini. The ribbon denotes the peptide backbone, side chains the cross-hatched areas. Six H bonds are indicated at the join between the monomers (Andersen et al., 1988) (B) Requirement of a dimer for single channel conductance (conductance tracing shown; drawing from O. Andersen) illustrating the mobility of the monomers in the monolayers and the possibility of forming hybrid channels (Andersen et al., 1988).

3.13 Summary

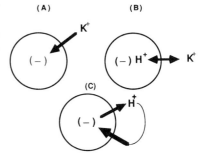

Figure 3.35. Inhibition of the formation of $\Delta\psi$ by K^+-valinomycin (A); of ΔpH, but not $\Delta\psi$, by H^+–K^+ exchange in the presence of nigericin (B); and by gramicidin or FCCP of ΔpH and/or $\Delta\psi$ generated by electrogenic H^+ translocation (C).

hydrogen bond at their formylated NH_2-termini to form an antiparallel dimer (Fig. 3.34A). The two halves of the dimer can diffuse independently in the two halves of the membrane bilayer, allowing the formation of hybrid dimers (Fig. 3.34B).

3.12.1 Prototype Ionophore Experiments

Valinomycin as a K^+-specific ionophore, nigericin and monensin as electroneutral K^+–H^+ and Na^+–H^+ exchange ionophores, and gramicidin or FCCP as H^+ ionophores, are frequently used to test for the existence of energy-linked events dependent on $\Delta\psi$ or ΔpH.

1. The presence of valinomycin and an excess of K^+ ($\sim$0.1 M) will inhibit activities dependent upon $\Delta\psi$, when the K^+ gradient is directed toward the negative side of the $\Delta\psi$ (Fig. 3.35A), or can be used to prevent formation of $\Delta\psi$ that will otherwise limit ion movement (Fig. 3.18C).
2. Activities dependent upon ΔpH will be inhibited by the electroneutral antiport activity of nigericin when the K^+ gradient is arranged to oppose the H^+ gradient (Fig. 3.35B). Since $H^+ \leftrightarrow K^+$ exchange does not affect $\Delta\psi$, activities dependent upon $\Delta\psi$ will not be affected. Subsequent addition of valinomycin will then inhibit the $\Delta\psi$ existing in the presence of nigericin. Exactly analogous to nigericin, monensin can be used to inhibit activities dependent upon a sodium gradient, ΔpNa.
3. The proton ionophores FCCP or gramicidin will inhibit activities dependent upon a ΔpH or a $\Delta\psi$ generated by energy-dependent H^+ translocation (Fig. 3.35C).

3.13 Summary

Chloroplasts, mitochondria, and Gram-negative bacteria are bounded by a pair of membranes, the outer membrane a shield against foreign macromolecules. The inner membrane contains the energy-transduction apparatus in mitochondria and bacteria, whereas in chloroplasts it is located in the thylakoid membrane. The mitochondrial inner membrane and thylakoid membrane lipids are characterized by a high degree of fatty acyl chain unsaturation (five double bonds/lipid), so that the viscosity of these membranes is low, allowing diffusion of proteins and quinone. The diffusion might be prevented by the high density of protein packing (protein:lipid = 3:1 by weight) except that much of this protein is extrinsic to the membrane. Mitochondria and chloroplasts import proteins ADP and orthophosphate, and export ATP, CO_2 fixation products and TCA cycle intermediates. The DNA coding capacity of plant mitochondria > chloroplasts > animal mitochondria. The similarities of membrane structure, and some of the intrinsic membrane protein complexes and gene sequences, are indicative of a common evolutionary background and divergence of chloroplasts and mitochondria from a common origin. The mechanism of energy storage in the respective membranes is then expected to be similar. The mitochondrial inner membrane contains the major electron transport and energy-transducing complexes: I and II (dehydrogenases), III (cytochrome b–c_1 complex), IV (cytochrome oxidase), ATP synthase, the ATP/ADP translocator, NADH–$NADP^+$ transhydrogenase, the extrinsic electron carrier cytochrome c, and ubiquinone. Among the principal and well studied proteins and protein complexes of thylakoid membrane are the light-harvesting proteins, the photosystem II, photosystem I, cytochrome b_6–f, and ATP synthase complexes, ferredoxin-$NADP^+$ reductase, the extrinsic electron carriers plastocyanin and ferredoxin, three extrinsic proteins of PS II, and plastoquinone.

The general mechanism for energy coupling that is able to explain the largest amount of data is the *chemiosmotic hypothesis*. Applied to mitochondria, bacterial and oxygenic photosynthesis, and *E. coli* membranes, (i) energy is stored in membranes in a ΔpH and $\Delta \psi$, and the stored energy is proportional to the proton electrochemical potential, $\Delta \tilde{\mu}_{H^+} = F \cdot \Delta \psi - 2.3RT \cdot \Delta pH$; (ii) the $\Delta \tilde{\mu}_{H^+}$ is formed by H^+ and e^- carriers and pumps; (iii) ATP synthesis is coupled to an H^+ flux from the $\tilde{\mu}_{H^+}$-positive side of the membrane, and ATP hydrolysis pumps H^+ across the membrane to the $\tilde{\mu}_{H^+}$-positive side, both reactions catalyzed by the multisubunit F_0F_1 ATPase. (iv) Uncouplers of energy transduction catalyze the equilibration of H^+ across the membrane bilayer. The chemiosmotic hypothesis has been tested by measurement of energy-linked H^+ movement across organelle or bacterial cell membranes, ATP synthesis or active transport driven by a $\Delta \tilde{\mu}_{H^+}$, and uncoupler-stimulated H^+ conductance in planar bilayers. The chemiosmotic formalism can also be applied to Na^+-linked energy transduction. There are some experimental data that are not readily explained by a simple application of the chemiosmotic hypothesis to a homogeneous population of membranes.

Radiolabeled or fluorescent weak acids or bases, or lipophilic ions, are used as extrinsic probes, respectively, of ΔpH and $\Delta \psi$. The carotenoid pigment in photosynthetic membranes has been used through the field-induced shift of its absorbance peak (Stark effect) as an intrinsic rapidly responding (< 1 nsec) probe of the $\Delta \psi$. Because of the specific capacitance of biological membranes,

~1 $\mu F/cm^2$, only a small H^+ or ion flux across the membrane is sufficient to generate a significant potential, e.g., 1 $H^+/10^4$ $Å^2$ generates a $\Delta\psi \simeq 100$ mV. Valinomycin is used to generate or dissipate K^+ diffusion potentials caused by ionophore-mediated movement of K^+ across the membrane along the concentration gradient. Valinomycin, a uniport ionophore carrier of Rb^+ ior K^+, and nigericin, a $K^+ \leftrightarrow H^+$ exchanger, are "carriers" with a characteristic turnover time of $\sim 10^4$ ions/s. Gramicidin is an ion "channel" ($H^+ > Rb^+ > K^+ > Na^+$) with a single channel conductance of approximately 10^7 s^{-1}.

An uncoupler protein that occurs naturally in brown fat mitochondria of animals adapted to a cold environment allows a large fraction of the free energy from electron transport to be released as heat instead of ATP. Addition of uncouplers or ADP and orthophosphate to mitochondria or chloroplasts in a state of high $\Delta\tilde{\mu}_{H^+}$ (static head) cause a large increase (3- to 10-fold) in electron transport rates associated with the decrease in $\Delta\tilde{\mu}_{H^+}$ associated with ATP synthesis, unless the organelles are already leaky to H^+ as occurs pathologically in Luft's disease. The H^+ flux and charge movement across the mitochondrial inner membrane associated with the synthesis of one ATP molecule is 3 H^+/ATP and 1 H^+/orthophosphate, and electrogenic exchange of 1 ADP and 1 ATP by the adenine nucleotide exchange translocator. The latter is the most abundant protein in nonthermogenic mitochondria. Unlike transport systems in bacteria, most mitochondrial and chloroplast transport systems involve exchange. ATP/ADP exchange, inhibited by carboxyatractyloside and bongkrekic acid which have been used to purify an M_r 30,000 translocator protein subunit, is the fastest transport system known in eukaryotic cells relying on respiration. Primary biliary cirrhosis of the liver is a disease characterized by antibodies generated against the adenine nucleotide translocator.

Problems

46. Calculate the pH gradient, ΔpH, for quenching of the probe, 9-aminoacridine, as shown, by illuminated chloroplasts if the chloroplast concentration is 20 μg Chl/ml, and the internal lumen volume is 50 μl H_2O/mg Chl.

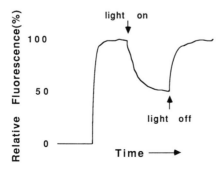

47. If the internal K^+ concentration of an *E. coli* cell (diameter 1 μm of an assumed spherical shape) is 0.2 M, calculate the fractional change in the internal K^+ concentration that will occur if K^+ efflux is used to establish a $\Delta\psi$ of -100 mV.
48. Using the data of section E, calculate the number of protons which would be extruded in an uncompensated manner from a spherical mitochondrion with a radius of 0.5 μm by an O_2 pulse which causes a potential of -100 mV to be generated across the mitochondrial membrane.
49. Calculate the diameter of the membrane vesicle population for which extrusion of 100 uncompensated charges will generate an electrical potential of 100 mV.
50. Calculate the approximate value of the band-shift arising from a $\Delta\psi$ of 160 mV across a 40 Å membrane bilayer of a carotenoid whose $\lambda_m = 500$ nm. Assume $E_0 = 2 \times 10^6$ V/cm, $\Delta\alpha = 4 \times 10^{-21}$ cm^3 and 1 V (esu) = 300 V (Standard International Units). [To check dimensions, remember that energy in esu has dimensions of (electric field)^2.]
51. Illumination of chloroplasts with a light intensity of 250 J/m^2-s generates a ΔpH of 3.35, $\Delta\psi = 0$, and a ratio of ATP to ADP and P_i for which the $\Delta G_{ATP} = 14.2$ kcal/mol (McCarty, 1978). Calculate the average number of protons required for the synthesis of one molecule of ATP (i.e., H^+/ATP).
52. (a) Calculate the basal rate of O_2 evolution by a chloroplast suspension, in μmol/mg chl/h, if chloroplasts at a concentration of 20 μg/mol chlorophyll in the absence of ADP and presence of P_i produce 30 nmol O_2/ml in 1 min. (b) Calculate the ATP/O ratio when addition of 120 μM ADP, assumed to be completely converted to ATP, causes a transient stimulation of the O_2 evolution rate by a factor of 3, which then abruptly decreases to the basal rate when the ADP is consumed.

PART II Components and Pathways for Electron Transport and H$^+$ Translocation

Chapter 4
Metalloproteins

4.1 Heme Proteins, Cytochromes *a* through *d*, and *o*

Cytochromes *a*–*d* were classified on the basis of characteristic absorbance maxima (Keilin and Keilin, 1966). The heme can be attached to the protein noncovalently (cytochromes *a*, *b*, and *d*) or covalently (cytochromes *c*). Most of the cytochromes in energy-transducing organelles are membrane-bound, although the *c*-type cytochromes are both bound and soluble. The heme prosthetic group of the *b*-type cytochromes, as well as myoglobin, hemoglobin, catalase, and peroxidase, is iron protoporphyrin IX (Fig. 4.1), and the heme of the other cytochrome types is derived from this compound. The four pyrrole nitrogens of the porphyrin serve as the in-plane ligands. The heme also has two more axial ligands perpendicular to the heme plane. These are usually histidyl residues (*bis*-histidyl coordination) in the *b*- and *a*-type cytochromes (Fig. 4.2A and C), one His ligand for cytochrome a_3 of the oxidase which binds O_2, and histidine-methionine for soluble *c*-type cytochromes (Fig. 4.2B).

The key to experimental studies of the redox properties of these cytochromes is that their absorbance spectra possess sharp features in the visible region that differ in the oxidized and reduced states (Fig. 4.3; Table 4.1). The reduced cytochromes have three prominent absorbance bands, α, β, γ (Soret), typically in the ratio 2:1:10 for the *b*- and *c*-type cytochromes. The α band is generally the most useful in experimental studies because: (i) it is narrow and pronounced in the reduced state but small and broad when oxidized, and (ii) for photosynthetic studies it absorbs in the green region of the spectrum where interference of the chlorophyll absorption bands is minimal (the "green window"). The oxidized cytochromes have a single prominent absorbance band (Soret) in the blue region and, for the *c*-type cytochromes with a

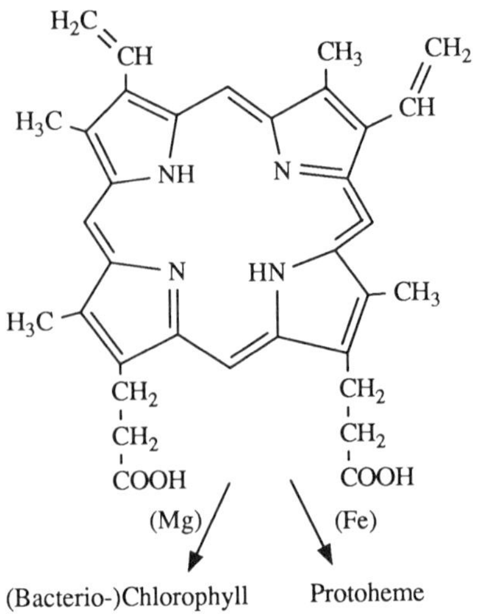

Figure 4.1. Structure of protoporphyrin IX, the precursor to protoheme, chlorophyll, and bacteriochlorophyll. The characteristic set of side chains on the four pyrrole rings is, reading clockwise from methyl (upper left, ring I, pos. 1), vinyl (I), methyl (II), vinyl (II), methyl (III), propionate (III), propionate (IV), and methyl (IV).

Figure 4.2. Heme structure. Modifications to the basic photoheme structure, and in the case of "siroheme" to its precursor, uroporphyrin, are indicated by heavy lines. (A) Protoheme is the prosthetic group for hemoglobin, catalase, peroxidase, cytochrome P-450, and the b-type cytochromes. (B) Heme c is the prosthetic group of most cytochromes that contain covalently bound heme and is characterized by addition of protein cysteinyl residues across the two vinyl side chains of protoheme at positions 2 and 4. (C) Variant heme c is the prosthetic group of a small number of protozoan mitochondrial cytochromes c and is characterized by a single thioether bond to the protein at position 4 and retention of a vinyl side chain at position 2. (D) Heme a is the prosthetic groups of the a-type oxidases and is characterized by hydroxyl and farnesyl additions to the vinyl side chain at position 2 and by oxidation of the 8-methyl group to a formyl side chain. (E) "Spirographic" heme is found in an annelid worm hemoglobin and is the only heme with a formyl substituent at position 2 instead of a vinyl. (F) "Siroheme" is the prosthetic group of sulfite and nitrite reductases. This iron containing isobacteriochlorin is a reductively dimethylated derivative of uroporphyrin, which is a precursor of protoheme. The dimethylated uroporphyrin itself is an intermediate in biosynthesis of vitamin B_{12} (Meyer and Kamen, 1982).

4.1 Heme Proteins, Cytochromes *a* through *d*, and *o*   143

A.

B.

C.

D.

E.

F.

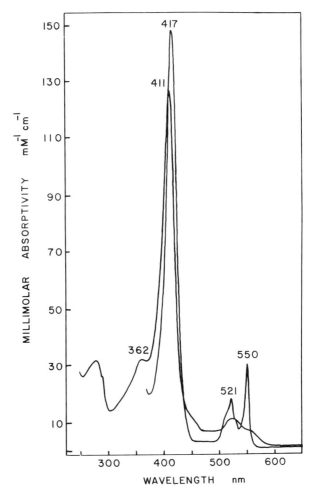

Figure 4.3. Visible absorption spectra of reduced (α, β, and Soret band peaks at 550, 521, and 417 nm) and oxidized cytochrome c (Meyer and Kamen, 1982).

Table 4.1. Absorption wavelength peaks of the reduced forms of the principal cytochrome classes

	Absorption band		
Heme protein	α (nm)	β (nm)	γ (Soret) (nm)
Cytochrome c	550–558	521–527	415–423
Cytochrome b	555–567	526–546	408–449
Cytochrome a	592–604	Absent	439–443

methionine ligand (see below), an absorbance band of much lower extinction (about 0.05 that of the reduced α-band) in the red region at 695 nm.

Instrumentation: Visible Light Difference Spectrophotometry

The relation between absorbance (A) and transmittance (T = ratio of light intensity transmitted by a sample to that incident on it) is

$$A = -\log_{10} T. \tag{1}$$

For small values (≤ 0.01) of ΔA,

$$\Delta A \simeq -\frac{1}{2.3}\frac{\Delta T}{T}. \tag{2}$$

The expected magnitude of ΔA can be calculated. For example, the stoichiometry of cytochrome c in beef heart mitochondria is approximately 0.2 nmol/mg membrane protein, the concentration, c, of 1 mg/ml membrane protein is 2×10^{-7} M, and the reduced-oxidized difference in the molar extinction coefficient ($\Delta \varepsilon_M$) at the reduced α-peak is 2×10^4 M^{-1} cm^{-1}. Therefore, $\Delta A = \Delta \varepsilon_M \cdot c = (2 \times 10^{-7}) \cdot (2 \times 10^4) = 4.0 \times 10^{-3}$ for a 1-cm pathlength. The noise level of the spectrophotometer is $\leq 10^{-4}$ absorbance units so that $\Delta A = 10^{-3}$ units can be readily measured. The absorbance changes associated with oxidation–reduction reactions in situ are small and occur in highly turbid and pigmented suspensions. Turbidity results from scattered light that is lost to the light detector and thus recorded as a higher absorbance. One experimental solution to this problem is to use a large (2-inch diameter) end-on photomultiplier tube as the detector, mounted close to the cuvette in order to collect a large solid angle of the scattered light, and a vertical light path so that the absorbance does not change as the sample settles (Norris and Butler, 1961).

Reduced minus oxidized spectra can be measured on a *split-beam* instrument (Fig. 4.4A) in which the beam from a single monochromater alternately scans each of the two cuvettes containing reduced and oxidized samples. The difference is detected by an amplifier tuned to the scanning frequency and phase. Difference spectra can also be measured on a single beam instrument with successive scans of the sample in different redox states stored in a computer.

Energy-transducing membranes undergo large changes in light scattering due to energy-linked ion movement and protein conformational changes. Correction for these scattering changes can often be handled by using a dual-wavelength spectrophotometer (Fig. 4.4B) to measure the absorbance change ΔA at the wavelength of interest, λ_1, relative to that at a nearby reference wavelength, λ_2, at which the ΔA absorbance change is small or zero (*iso*sbestic wavelength) and most of the apparent ΔA arises from the scattering change. An isosbestic wavelength in the α-band region, 540–542 nm, for the absorbance of reduced and oxidized cytochrome c (Fig. 4.3) is commonly used as a reference wavelength for c-type cytochromes. The redox response in respiratory membranes is initiated by addition of substrate, reductant or oxidant. Redox changes in photosynthetic membranes are also initiated by an actinic (photosynthetically active) light, oriented orthogonally to the measuring beam (Fig. 4.4C). The actinic light source may be a short xenon (1–10 μs duration) or laser (nano- or picoseconds) light flash. With appropriate optical and electronic filtering, the actinic light perturbs the redox state of the sample, but does not influence the detector, so that the kinetic response of the oxidation–reduction can be determined. Measurements of cytochrome difference spectra are often made at low (liquid nitrogen, 77 K) temperature, where band-widths are narrower, in order to better resolve neighboring components (Fig. 4.4D).

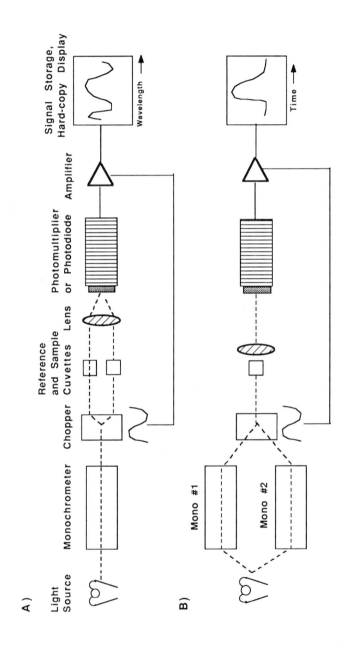

4.1 Heme Proteins, Cytochromes *a* through *d*, and *o*

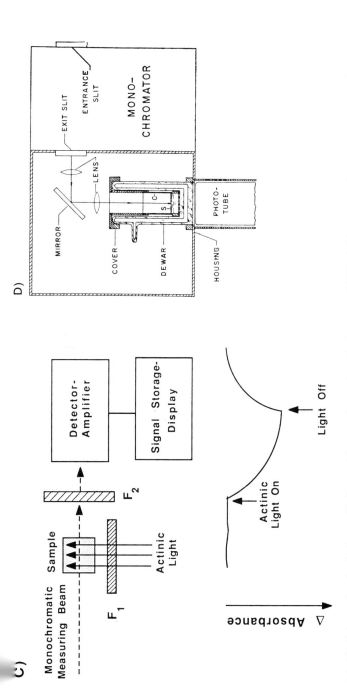

Figure 4.4. (A) Schematic of split-beam (2 chopped beams, 1 wavelength, 2 cuvettes) UV-visible spectrophotometer, and (B) dual wavelength (2 chopped beams, 2 wavelengths, 1 cuvette) spectrophotometer. Amplifier-detector is tuned to chopping frequency. (C) Measurement of light-induced absorbance changes with the wavelength band of the actinic light defined by optical filters F_1 and F_2, used to block possible artifactual fluorescence from the sample. Time course shown of light-induced absorbance change; submillisecond data usually processed by signal averager. (D) Monochrometer, optics, dewar, and photomultiplier tube for measurement of absorbance changes in turbid samples at 77 K; vertical light path used to minimize sample settling and photomultiplier tube mounted close to sample to minimize loss of scattered light (Norris and Butler, 1961. From IRE Transactions on Bio-Medical Electronics, *BME* vol. 8, pp. 153–157, © 1961 IRE (now IEEE)).

4.1.1 Cytochrome Oxidases

Cytochrome oxidase, ferrocytochrome $c:O_2$ oxidoreductase, is the terminal electron carrier chain of mitochondria and some aerobic bacteria that oxidize cytochrome c and reduce O_2 to H_2O. It is a multisubunit metalloprotein complex containing two molecules of heme a and two copper atoms. Heme a differs from heme b by having the vinyl group in the 2-position replaced by the hydroxyalkyl,

$$\underset{\underset{\displaystyle OH-CH}{\displaystyle |}}{(CH_2)_3} - \underset{\displaystyle CH_3}{\overset{\displaystyle CH_3}{|}} - (CH_2)_3 - \underset{\displaystyle CH}{\overset{\displaystyle CH_3}{|}} - (CH_2)_3 - \underset{\displaystyle CH_3}{\overset{\displaystyle CH_3}{|}}$$

and oxidation of the 8-methyl group to a formyl side chain. The presence of the formyl group results in a shift of the spectral peaks of the reduced cytochrome to longer wavelength (Table 4.1). The protein complex can contain as few as 2–3 subunits in prokaryotes (Fee et al., 1986), but 9–13 in eukaryotes (Capaldi, 1982). The proposal for two heme centers in the oxidase was originally based on spectral studies of differential effects of inhibitors (e.g., CO, CN^-), oxidants (O_2 and ferricyanide), and reductants on the Soret (~ 445 nm) and α (~ 605 nm) absorbance bands of the oxidase, as well as the finding that the stoichiometry for the binding of CO is 1 CO: 2 heme a. The notation heme a_3 was given to the component that reacted directly with O_2 and the inhibitors HCN or CO (Keilin and Keilin, 1966). The contribution of the a_3 and a components to the absorption peaks in the visible absorption spectrum, along with the electron paramagnetic resonance (EPR) spectral peaks is shown in Table 4.2. The high spin character of oxidized cytochrome a_3 also indicates that either the iron of heme a_3 is 5-coordinate or one of the axial ligands is weak enough that it could be displaced by a ligand such as O_2.

Table 4.2. Spectral properties of metal centers of cytochrome oxidase

Heme	$\Delta\varepsilon_m (mM^{-1} cm^{-1})^a$	g values (EPR)[b]
a and a_3		
Fe^{2+} Soret band	150 (445–460 nm), 70% a_3	—
Fe^{2+} α band	25 (605–630 nm), 25% a_3	—
Fe^{3+} a	—	3, 2, 1.5 (Low spin)
Fe^{3+} a_3	—	6 (High spin)
Copper		
Cu_A	2 (830 nm)	2
Cu_B	Silent	Silent

[a] Differential millimolar extinction coefficient between peak and reference wavelengths.
[b] see, section 4.9.1.1
From Wikström and Saraste (1984).

4.1 Heme Proteins, Cytochromes a through d, and o

Table 4.3. Polypeptides of bovine cytochrome oxidase[a]

Subunit no.	MW $\times 10^{-3}$
I[b,c]	57.0 (m)[d]
II	26.0 (m)
III	29.9 (m)
IV	17.2
V or Va	12.4
Vb	10.7
VIa	9.4
VIb	8.5
VIc	10.1
VIIa	5.5
VIIb	5.0
VIIc	6.2

[a] Maximum oxidase activity, 400 e^-/mol a-a_3-s, pH 7.0, 25°C.
[b] Notation is from Kadenbach and Merle (1981).
[c] The stoichiometry of all polypeptides except possibly SU III and the smallest SU is one per monomer (Wikström and Saraste, 1984).
[d] (m), mitochondrial-encoded; all others of cytoplasmic origin.
From Capaldi (1982) and Wikström et al. (1985).

The two largest subunits (subunits I, II) of the mammalian mitochrondrial cytochrome oxidase (Table 4.3) are involved in binding hemes a and a_3 and the two Cu. Subunit III may have a role in facilitating H^+ translocation. These three subunits are encoded by mitochondrial DNA (Fig. 3.2C). All of the other subunits are cytoplasmically encoded. The oxidases from lower eukaryotes (e.g., yeast) include six additional subunits made in the cytoplasm, and the higher eukaryotes (e.g., bovine, human), 10 more subunits (Zhang et al., 1988).

The mitochondrial-encoded subunits I–III seem to be present in all tissues in immunologically similar forms. The 10 nuclear-encoded subunits show immunological differences between two or more tissues, particularly between liver, kidney, brain, heart, or skeletal muscle, suggesting that at least some of these may be tissue-specific isozymes. Nuclear-encoded subunits from fetal and adult tissues showed immunological differences in liver, heart, and skeletal muscle. Topographical studies indicate that some small nuclear encoded subunits are close to the catalytic subunits, suggesting the possibility of regulatory functions, one of which could affect oxidase affinity for cytochrome c (Jarausch and Kadenbach, 1985a,b). The structure of cytochrome oxidase will be discussed further in Chap. 7 in connection with its proton-pumping function.

Plant mitochondria and bacteria have alternate cytochrome oxidases that branch from the quinone pool, whose purpose in the mitochrondrial system (Chap. 5, section 5.1) is to provide a pathway for heat generation, and in the latter (e.g., *E. coli*) to efficiently utilize O_2 at different ambient levels.

4.1.2 Cytochromes o and d

Cytochrome o, with protoheme as the prosthetic group, is a common terminal oxidase in bacteria where it may be present as the sole oxidase or as one of the oxidases in a branched respiratory chain (Poole, 1983; Wood, 1984). The presence of cytochrome o can be detected by the Soret band (410–420 nm) absorbance of its reduced, CO-ligated form relative to that of the reduced cytochrome in the absence of CO. Cytochrome d is often present as an alternative oxidase in Gram-negative bacteria that can adapt to environments where the surrounding O_2 content can vary appreciably. The structure of the metal-free methyl ester of heme d is shown in Fig. 4.5.

The cytochrome d and o oxidase complexes of *Escherichia coli* (Table 4.4) function, respectively, under conditions of low (e.g., stationary phase) and high O_2 tension (Fig. 3.5B). When the cells are grown under high aeration to early log phase, the cytochrome o oxidase complex (containing cytochromes b_{555} and b_{562}) is present in the cytoplasmic membrane (Ingeldew and Poole, 1984). When the cells are grown under O_2-limited conditions, the cytochrome d terminal oxidase complex (cytochrome b_{558}, b_{595}, and d) is induced (Au et al., 1985). Neither complex is used when the bacteria are grown fermentatively. The mechanism of $\Delta\tilde{\mu}_{H^+}$ generation by the cytochrome o oxidase is known to be different from the H^+ pump system of the a-a_3 cytochrome oxidase (Chap. 7, section 3). Turnover of o oxidase generates a $\Delta\psi$, inside negative, due to vectorial electron flow from the outer to the inner membrane surface. The ΔpH inside alkaline arises from consumption and release of H^+ at the inner and outer membrane surfaces respectively (Matsushita et al., 1984).

The structure of cytochrome d appears to be an $\alpha\beta$ dimer containing two b (b_{558} and b_{595}) and two d hemes accounting for the four Fe, in a protein having a molecular weight of $\sim 100,000$. The α ($M_r = 57,000$) subunit contains the low-spin cytochrome b_{558} responsible for quinol oxidation. The β

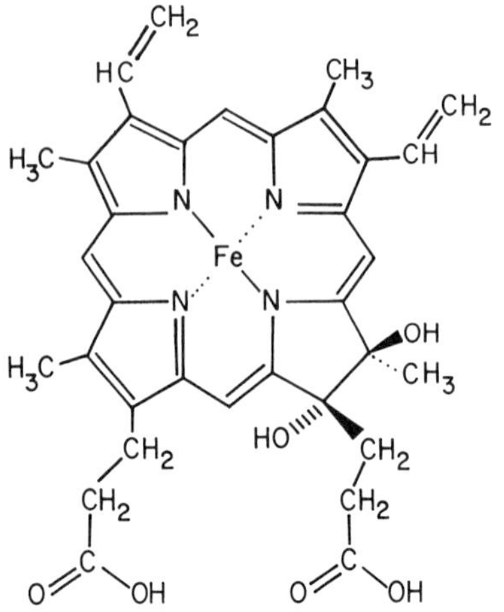

Figure 4.5. Proposed structure of chlorin d, the metal free methyl ester derived from heme d (Anraku and Gennis, 1987; *cf*. Timkovich et al., 1985).

4.2 Occurrence of b Cytochromes 151

Table 4.4. Properties of E. coli terminal oxidase complexes

	Cytochrome	
	o	d
1. Polypeptides	53, 36(24, 17) kDa	57 (α) 43 (β) kDa
2. Protein	n.k.	~100 kDa
3. Position on E. coli map	10.2'	16.5'
4. Metal content	20 nmol Fe, 17 nmol Cu per mg protein	Fe, 4; Cu, none
5. Heme content	Heme b, 2.0 per 100 kDa; 2 Cu	Heme b, 2.0 (b_{558}, b_{595}), heme d, 2.0
6. Reduced α-band maximum	555, 562 nm (77 K)	558, 595, 628 nm (77 K)
7. Midpoint potential	$E_{m7.4} = 125$ mV (pH-dependent)	$E_{m7} = +160$ mV (b_{558}); $= +260$ mV (d)

n.k., Not known.
From Anraku and Gennis (1987).

($M_r = 43,000$) subunit contains the cytochrome b_{595} and two cytochrome d hemes. Using antibodies to the E. coli cytochrome d complex, a related complex has been found in a variety of Gram-negative bacteria: *Serratia marcescens*, *Salmonella typhimurium*, *Klebsiella pneumoniae*, *Photobacterium phosphoreum*, and *Azotobacter vinelandii* (Kranz and Gennis, 1985).

4.2 Occurrence of b Cytochromes

The protoheme-containing b cytochromes in energy-transducing membranes are intrinsic membrane proteins, as in the case of E. coli mentioned above. The presence of a b cytochrome has often been inferred from the existence of a reduced α-band absorbance maximum in the wavelength range 560–566 nm. The number of b cytochromes components inferred from spectrophotometric and electrochemical data often has exceeded the number of protein components that could be biochemically isolated and purified.

A b-type cytochrome functions in the bc_1 complex of mitochondria and chromatophores and the $b_6 f$ complex of chloroplasts (Widger et al., 1984). The cytochrome is organelle-encoded in the case of mitochondria and chloroplasts, and coded by one of the genes in an operon for the bc_1 complex in photosynthetic bacteria. These complexes have a central role in all of these membranes. They provide the only pathway for transferring electrons from quinol on the reducing side of both respiratory and photosynthetic electron transport chains to plastocyanin or a c-type cytochrome on the oxidizing side (Fig. 4.6), while coupling electron flow to H^+ translocation (Chap. 7, section 4). The intramolecular heme cross-linking of these b cytochromes, as well as subunit I of cytochrome oxidase, illustrates

A. Chloroplasts

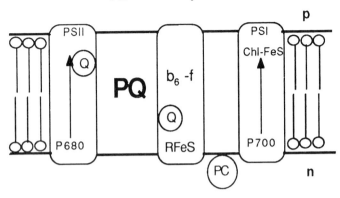

B. Mitochondria

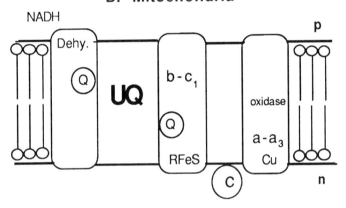

Figure 4.6. Intermediate position in (A) chloroplasts and (b) mitochondria of the cytochrome b_6–f or b–c_1 complex in the electron transport chain. The three transmembrane complexes in each pathway are shown. p and n sides of the membrane have (+) and (−) values of the $\tilde{\mu}_{H^+}$.

an emerging structural motif for membrane-bound cytochromes (Chap 7, sections 7.2 and 7.4). The membrane-bound cytochrome b_{559} that is part of the chloroplast photosystem II (PSII) reaction center also has a heme crosslinked structure, but it is intermolecular, resulting in a dimer (Chap. 6, section 6.7).

4.3 Structure of Cytochrome c

Most of the cytochromes in energy transducing membranes are tightly bound to the membrane and can be extracted only with detergent. In contrast to these intrinsic membrane proteins, there is a ubiquitous group of small (85–

4.3 Structure of Cytochrome c

Table 4.5. Amino acid sequence alignment of photosynthetic bacterial cytochromes c_2 and cytochrome c from *Paracoccus denitrificans* and horse

```
                       HEME————————————————————
                        |
1.   QEGDPEAGAKAFNQ-COTCHVIVDDSGTTIAGRNAKTGPNLYGVVGRTAGTQADFKGYGEGMKEAGAKG-
2.   QDGDAADGEKEFNK-CKACHMIQAPDGTDII-KGGKTGPNLYGVVGRKIASEEGFK-YGEGILEVAEKNP
3.   GDAAKGEKEFNK-CKTCHSIIAPDGTEIV-KGAKTGPNLYGVVGRTAGTYPEFK-YKDSIVALGASG-
4.   EGDAAAGEKVSKK-CLACHTFDQGG------ANKVGPNLFGVFENTAAHKDNYA-YSESYTEMKAKG-
5.   QDAASGEQVFKQ-CLVCHSIGPGA-------KNKVGPVLNGLFGRHSGTIEGFS-YSDANKN---SG-
6.   GDVEKGKKIFVQKCAQCHTVEKGG-------KHKTGPNLHGLFGRKTGQAPGFT-YTDANKN---KG-
              10                   30
```

```
     (HEME cont.)
                                    |
1.   -LAWDEEHFVQYVQDPTKFLKEYTGDAK------AKGKM--TFK-LKKEADAHNIWAYLQQVAVRP
2.   DLTWTEADLIEYVTDPKPWLVKMTDDKG------AKTKM--TFT-MGK--NQADVVAFLAQNSPDAGGDEAA
3.   -FAWTEEDIATYVKDPGAFLKEKTDDKK------AKSGM--AFT-LAK--GGEDVAAYLASVVK
4.   -LTWTEANLAAYVKDPKAFVLEKSGDPK------AKSKM--TFK-LTKDDEIENVIAYLKTLK
5.   -ITWTEEVFREYIRDPKAKI------PG--------TKM--IFAGIKDEQKVSDLIAYLKQFNADGSKK
6.   -ITWKEETLMEYLENPKKYI------PG--------TKM--IFAGIKKKTEREDLIAYLKKATNE
              60                   80              100
```

1. *Rb. sphaeroides*; 2. *P. denitrificans*; 3. *Rb. capsulatus*; 4. *R. rubrum*; 5. *Rps. viridis*; 6. Horse. The positions of insertions and deletions have been confirmed by three-dimensional structure comparisons for horse cytochrome c and *R. rubrum* and *P. denitrificans* cytochromes c_2. The numbering is based on the horse cytochrome sequence. Functionally equivalent residues are connected by lines between the sequences. The heme is bound hear the N-terminus to a pair of cysteins at positions 14 and 17, and the heme iron is ligated to His-18 and Met-80. The Met-80 sulfur is H-bonded to Tyr-67. One of the heme propionates is H-bonded to Arg-38, Trp-59, and Try-48. His-18 is H-bonded to Pro-30. Less highly conserved are the lysines 13, 27, 72, 73, 79, 86, and 87, which define an area for binding interactions with bovine cytochrome oxidase and reductase (Meyer and Kamen, 1982).

135 amino acids, MW = 10–15,000) high-potential ($E_m \gtrsim +200$ mV) c-type cytochromes that are weakly bound to the membrane (peripheral membrane proteins) and are water-soluble. These cytochromes contain a single, covalently bound heme c attached via two thioether linkages to a single polypeptide chain. Because soluble proteins can generally be purified far more easily than those bound intrinsically to the membrane, this family of cytochromes has been the most widely studied. Amino acid sequences are known for at least 100 cytochromes c (e.g., six are shown in Table 4.5), and high-resolution crystal structures are available for bacterial, algal, and eukaryotic cytochromes c. The single heme group is covalently attached at positions 2 and 4, as shown below, to two cysteinyl (R) residues

$$\begin{array}{c} \text{H} \\ | \\ -\text{C}-\text{CH}_3 \\ | \\ \text{SR} \end{array}$$

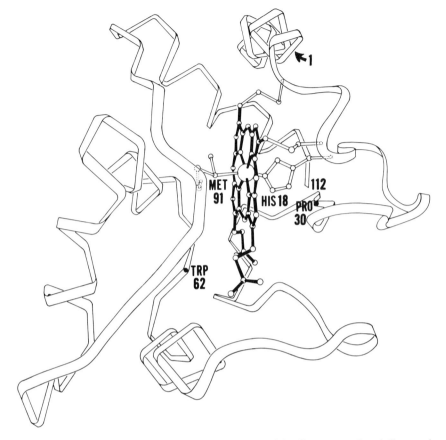

Figure 4.7. Structure of cytochrome c_2. Structure originally presented and discussed in Salemme (1977). [From Cramer and Crofts (1982).]

in a conserved (Cys-X-Y-Cys-His) sequence near the NH_2-terminus of the polypeptide. A methionine sulfur and histidine nitrogen serve as the two axial ligands for the heme iron (Fig. 4.7).

The heme of the eukaryotic mitochondrial cytochromes is located in a hydrophobic crevice surrounded by nonpolar amino acid residues, with only the front edge, including part of porphyrin ring II and the methine bridge between rings II and III, exposed to the solvent. In addition to the covalent thioether bonds and the bonds to the two axial ligands, the heme is positioned by hydrogen bonds from a propionic acid residue to Tyr 48 and Trp 59. Amino acid residues that are critical for function are expected to be conserved in evolution. The histidine and methionine that hold the heme in place and serve as axial ligands have been found to be invariant during evolution, as have most of the nonpolar amino acids that provide the hydrophobic contacts for the solvent-inaccessible portion of the heme (Table 4.6).

4.3 Structure of Cytochrome c

Table 4.6. Highly or totally conserved amino acids in cytochrome c

Amino acid	Number in mammalian cytochrome c	Role
Cys	14, 17	Heme attachment
His	18	Heme ligand
Met	80	Heme ligand
Tyr	48	Heme hydrogen bond
Tryp	59	Heme hydrogen bond
Leu	32	Heme neighbor
Phe	82	Heme neighbor
Gly	6, 29, 34	Tight region
Pro	30, 71	Sharp bend
Lys	8, 13, 72, 73, 86, 87	(+) charged ring around heme; binding to electron transfer partners

The structures of cytochrome c from the aerobic bacteria (*Pseudomonas aeruginosa, Paracoccus denitrificans*) and the photosynthetic bacteria (*Chlorobium limicola* and *Rhodospirillum rubrum*) are examples of three subgroups of the cytochrome c family. (i) The short (S) cytochromes c (*Pseudomonas* and *Chlorobium*) lack a loop (residues 37–59 in the tuna numbering system) that forms the bottom of the molecule in the tuna structure. The medium (M) cytochromes c, like that from tuna, are the type found in mitochondria and purple photosynthetic bacteria such as *R. rubrum*. The large (L) cytochromes (e.g., that from *P. denitrificans*) have an additional segment in the region of segment 54–77 and, in some cases, also near residue 23 (Dickerson, 1980a,b). The similar size of the cytochrome c of mitochondria and some of the purple photosynthetic bacteria has suggested that the latter may be an evolutionary precursor (Dickerson, 1980b).

4.3.1 Flexibility of Cytochrome c

The structure of a protein derived from x-ray analysis of a crystal emphasizes a static view of the protein. High-resolution proton nuclear magnetic resonance (NMR) spectra, on the other hand, can include information on the motion of the individual proton-containing groups. Such information is included in the proton NMR spectrum of cytochrome c, which shows that the mobility of the different groups varies across the cytochrome c molecule (Fig. 4.8). Of the aromatic residues, Phe-82 and Tyr-74 flip rapidly, but residues 10, 46, and 48 move slowly. The motions of this protein are certainly central to its function, although protein structure analysis has not proceeded to the point where the consequences for function of small (~ 1 Å) intramolecular motions are understood.

Figure 4.8. Mobility map of cytochrome c. The heme is seen on the vertical dark rectangle coordinated by the methionine sulfur on the right with two cysteine sulfurs covalently bound below to heme side chain vinyls, 2 and 4. The filled-in regions are relatively immobile and the cross-hatched more mobile (Williams, 1985).

4.3.2 The Multiheme Cytochrome c_3

Cytochrome c_3 is found in anaerobic sulfate- and sulfite-reducing bacteria such as the genus *Desulfovibrio* (Santos et al., 1984). The interesting structural feature of this cytochrome is that it contains four hemes arranged in approximately perpendicular pairs in a protein with a molecular weight of 13,000 (Fig. 4.9), a polypeptide of approximately the same size used to package the single heme of mammalian cytochrome c. The amino acid sequences of cytochrome c_3 isolated from a half-dozen different species are known, and x-ray diffraction data exist to 1.8 Å resolution for the cytochrome from *D. vulgaris* (Higuchi et al., 1984) and to 2.5 Å for *D. desulfuricans* (Pierrot et al., 1982). The linkage of the heme to the protein is through *bis*-histidyl axial ligation, as contrasted with the His, Met ligands in most of the monoheme soluble c-type cytochromes. The additional covalent thioether linkages to the protein of the two cysteines are as in monoheme cytochrome c.

The four cytochrome c_3 hemes have unusually low potentials, partly because of the *bis*-histidine coordination of the heme (Chap. 2, section 7). The E_m values of the four hemes are: -235 mV, -235 mV, -310 mV, and -315 mV (*D. Gigas*); -285 mV, -310 mV, -320 mV, and -325 mV (*D. vulgaris*), and -165 mV, -305 mV, -365 mV, and -400 mV (*D. desul-*

4.3 Structure of Cytochrome c

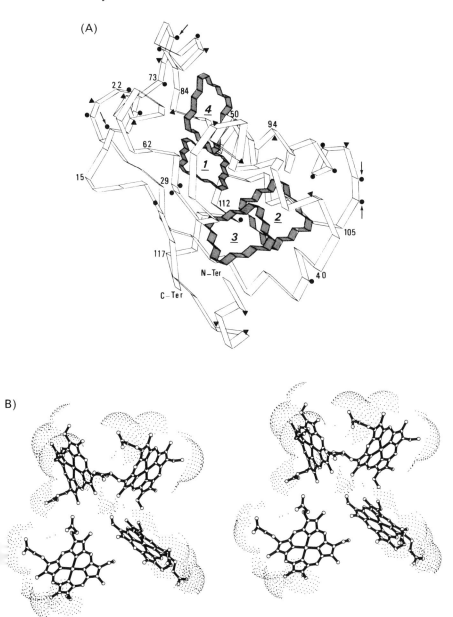

Figure 4.9. Models of cytochrome c_3 from solved crystal structures of *Desulfovibrio desulfuricans*. (A) Arrangement of hemes (in bold outline, and numbered *1–4*), and basic (o), acidic (△), and invariant (→) amino acids. To conform with the coordinates in the Protein Data Bank, heme 1 should be changed to 4, 2 → 3, 3 → 1, and 4 → 2 (Pierrot et al., 1982). (B) Model of cytochrome c_3 and heme arrangement from structure obtained at 1.8 Å resolution. (Figure contributed by Y. Higuchi, derived from Higuchi et al., 1984.)

furicans). Titration of the midpoint potentials and their pH dependence in the *D. gigas* cytochrome, indicated that the heme groups interact so that the E_m of any one heme depends on the redox state of its neighbors (Santos et al., 1984). The purpose of the heme-heme interaction and the presence of the four hemes in cytochrome c_3 is presumably related to the physiological role of the cytochrome which is not well understood. The cytochrome may utilize the perpendicular arrangement of the hemes for the purpose of storing reducing equivalents that are obtained from H_2 in the periplasm of *D. desulfuricans*. Electrons would then be donated to the electron transport chain in the membrane which, in turn, provides electrons in the cytoplasm for the reduction of sulfate (to sulfite via adenosine phosphosulfate, $E_{m7} = -60$ mV) and sulfite (to sulfide, using $6\,e^-, 6\,H^+, E_{m7} = -116$ mV). H_2 would be regenerated in the cytoplasm via ferredoxin and the oxidation of pyruvate and lactate (Pettigrew and Moore, 1987; Dolla and Bruschi, 1988).

4.4 Structure–Function in Mitochondrial Cytochrome *c*

Structural features of mitochondrial cytochrome *c* relevant to the mechanism of electron transfer are: (a) the exposure of a heme edge to the solvent; (b) a highly conserved positive charge distribution about this exposed heme edge (Table 4.5). A role of cytochrome lysine residues in stabilizing catalytically active cytochrome *c*·cytochrome oxidase and cytochrome *c*·cytochrome bc_1 complexes through electrostatic interaction has been documented by: (i) chemical modification of specific lysines, and (ii) differential labeling of lysines in the presence and absence of the reaction partner.

4.4.1 Specific Modification of Cytochrome Lysines

Selective and specific chemical modification of cytochrome *c* lysines surrounding the heme caused the loss of cytochrome *c* reactivity with both its oxidase and reductase (Brautigan et al., 1978). Major studies on the effects of such specific modification of cytochrome *c* lysines have been carried out with the reagent 4-chloro-3,5-dinitrobenzoic acid (CDNP) and *n*-trifluoromethyl-phenylcarbamoyl (CF_3 PhNHCO-) derivatives. CDNP selectively modified individual lysines centered around the β carbon of phenylalanine 82 (Fig. 4.10). The resulting inhibition of the binding to cytochrome oxidase correlated with the order of elution from a negatively charged carboxymethyl-(CM) cellulose column (Table 4.7), implying that CDNP modification has a systematic effect on the net or accessible positive charge on the "front" surface of cytochrome *c*.

Lysine modification to produce the (CF_3 PhNHCO-) derivatives changed the charge on a single lysine residue from +1 to 0, but did not significantly alter the E_m or visible absorbance spectrum (Smith et al., 1980). The (CF_3 PhNHCO-) derivatives behaved similarly to those affected by CDNP, and caused significant inhibition of the reaction of the cytochrome with

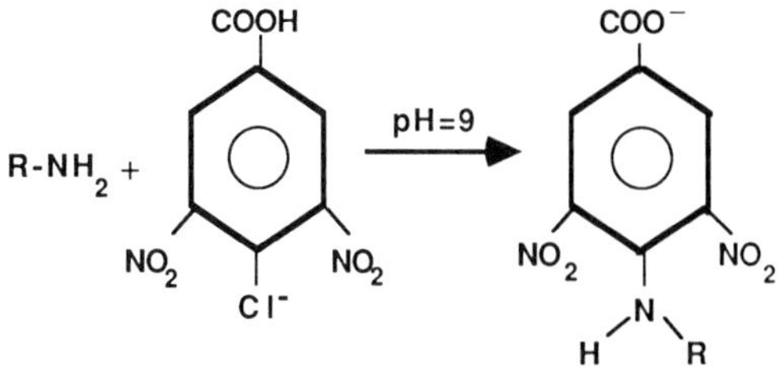

4.4 Structure–Function in Mitochondiral Cytochrome c

Figure 4.10. Reactant and product structures for the reaction of 4-chloro-3,5-dinitrobenzoic acid with ε-amino groups of cytochrome c lysines to form the stable, negatively charged, yellow CDNP-lysyl product. R represents the remainder of the protein which contains an N-acetylamino terminus and no free cysteine. (Adapted from Brautigan et al., 1978.)

Table 4.7. Binding of modified cytochrome c to oxidase and to CM-cellulose

Residue modified[a]	Surface distance (Å) from β carbon of Phe-82	K_D^b (M)	Order of elution from CM-cellulose
13	5	1.7×10^{-6}	1
72	9	8.6×10^{-7}	2
87	14	4.3×10^{-7}	3
3	17	2.4×10^{-7}	4
27	18	1.9×10^{-7}	5
39	32	3.4×10^{-8}	6
60	35	3.0×10^{-8}	7
22	36	3.6×10^{-8}	8
99	38	—	
Unmodified protein		2.4×10^{-8}	9

[a] CDNP-modified.
[b] Kinetically determined dissociation constant for reaction with cytochrome c oxidase.
From Ferguson-Miller et al. (1978).

cytochrome oxidase and its reductase, the cytochrome bc_1 complex. Modification of lysine residues that were more distant from the front heme edge had relatively little effect.

4.4.2 Differential Labeling of Lysines in the Presence and Absence of Reaction Partners

Acetylation and labeling of the lysyl residues attached to the protein, R, was accomplished with the [^{3}H]-labeled amino group reagent, acetic anhydride:

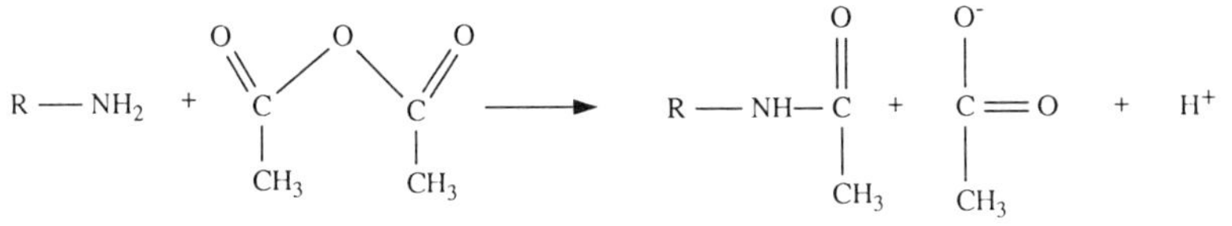

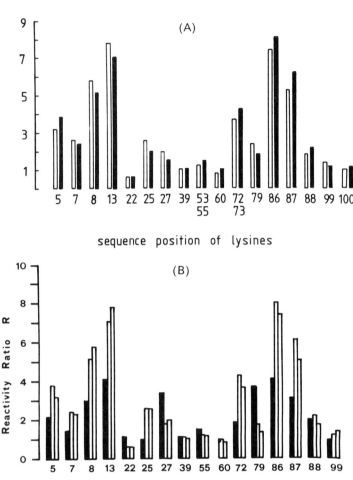

Figure 4.11. Differential chemical modification of cytochrome c. (A) The height of each bar represents the reactivity with respect to acetylation of a Lys residue of cytochrome c divided by the reactivity of the same residue when cytochrome c was complexed to cytochrome c oxidase (open columns) or cytochrome bc_1 complex (solid columns) (Rieder and Bosshard, 1980). (B) Comparison of reactivity ratios obtained by differential chemical modification of equine cytochrome c in the presence and absence of flavocytochrome c_{552} (filled columns), bc_1 complex (left open column), or cytochrome oxidase (right open column). Data for mitochondrial redox partners (Rieder and Bosshard, 1980; Margoliash and Bosshard, 1983). R-values for Lys 55, 72, and 99 are average values for Lys 53 + 55, 72 + 73, and 99 + 100, respectively (Bosshard et al., 1986). Because of variation in lysine reactivity, the labeling was normalized to the complete labeling of the denatured cytochrome with [^{14}C]acetic anhydride. The ^{3}H:^{14}C ratio for free and masked cytochromes expressed the normalized reactivity of the lysines. The quotient of the normalized reactivity for the free and masked cytochrome was defined as the shielding factor, R (Rieder and Bosshard, 1980).

4.4 Structure–Function in Mitochondiral Cytochrome c

The determination of the region of the cytochrome c protein that was occluded by reaction with reductase or oxidase was accomplished by comparing the labeling by small amounts of ^{3}H-labeled acetic anhydride of (i) free cytochrome c and (ii) cytochrome c masked by the reaction partner.

The position of the labeled lysines was determined by proteolytic digestion of the cytochrome c to small peptides and subsequent determination of the peptide sequences by Edman degradation. The distribution of the shielding factor "R" along the cytochrome c polypeptide for the reaction of cytochrome c with different electron transport partners, (e.g., cytochrome c oxidase and cytochrome c reductase (the bc_1 complex)) is shown (Fig. 4.11), along with a schematic summary of the relative shielding of the lysines viewed from the "front" face of horse heart cytochrome c, showing the exposed heme edge (Fig. 4.12). Complex formation between cytochrome c and at least six reaction partners (cytochrome oxidase, cytochrome bc_1 complex, cytochrome c peroxidase, cytochrome b_5, sulfite oxidase, and the photosynthetic reaction center of *Rb. sphaeroides*) has been shown to be stabilized by electrostatic forces (Margoliash and Bosshard, 1983), as has complex formation between the photosynthetic cytochrome c_2 and the reaction center and bc_1 complex of the photosynthetic bacteria *Rb. sphaeroides* (Hall et al., 1987a,b) and *R. rubrum* (Bosshard et al., 1987; Hall et al., 1987c,d; Van der Waal et al., 1987).

All of these chemical modification experiments indicate that electron transfer to and from cytochrome c occurs through the exposed front heme edge of cytochrome c, with the (+) charged lysine residues shown in Fig. 4.12 providing the main ionic attraction to the acceptor and/or donor redox protein (Rieder and Bosshard, 1980; Margoliash and Bosshard, 1983).

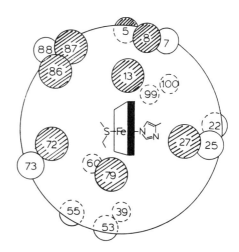

Figure 4.12. A schematic diagram of equine cytochrome c viewed from the front of the heme crevice. The approximate positions of the β-carbons of the lysine residues are indicated by closed and dashed circles for residues located towards the front and back of the molecule, respectively. Differential chemical modification indicates that complex formation with different redox partners protects the same residues on the top and front side of cytochrome c (hatched circles). However, the degree of protection of lysine residues differs somewhat, depending on the redox partners, indicating slightly different modes of binding. (Figures 11 and 12 contributed by H. Bosshard.)

Although this arrangement of lysines and the exposed heme edge is found in many cytochromes c and appears to be important in the docking of cytochromes c to several different reaction partners, differences of orientation have been found among the c cytochromes: (i) The heme orientation relative to the photosynthetic reaction center is tilted by 7–8° closer to the membrane normal and rotated by 32° for the photosynthetic cytochrome c_2 compared to mammalian cytochrome c. The difference in orientation, implying some difference in orientation of the bound heme, may explain why electron transfer from cytochrome c_2 to the reaction center is 20-fold faster for c_2 than mitochondrial cytochrome c (Tiede, 1987). (ii) The domain of cytochrome c acting as an acceptor for flavocytochrome c_{552} of the photosynthetic bacterium *Chromatium vinosum* extends to the right of the heme edge and involves fewer ionic bonds than with other redox partners of cytochrome c (Bosshard et al., 1986).

4.5 Residues of Reaction Partners That Are Complementary to Cytochrome c Lysines

The nature of the binding sites complementary to the front face lysines of cytochrome c has been investigated for (i) cytochrome b_5, (ii) cytochrome c peroxidase (CCP), and (iii) cytochrome c_1.

4.5.1 Cytochrome b_5

The reduction by NADH of a cytochrome b_5-like heme protein located in the outer mitochondrial membrane and its subsequent oxidation by cytochrome c allows cytochrome c to function as an electron shuttle in an intermembrane pathway from exogenous NADH to cytochrome oxidase (Ito, 1980; Bernardi and Azzone, 1981). In a model complex derived from crystal structure data of cytochromes c and b_5 (Salemme, 1977), negatively charged carboxylate groups surrounding the cytochrome b_5 heme were found to interact electrostatically with the lysines around the parallel heme of cytochrome c. A role of the cytochrome b_5 heme propionic acids in the complex formation was indicated by an alternative interaction geometry, involving different amino acids and noncoplanar orientation of the hemes, using dimethyl ester-substituted cytochrome b_5 in which the charges on the two propionic acids are eliminated (Mauk and Mauk, 1986).

4.5.2 Cytochrome c peroxidase

Cytochrome c peroxidase is found in the intermembrane space, between the inner and outer membranes of mitochondria from aerobically grown yeast.

4.5 Residues of Reaction Partners That Are Complementary

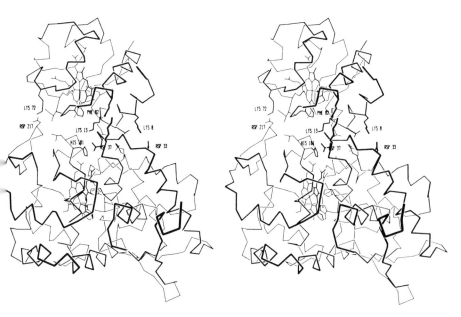

Figure 4.13. Stereoscopic view of cytochrome c peroxidase–cytochrome c complex using α-carbon backbone models and the 294-residue CCP structure (Finzel et al., 1984; Poulos and Finzel, 1984). "Front" and "back" are bold and light. Heme of cytochrome c can be seen near the top, along with its Phe-82, Lys-8, Lys-13 and Lys-72. Heme of CCP with its proximal ligand His-175 is near the bottom on the left and residues Asp-33, Asp-37, Asp-217, and His-181 are designated. (Figure contributed by T. Poulos.)

CCP catalyzes the reduction of hydrogen peroxide by ferrocytochrome c. The crystal structures of cytochrome c (e.g., tuna, Takano and Dickerson, 1980) and of CCP (e.g., yeast, Finzel et al., 1984) were used to locate a group of negatively charged aspartate residues around the CCP heme that interacts with cytochrome c (Fig. 4.13, left). The role of some of these residues in electrostatic interactions important to formation of the electron transfer complex has been confirmed by cross-linking (Bisson and Capaldi, 1981; Waldemeyer and Bosshard, 1985) and differential modification in the presence and absence of cytochrome c (Bechtold and Bosshard, 1985). When contour lines of the average electrostatic potential energy of the center of mass of cytochrome c are drawn around cytochrome c peroxidase, it can be seen that the electrostatic interactions allow a larger number of successful encounter complexes (Northrup et al., 1988).

The two hemes of cytochrome c and CCP are also parallel, but slightly displaced, with edge-to-edge and Fe–Fe distances of 16.5 Å and 24.5 Å, respectively. The long-range transfer may be facilitated in the cytochrome $c \cdot$ CCP case by the proximity of aromatic residues. The largest distance between any two aromatic groups in the bridged structure, much shorter than 16–17 Å, is the 5 Å between Phe 87 (analogous to the Phe-82 of mammalian cytochrome c described above) of yeast cytochrome c and its heme ring. The model suggested that Phe-87 of the cytochrome and His-181 of the CCP could mediate electron transfer between the hemes, consistent with cytochrome c protecting His-181 of CCP against photooxidative destruction (Bosshard et al., 1984). A critical role of the aromatic Phe-87 in this electron transfer reaction was suggested by inhibition of the electron transfer rate by a factor of 10^4 when this phenylalanine was changed to a Ser or Gly residue by directed mutagenesis (Liang et al., 1987) (see Chap. 2, section 2.10.5) The mutated genes were transferred into a yeast strain that lacked the cytochrome isozymes and was unable to grow on a nonfermentable carbon source, an ability that was acquired after transfer of the mutated genes. The purified mutant proteins possessed spectra similar to wild-type. The question that often arises with mutagenesis experiments is whether or not the mutation caused secondary or additional changes in protein structure that were not anticipated and, in this case, might affect the electron transfer rate.

4.5.3 Identity of Complementary Acidic Residues

Differential labeling was also used to identify acidic residues on cytochrome c_1 of the cytochrome bc_1 complex that are involved in binding to cytochrome c (Stonehuerner et al., 1985). The only region of cytochrome c_1 that was protected by cytochrome c from carboxyl group labeling by [1-ethyl-3-(3-[^{14}C] trimethyl-aminopropyl) carbodiimide] (ETC) was that spanning residues 63–81 containing seven acidic residues, Glu-66, Glu-67, Glu-69, Asp-72, Glu-76, Asp-77, and Glu-79. The involvement of an acidic sequence 165–174 of cytochrome c_1 was indicated from cross-linking to arylazidolysine 13 of cytochrome c (Bröger et al., 1983).

4.6 Diffusion and Orientability of Cytochrome c

The above considerations describe the complementary charge interactions between cytochrome c and its reductant and oxidants that lead to a productive electron transfer complex. The use of the same front face for oxidation and reduction in reactions that are fast and almost diffusion-controlled implies that the cytochrome must rapidly diffuse in order to sequentially collide and react with its reductant and oxidant. Three types of diffusional motion have been proposed (Margoliash and Bosshard, 1983): (i) A limited, rotational motion of cytochrome c confined to a cavity between the cytochrome bc_1 complex and cytochrome oxidase. (ii) Two-dimensional translational diffusion along the membrane surface between cytochrome bc_1 complex and cytochrome oxidase centers. (iii) Three-dimensional diffusion in which cytochrome

c dissociates from the b–c_1 complex and then moves in the aqueous space between the inner and outer mitochondrial membranes prior to reassociation with a cytochrome oxidase. The existence of both rotational (rotational relaxation times, 300 μs, 6 ms; Dixit et al., 1982) and translational (diffusion constant $D \simeq 10^{-9}$ cm$^2 \cdot$s^{-1}; Gupte et al., 1984; Hochman et al., 1985) motion of cytochrome c in mitochondrial and artificial membranes has been documented. The translational diffusion and oxidation is inhibited, particularly at low ionic strength, by the presence of acidic phospholipids to which the positively charged heme face of the cytochrome can bind electrostatically (Overfield and Wraight, 1980).

Because the surface area of the cytochrome c heme that is accessible to water molecules occupies only 0.6% of the total surface area of the protein (Koppenol and Margoliash, 1982), there must be a mechanism to properly orient the cytochrome toward its reductant and oxidant. In fact, the populations of cytochrome c and photosynthetic cytochrome c_2 molecules that transfer electrons rapidly to photosynthetic reaction centers were found to be optically dichroic and therefore oriented (Tiede, 1987). It has been proposed that the large electrical dipole moment of cytochrome c, with positive charges on the front face and negative on the back, can orient its reaction partners and be oriented by them (Koppenol and Margoliash, 1982). The dipole moment of cytochrome c was estimated to be 300–325 debye units (1 debye = 10^{-18} esu·cm = 3.3×10^{-30} coulomb-meter; 5 debye equals one electronic charge times an interatomic distance). The positive end of the dipole moment coincides with Phe-82 of mammalian cytochrome c at the center of the reactive lysines in the front face.

4.7 Membrane-Bound c-Type Cytochromes: Cytochromes c_1 and f

Until the advent of molecular biology and sequencing at the level of the gene, it was difficult to obtain complete amino acid sequences of membrane-bound proteins. The amino sequence of mitochondrial cytochrome c_1 has been determined for mitochondria (Wakabayashi, 1982; Sadler et al., 1984) and at the level of the gene for the photosynthetic bacterium, *Rb. capsulatus* (Gabellini and Sebald, 1986; Davidson and Daldal, 1987a). The sequence of cytochrome f has been obtained from the nucleotide sequence of spinach (Alt and Herrmann, 1984) and pea (Willey et al., 1984) chloroplast DNA (Tables 4.8 and 4.9).

The degree of homology of the mitochondrial and photosynthetic bacterial cytochrome c_1 sequences with that of cytochrome f is small (<5%) outside of the Cys-X-Y-Cys-His heme ligand sequence (residues 56–60 in the chloroplast sequence). The identity of the second ligand is probably a lysine in cytochrome f (Table 4.8), an unusual heme ligand. A 20–21 residue sequence of nonpolar amino acids near the COOH-terminus is common to all of the

Table 4.8. Properties of the amino acid sequences of cytochrome c_1 and f

Source	No. of amino acids	MW	Heme ligands[a]	Leader sequence (No. of residues)	Position of hydrophobic sequence
Yeast mitochondria	248	27,000	His-44:	+(61)	212–226
Rb. capsulatus	259[b]	28,300	His-38; Met-184(?)	+(21)	228–248
Pea chloroplasts	285[c]	31,100	His-25; Lys-145[e]	+(35)	251–270
Spinach chloroplasts	285[d]	31,300	His-25; Lys-145	+(35)	251–270

[a] Residue numbering is derived from Table 4.9 for the mature protein.
[b] Gabellini and Sebald (1986); Davidson and Daldal (1987a,b).
[c] Willey et al. (1984).
[d] Alt and Herrmann (1984).
[e] Lys implicated by spectroscopy (Siedow et al., 1980; Davis et al., 1988; the latter work suggests Lys from sequence comparisons).

Table 4.9. Amino acid sequences of cytochromes c_1 and f

```
              10        20        30  ↓     40        50        60        70
f :   MQTINTFSWIKEQITRSIFISLILYIITRSSIANAYPIFAQQGYENPREATGRIVCANCHLANKPVDIEVPQAVLPI
c₁:   MKKLLISAVSALVLGSGAALANSNVQDHAFSFEGIFGKFDQAQLRRGFQVYSEVCSTCHGMKFVPIRTLSDDGPC
                                  ↑

              90       100       110       120       130       140       150
f :   EAVVRIPYDMQLKQVLANGKKGGLNVGAVLILPEGFKIAPPDRIPGEMAEEVGDLSFQSYFPNKVNILVIGPVPGKK
c₁:   TFVREYAAGLDTIIDKDSGEERDRKETDMFPTRVGDGMGPDLSVMAKARAGFSGPAGSGMNQLFKLIGGPEYRYRY

             170       180       190       200       210       220       230
f :   ITFPILAPDPATKKDVHFLKYPIYVGGNRGRGQIYPDGSKSNNTVYNSFATGIVKKIVRKEKGGYEINIADASDER
c₁:   PEENPACAPEGIDGYYYNEVFQVGGVPDTCKDAAGIKTTHGSWAQMPPALFDDLVTYEDGTDATVDQMDQDVASFL

             250       260       270       280       290       300       310
f :   IIPRGPELLVSEGESIKLDQPLTSNPNVGGFGQGDAEVVLQDPLRIQGLLFFFASVILAQIFLVLKKKQFEKVQLS
c₁:   EPKLVARKQMGLVAVVMLGLLSVMLYLTNKRLWAPYKRQKA
```

[a] NH_2-termini of mature f and c_1 polypeptides (marked by arrows) are Tyr-35 (Alam and Krogmann, and Asn-22 (Gabellini and Sebald, 1986). Sequences of f and c_1 are from spinach chloroplasts (Alt Hermann, 1984) and Rb. capsulatus (Gabellini and Sebald, 1986; Davidson and Daldal, 1987a). hydrophobic domains near the —COOH terminus are underlined, as is the conserved segment including the proximal heme histidine ligand.

4.7 Membrane-Bound c-Type Cytochromes: Cytochromes c_1 and f 167

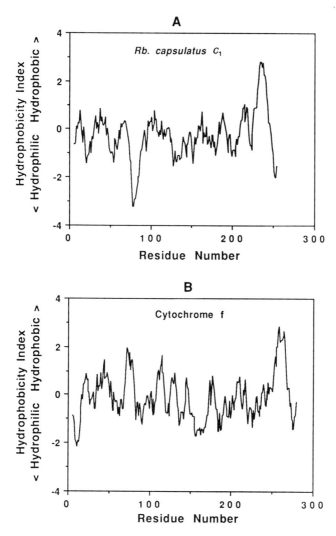

Figure 4.14. Hydrophobicity plot from sequence data of Table 4.8 for (A) cytochrome c_1 from *Rb. capsulatus* and (B) spinach chloroplast cytochrome f. The residue numbering begins at the NH_2-terminus of the mature protein. The hydropathy index was averaged over a window of 11 amino acids at each residue using the data base of Kyte and Doolittle (1982) (Appendix III).

sequenced cytochromes c_1 and f, as seen from the sequence comparison of mitochondrial and *Rb. capsulatus* cytochrome c_1 and spinach chloroplast cytochrome f (Table 4.9), and the hydropathy plots (Fig. 4.14). This hydrophobic peptide segment is likely to be a transmembrane α-helical domain involved in anchoring the cytochrome to the membrane.

The topography of membrane proteins can be probed with proteases that

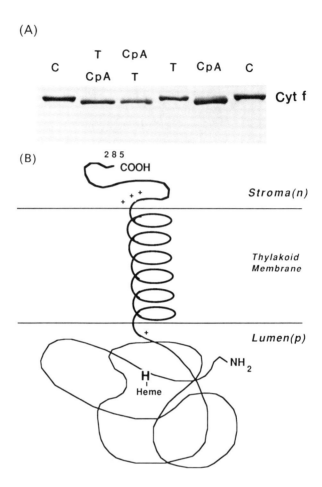

Figure 4.15. (A) Heme-stained SDS-PAGE gel of cytochrome f (top lanes) and cytochrome b_6 (bottom lanes) thylakoid membranes exposed to carboxypeptidase A (CpA) and trypsin (T). The untreated control is labeled C. The small decrease in molecular weight ($\Delta M_r = -2,000$) of cyt f after CpA treatment shows that the COOH-terminal end of the protein is on the stromal side of the membrane, as shown in (B). Cytochrome b_6 is also partially cleaved by trypsin after a previous treatment with CpA (Szczepaniak et al., 1989). (B) Topography of pea or spinach cytochrome f predicted from sequence and protease accessibility. The histidine ligand to the heme is residue 25. The lumenal location of the NH_2-terminus of cyt f is unusual among thylakoid membrane proteins (adapted from Willey et al., 1984).

4.8 Copper Proteins: Plastocyanin

react with specific surface-exposed groups. For example, the proteolytic enzymes trypsin and V8 protease will cut, respectively, on the COOH-side of Lys and Arg, and at Glu and Asp. The proteolysis products can be detected using (i) antibodies to the whole protein or a specific peptide ("Western Blot"). (ii) In the case of heme proteins, the protein products containing bound heme can be detected by using a dye assay for heme peroxidase activity. The model for cytochrome f shown in Fig. 4.15B is inferred from: (i) the selective ability of carboxypeptidase added to thylakoids and right-side-out membranes to cause a shift of about 2,000 in relative molecular weight indicating that approximately 15 residues near the COOH-end of the protein are exposed on the stromal side of the membrane (Fig. 4.15A); (ii) trypsin, but not carboxypeptidase, added to inside-out or detergent-permeabilized membranes caused the loss of the cytochrome band from the gel (Szczepaniak et al., 1989), indicating that most of the cytochrome f mass is exposed on the lumen side. The one residue known to function as a heme ligand of cytochrome f, His 25, would then be in the polar domain of the polypeptide on the lumen side of the membrane. Cytochrome f or c_1 functions in the sequence (Rieske [2Fe-2S] center) → (cyt f or c_1) → (plastocyanin or cyt c). These carriers form a group of high potential (0.3 V $\lesssim E_m \lesssim$ 0.4 V) electron donors that supply electrons to the plant photosynthetic reaction center, to cytochrome oxidase, or to the bacterial photosynthetic reaction center.

The diffusible electron acceptors of cytochromes c_1 and f, respectively, are the cytochrome c of mitochondria and c_2 of chromatophores, and the acidic copper protein, plastocyanin, which have been called "distributive" carriers, and usually provide the bridge between the cytochrome $b_6 f$ or bc_1 complex and the oxidase or reaction center. The locations of cytochromes c, c_2, and plastocyanin are, respectively, on the cytoplasmic, periplasmic, and lumenal sides of the membrane, which is defined as the $\tilde{\mu}_{H^+}$-positive, or "p"-side. The presence of this high potential loosely bound distributive carrier is almost universal, but not quite so. Cytochrome c_2-deficient mutants of *Rb. capsulatus* can be obtained that are photosynthetically competent (Daldal et al., 1986). No alternate photooxidizable carrier has been found, suggesting the possibility of a structural modification of the membrane protein organization in this mutant resulting in a "super-complex" between the reaction center and the cytochrome bc_1 complex.

4.8 Copper Proteins: Plastocyanin

Plastocyanin is a nuclear-encoded water-soluble protein consisting of a single polypeptide of molecular weight 10,500 that coordinates a single copper atom. The protein is characterized by its blue color in the oxidized state, a broad absorption peak centered at about 600 nm ($\varepsilon_M \simeq$ 5,000 M^{-1} cm^{-1}), and an E_m = +370 mV that allows it to connect cytochrome f and the chloroplast reaction center P700. The crystal structure of oxidized poplar plastocyanin

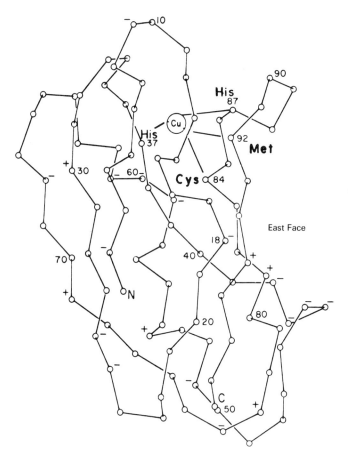

Figure 4.16. Structure of spinach plastocyanin. The figure was redrawn by D. W. Krogmann, using the original figure of Coleman and co-workers (1978) (from Cramer and Crofts, 1982). Negatively charged molecules such as ferricyanide are thought to bind at the top of the molecule near His-87. Positively charged residues from the plastocyanin reaction partners, cytochrome f and the P700 reaction center, could form electrostatically stabilized complexes with a conserved negative patch, residues 42–45, on the "east face" of the plastocyanin molecule (Anderson et al., 1985). The copper is coordinated by the imidazole nitrogen atoms of His-37 and His-87, the sulfur of Cys-84, and the sulfur of Met-92.

has been solved to a 1.6 Å resolution (Guss and Freeman, 1983), for which a representation is shown in Fig. 4.16. A solution structure has also been obtained by NMR spectroscopy for plastocyanin from the green alga *Scenedesmus obliquus* (Moore et al., 1988). Algae such as *Chlamydomonas reinhardtii* or *Porphyridium boryanum*, or cyanobacteria such as *Anabaena variabilis*, can use either plastocyanin or a *c*-type cytochrome as the "distributive" carrier depending on the metal content of the growth medium. This physiological flexibility is important because copper precipitates in a reducing environment, and iron becomes an insoluble ferric hydroxide under oxidizing conditions (Wood, 1978). The flexibility also occurs in the ability of flavodoxin to substitute for ferredoxin when iron is in limited supply in some algae. The molecular basis for the metal-regulated gene expression of cytochrome c-552 and plastocyanin in *C. reinhardtii* is that the plastocyanin apoprotein is degraded in the absence of Cu (Merchant and Bogorad, 1986a), and that the messenger RNA of the cytochrome and the plastocyanin, respectively, accumulates only in the absence and presence of Cu (Merchant and Bogorad, 1986b).

4.9 Iron-Sulfur Proteins

Membrane-bound electron transport chains contain an excess or similar amount of nonheme compared to heme iron. For example, beef heart mitochondria typically contain 3.0–3.5 nmol nonheme iron/mg protein compared to 2.5–3.0 nmol heme iron/mg protein. The nonheme iron present in the electron transfer chains of mitochondria, bacteria, and chloroplasts is in the form of protein-bound iron-sulfur centers (Fig. 4.17) which can contain 2, 4, and also 3 Fe per center. In the case of the 2Fe and 4Fe centers, for which

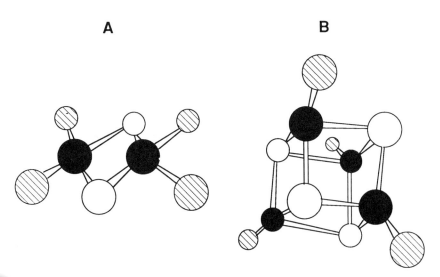

Figure 4.17. Structure of iron-sulfur linkage characteristic of (A) [2Fe-2S] (binuclear), and (B) [4Fe-4S] (tetranuclear) clusters originally found in plant and bacterial ferredoxins, respectively. The 2Fe and 4Fe clusters are each tetrahedrally coordinated to sulfur, with an Fe-Fe separation of approximately 2.7 Å. Four of the S atoms in the cluster are contributed by protein Cys residues. Fe (black); acid-labile sulfur (open); cysteine sulfur (hatched).

there is more information, the centers contain equimolar amounts of acid-labile inorganic sulfide, leading to the nomenclature [2Fe-2S] and [4Fe-4S] centers. A [3Fe-4S] complex has been defined which is an apo-iron cube with an average Fe-Fe distance of 2.7 Å, about the same as for the 2Fe and 4Fe centers (Stout et al., 1988; George and George, 1988).

It was shown that acid-labile sulfide was inorganic sulfide and not derived from a sulfur-containing amino acid by regeneration of the native protein by addition of ferrous iron and inorganic sulfide to *Clostridial* apoprotein (Malkin and Rabinowitz, 1966). The essential role of cysteine residues is illustrated by the high degree of conservation of four cysteine residues in a number of [2Fe-2S] and [4Fe-4S] cluster-containing ferredoxin sequences (Hall et al., 1975). X-ray analysis of crystal structures showed that each iron in the [2Fe-2S] cluster is in an approximately tetrahedral coordination geometry, surrounded by four sulfur ligands, two inorganic sulfides and two cysteinyl sulfurs. The [4Fe-4S] clusters found in bacterial ferredoxins are approximately cubic, with each iron again surrounded by four sulfur ligands, three as organic sulfide and one cysteinyl sulfur. The structure shown in Fig. 4.17 for the electron-carrying prosthetic group of the [2Fe-2S] ferredoxins was based on these data, spectroscopic measurements, and x-ray crystal structures of the [2Fe-2S] ferredoxin from the cyanobacterium *Spirulina platensis* (Tsukihara et al., 1981).

4.9.1 Ferredoxins

The first soluble iron-sulfur protein was discovered in the obligate anaerobe *Clostridium pasteurianum* (Mortenson et al., 1962), where it serves as a low potential electron donor for H_2 production and N_2 reduction. The protein was named *ferredoxin* and this name has been widely used as the generic term for a large number of soluble iron-sulfur proteins. In brief, the characteristics of the low ($\leq 12,000$)-molecular-weight ferredoxins are: (i) equimolar nonheme iron and acid-labile sulfide (released as H_2S when the proteins are acidified); (ii) a relatively negative oxidation–reduction potential ($E_{m7} \leq -200$ mV); (iii) diamagnetism of the protein (magnetic spin number, $S = 0$) in the oxidized state, and a paramagnetic ($S = 1/2$) state upon one-electron reduction.

To reconcile the observation that oxidized ferredoxins show no net unpaired electron spin with evidence (e.g., from Mössbauer spectroscopy) that both irons in oxidized [2Fe-2S] centers are high spin ($S = 5/2$), it was necessary to assume that the two high-spin Fe^{3+} irons are antiferromagnetically coupled to produce a net $S = 0$ spin state at low temperatures. One-electron reduction of the cluster produces a high-spin Fe^{2+} ($S = 2$) along with the high-spin Fe^{3+} ($S = 5/2$), and the antiferromagnetic coupling produces a net spin ($S = 1/2$) delocalized over the reduced cluster. Because the optical absorbance changes that occur during oxidation–reduction of ferredoxins are relatively small and centered in a spectral region where they can be obscured

4.9 Iron-Sulfur Proteins

by larger cytochrome absorbance changes, it is usually necessary to use EPR spectroscopy to follow redox reactions of iron-sulfur proteins in membranes.

Instrumentation: Electron Paramagnetic Resonance Spectrometry

The unpaired electron has a magnetic moment $\vec{\mu}$ that is proportional to the spin angular momentum $\vec{S}$. The spin angular momentum is usually expressed in units of $\hbar$, Planck's constant divided by 2π. The ratio between the magnetic moment and angular momentum is γ, the gyromagnetic ratio.

Then

$$|\vec{\mu}| = \gamma \hbar |\vec{S}| \tag{3}$$

$$\gamma = -g \frac{e}{2mc}, \tag{4}$$

where

$g = 2.00232$ for a free electron; g values of valence electron are different (Table 4.10), and reflect the magnetic environment, and the sign is negative because

e = the electronic charge is negative. Also,

m = the electron mass,

c = the speed of light (3×10^{10} cm/s, 3×10^8 m/s),

and

$$\hbar = h/2\pi = \frac{6.63}{2\pi} \times 10^{-27} \text{ erg} \cdot \text{s}^{-1}, \quad \text{or} \quad \frac{6.63}{2\pi} \times 10^{-34} \text{ J} \cdot \text{s}^{-1}.$$

As an example of net unpaired electron spin, oxidized ferricytochrome generally has two spin states, "high" and "low" which are a consequence of the ferric iron (atomic number, 26) having five electrons in the 3d orbital that can pair off (Feher, 1969) as:

$\uparrow\downarrow$ $\uparrow\downarrow$ $\uparrow$, net spin 1/2. "low spin", $g \simeq 3.0$

or:

$\uparrow\downarrow$ $\uparrow\uparrow$ $\uparrow$, net spin 5/2. "high spin", $g \simeq 6.0$.

The proportionality factor between $\vec{\mu} = -g\beta\vec{S}$, and $\vec{S}$, is $g\beta$, where

$$\beta \equiv e\hbar/2mc, \tag{5}$$

is the Bohr magneton and equals 9.27×10^{-21} erg-gauss^{-1}, or 9.27×10^{-24} J/Tesla.

The energy, E, of the component with magnetic moment $\vec{\mu}$ oriented at an angle Θ to the magnetic field $\vec{H}$ is:

Table 4.10. EPR properties of some metalloproteins

Metal ion	Electronic configuration	Spin state	Example	g-values[a]	Nuclear spin	Temp (°
Fe(III)	$3d^5$	1/2	Cytochromes c	3.8–0.5	0	<100
Fe(III), Fe(II) pair	$3d^5, 3d^6$	1/2	Spinach ferredoxin	1.7–2.1	0	<50
3 Fe(III), Fe(II) tetrad	$(3d^5)_3 3d^6$	1/2	HiPIP	2–2.2	0	<50
Fe(III), 3 Fe(II) tetrad	$3d^5(3d^6)_3$	1/2	Bacterial ferredoxin	1.7–2.1	0	<50
Co(II)	$3d^7$	1/2	Cobalamin	2.0–2.3	7/2	<120
Ni(III)	$3d^7$	1/2	Hydrogenase	2–2.2	0	<5
Cu(II)	$3d^9$	1/2	Plastocyanin	2–2.4	3/2	<30

[a] The ranges shown encompass both g-anisotropy in a given system and the variation among proteins. From Palmer (1985).

$$E = -\vec{\mu} \cdot \vec{H} = -\mu H \cos \Theta, \tag{6}$$

$$= g\beta \vec{S} \cdot \vec{H} = g\beta S H \cos \Theta \tag{7}$$

When the electron is in a local environment with a magnetic field that varies with direction, the g value will also be direction-dependent, splitting into g_z, parallel to the test field, and g_x, g_y values perpendicular to the test field. The EPR properties of different metal ions found in biochemical systems are shown in Table 4.10.

The energy of a magnetic moment in a magnetic field H is $E = -\vec{\mu} \cdot \vec{H}$, indicating that the energy is minimized when the dipole is aligned along the direction of the field. The spin vector is quantized, and can have the value $\pm 1/2$, depending on whether the spin vector is aligned parallel (↑) or antiparallel (↓) to the field. The field axis is usually defined as the z-axis. So $S_z = \pm 1/2$, for which $\cos \Theta$ in Eq. 7 $= \pm 1$.

$$\vec{H} \equiv \vec{H}_z; \quad (\uparrow) S_z = +1/2, \quad (\downarrow) S_z = -1/2$$

The energy, E, of the electron in the magnetic field is then split. From (7),

$$E = +\frac{g\beta H}{2}, \quad \text{for } S_z = +1/2, \tag{8a}$$

$$E = -\frac{g\beta H}{2}, \quad \text{for } S_z = -1/2. \tag{8b}$$

That is, the state with the spin vector parallel to the magnetic field will occupy the higher energy state.

The energy splitting as a function of magnetic field H will vary as in Fig. 4.18.

4.9 Iron-Sulfur Proteins

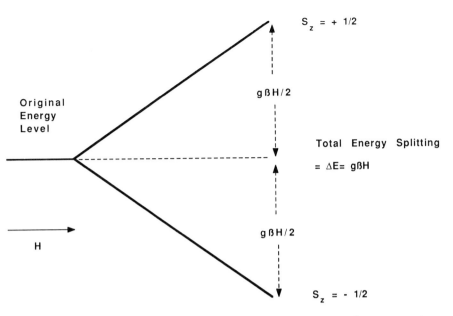

Figure 4.18. Energy splitting as a function of the magnetic field, H, of two states of a free electron, with $S_z = \pm 1/2$.

The magnetization vector, due to the spin in this case, precesses around the static magnetic field, H_z, at the Larmor frequency, $\gamma H_z |S_z|$.

It is possible to "flip" the spin from $S_z = -1/2$ to $S_z = +1/2$ by applying an amount of energy $h\nu = g\beta H$ in a radio frequency field whose magnetic vector is aligned perpendicular to the orientation of the static field (Fig. 4.19), rotates at the Larmor frequency so as to remain synchronously aligned with the magnetization vector, and results in the spin flip.

The frequency ν of the microwave radiation required to produce the transition is calculated from the resonant condition in which the energy, $h\nu$, in a photon of the microwave radiation equals the energy difference between lower and upper states in Fig. 4.18.

$$h\nu = g\beta H. \quad (9)$$

If $H \approx 3,300$ gauss (x-band), then:

$$\nu = \frac{g\beta H}{h} = \frac{2(9.3 \times 10^{-21})(3.3 \times 10^3)}{(6.63 \times 10^{-27})}$$

$= 9.26 \times 10^9$ cycles per second, 9.26 gigahertz, a microwave frequency.

The usual electron paramagnetic resonance (EPR) or electron spin resonance (ESR) spectrometer (Fig. 4.20A) operates at fixed frequency and the magnetic field is varied. The detection is improved by modulating the magnetic field. If the magnetic field modulation is smaller than the line width of

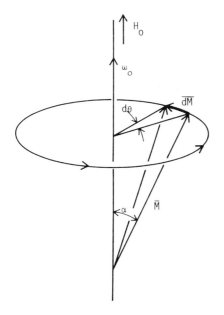

Figure 4.19. Precession of electron spin around magnetic field H_o. In the presence of a magnetic field, the magnetization vector, M, of the electron precesses around the z component of the field, H_z, at the Larmor frequency ω_0 (Feher, 1969).

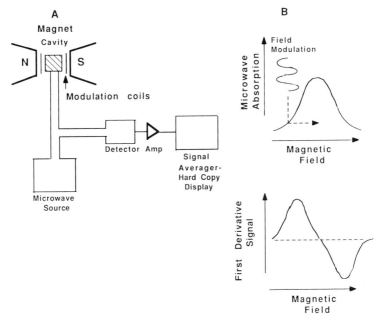

Figure 4.20. (A) Components of EPR spectrometer. In analogy to the optical spectrometer (Fig. 4.4), a monochromatic beam of microwave radiation passes through the sample (in the case of EPR in a magnetic field) and is detected by a silicon crystal. (B) Absorption profile of microwave energy showing field modulation of the magnetic field which is scanned, and resulting detection of the first derivative signal.

the absorption signal (Fig. 4.20B), then the detected signal will appear as the first derivative of the absorption profile if the detector-amplifier is tuned to the frequency and phase of the modulated signal. This technique of phase-sensitive detection is very common in spectroscopic measurements because it allows detection and measurement of a small signal modulated at a predetermined frequency in a large amplitude background.

Very similar techniques are used to study nuclear magnetic resonance (NMR). Since the elemental nuclear mass is typically $\sim 2{,}000$ times that of the electron, the frequency of the absorbed electromagnetic radiation would be $\sim 1/2{,}000$ as large for the same magnetic field (Eq. 9). The actual magnetic field used ($\sim 10^5$ gauss) in modern NMR spectrometers is designed to be large, using superconducting magnets in order to obtain greater spectral resolution.

4.9.2 High-Potential Iron-Sulfur Proteins ("HiPIP")

A number of soluble iron-sulfur proteins, the best characterized being the high potential iron protein (HiPIP) from the photosynthetic bacterium *Chromatium vinosum*, differed from the ferredoxins in that they were diamagnetic ($S = 0$) in the reduced form but became paramagnetic ($S = 1/2$) on one-electron oxidation. *C. vinosum* HiPIP is much more electropositive ($E_{m7} = +350$ mV) than typical ferredoxins. *C. vinosum* HiPIP contains a [4Fe-4S] (Cys)$_4$ cluster identical in three-dimensional structure to those of the [4Fe-4S] ferredoxins (Dus et al., 1973). The "three-state hypothesis," in which it was proposed that the [4Fe-4S] clusters in iron-sulfur proteins could exist in three oxidation states (Carter et al., 1972; Yoch and Carrithers, 1979), was put forward to reconcile these data.

$$[4Fe\text{-}4S(SR)_4]^{3-} \underset{+1e^-}{\overset{-1e^-}{\rightleftarrows}} [4Fe\text{-}4S(SR)_4]^{-2} \underset{+1e^-}{\overset{-1e^-}{\rightleftarrows}} [4Fe\text{-}4S(SR)_4]^{-1}$$

$$Fd_{red} \rightleftarrows Fd_{ox} \rightleftarrows Fd_{super\text{-}ox}$$

$$HiPIP_{super\text{-}red} \rightleftarrows HiPIP_{red} \rightleftarrows HiPIP_{ox}$$

$$3Fe(II) + 1Fe(III) \qquad 2Fe(II) + 2Fe(III) \qquad 1Fe(II) + 3Fe(III)$$

where (—SR) indicates the cysteinyl groups contributed by the proteins. The formal charges attributed to the individual species in this model are (-2) for inorganic sulfide, (-1) for RS$^-$, and $+3$ and $+2$ for ferric and ferrous iron, respectively. The hypothesis predicts that the cluster oxidation states [formally equivalent to 2Fe(II) + 2Fe(III)] found in oxidized ferredoxin and reduced HiPIP are identical, explaining their similar magnetic properties (EPR silent due to a net $S = 0$ ground state arising from anti-ferromagnetic coupling of the irons through bridging sulfur ligands). From this $S = 0$ state, either gain of an electron during reduction, or loss of an electron during oxidation, would produce a net $S = 1/2$, EPR-detectable species. Only the $[4\text{Fe-4S(SR)}_4]^{-3}$ and $[4\text{Fe-4S(SR)}_4]^{-2}$ states would be observed in the native form of HiPIP while only the $[4\text{Fe-4S(SR)}_4]^{-2}$ and $[4\text{Fe-4S(SR)}_4]^{-1}$ states would be accessible in the ferredoxins due to conformational constraints (Cammack, 1973).

4.9.3 Ferredoxin in Oxygenic Photosynthesis

Ferredoxin has a central function in photosynthetic noncyclic reduction of $NADP^+$ and photosystem I cyclic phosphorylation (Arnon, 1984; see below, Chap. 6, section 6.9). There are two ferredoxin isozymes in spinach chloroplasts. Ferredoxin I (MW = 10,485, $E_{m7} = -420$ mV) isolated from spinach chloroplasts contains a single [2Fe-2S] cluster, similar in structure to that deduced from x-ray crystallographic data for the functionally similar [2Fe-2S] protein isolated from the cyanobacterium *S. platensis* (Tsukihara et al., 1981). The gene has also been cloned and sequenced in cyanobacteria (Alam et al., 1986). The highly homologous ferredoxin II (MW = 10,335) may predominate during early stages of leaf development (Takahashi et al., 1983). The reduction of $NADP^+$ by PS I in plant photosynthesis is mediated, in addition to ferredoxin, by the peripheral membrane-protein, ferredoxin: $NADP^+$ oxidoreductase, an FAD-containing (Fig. 4.21) enzyme often referred to by the acronym FNR (Ferredoxin: $NADP^+$ oxidoreductase), which is associated with an M_r 17,500 intrinsic membrane protein (Vallejos et al., 1984). The FNR from spinach chloroplasts, consisting of 314 amino acids, has been crystallized and the structure solved at a resolution of 3.7 Å (Sheriff and Herriott, 1981).

FNR consists of two structural domains, as anticipated for a molecule binding $NADP^+$ and FAD: (i) An $NADP^+$-binding domain in the carboxyl-terminal end of the first strand of a four-stranded parallel β sheet with several adjacent α-helices is similar to the "nucleotide-binding fold" defined by Rossmann et al. (1974). This site is exposed to the solvent. Affinity labeling of the enzyme with an $NADP^+$ analog identified a series of amino acids (Gly-Glu-Lys-Met-Tyr-Ile-Glu-Thr-Arg, corresponding to residues 242–250 in the spinach enzyme) at the $NADP^+$-binding site, with Lys-244 the actual site of attachment of the nucleotide analog (Chan et al., 1985). (ii) From the structure, and sequence identities between the chloroplast enzyme (Table 4.11) and the cyanobacterium, *Spirulina*, the location of the tightly ($K_d = 3 \times 10^{-9}$ M), noncovalently bound FAD prosthetic group appears to lie in a region encompassing amino acid residues 86–99 in the spinach enzyme (Sheriff and Herriott, 1981). Lys-85 and/or Lys-88 in this region provide a major site of electrostatic interaction with the carboxylates of Glu 92–94 of spinach ferredoxin that results in 1:1 complex formation (Vieira and Davis, 1986; Zanetti et al., 1988).

The K_d of the complex is $<2 \times 10^{-8}$ M in 0.05 M buffer, with almost complete dissociation occurring in 0.15 M NaCl (Ricard et al., 1980; Batie and

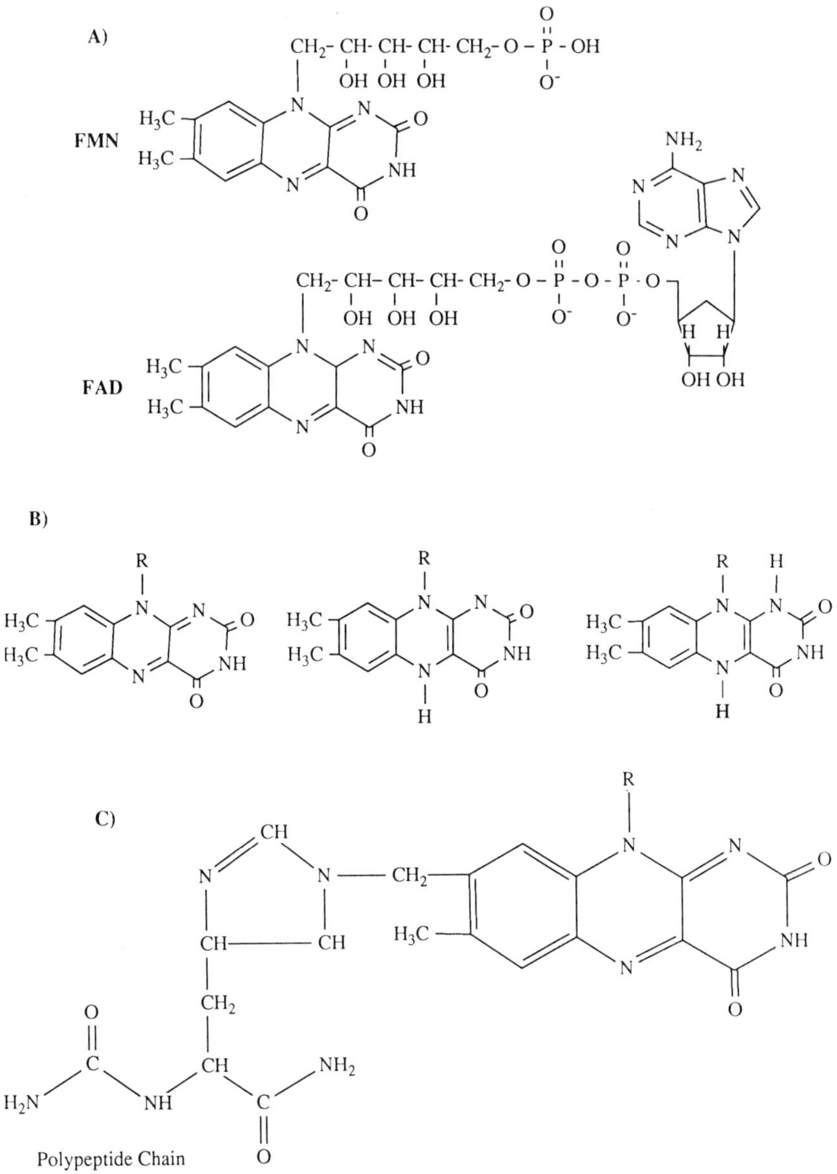

4.9 Iron-Sulfur Proteins

Figure 4.21. (A) Flavin mononucleotide (FMN, riboflavin phosphate) and flavin adenine dinucleotide (FAD); (B) Oxidation-reduction states of flavin isoalloxazine ring; left, oxidized; middle, semiquinone; right, reduced; λ_{max} of oxidized flavin, 450–470 nm, "blue" neutral semiquinone and "red" semiquinone anion detectable by visible and EPR absorbance spectra. Flavin must be able to utilize semiquinone state when accepting electrons from an obligatory one-electron donor such as ferredoxin. It must also be able to then transfer $2e^-$ to $NADP^+$. (C) 8α-(N-3-histidyl) covalent riboflavin linkage found in succinate dehydrogenase (section 4.11).

Table 4.11. Sequence of spinach chloroplast FNR. The underlined regions, 86–99, and 242–250, encompass the FAD- and NADP$^+$-binding domains, respectively.

1[a]	10	20	30	40	50	60	70	80

QIASDVEAPPPAPAKVEKHSKKMEEGITVNKFKPKTPYVGRCLLNTKITGDDAPGETWHMVFS-HEGEIPYREGQSVGVIP

	90	100	110	120	130	140	150	160

DGEDK<u>NGKPHKLRLYSIAS</u>SALGDFGDAKSVSLCYKRLIYTN-DAGETIKGVCSNFLCDLKPGAEVKLTGPVGKEMLMPKD

	170	180	190	200	210	220	230	240

PNATIIMLGTGTGIAPFRSFLWKMFFEKHDDYKFNGLAWLFLGVPTSSSLLYKEEFEKMKEKAPDNFRLDFAVSREQTNE

	250	260	270	280	290	300	310	314

K<u>GEKMYIQTR</u>MAQYAVELWEMLKKDNTYVYMCGLKGMEKGIDDIMVSLAAAEGIDWIEYKRQLKKAEQWNVEVY

[a] NH$_2$-terminal residue found to be isoleucine (Zanetti et al., 1988).
From Karplus et al. (1984).

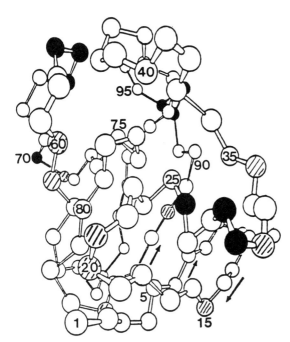

Figure 4.22. Location (in black) of carboxylate residues in spinach ferredoxin that are involved in electrostatic complex formation with FNR. Only α-carbons are shown. Other carboxylates are cross-hatched. The [2Fe-2S] center is attached at Cys residues 39, 44, 47, and 77 (Vieira and Davis, 1986). The major site is probably Glu 92–94 (Zanetti et al., 1988).

Kamin, 1984). The E_m of the FNR and ferredoxin changed by +25 mV and −80 mV, in the complex (Smith et al., 1981; Batie and Kamin, 1984). Fd:FNR complex formation also affected the NMR signals of three ferredoxin glutamate residues in the ^{13}C-NMR spectrum of 20% ^{13}C-enriched ferredoxin from the cyanobacterium *Anabaena variabilis* (Chan et al., 1983), which are probably three of the group Asp-84, Glu-88, Glu-92, Glu-93, and Glu-94 in the spinach ferredoxin that were implicated from cross-linking studies in complex formation with FNR (Fig. 4.22).

Chemical modification of carboxyl residues on ferredoxin does not affect the ability of ferredoxin to accept electrons from the reducing side of PS I but does interfere with ferredoxin binding to FNR and other ferredoxin-dependent chloroplast enzymes (Vieira and Davis, 1986; Hirasawa et al., 1986), indicating different oxidation–reduction sites on the protein, a situation different from that operating in the oxidation and reduction of cytochrome c (section 4.4). Similarly, FNR appears to contain separate binding domains for ferredoxin and $NADP^+$, as indicated by chemical modification studies of FNR Arg and Lys residues (Zanetti, 1976), and from the observation that a ternary ferredoxin · FNR · $NADP^+$ complex can be formed (Batie and Kamin, 1984).

4.10 Membrane-Bound Iron-Sulfur Proteins

4.10.1 The "Rieske" High-Potential [2Fe-2S] Protein

The electron transport chains of mitochondria, chloroplasts, and chromatophores contain a high-potential ($E_{m7} \simeq +250$ to $+300$ mV) membrane-bound [2Fe-2S] protein with an M_r value of 26,000, a broad visible absorption spectrum with peaks of the oxidized proteins at 575, 460, and 315 nm, and a $g_y = 1.90$ EPR signal in the reduced form instead of the $g_y = 1.93$–1.94 signal of reduced ferredoxin (Rieske, 1976). The Rieske protein is diamagnetic ($S = 0$) in the oxidized form and paramagnetic ($S = 1/2$) when reduced, yielding an EPR spectrum of the reduced protein with g values of 2.02–2.04, 1.89–1.90, and 1.79–1.81. The purified Rieske protein is required for electron transfer from succinate or ubiquinol to cytochrome c_1, or from plastoquinol to cytochrome f, and could reconstitute this activity and others associated with the cytochrome bc_1 complex when added back to the bc_1 complex depleted of the Rieske protein (Trumpower, 1981b).

Coordination of this group of [2Fe-2S] proteins does not involve four Cys sulfurs. Chemical analysis of the M_r 20,000 protein from the thermophilic bacterium *Thermus thermophilus* showed the presence of four nonheme iron and four acid-labile sulfides, so that the *T. thermophilus* protein contains two [2Fe-2S] clusters in contrast to the single cluster in mitochondria, chloroplasts, and chromatophores. However, amino acid analyses showed that the protein contains only four cysteine residues, whereas one would have expected eight, four Cys per [2Fe-2S] cluster. Therefore, each [2Fe-2S] cluster must contain two nonsulfur iron ligands, which may be histidine residues (Cline et al., 1985; Kuila and Fee, 1986). A similar role of nitrogen ligands exists for the yeast mitochondrial Rieske center (Telser et al., 1987). The amino acid sequence of the [2Fe-2S] Rieske protein from mitochondria of the fungus *Neurospora crassa* and from chromatophores of the photosynthetic bacterium *Rb. capsulatus* that each contain a single [2Fe-2S] center are compared in Table 4.13. The polypeptides are 199 and 191 amino acids in length. There

Table 4.12. Comparison of amino acid sequences of the Rieske iron-sulfur proteins of
Rb. capsulatus and *Neurospora crassa*[a] mitochondria

```
              10         20         30         40         50         60         70         80
RbC: MSHAEDNAGTRRDFLYHATAATGVVVTGAAVWPLINQMNASADVKAMSSIFVDVSAVEVGTQLTVKWRGKPVFIRRDEH
NC:  SKAPPSTNMLFSYFMVGTMGAITAAGAKSTIQEFLKNMSASADVLAMKVEVDLNAIPEGKNVIIKWRGKPVFIRHRTPA

              90        100        110        120        130        140        150        1
RbC: DIELARSVPLGALRDTSAENANKPGAEATDENRSLAAFDGTNTGEWLVMLGVCTHLGCVPMGDKSGDFGGWFCPCHGS
NC:  EIEEANKVNVATLRD...      ...PETDADRVKKP...    ..EMLVMLGVCTHLGCVPIGE.AGDYGGWFCPCHGS

              170        180        190
RbC: DSAGRIRKGPAPRRNLDIPVAAFDETTIKLG
NC:  DISGRIRKGPAPLNLEIPLYEFPEEGKLVIG
```

[a] NC, Not shown, 57 residues before first residue, Ser, that consist of a 32-residue leader sequence and 2 residues of the 199-residue protein. The four Cys and three His residues that are identical between the two sequences are underlined. Sequences predicted for the Rieske ISp from bovine mitochondria (Schägg et al., 1987) and chloroplasts (Steppuhn et al., 1987) have also been reported.
From Harnischet et al. (1985); Gabellini and Sebald (1986); Davidson and Daldal (1987a).

are four cysteines, Cys-133, 138, 153, and 155, which are the only Cys residues in the proteins, and are conserved between these sequences and others. Two of these Cys must be involved in coordination of the Fe atom. The identity of the other two nitrogenous ligands is not known at present. They are presumably two of the three conserved histidines near the cysteines, which are underlined in Table 4.13.

4.10.2 The Membrane-Bound FeS-Flavoprotein, NADH:Ubiquinone Oxidoreductase (Complex I)

The oxidation of a large number of substrates (including the Krebs cycle intermediates isocitrate, α-ketoglutarate, and L-malate) by NAD^+-specific enzymes results in the presence of a continuing supply of NADH in the mitochondrial matrix in eukaryotic cells. An important function of the mitochondrial electron transfer chain is to oxidize this NADH back to NAD^+, ensuring an adequate supply of oxidant so that NAD^+-requiring oxidation reactions do not become oxidant-limited. NADH oxidation also initiates the highly exergonic respiratory electron transfer to O_2 that provides the energy for the formation of most of the cell's ATP supply. This oxidation of NADH is catalyzed by complex I, the FMN-containing NADH:ubiquinone oxidoreductase, a membrane-bound, multisubunit enzyme complex with the largest number of subunits, approximately 25, of any complex in the respiratory chain (Table 4.13). The function of many of these subunits is unknown, and some may be involved in the binding of TCA cycle dehydrogenases, e.g., malate dehydrogenase, α-ketoglutarate dehydrogenase, and pyruvate dehydrogenase, to complex I (Srere, 1987). Complex I from human mitochondria contains six of the thirteen polypeptides that are encoded by the mitochon-

4.10 Membrane-Bound Iron-Sulfur Proteins   183

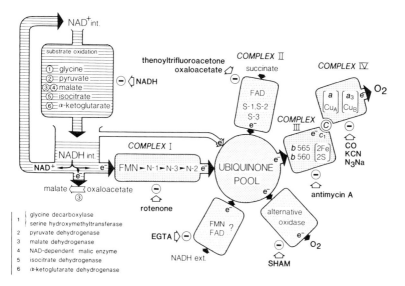

Figure 4.23. The role of a central quinone pool in connecting dehydrogenases, the cytochrome b–c_1 complex (complex III), and oxidases in plant mitochondria. Inhibitors acting at specific sites in the respiratory chain, most of which are quinone analogs (Chap. 5, section 5.1), except for those acting on the cytochrome oxidase, are shown (Douce and Neuberger, 1987).

drial DNA (Chomyn et al., 1985). Kinetic and extraction/reconstitution studies have shown ubiquinone to have a central role as the electron acceptor for Complex I and other mitochondrial dehydrogenases (Fig. 4.23), and that the ubiquinol formed from ubiquinone reduction serves as the electron donor to the next segment of the respiratory chain, the cytochrome bc_1 complex or complex III, and in the case of plant mitochondria, an alternate oxidase that branches from the quinone pool. As predicted from the scheme shown in Fig. 4.23, respiratory activity can be reconstituted from the combination of complex I (NADH dehydrogenase) or complex II (succinate dehydrogenase) and ubiquinone, complex III (cytochrome bc_1), cytochrome c, and complex IV (cytochrome oxidase).

The large number of iron-sulfur centers in complex I and the fact that the different centers have similar EPR spectra have made it difficult to analyze the EPR spectra of these centers in complex I. However, investigation of the temperature, power, and redox dependencies of the spectra, as well as the use of spectral simulation, indicated the presence of at least five active iron-sulfur centers in the mitochondrial NADH:ubiquinone oxidoreductase (Table 4.14), all of which are of the "ferredoxin-type" in that they are diamagnetic ($S = 0$) and EPR-silent in the oxidized form, and paramagnetic ($S = 1/2$) upon reduction by one electron.

$(NH_4)_2SO_4$ fractionation of complex I after treatment with a chaotropic

Table 4.13. Properties of mitochondrial complex I from beef heart mitochondria

Monomer molecular weight: $7-8 \times 10^5$
Number of polypeptides: ~ 25 (~ 1 copy per complex; M_r values in Table 4.15)
Flavin, FMN: 1.0–1.3 nmol/mg protein
Nonheme Fe or acid-labile S: 22–24 per FMN.
UQ/FMN: 2–4
Number of active EPR-detectable FeS centers: 5 (see Table 4.14)
E_{m8}(mV), FMN/FMNH$_2$, -320;
    FMN/FMNH·, -380
    FMNH·/FMNH$_2$; -260
Cytochrome content: <0.1 nmol/mg protein
Lipid content: 0.25 µmol lipid P_i/mg protein
                  200–300 molecules/FMN
                  20 cardiolipid molecules/FMN
Inhibitors: Piericidin A ($K_D = 10^{-7}$ M); rotenone (except *C. utilis*) (Fig. 4.24).
Structural heterogeneity, possibility of dimer (Ragan, 1987).
Function in H$^+$ translocation (Chap. 7).

From Hatefi (1985); Ragan (1985).

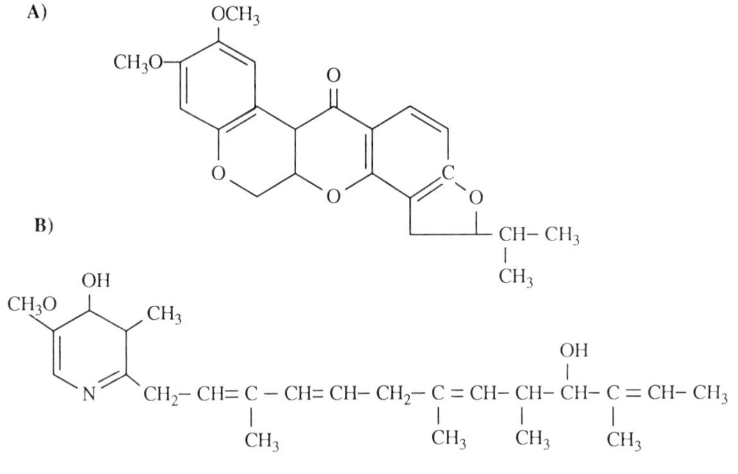

Figure 4.24. Structural formulae of the inhibitors of complex I, (A) rotenone, an insecticide, and (B) piericidin, a structural analog of ubiquinone. The binding site is on the M_r 33,000 polypeptide of the H subcomplex.

agent resulted in the isolation of three defined subcomplexes, two water soluble, flavin-containing (Fp) and iron-enriched (Ip), and the third hydrophobic (H). Each fraction contains at least one of the active iron-sulfur centers (Table 4.15). The Fp fraction is thought to contain the N1b and N3 centers, along with most or all of the bound FMN, and the NADH binding site attached to an M_r 51,000 polypeptide. The midpoint potential data indicate

4.10 Membrane-Bound Iron-Sulfur Proteins 185

Table 4.14. Classification of EPR-detectable FeS centers active in complex I

A.

Designation Ohnishi[a]	g Values			Concentration (unpaired e$^-$/FMN)	Cluster structure
	g_z	g_y	g_x		
N1a	2.03	1.95	1.91	1/2	[2Fe-2S]
N1a	2.02	1.94	1.92	1/2-1	[2Fe-2S]
N2	2.05	~1.92	~1.92	1	[4Fe-4S]
N3	2.04	~1.93	~1.86	1	[4Fe-4S]
N4	2.10	~1.93	~1.88	1	[4Fe-4S]

B.

Center	E_{m7} (mV)	n-Value	$\Delta E_m/\mathrm{pH}$ (mV)
N1a	−370	1	−60
N1b	−250	1	0
N2	−20	1	−60
N3	−255	1	0
N4	−255	1	0

[a] Nomenclature of Ohnishi (1979) and Ohnishi et al. (1981); components that have been detected spectoscopically, but have a stoichimetry < 1/2 are omitted from this table.
From Ohnishi (1979); Beinert and Albracht (1982); Wikström and Saraste (1984); Ragan (1985).

Table 4.15. Fractionation of iron-sulfur centers of complex I

Fraction	Subunit ($M_r \times 10^{-3}$)	Cluster structure	FeS Center	Fe (mole/polypeptide)
Fp	51 (FMN, NADH active site)	[4Fe-4S]	N3	6
	24	[2Fe-2S]	N1b	
Ip	75	[2Fe-2S]	N1a	10
	49	[2Fe-2S]	N4	
	30 + 13	[2Fe-2S] or [4Fe-4S]	? or N4	
H	33 and	[2Fe-2S]	?	6
	approx. five others	[4Fe-4S]	N2	

[a] From Ragan (1985, 1987).

that the N2 center associated with subcomplex H should be reduced most readily. At 20°C, however, the kinetics of reduction of all four centers occurs within 10 ms, and the sequence of reduction has not been resolved. On the basis of its relatively positive E_m values, center N2 is the likely donor to the ubiquinone.

A role for the iron-sulfur centers in energy conservation is implied by the loss in the yeast *C. utilis*, and the bacterium *P. denitrificans* when grown on sulfate-limited medium, of both the iron-sulfur center EPR signals and of ATP synthesis associated with electron flow through complex I.

4.10.3 Topography of Complex I and a Possible Electron Transport Pathway

The arrangement of the polypeptides of membrane proteins can be probed with hydrophilic and hydrophobic agents that react, respectively, with polypeptide groups exposed to the aqueous phase or buried in the protein and/or inserted into the membrane bilayer. Common hydrophilic agents are diazobenzenesulfonate (DABS), lactoperoxidase-mediated iodination, and nitroazidophenyltaurine (NAP-taurine). Lipid soluble probes include [3-(trifluoromethyl)-3-(*m*-iodophenyl)diazirine] TID and arylazidophosphatidylcholine. None of the three subunits of the Fp fraction were labeled by any probe and these were inferred to be buried within the membrane. All of the other subunits of the Ip and H fractions defined in Table 4.15 were labeled by DABS, NAP-taurine, or hydrophobic probes. A topographical model of the complex derived from many such probe experiments as well as chemical cross-linking, is shown in Fig. 4.25. NADH would bind in a hydrophobic pocket on the large subunit of the Fp fragment, which is close to the largest subunit of Ip. Electron transfer would occur from (i) NADH to FMN near the NADH binding site and then to centers N3, N1b on the same Fp subunit, (ii) to the N1a and N4 centers of Ip, (iii) to the high potential N2 center that is located in the H domain, and then (iv) via semiquinone intermediates to the ubiquinone pool. The hydrophobic H domain contains the six mitochondrial-encoded polypeptides. One of these, a 33-kDa polypeptide on which the rotenone inhibitor binding site has been located, may contain the site for ubiquinone reduction (Ragan, 1987),

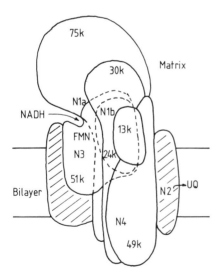

Figure 4.25. Topographical model of complex I (Ragan, 1987); Fp contains $M_r \simeq 51,000$ and 24,000 polypeptides and the FMN binding site; Ip (iron-sulfur protein) contains the M_r 75, 49, 30, and 13,000 polypeptides. The assignments of the iron-sulfur clusters are indicated, and the H domain is defined by the shaded area.

4.11 The Membrane-Bound FeS-Flavoprotein, Succinate: Ubiquinone Oxidoreductase (Complex II)

This intrinsic, multiprotein complex of the inner mitochondrial membrane is responsible through its succinate dehydrogenase activity for the oxidation of succinate to fumarate in the matrix during the operation of the Krebs cycle. Succinate dehydrogenase (SDH) is the major component of complex II and the only enzyme of the Krebs cycle that is membrane-bound. As expected from the relatively small ΔE_m that exists between the succinate/fumarate couple ($E_{m7} = +30$ mV) and the UQ-10/UQH$_2$-10 couple ($E_{m7} = +65$ mV; Urban and Klingenberg, 1969), ATP synthesis is not associated with electron flow through Complex II to ubiquinone. The resulting ubiquinol becomes part of the pool which can serve as an electron donor to the bc_1 complex and a proton carrier across the membrane bilayer (Chap. 7). Complex II is located on the cytoplasmic or inner membrane in bacteria and mitochondria, respectively. Mitochondrial complex II consists of four polypeptides: M_r 70,000 and 27,000 (containing the SDH), 15,000 and 13,000. The two smaller polypeptides are a membrane anchoring fraction that contain a cytochrome b (b_{560}) and also bind quinone ("QP-S"). Addition of the two smaller subunits can convert succinate dehydrogenase to succinate-ubiquinone reductase. The SDH fraction is water-soluble and contains one bound FAD, and 7–8 equivalents of nonheme iron and acid-labile sulfide (Table 4.16). The single FAD of SDH is covalently attached, via an 8α-histidyl linkage, to the M_r 70,000 subunit of succinic dehydrogenase (Fig. 4.21C). The redox components detectable in the mitochondrial complex II are summarized in Table 4.17. SDH is organized in the Gram-positive bacterium *Bacillus subtilis* as an operon containing the genes for a b-type cytochrome and an iron-sulfur protein: sdhA (cytochrome b; MW = 22,900), sdhB (flavoprotein; MW = 65,000) and sdhC (iron-sulfur protein; MW = 28,000). The cytochrome b appears essential for the anchoring of the complex to the membrane (Magnusson et al., 1985).

Table 4.16. Properties of bovine heart succinate dehydrogenase and its subunits

Molecule	Molecular weight	FAD/ polypeptide	Fe/ polypeptide[a]	Acid labile sulfide/ polypeptide
Succinate dehydrogenase	97,000 ± 4%	1	7–8	7–8
Large subunit (Fp)	70,000 ± 7%	1	4	4
Small subunit (Ip)	27,000 ± 5%	0	3–4	3–4

[a] More recent data indicate that all of the iron-sulfur centers are on the small subunit (Fig. 4.26). From Hatefi et al. (1985).

Table 4.17. Redox components of bovine heart mitochondrial complex II

Component	E_{m7} (mV)	g Values	Stoichiometry:FAD	Reference
FAD/FADH$_2$	0 to −90	—	—	Gutman et al., (1980)
FAD/FAD$^{\dot-}$	−130	2.00 (FAD$^{\dot-}$)	—	
FAD$^{\dot-}$/FADH$_2$ (pK = 7.7)	−30		—	Ohnishi et al. (1981)
[2Fe-2S] center S-1 (succinate-reducible)	0	1.93 (cntr.)	1:1	Ohnishi et al. (1976)
[2Fe-2S] or [4Fe-4S] center S-2 (not succinate-reducible)	−270	Complex interaction with S-1	1:1	Ohnishi et al. (1976); Albracht, (1980); Maguire et al. (198?)
S-3, 3Fe-XS (X = 3,4) center[a] (M^r 28,000)	+130	2.014 (cntr.)	1:1	Ackrell et al. (1984)
Two bound quinones detected as ubiquinone radicals pK = 6.5, 8.0), 7.5–8.0 Å apart which are eliminated by TTFA[b] (M$_r$ 15, 13,000)	+80, +140 (pH 7.4), Q/Q$^{\dot{}}$ and Q$^{\dot{}}$/QH$_2$	2.04, 1.99 (Oxidized)	—	Ingledew and Ohnishi (1977); Salerno and Ohnishi (1980); Yu and Yu (198?)
b_{560} (M$_r$ 15–13,000).	−185	—	—	Hatefi and Galan? (1980); Yu et al (1987)

[a] Recently discovered iron-sulfur cluster that is paramagnetic in the oxidized ($S = 1/2$), where is ca? detected, and reduced ($S = 2$ ground state) states.
[b] Thenoyltrifluoroacetone.

4.11.1 Topography of Complex II and Mechanism of Electron Flow from Succinate to Ubiquinone

Based on amino acid sequence data of the complex II components, and data on spin-coupling of three iron-sulfur clusters, S-1, S-2, and S-3, a topographical model of complex II has been proposed (Fig. 4.26). Succinate, produced in the mitochondrial matrix, would be bound to an active site near the covalently bound FAD that is present on the membrane-bound 70-kDa subunit and accessible from the matrix space. Succinate can be oxidized to fumarate in a two electron step by FAD. The details of the subsequent pathway are not well established, but the relatively high stability of the FAD semiquinone state suggests that one pathway of electron transfer from FADH$_2$ could occur by two, sequential one-electron steps involving FeS centers S-1 and S-3 (Yu et al., 1987). The role of center S-2 is less clear. Reduced centers S-3 and S-1 would reduce the ubisemiquinone pair attached to the

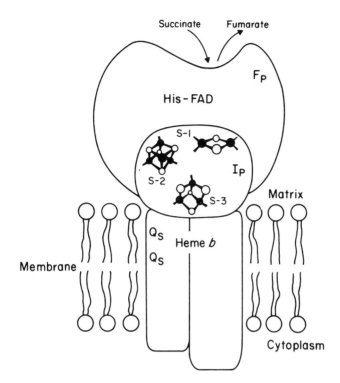

Figure 4.26. Schematic representation of the topography of succinate dehydrogenase and cytochrome b in complex II, showing the covalently bound FAD, iron-sulfur clusters S-1, S-2, and S-3 on the small subunit Ip, heme b, and a bound pair of ubiquinone molecules (Ohnishi, 1987).

QP-S quinone-binding protein, either in parallel or in series, with subsequent release of the semiquinones, and formation of ubiquinol by semiquinone dismutation (Chap. 2, Eq. 17d). The specific inhibitor of complex II, thenoyl-trifluoroacetone (TTFA), inhibits electron flow between center S-3 and the ubiquinone pool, perhaps at the binding site of the specialized semiquinone pair on QP-S (Ingledew and Ohnishi, 1977).

4.12 Summary

Cytochromes a–d have been classified on the basis of their characteristic visible absorbance spectra. The heme, derived in all cases from protoporphyrin IX, is attached non-covalently to cytochromes a, b, and d, and covalently to cytochrome c. The optical spectra, commonly used to assay changes in redox state, have 3 major absorbance bands, the α, β, and γ bands, typically in a ratio of 2:1:10 for the b and c cytochromes. Particularly for the pigmented photosynthetic membranes, redox changes are easiest to measure in the α band. The use of isosbestic reference wavelengths, and efficient collection of scattered light from turbid and optically dense organelle or cell suspensions, allows measurement of small ($\Delta A \sim 10^{-4}-10^{-3}$) absorbance changes associated with physiological oxidation-reduction events.

Cytochrome oxidase is a multisubunit metalloprotein complex containing two molecules of heme a and two copper atoms, two polypeptide subunits in some prokaryotes and 9-13 in eukaryotes, of which only the three largest are encoded by mitochondrial DNA. Some of the smaller nuclear-encoded subunits in mammalian oxidase are tissue-specific. Plant mitochondria and membranes from bacteria such as $E.$ $coli$ have alternate cytochrome oxidases. The cytochrome d (containing b and d hemes) and o (containing b heme and copper) of the $E.$ $coli$ membrane function, respectively, under conditions of low (e.g., stationary phase) and high O_2 tension. Cytochrome b functions in a membrane-bound complex in mitochondria, chloroplasts, photosynthetic bacteria, and some obligately aerobic bacteria in the center of the electron transport chains, in the bc_1 complex between the oxidase and dehydrogenase complexes of the respiratory chain, and in the $b_6 f$ complex between PS II and PS I in oxygenic photosynthesis. The intramolecular cross-linking by heme of transmembrane peptides in these b cytochromes, as well as in subunit I of cytochrome oxidase, is a structural motif. In contrast to the membrane-bound cytochromes, the small (~ 100 amino acids) high potential c-type cytochromes are water-soluble and only weakly bound to the membrane, as is the copper protein plastocyanin. These proteins provide a diffusible redox link between the central cytochrome b complex and the oxidase or PS I. The histidine heme ligand of cytochrome c near the NH_2 terminus occurs in a diagnostic sequence, Cys-X-Y-His-Cys, in which the two Cys provide the covalent coordination to the heme group at side chain positions 2 and 4. Cytochrome c_3 is a small protein of about the same size, but with four hemes that function in the pathway from ferredoxin to hydrogenase and sulfite reduction. Because it is relatively easy to isolate the water soluble mitochondrial cytochrome c, its structure-function relationships have been studied extensively. Structural features found critical to function are (i) the exposure of a heme edge to the external solvent. (ii) A highly conserved positive charge distribution around this exposed heme edge provides electrostatic interaction with negatively charged residues on the reaction partners. The critical Lys residues in cytochrome c from horse, Lys 8, 13, 27, 72, 73, 86, 87, could be identified by chemical modification and assay of resulting activity and effective charge of the protein, and by differential labeling in the presence and absence of the reaction partners of the cytochrome. Complementary acidic carboxylate residues provide the other half of the electrostatic interaction in cytochrome b_5 and cytochrome c peroxidase, with which cytochrome c can form electron transfer complexes. This interaction is frequently found when either the donor or acceptor is water-soluble. Cytochromes c_1 and f are membrane-bound c-type cytochromes in the central cytochrome bc_1 or $b_6 f$ complexes. These

4.12 Summary

proteins have little primary sequence identity, but have one membrane spanning α-helix near the COOH-terminus which is on the n-side of the membrane. (In order to have a common notation to apply to chloroplast, chromatophore, and mitochondrial membranes, the membrane sidedness is defined as "n" and "p" corresponding to the sides with relatively negative and positive values of the $\tilde{\mu}_{H^+}$].

Electron transport chains contain an excess or comparable amount of nonheme compared to heme Fe, ~ 3.0 nmol/mg protein of each in mitochondrial membranes of bovine heart. The non-heme Fe is in the form of protein-bound iron-sulfur centers that can contain 2, 4, and also 3 Fe/center. The 2Fe and 4Fe centers are equimolar with acid-labile sulfide, which is inorganic sulfide, leading to the notation of [2Fe-2S] and [4Fe-4S] centers. In each case, the Fe is surrounded by four sulfur ligands, two inorganic sulfides and two cysteine residues in the case of the [2Fe-2S] ferredoxins. Ferredoxin was the name given to the water-soluble iron-sulfur protein discovered in *Clostridium pasteurianum*. Ferredoxins are characterized by their low molecular weight ($\leq 12,000$), equimolar nonheme iron and acid-labile sulfide, negative E_m (≤ -200 mV), diamagnetism ($S = 0$) in the oxidized state, and paramagnetic $S = 1/2$ state on one-electron reduction. Because of their paramagnetism and broad optical absorbance bands, iron-sulfur proteins are commonly studied by electron paramagnetic resonance (EPR) spectroscopy. Ferredoxin-NADP$^+$-reductase (FNR), an FAD-enzyme, functions in the NADP$^+$ reduction pathway associated with photosystem I of oxygenic photosynthesis. The interaction between ferredoxin and FNR is also electrostatic, involving positively charged Lys residues on FNR and negatively charged carboxylates on ferredoxin. A membrane-bound [2Fe-2S] "Rieske" protein is found in the ubiquitous cytochrome bc_1 and $b_6 f$ complexes. It has the unique properties of a high redox potential ($E_m = +250$–300 mV) and ligation by at least one His and 2–3 Cys instead of 4. Complexes I and II of the mitochondrial respiratory chain, the FMN-containing NADH: and FAD-succinate: ubiquinone oxidoreductases, contain many iron-sulfur centers, approximately 24 and 15, respectively. In complex I, five active iron-sulfur centers can be distinguished by EPR spectroscopy and redox titrations, and 3 subcomplexes can be separated: (i) "Fp" containing at least two polypeptide components, two of the active iron-sulfur centers, the FMN, and the active site; (ii) the Fe-enriched "Ip" subcomplex containing two active iron-sulfur centers; (iii) the hydrophobic subcomplex "H" contains the iron-sulfur center with the most positive E_m that is the likely donor to ubiquinone. In human mitochondria, it contains 6 of the 13 mitochondrial-encoded polypeptides. Mitochondrial complex II consists of 3–4 polypeptides, with the FAD covalently attached to an M_r 65–70,000 subunit near the active site for binding of succinate where it undergoes a 2e$^-$ oxidation by FAD to fumarate. Three iron-sulfur centers localized in an M_r 27–28,000 subunit are involved in the oxidative pathway to quinone which is bound to one or two low-molecular-weight polypeptides.

Problems

53. Calculate the absorbance change at 563 nm, relative to a 575 nm reference wavelength, of the chloroplast cytochrome b_6 when it is reduced 50% by a light flash in a suspension of chloroplasts containing 20 μg chlorophyll/ml. The oxidized minus reduced millimolar extinction coefficient (563–575 nm) and heme stoichiometry of the cytochrome are 17 mM^{-1} cm^{-1} and 3 nmol/mg chlorophyll, respectively.
54. Calculate the g value of (a) the high-spin cytochrome that shows a peak in microwave absorption at a magnetic field of 1,100 gauss, using a microwave frequency of 9.13 GHz (9.13 × 19^9 Hz), and (b) of a low spin cytochrome that has a peak at 2,000 gauss. (c) Calculate the magnetic field at which the peak of the free radical of the chloroplast reaction center P700 free radical, $g = 2.002$, would be detected under the above conditions.
55. Calculate (a) the value of the Bohr magneton for a proton ($m_p = 1836\, m_e$) and (b) the frequency at which a high-field proton NMR instrument will operate if the imposed magnetic field is 10^5 gauss (10 Tesla), the proton spin is 1/2 (units of $\hbar$), and the g value is 2.79.
56. (a) Calculate the energy change $U = -\vec{\mu}\cdot\vec{E} = -\mu E \cos\Theta$ involved in orienting ferrocytochrome c through an angle of 60° to the reaction site on the surface of cytochrome oxidase where at equilibrium it would be parallel to the electric field vector at the surface of the oxidase. (b) How much larger is this energy than thermal energy, kT, that would cause disorientation? The dipole moment, $\vec{\mu}$, of the cytochrome is 310 debye and the electric field strength at the surface of the oxidase is 3×10^7 V/m (Koppenol and Margoliash, 1982).
57. (a) Calculate the E_m value for the $n = 2$ FAD/FADH$_2$ couple, using the values in Table 4.17 for the two flavin semiquinone reactions. (b) What is the semiquinone formation constant of the FAD flavin?

Chapter 5
The Quinone Connection

5.1 Structures, Stoichiometry, Pools, and Branch Points

5.1.1 Structures and Stoichiometry

The concept that the long-chain quinones in energy-transducing membranes (Fig. 5.1A and B) act as mobile carriers of electrons and protons is partly based on the high concentration of quinone in all energy-transducing membranes (Table 5.1). This quinone can be readily extracted with organic solvent. In spite of the apparent excess of quinone in all these systems, most of it is needed to maintain a normal respiratory flow, because partial extraction of mitochondrial quinone results in loss of activity (Schneider et al., 1985). Quinone analogue inhibitors block electron transfer by competing for the quinone binding sites located on electron transport protein complexes (Fig. 5.1C). The compounds 3-(3,4-dichlorophenyl)-1,1-dimethylurea (DCMU) and atrazine inhibit plant and algal photosynthetic electron transport at the site (the quinone-binding D1 polypeptide) of reduction of the plastoquinone pool by photosystem II (PS II). Piericidin and rotenone inhibit at the site of reduction of the ubiquinone (UQ) pool on the NADH dehydrogenase (Fig. 4.25), and several inhibit effectively on the p-side of the membrane (toward which protons are translocated, positive $\tilde{\mu}_{H^+}$) at the cytochrome $b_6 f$ (2,5-dibromo-3-methyl-6-isopropylbenzoquinone, stigmatellin) or bc_1 (stigmatellin, 5-n-undecyl-6-hydroxy-4,7-dioxobenzothiazole, myxathiazol) complexes, preventing oxidation of the plasto- or ubiquinone pool. Antimycin A inhibits the bc_1 complex on the n-side of the membrane.

5: The Quinone Connection

(A), (B), C.1, C.2, C.3, C.4, C.5, C.6, C.7

5.1 Structures, Stoichiometry, Pools, and Branch Points

Table 5.1. Quinone pool sizes

Membrane system	Quinone[a]	Pool size
Chloroplasts	PQ-9	40[b]
Cyanobacteria		
A. nidulans	PQ-9	50[b]
A. variabilis	PQ-9	7
Photosynthetic bacteria		
Rhodospirillaceae		
photosynthetic	UQ-10	25–40[b]
aerobic	UQ-10	5–10[c]
C. vinosum	UQ-7	15[b]
	MQ-7	6
Mitochondria		
Mammals	UQ-10	6–8[d]
Plants (Arum)	UQ-10	7
Yeast	UQ-6	36
Enterobacteria	UQ-8 (log ϕ)	6–8[c]
	MQ-8 (stat. ϕ)	6–8
Myobacterium phlei	MQ-9	45[d]
B. megatherium	MQ-7	5–10[c]
P. vulgaris	MQ-6	21[c]

[a] PQ, UQ, MQ, plasto-, ubi-, menaquinone.
[b] Normalized to P700 or P870 reaction center; the actual number of PQ molecules turning over rapidly in the electron transport is 3–6.
[c] Normalized to cytochrome b.
[d] Normalized to cytochrome aa_3.
From Hauska and Hurt (1982).

Figure 5.1. (A) Structure of ubiquinone and plastoquinone with substituent groups and isoprenoid chain length n. The quinone in mammalian and eukaryotic mitochondria (Crane et al., 1957) is usually UQ-10 (Crane, 1977; Crane and Barr, 1985), next most frequently UQ-9, and UQ-6 in shark liver. UQ-6 is also common in yeast mitochondria. UQ-9 and UQ-10 form the bulk quinone in the photosynthetic bacteria, *Rps. viridis* and *Rb. sphaeroides*, respectively, and UQ-8 is used in *E. coli*.

	R_1	R_2	R_3	n	
UQ	CH_3	CH_3O	CH_3O	10	(eukaryotic mitochondria)
PQ	H	CH_3	CH_3	9	

(B) Structure of menaquinone found in some bacteria such as *B. megatherium* and *V. succinogenes*. $n = 9$ in the bound quinone Q_A of the purple photosynthetic bacterium *Rps. viridis*. (C) Quinone-like inhibitors: (1) antimycin, R = *n*-hexyl or *n*-butyl for A_1 and A_3 derivatives; (2) atrazine; (3) DCMU; (4) DBMIB; (5) UHDBT; (6) stigmatellin; (7) myxathiazol; (8) also, piericidin (Fig. 4.24B).

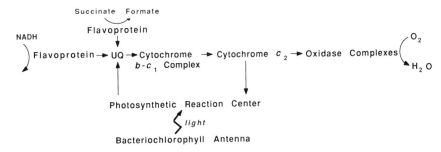

Figure 5.2. Ubiquinone and cyt c_2 as branch points for respiration and photosynthesis in *Rb. capsulatus*. The arrows show the direction of electron flow.

5.1.2 The Quinone Pool Is a Branch Point for Electron Transport

The many structural similarities between the membranes of mitochondria, photosynthetic bacteria, chloroplast thylakoids, and cyanobacteria and also between some of their intrinsic proteins imply an overlapping evolutionary background. Indeed, the electron transport chains of photosynthetic bacteria and cyanobacteria can be used for both respiration and photosynthesis (Binder, 1982), with the long-chain ubiquinone and plastoquinone able to function, along with a c-type cytochrome or plastocyanin, as mobile or distributive electron carriers at branch points of the chains (Fig. 5.2). The ubiquinone pool has a central branching role mediating between several dehydrogenases and the $b-c_1$ complex in mammalian mitochondria and as a branching pool between several dehydrogenases and two alternative oxidases in *E. coli* membranes (Fig. 3.5B). Although the ubiquinone pool can interact with many different dehydrogenases in the respiratory chain, it behaves kinetically like a homogeneous pool, its oxidation and reduction following first-order kinetics (Kröger and Klingenberg, 1973). Plastoquinone connects many electron transport chains on the donor and acceptor sides of the pool in the choroplast electron transport chain (Siggel et al., 1972).

The Branched-Chain in Plant Mitochondria Generates Heat

The structure of the plant mitochondrial respiratory chain is similar to that of mammalian mitochondria except for additional substrate oxidation pathways and, like *E. coli*, an alternate oxidase. Unlike mammalian mitochondria which cannot oxidize exogenous NADH, but must transport external NADH through a shuttle system, plant mitochondria can oxidize both exogenous and endogenous NADH. The route of oxidation of the exogenous NADH, however, bypasses the first (dehydrogenase) site of H^+ translocation and phosphorylation and is not sensitive to rotenone or piericidin, the two usual inhibitors of NADH dehydrogenase. The external NADH is oxidized by a flavoprotein dehydrogenase on the outer surface of the inner plant mitochondrial membrane, which then donates electrons and protons directly to the ubiquinone pool. Practically all plant tissues show a residual respiration in the presence of the usual mammalian cytochrome oxidase inhibitors, carbon monoxide, cyanide, or azide (Table 5.2). Some plant mitochondria are known to have an alternative pathway to molecular O_2 that is resistant to usual respiratory inhibitors such as cyanide, azide, carbon monoxide, and antimycin A, although it is sensitive to hydroxamic acids [e.g., salicylhydroxamic acid (SHAM)].

The branch point to the alternate oxidase appears to occur in the region of the quinone pool, because oxidation of succinate and exogenous NADH is inhibitor resistant, but oxidation of reduced cytochrome c (Fig. 5.3) is sensitive to inhibitors of the oxidase. The pathway through the alternate oxidase is not coupled to $\Delta\tilde{\mu}_{H^+}$ generation and ATP synthesis, but bypasses the cytochrome b–c_1 complex (Fig. 5.3) and instead releases heat energy. The heat evolution protects the flower-bearing tissue (e.g., skunk cabbage spadix) from extreme

Table 5.2. Resistance of plant mitochondria to potassium cyanide

Species	Tissue	Cyanide resistance (%)
Cotton	Roots	22
Bean	Roots	41
Spinach	Roots	34
Wheat	Roots	35
Maize	Roots	32
Pea	Leaves	30
Spinach	Leaves	27

[a] Mitochondrial substrates were 10 mM malate plus 10 mM succinate; KCN (0.2 mM) was added in the presence of ADP. The cyanide-resistant O_2 uptake was inhibited by SHAM. After Douce (1985).

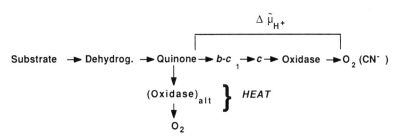

Figure 5.3. Branched electron transport chain to energy ($\Delta\tilde{\mu}_{H^+}$)-conserving and CN^--sensitive oxidase and the heat-generating alternate (alt) oxidase in plant mitochondria.

cold. It is known that in *Arum* lilies heat volatilizes odor-forming amines that serve as insect attractants and thus promote pollination.

5.2 Reconstitution of Quinone Function Requires Q_n with $n \geq 3$

Reconstitution studies in which physiological membranes have been depleted of almost all quinone, and function restored by addition of Q with variable n, have been helpful in elucidating Q topography and function. In general, Q_n with $n \geq 3$ is required to restore or establish function: (i) UQ_8–UQ_{10} was much more efficient than UQ_0–UQ_2 in restoring NADH oxidase activity to extracted heart mitochondria (Crane, 1977); (ii) phosphorylation and electron transport were restored to depleted chromatophore membranes by UQ_3 and UQ_{10}, and UQ_1 or UQ_2 were much less effective (Baccarini-Melandri et al., 1980); (iii) UQ_3–UQ_8 reconstituted NADH oxidase activity, whereas UQ_1–UQ_2 were relatively inactive, although D-lactate-driven proline transport was restored by lower concentrations of UQ_1 than UQ_8 (Stroobant and Kaback, 1979). Thus, lipophilic quinone is required for function, although the above experiments do not explain the physiological requirement for very large $n = 8$–10 rather than $n = 3$ for the isoprenoid chain number.

5.3 The Quinone Pool Is Located Near the Center of the Membrane Bilayer

The location of the quinone pool was inferred from (i) inaccessibility of UQ_{10} in mitochondrial membranes to hydrophilic donors and acceptors (Crane, 1977), (ii) inaccessibility of UQ_{10} to the short-range reductant borohydride and small perturbation of the liposome phospholipid phase transition by long-chain quinone (Fig. 5.4) (Ulrich et al., 1985); (iii) calorimetry and fluorescence probe measurements (Katsikas and Quinn, 1982) indicating that long-chain but not short-chain quinone is in the central low viscosity (approx. 0.9 poise, Schneider et al., 1985) region of the membrane bilayer. It was also inferred that the long-chain quinone resides in the center of the lipid bilayer in concentrated domains where the lifetime of the semiquinone would be small, because short-chain UQ_1 incorporated into liposomes, but not long-chain UQ_{10}, could be reduced to the semiquinone state by externally added reductant (Futami and Hauska, 1979). Analysis of quinone motion and the quinone–lipid system by nuclear magnetic resonance (NMR) spectroscopy has thus far not come to a consensus on the details of the quinone–lipid interaction (Stidham et al., 1984; Ulrich et al., 1985; Cornell et al., 1987).

The pool of lipophilic long-chain quinones has the common property in mitochondria, chloroplasts, and bacteria of mediating electron transport and H^+ translocation from the reducing end of the chain, dehydrogenases in

5.4 The Quinone Connection Across the Center of the Membrane

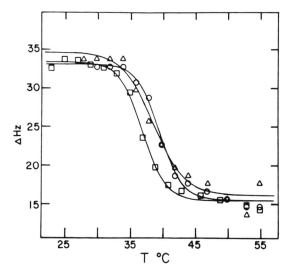

Figure 5.4. Lipid phase transition of dipalmitoylphosphatidylcholine (DPPC) vesicles containing 0.05 mol UQ: mol lipid. The phase transition was measured by ^{1}H-NMR, using the temperature dependence of the chemical shift of the inner choline methyl head group protons. (○) DPPC vesicles, $T_m = 40°C$, (△) vesicles plus UQ_{10}, $T_m = 39°C$; (□) vesicles plus UQ_2, $T_m = 36°C$. (Reprinted with permission from Ulrich et al., *Biochemistry.* Copyright 1985 American Chemical Society.)

respiratory chains or the photosynthetic PS II, to the cytochrome b–c_1 and b_6–f complexes (Figs. 5.2, 5.3, and 5.5). The rate-limiting step is quinol (QH_2) oxidation and is approximately 10 ms in oxygenic photosynthesis.

5.4 The Quinone Connection Across the Center of the Membrane

The special ability of long-chain quinones to reconstitute energy transduction in respiratory and photosynthetic membranes correlates with the tendency of these quinones to reside near the center of the membrane bilayer. The large amount of lipophilic quinones in energy transducing membranes, as well as the protonation–deprotonation associated with reduction–oxidation reactions of quinones (Chap. 2), led to the proposal in the original formulation of the chemiosmotic hypothesis for a central role of quinone in the mechanism of H^+ translocation. It was envisioned that transmembrane "flip-flop" of these quinones, together with reduction at one surface and oxidation at the other (Mitchell, 1966), could provide a mechanism for the transmembrane movement of protons associated with energy transduction (Chap. 3). The bulky structure of all *trans*-UQ_{10} (Fig. 5.6A) or PQ_9 (Fig. 5.6B) made it difficult, however, to conceive of a completely transmembrane "flip-flop" capability of

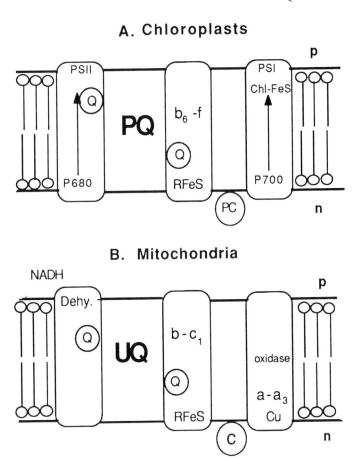

Figure 5.5. Common features of the quinone connection through a central pool (PQ or UQ) in (A) photosynthetic and (B) respiratory electron transport systems. The quinone binding sites are denoted by Q on the PS II or dehydrogenase complexes on the reducing side of the pool, and on the b_6–f or b–c_1 complexes on the oxidizing side. PC, plastocyanin; RFeS, Rieske iron-sulfur protein; C, cytochrome c.

these quinones. X-ray-diffraction analysis of the photosynthetic reaction center shows the quinone bound to the reaction center to be oriented with most of the long isoprenoid chain extended in the plane of the membrane bilayer (see Fig. 6.19C).

The occurrence of transmembrane "flip-flop" of UQ-10 was tested in synthetic DPPC (16-carbon fatty acyl chain, dipalmitoylphosphatidylcholine) liposomes through the ability of UQ-10 incorporated in the bilayer to mediate electron transfer from reductant (short-range borohydride, BH_4^-, or longer range, dithionite) to ferricyanide encapsulated in the membrane interior. The long-chain UQ-10 was the least efficient mediator of reduction (Fig. 5.7A) of the encapsulated ferricyanide by BH_4^- (Fig. 5.7B), showing that

5.4 The Quinone Connection Across the Center of the Membrane

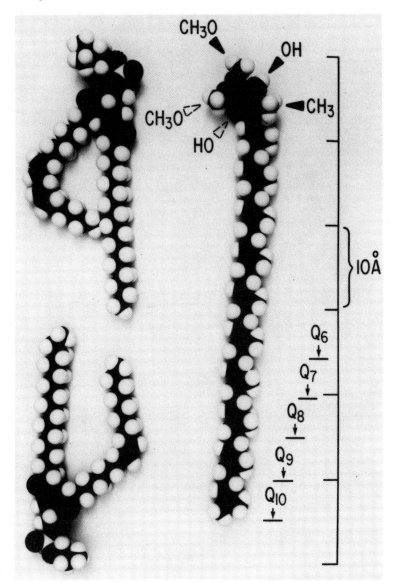

Figure 5.6. (A) UQ_{10} in the *trans*-conformation, compared to phospholipids: 18:0, 18:2, phosphatidylcholine (upper left) and 16:0, 18:1 phosphatidylethanolamine (lower left) (Trumpower, 1981a). (B) Space-filling models of the chloroplast glycolipid monogalactosyldiacylglycerol and PQ_A (PQ_9) (Millner and Barber, 1984).

MGDG

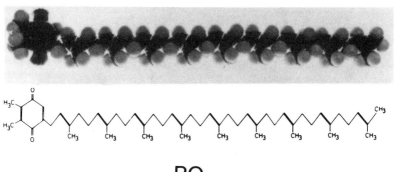

PQ_A

B

Figure 5.6 (*continued*)

Figure 5.7. (A) Quinone-facilitated reduction of ferricyanide encapsulated in UQ-DPPC-liposomes by sodium borohydride or dithionite (Girvin, 1985; *cf.*, Ulrich et al., 1985). (B) Time dependence of reduction by borohydride (BH_4^-) of ferricyanide encapsulated in UQ-liposomes. Borohydride (3 min, pH 8.8) added to DPPC liposomes containing 0.01 mol % UQ_n. (C) Time dependence of reduction by dithionite of ferricyanide encapsulated in UQ-liposomes. The order of quinone efficacy for ferricyanide reduction by dithionite, which can reduce UQ_{10} over longer distances, was opposite to that obtained with BH_4^- (Girvin, 1985; also see Ulrich et al., 1985; Futami et al., 1979).

5.4 The Quinone Connection Across the Center of the Membrane

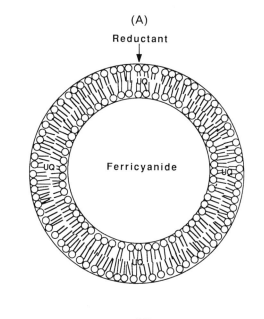

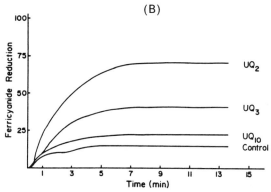

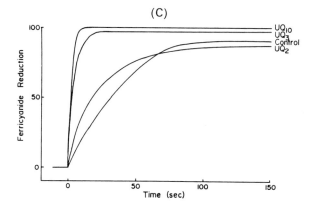

the UQ-10 does not "flip-flop" efficiently across a membrane bilayer of physiological thickness (≥ 16 carbon fatty acid chains). The ability of the longer range reductant dithionite, which transfers electrons instead of hydride ions, to act much more rapidly and to use UQ-10 more efficiently than shorter chain quinones (Fig. 5.7C) proved that the UQ-10 functions within the bilayer, and that it catalyzed transmembrane electron transport. Its ability to act efficiently with dithionite but not BH_4^- is consistent with a position near the bilayer center removed from the aqueous interface. For membranes of physiological thickness, it was inferred from the data of Fig. 5.7B and C that long-chain mobile quinones facilitate H^+ conduction across the central region of the membrane bilayer (compare with Fig. 5.5) on either side of which reside quinone binding sites in proteins (e.g., NADH or succinate dehydrogenase, bacterial or photosystem II reaction centers, or the cytochrome $b-c_1$ and b_6-f complexes) that complete the transmembrane H^+ pathway.

5.5 Quinone Lateral Mobility

The low viscosity of mitochondria and chloroplast membranes due to their highly unsaturated lipid chains (Tables 3.5 and 3.6), and the particularly fluid nature of the center of the bilayer, allows the possibility of lateral diffusion of intrinsic proteins and the quinone pool (Hackenbrock, 1981; Barber, 1983). Values of the lateral diffusion coefficients of long-chain quinone for the electron transfer chains of mitochondria, chloroplasts, and chromatophores were obtained from (i) measurements of electron transport activities in phospholipid-enriched membranes and (ii) measurement of quinone diffusion coefficients in liposomes (Gupte et al., 1984; Fato et al., 1986; Blackwell et al., 1987). Incorporation of exogenous phospholipid into the mitochondrial inner membrane resulted in an increase in the spacing between integral proteins observed by freeze-fracture electron microscopy (Schneider et al., 1985). The inhibition of electron transport from succinate or NADH by dilution of the electron transport complexes in the lipid would be expected if these reactions do not occur within a single complex, but rather by diffusion and collision. In fact, the rates of electron transport from NADH and succinate to cytochrome c decreased by more than 10-fold when the intrinsic cytochrome concentration was diluted 4-fold by added phospholipid (Schneider et al., 1985).

Data on diffusion constants of quinone in natural and artificial membranes obtained by the techniques of (i) fluorescence recovery after photobleaching ("FRAP") of labeled quinone and (ii) collisional quenching of fluorescence by quinone (Fato et al., 1986; Blackwell et al., 1987) are summarized in Table 5.3. The diffusion coefficient, D, required to allow a net displacement over a distance, d, on a two-dimensional surface, in time t, is:

$$d = \sqrt{\bar{x}^2 + \bar{y}^2} = \sqrt{4Dt} \qquad (1)$$

5.5 Quinone Lateral Mobility

Table 5.3. Diffusion coefficients (D_{UQ}) of ubiquinone in mitochondrial membranes and liposomes

Quinone	D_{UQ} (cm^2 s^{-1})	Technique	Reference
Q_oC_{10}NBDHA[a]		FRAP	Gupte et al. (1984)
Megamitochondria[b]	$3.7\,(\pm 1) \times 10^{-9}$		
Fused mitochondria[c]	$2.3\,(\pm 1.1) \times 10^{-9}$		
PQ-9, liposomes	2×10^{-7}	Quenching of pyrene fluorescence	Blackwell et al. (1987)

[a] Fluorescent ubiquinone analog with the 10-carbon side chain of N-4-nitrobenz-2-oxa-1,3-diazole-hexanoic acid.
[b] Giant mitochondria ($\sim 15\ \mu$m diameter, made from mice fed with the drug cuprizone).
[c] Large mitochondria made by fusion of mitochondrial inner membranes in the presence of high (10 mM) Ca^{2+} (Gupte et al., 1984).

(Villars and Benedek, 1974). Thus, in 10 ms a molecule with a diffusion coefficient of 10^{-8} cm$^2 \cdot$s^{-1} will undergo a root mean square net displacement of 2,000 Å from its starting point.

The value of 2×10^{-7} cm^2 s^{-1} for the diffusion coefficient obtained by quinone quenching of fluorescence in liposomes is about two orders of magnitude larger than the values determined for the fluorescent decylubiquinone analog in giant mitochondria by the FRAP technique (Fig. 5.8). The discrepancy may be due to (i) the distance scale of the FRAP diffusion measurement being larger ($\sim 1\ \mu$m) than that of the fluorescence quenching (≥ 10 Å) or (ii) different locations of the ubiquinone analog and the natural ubiquinone in the membrane corresponding to different effective viscosities for translational diffusion. As noted above, the shorter chain analog has a tendency to reside near the lipid head group, whereas much of the long-chain UQ-10 preferentially resides near the center of the bilayer, where the effective viscosity of the membrane is much smaller than that in the lipid head group region. (iii) The absence of protein in the liposomes may allow a larger average mean free path of the quinone between collisions.

The lateral diffusion coefficients of the other fluorescent-labeled electron transport components measured by the FRAP technique are summarized in Table 5.4. Except for ubiquinone, a fluorescent-labeled antibody was attached to each component. Diffusion of the peripheral cytochrome c is dependent on ionic strength, the positively charged protein binding more tightly to the negatively charged membrane surface at low ionic strength (Overfield and Wraight, 1980). The membrane surface is thus electrically "sticky" for cytochrome c. The higher diffusion of plastocyanin, also a peripheral protein, may be due to its acidic (negatively charged) nature. The minimum diffusion coefficient that is needed for an electron transfer model based completely on random diffusion has been estimated to be 1–2×10^{-9} cm^2 s^{-1} (Hochman et

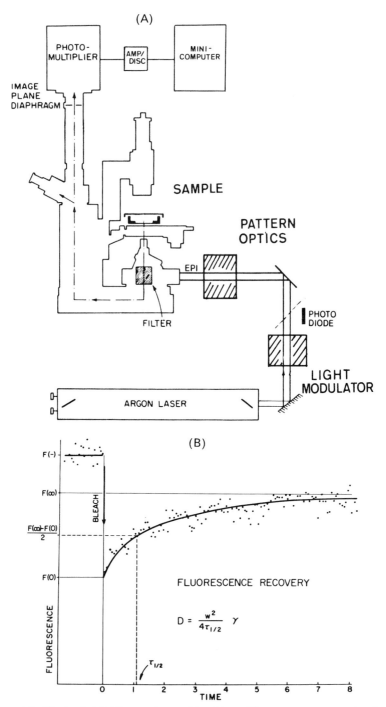

Figure 5.8. The basic FRAP experiment: Bleaching of fluorescent macromolecules in a spot ($\geq 1\ \mu m^2$) on a membrane surface. (A) Optics and electronics of a FRAP system. Argon laser is focused in small spot on microscope stage. Bleaching of spot detected by photomultiplier tube. (Drawing by D. Axelrod, after Axelrod et al., 1976.) (B) Time course, in seconds, of fluorescence recovery due to diffusion of neighboring molecules into the bleached spot after the laser flash.

5.5 Quinone Lateral Mobility

Table 5.4. Lateral diffusion coefficients for mitochondrial inner membrane and thylakoid

Membrane	Component	Ionic condition	D, cm^2/s	Reference
Megamitochondria	Ubiquinone	0.3 mM Buffer	3.7 ($\pm$1.0) $\times$ 10^{-9}	Gupte et al. (1984)
Megamitochondria	Cytochrome b–c_1	10 mM Buffer/50 mM NaCl	3.6 ($\pm$2.7) $\times$ 10^{-10}	Gupte et al. (1984)
Fused mitochondria	Ubiquinone	0.3 mM Buffer	2.3 ($\pm$1.1) $\times$ 10^{-9}	Gupte et al. (1984)
Fused mitochondria	Cytochrome b–c_1	0.3 mM Buffer	4.8 ($\pm$1.9) $\times$ 10^{-10}	Gupte et al. (1984)
Fused mitochondria	Cytochrome b–c_1	10 mM Buffer/50 mM NaCl	4.2 ($\pm$1.9) $\times$ 10^{-10}	Gupte et al. (1984)
Fused mitochondria	Cytochrome oxidase	10 mM Buffer/50 mM NaCl	3.6 ($\pm$1.7) $\times$ 10^{-10}	Gupte et al. (1984)
Megamitochondria	Cytochrome oxidase	42 mM Mannitol/8 mM HEPES	1.5 ($\pm$1.0) $\times$ 10^{-10}	Hochman et al. (1985)
Fused mitochondria	Cytochrome c	0.3 mM Buffer	5.9 ($\pm$2.2) $\times$ 10^{-11}	Gupte et al. (1984)
Fused mitochondria	Cytochrome c	10 mM Buffer	2.7 ($\pm$1.3) $\times$ 10^{-10}	Gupte et al. (1984)
Fused mitochondria	Cytochrome c	25 mM Buffer	1.9 ($\pm$0.6) $\times$ 10^{-9}	Gupte et al. (1984)
Megamitochondria	Cytochrome c	42 mM Mannitol/8 mM HEPES	3.5 ($\pm$1.5) $\times$ 10^{-10}	Hochman et al. (1985)
Liposomes	Plastocyanin	10 mM Buffer/15 mM NaCl	5 $\times$ 10^{-8}	Fragata et al. (1984)

al., 1985). Most of the entries in Table 5.4 have diffusion coefficients smaller than this threshold value. Only ubiquinone and cytochrome c at high ionic strength exceed it. Because cytochrome c forms stable complexes with other proteins (Chap. 4), quinone binds to proteins (section 5.7), and the cytochrome $b-c_1$ and oxidase complexes can be isolated together as a super complex from the bacterium *P. dentrificans* (Berry and Trumpower, 1985), it seems likely that physiological electron transfer occurs both through a process of random diffusion and also within complexes that have a lifetime long enough to allow electron transfer (Hochman et al., 1985).

5.6 The Segregation of Electron Transport Components in Thylakoids Requires Lateral Mobility of Quinone

The ratio of stacked to unstacked (grana to stroma) membrane regions in higher plant chloroplasts is variable (Staehelin, 1986). Qualitatively, there is an inverse relationship between the amount of thylakoid membrane stacking and the intensity of light used for growth. Plants grown under low light have larger diameter grana with more stacked membranes than those grown under high light [e.g., *Zea* mays mesophyll chloroplasts: (i) high light, 61% stacked, average granum diameter, 0.35 μm; (ii) low light, 73%, 0.43 μm (Staehelin, 1986)]. Shade plants of tropical rain forests have an even greater ratio of stacked to unstacked membranes, with grana stacks consisting of as many as 100 lamellae.

The distribution of chloroplast membrane mass between the internal grana and stromal lamellae membranes in spinach chloroplasts is typically 75–85% grana and 15–25% stroma (Fig. 5.9A). A three-dimensional view of the thylakoid membrane architecture shows helically arranged stromal membranes that connect to the cylindrical stacks of discs in the grana (Fig. 5.9B). The light-harvesting and electron transport components associated with PS II and I in these chloroplasts are distributed unevenly between grana and stroma fractions. Immunogold labeling of the chlorophyll-binding CP43 protein closely associated with PS II shows the label to be concentrated in the grana stacks (Fig. 5.9C), whereas gold label associated with antibody to the P700 PS I reaction center seems to be excluded from the grana regions (Fig. 5.9D). Direct contact between stacked containing tetrameric protein particles and unstacked regions containing smaller nontetrameric particles is shown in the freeze-fracture electron micrograph of barley chloroplasts (Fig. 5.9E). The white areas represent infolding of the membranes, which are continuous with the grana disc, also seen as pockets around the stacked region in Fig. 5.9F.

The distribution of PS II and PS I components in grana and stroma has been studied in appressed membranes that are derived from the grana stacks minus the end membranes, and the "stromal membrane" fraction that consists of end membranes from the grana stacks together with the isolated stroma lamellae. Freeze-etch microscopy (Fig. 5.10A and B), along with the thin

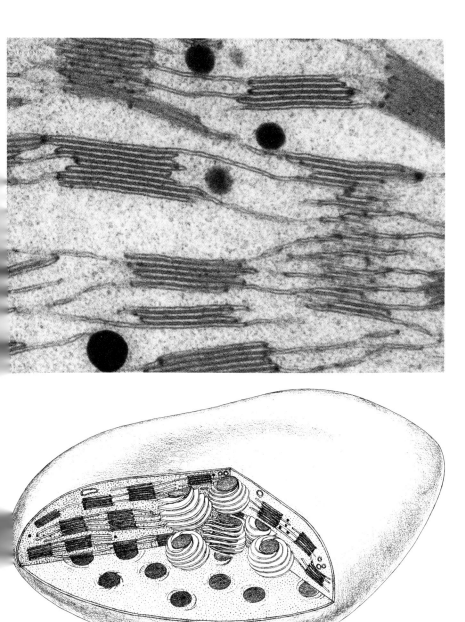

Figure 5.9. (A) Thin section through a portion of a spinach chloroplast showing interconnected appressed grana and unstacked stroma thylakoid membranes. The flattened sac-like structure of individual thylakoids can be noted; × 80,000 (Staehelin, 1986). (B) Diagram of thylakoid membrane architecture based on electron micrographs of thin sections. Grana stacks can be seen to arise from two-dimensional cross-sectioned view of cylindrically shaped stacks of discs that are interconnected; stromal unappressed membranes arise from helically arranged membrane sheets with fret-like connections to the discs (Ort, 1986).

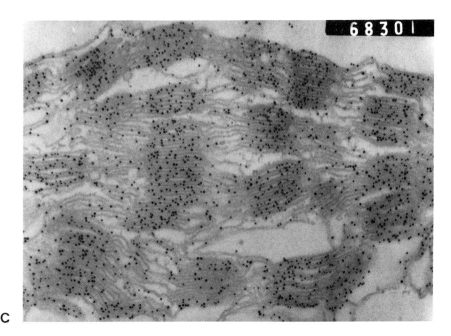

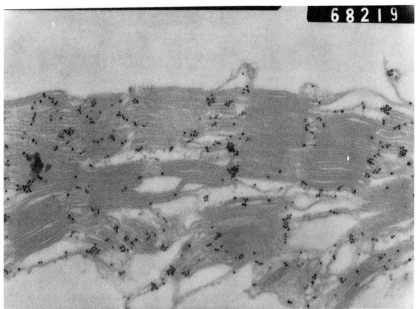

Figure 5.9C,D. Immunogold labeling of (C) chlorophyll binding protein; α-CP43, chlorophyll binding protein, associated with PS II reaction center, and (D) P700 reaction center protein of PS I; from wild-type barley seedlings. × 60,000; reproduced at 60%.

5.6 The Segregation of Electron Transport Components 211

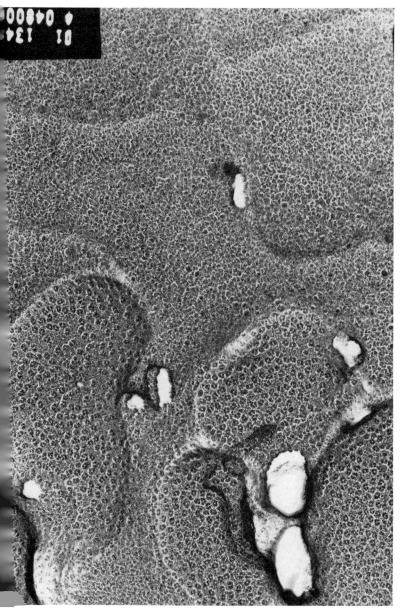

Figure 5.9E,F. E, Freeze-fracture electron micrographs of barley chloroplasts. (E) magnification, ×200,000; reproduced at 64%; 13 mm = 0.1 μm; (F) magnification, ×150,000, reproduced at 80%; 24 mm = 0.2 μm. White areas in (E) represent infolding of the membrane, continuous with grana disc, also seen as pockets around the stacks in (F). (C–F) From D. Simpson; compare with Simpson (1979).

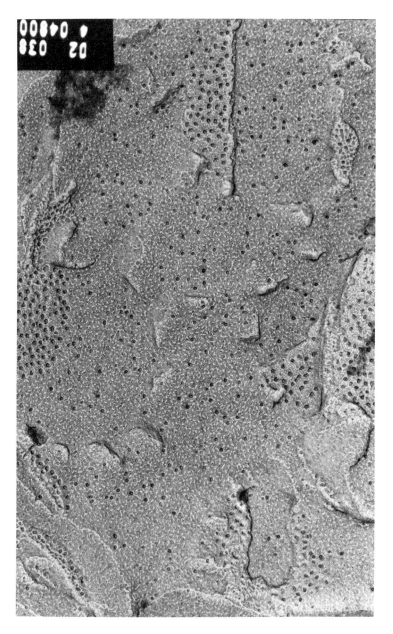

Figure 5.9 F *(continued)*

5.6 The Segregation of Electron Transport Components

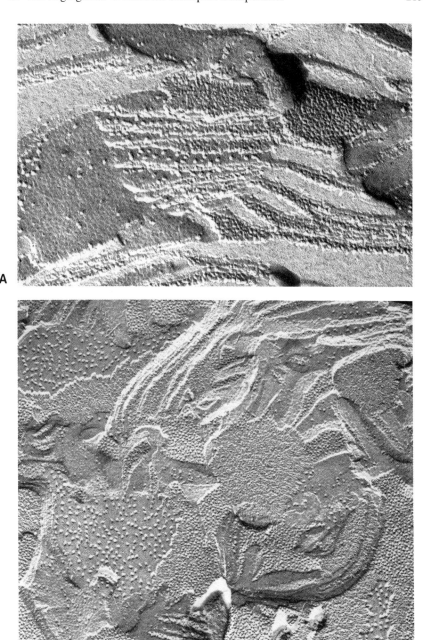

Figure 5.10. (A) Freeze-fracture electron micrograph of a single granum and associated stromal membranes from a pea chloroplast, showing the continuity of the stromal membranes where stroma and grana cross (arrowheads). ×120,000 (reproduced at 90%). (Staehelin, 1986.) (B) Freeze-fracture of pea chloroplasts, in which the angled stroma resemble the blades of a windmill, with the central grana corresponding to the hub. ×65,000 (reproduced at 90%). (Staehelin, 1986.)

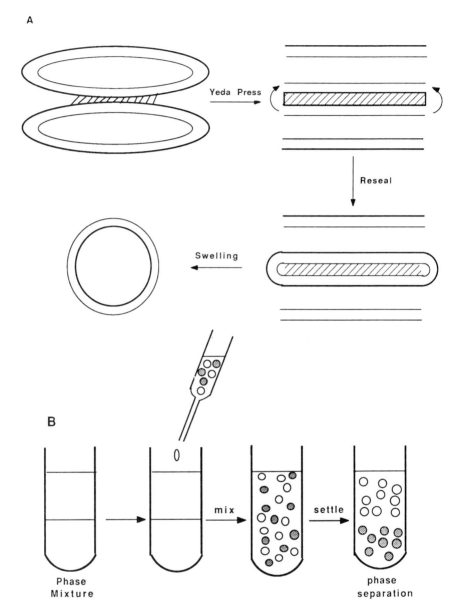

Figure 5.11. Three-step (A–C) procedure for phase separation of inside-out and right-side-out chloroplast membrane vesicles. (A) Depiction of fractionation by high-pressure cell of thylakoid membranes into inside-out appressed and stromal fractions. Dots represent attractive forces between membranes. (Adapted from Andersson et al., 1980.) (B) Partition of right-side out (○) and inside-out (●) thylakoid membranes in a two-phase Dextran T500-PEG 4000 system. (Adapted from Larsson, 1983.) (C) Three centrifugation step separation procedure for isolation from spinach chloroplasts of inside-out and right-side-out thylakoid vesicles in the bottom (B) and top (T) phases.

5.6 The Segregation of Electron Transport Components

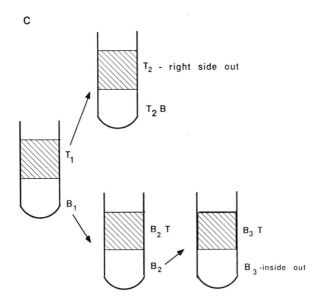

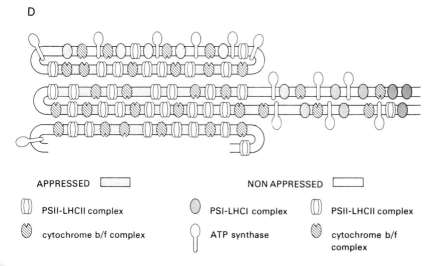

respectively, that are enriched in dextran and PEG (adapted from Larsson, 1983) (D) Proposed distribution of electron transport complexes in thylakoid membranes. (Note the difference in composition of grana and stromal membrane regions.) Preferential illumination of PS II and reduction of the PQ pool leads to a greater extent of phosphorylation of LHC II and net migration from grana to stroma (see Chap. 6.4.2). The ATPase is exclusively in the stromal fraction (Staehelin, 1986). The structural separation of PS II and ATPase argues for delocalization of the H^+ coupling (Table 3.11). (Figure contributed by J. Anderson.)

Table 5.5. Relative distribution of thylakoid membrane components

Component	Appressed (grana) membranes (percent)	Stromal and "end region" membranes (percent)
PS II	75–85	15–25
PS I	<15–30[b]	70–85
	0.67–0.9	0.1–0.33
Chl a/b LHC[a]	More (70–90), dependent on LHC kinase[c]	Less (10–30) dependent on LHC kinase[c]
ATP synthase	0	100

[a] Chlorophyll a/b light-harvesting complex.
[b] Atta-Asafo-Adjei and Dilley (1985).
[c] See Chap. 6.4.2.
From Staehelin (1986).

section pictures of Fig. 5.9A–C, show the grana and stromal membranes, which share a common lumenal space. The initial step in the membrane fractionation between appressed membranes and the stromal lamellae is breakage of the appressed membranes in a pressure cell (Fig. 5.11A), resulting in formation of (i) larger inside-out thylakoid membranes and (ii) the "stromal membrane" fraction that includes end regions of the appressed membranes (Åkerlund and Andersson, 1983). The two membrane populations can be separated in a two-phase system, in which membranes are separated by differences in membrane surface charge and other surface properties related to hydrophobicity (Fig. 5.11B). The inside-out membranes are located in the bottom phase enriched in $>5\%$ Dextran T500 ($\overline{MW} = 5 \times 10^5$), and the right-side-out membranes in the top phase enriched in $>5\%$ polyethylene glycol 4000 ($\overline{MW} = 3{,}100$–$3{,}300$) (Fig. 5.11C). Lateral segregation of PS II and PS I reaction center complexes (Figs. 5.9C, D; Table 5.5) in the ratio PS II:PS I $\simeq 5{:}1$ and $\simeq 1{:}5$ in grana and stroma, respectively (Anderson, 1981; Anderson and Haehnel, 1982; Anderson and Melis, 1983; Atta-Asafo-Adjei and Dilley, 1985; Staehelin, 1986) implies that mobile carriers such as plastoquinone and plastocyanin must diffuse over relatively large distances in order to transfer electrons from PS II to PS I.

PS II has two components, the larger fraction PS IIα consisting of larger (10–17 nm) particles seen in freeze-etch electron microscopy located in the appressed granal membranes and the smaller fraction PS IIβ made of smaller (~ 11 nm) particles located in the stroma. PS I is located in the stromal fraction. It can be seen in the end membranes of the grana where it can readily contact ferredoxin, the electron acceptor for PS I located in the stroma or on the outside of the thylakoid membrane (Fig. 5.11D). The model for nonrandom distribution of PS II and I that requires the shortest distance of carrier

Table 5.6. Calculated root-mean-square distance of diffusion of plastoquinone and plastocyanin during reduction and oxidation; $d = \sqrt{4Dt}$

Component	$t_{1/2}$ (s)	D (cm² s⁻¹)	Distance (Å)
A. Plastoquinone	10^{-2} (o)	3×10^{-9}	1,100
	5×10^{-4} (r)	3×10^{-9}	250
	10^{-2} (o)	2×10^{-7}	10,000
	5×10^{-4} (r)	2×10^{-7}	2,500
B. Plastocyanin	3×10^{-4} (r and o)	5×10^{-8}	750

o, oxidized; r, reduced.

diffusion between PS II and PS I would involve diffusion of quinone and plastocyanin from the center to the periphery of the grana stacks. However, in a high-light plant, there is also presumably a diffusive connection to the stroma lamellae.

The cytochrome b_6–f complexes have a slightly higher density in the grana compared to the stroma fraction (Staehelin, 1986). For 3:2 and 4:1 ratios of grana:stroma areas, approximately two-thirds and 0.8–0.9 of the b_6–f complexes, respectively, would be present in the stacked membranes (Staehelin, 1986).

The distance requirement in this model for the mobile carriers, plastoquinone and plastocyanin, in the plane of the ~0.5 μm diameter grana discs does not pose a serious constraint for a diffusion-limited mechanism, given the current estimates of their diffusion constants (Table 5.6), if the average distance between PS II centers and b_6–f complexes is less than 200 Å (Whitmarsh, 1986). The sufficiency of a diffusive step from cytochrome f to P700, which would be traversed by plastocyanin, is more of a problem if P700 is constrained to be in the edges of the grana stacks and external lamellae, because the characteristic distance for plastocyanin diffusion is 750 Å (Table 5.6B).

5.7 Quinone-Binding Proteins

The protein-bound quinone is a small fraction (5–30%, depending on total pool size) of the total quinone present in the mitochondrial inner membrane and thylakoids. The quinone-proteins have been inferred from the presence of semiquinone free-radical electron paramagnetic resonance (EPR) signals or optical absorbance changes associated with redox events in the membrane, or by covalent labeling of protein components with azido-labeled quinone analogs as photoaffinity probes (Fig. 5.12). The quinol, quinone, and semiquinone compounds whose spectra are shown (Fig. 5.13), and whose thermo-

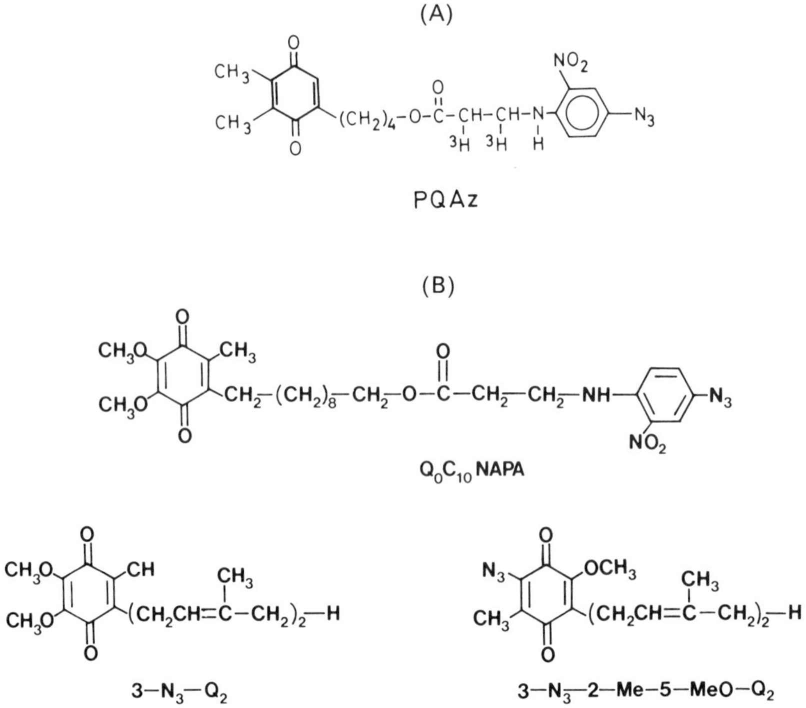

(A)

PQAz

(B)

Q_0C_{10} NAPA

$3-N_3-Q_2$

$3-N_3-2-Me-5-MeO-Q_2$

Figure 5.12. Structure of quinone photoaffinity labels. (A) Plastoquinone azide used to label cytochrome b_6 and the Rieske iron-sulfur protein in the chloroplast b_6-f complex (Oettmeier et al., 1982). (B) Azido- and arylazido-Q derivatives used to label the quinone binding sites of the $b-c_1$ complex of mitochondria and photosynthetic bacterium Rb. sphaeroides, and cytochrome d oxidase of E. coli (Yu and Yu, 1987b). Azido compounds. $R-N_3$, typically absorb light in the near-ultraviolet (e.g., 390 nm) through which they are converted to highly reactive nitrenes $[R-N_3 \xrightarrow{hv} R-N: + N_2]$.

dynamic relationships were discussed in Chap. 2, are important in studies of the photosynthetic reaction centers and the cytochrome $b-c_1$ and b_6-f complexes.

The protein subunits labeled as quinone-binding by the quinone photoaffinity probes or identified by x-ray analysis of the crystallized bacterial reaction center are summarized in Table 5.7.

EPR signals have been detected for the semiquinone species, $Q^{\dot-}$ or QH', for the protein-bound quinones listed in Table 5.8. The $Q_s^{\dot-}$ signal is associated with the M_r 13–15,000 quinone-binding polypeptides of mitochondrial complex II (Table 4.17). Two different ubisemiquinone EPR signals have been associated with the mitochondrial $b-c_1$ complex arising from semiquinone components, $Q_{in}^{\dot-}$ and $Q_{out}^{\dot-}$, on the inner and outer sides (n and p sides,

5.7 Quinone-Binding Proteins

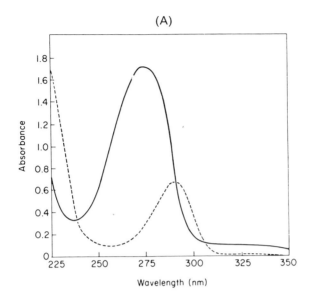

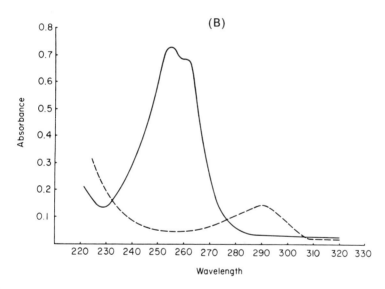

Figure 5.13. (A) and (B) Absorption spectra of UQ-10 and plastoquinone-9, respectively (Barr and Crane, 1971). Oxidized (—); reduced (- - -). (C) Semiquinone anion radical-quinone difference spectra: (a) plastosemiquinone-9 anion minus plastoquinone-9, (b) ubisemiquinone-10 minus UQ-10, (c) ubisemiquinone-0 minus UQ-0 (Bensasson and Land, 1973).

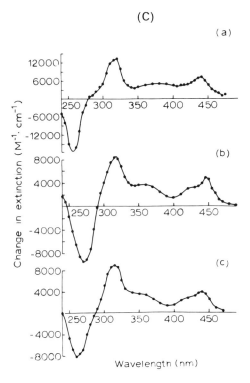

Figure 5.13 (continued)

corresponding to negative and positive $\tilde{\mu}_{H^+}$) of the mitochondrial inner membrane (de Vries et al., 1980, 1981; Ohnishi and Trumpower, 1980). $Q_{in}^{\doteq}$ or $Q_n^{\doteq}$ is a relatively stable signal (0.4/cyt c_1, $K_s = 5 \times 10^{-2}$) that is eliminated by the inhibitor antimycin A (Fig. 5.1C.1) whose site of action is on the matrix or n-side of the membrane (see Table 7.5A). An unstable, antimycin-insensitive signal, $Q_{out}^{\doteq}$ or $Q_p^{\doteq}$, on the cytoplasmic side of the mitochondrial membrane has been detected during oxidant-induced reduction of cytochrome b (Chap. 7). Association of $Q_p^{\doteq}$ and $Q_n^{\doteq}$ with the inhibitor sites on the cytoplasmic and matrix sides of the membrane is implied by the inhibitor 2,3-dimercaptopropanol (British Anti-Lewisite) eliminating both the Rieske center and Q_p^- EPR signals in the presence of oxygen, but not affecting the $Q_{in}^{\doteq}$ signal (de Vries et al., 1981). An antimycin-sensitive ubisemiquinone, $Q_n^{\doteq}$, has also been found in the cytochrome b–c_1 complex of *Rb. sphaeroides* chromatophores (Table 5.8). The EPR properties of semiquinones detected in the respiratory chains of mitochondria and *P. denitrificans*, and the bacterial photosynthetic reaction centers, are summarized in Table 5.8.

5.7 Quinone-Binding Proteins

Table 5.7. Proteins inferred to contain a quinone binding site from quinone photoaffinity labeling experiments

Protein or polypeptide	Function or activity of bound quinones	Reference
Mitochondrial M_r 13-15,000, QP-S	Succinate-ubiquinone reductase	Table 4.16 (see above)
Mitochondrial M_r 37,000 Cyt b and M_r 17,000 of cytochrome $b-c_1$ complex (complex III). Labeling of M_r 17,000 dependent on phospholipid.	Electron donation of UQH_2 to ISp and of $UQ^{\dot{-}}$ to heme b_p of cytochrome b-566.	Yu and Yu (1987b)
Chloroplast M_r 23,000 cyt b_6, M_r 20,000 ISp of cytochrome b_6-f complex. [Different result obtained when quinone labeled on ring with azido group instead of an isoprenoid side chain, and b_6-f complex was depleted of lipid. In this case only M_r 17,000 subunit IV was labeled].	Electron donation of PQH_2 to ISp and of $PQ^{\dot{-}}$ to heme b_p of cytochrome b_6.	Oettmeier (1982); Doyle et. al., (1989); Chap. 7.4
Rb. sphaeroides M_r 43,000 cyt b and M_r 12,000 polypeptide of chromatophore cytochrome $b-c_1$ complex. Labeling increased after phospholipase A_2 digestion indicating lipid involvement. Labeling not decreased by pretreatment with putative quinone analogs HQNO, UHDBT, myxothiazol, or antimycin.	Electron donation of UQH_2 to ISp and/ or of $UQ^{\dot{-}}$ to heme b_p of cytochrome b-566.	Yu and Yu (1987); Chap. 7.4.
E. coli M_r 57,000 α subunit of cytochrome d terminal oxidase.	Acceptor side for binding of UQ_8H_2 to low K_m oxidase of E. coli.	Yang et al (1986); Fig. 3.5B, Table 4.4
PS II reaction center polypeptides.	Interface between secondary electron acceptor of PS II and PQ pool.	Pfister et al. (1981); Steinback et al. (1981)
M and L subunits of bacterial photosynthetic reaction centers of Rps. viridis and Rb. sphaeroides.	Secondary electron acceptor on 200–300 ps time scale; H^+ acceptor.	Marinetti et al. (1979); de Vitry and Diner (1984)

iron-sulfur protein.

Table 5.8. EPR properties of semiquinones detected in bovine mitochondria (Bowyer and Ohnishi, 1985), *Paracoccus denitrificans* (Meinhardt et al., 1987), and chromatophores (Robertson et al., 1984)

Source	Radical species	g Value	Line width (Gauss)	K_s (pH 7)	E_{m7} [$QH_2/Q^{\cdot -}$] (mV)
A. Mitochondria					
1. Succinate → UQ	$Q_s^{\cdot -}$	2.005	12	10	135
2. Cyt b–c_1	$Q_n^{\cdot -}$ (anti A-sens)[a]	2.005	9.5–10	5×10^{-2}	80
	$Q_p^{\cdot -}$	2.005	8.3	Very small	ND
B. *P. denitrificans*					
Cyt b–c_1 complex	$Q_n^{\cdot -}$	2.004	10	2.2×10^{-1}	42 (pH 8.5)
C. Chromatophores					
1. b–c_1 Complex (antimycin-sensitive)	$Q_n^{\cdot -}$	2.005	8	10^{-2} (pH 7.4)	140 (pH 7.4)
2. Reaction center	$Q_A^{\cdot -} Fe^{2+}$	1.84 (peak) 1.68 (Trough)	broad	Very stable	−45 (Q/Q$^{\cdot -}$)
		2.0046 (−Fe)	8.1	—	—
	$Q_A Fe^{2+} Q_B^{\cdot -}$	1.84 (Peak) 1.63 (Trough)	broad	Stable	40 (Q/Q$^{\cdot -}$)

ND, Not determined.

[a] Q_n and Q_p denote quinone binding sites on the sides of the membrane that have relatively negative and positive values of $\tilde{\mu}_{H^+}$, respectively; these are the matrix and cytoplasmic sides of the mitochondrial inner membrane, and the cytoplasmic and periplasmic sides of the bacterial inner membrane.

5.8 Quinone Electron Acceptors in Photosynthetic Reaction Centers

The primary photochemistry leading to charge separation in photosynthesis starts with absorption of a photon leading to the excited state of the reaction center (Chap. 2, section 2.9), from which the electron is transferred to a bacteriopheophytin (BPh) or pheophytin (Ph) [$E_m = -0.6$ V] in ~3 ps in the bacterial and oxygenic photosynthetic systems, respectively. To prevent excited state decay or back-reaction, the electron must be transferred to a higher potential acceptor in a time $<10^{-9}$ s. The acceptor is the bound ubi-, mena-, or plastoquinone, Q_A, whose one-electron reduction generates $Q_A^{\pm}$ (Table 5.8). $Q_A^{\pm}$ cannot be trapped on the first two flashes unless an inhibitor is present because the electrons flow downhill to Q_B (positive ΔE_m, Table 5.8C), as shown in Fig. 5.14. A broad $g = 1.84$ EPR signal arising from a $Q_A^{\pm}$ signal broadened by a neighboring Fe^{2+} atom is induced by illumination and is formed after three light flashes under conditions where doubly reduced Q_B is trapped, thereby blocking the oxidation of $Q_A^{\pm}$ (Fig. 5.14C). When the bound Q_B is oxidized, the electron is transferred from $Q_A^{\pm}$ to Q_B (ox) in 0.1–0.5 ms. The anionic semiquinone, $Q_B^{\pm}$, is the species detected after one light flash (Fig. 5.14A) as a $g = 1.84$ ($g = 1.69$ trough). It is also seen as an absorption band near 320 nm in the light minus dark difference spectrum which resembles the chemically generated semiquinone difference spectrum (Fig. 5.13). Although a Fe^{2+} is situated structurally in the path between Q_A and Q_B in both the bacterial and the plant PS II reaction centers, it does not undergo obligatory redox changes. However, it may facilitate electron transfer to Q_B by catalyzing the required H^+ transfer (Petrouleas and Diner, 1987). Removal of the metal from the bacterial reaction center decreases the transfer rate from Q_A to Q_B and the quantum yield of $Q_A^{\pm}$ formation by a factor of two (Kirmaier et al., 1986).

The mechanism of quinone reduction and the structures involved in quinone binding appear to be similar in bacterial reaction centers and those of the plant and algal PS II. The $Q_B^{\pm}$ semiquinone is stabilized through binding to the L (light) or D1 polypeptide of the bacterial or PS II reaction center, respectively. The E_m values of approximately $+40$ mV for the $Q_B/Q_B^{\pm}$ couple and -40 mV for $Q_B^{\pm}/Q_BH_2$ correspond to a stability constant (Chap. 2) of 20 (Rutherford and Evans, 1980). A second flash results in a second electron transfer through the $Q_A^{\pm}$ species forming the doubly reduced quinol species, Q_BH_2, which is loosely bound and rapidly released ($t_{1/2} < 1$ ms) into the quinone pool, where it is oxidized by the cytochrome b–c_1 or b_6–f complex. Thus, the reduction of Q_B shows a two-flash periodicity, and the Q_B reduction system in both chloroplast and chromatophore systems is called a "two-electron gate" (Wraight, 1982), in which binding of a Q from the pool at the Q_B site and release of Q_BH_2 to the pool take place on a submillisecond time scale (Crofts et al., 1984). The light-induced reduction of the primary and secondary bound quinone acceptors can be summarized as shown in Table 5.9.

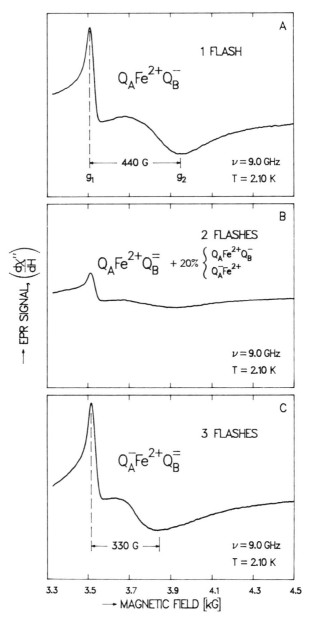

Figure 5.14. EPR signals of Q_A-Fe-Q_B semiquinone species from bacterial reaction centers of the *Rb. sphaeroides* carotenoidless mutant R-26, frozen after 1, 2, or 3 laser flashes, at room temperature in the presence of reduced cytochrome c (Reproduced from Butler et al., the *Biophysical Journal*, 1984, 45, pp 947–973 by copyright permission of the Biophysical Society.) (Also see Okamura et al., 1982).

5.8 Quinone Electron Acceptors in Photosynthetic Reaction Centers

Table 5.9. The two-electron gate in bacterial reaction centers: redox reactions and H^+ uptake of quinone acceptors

	Notes[a]
(1) $Q_A R_A - Q_B R_B \xrightarrow{h\nu} Q_A^{\cdot -} R_A - Q_B R_B$ $\to Q_A R_A - Q_B^{\cdot -} R_B;$	(1) First flash, e^- transfer to Q_A without protonation, with effective $E_m \simeq -150$ mV, e^- on Q_B after 0.5 ms, spectra indicate $Q_B^{\cdot -}$ mostly unprotonated. Fe^{2+} between Q_A and Q_B does not affect e^- transfer, but does alter EPR signal of $Q_A^{\cdot -}$, $Q_B^{\cdot -}$; R_A, R_B, are protonable amino acids associated with Q_A, Q_B.
(2) $Q_A R_A - Q_B^{\cdot -} R_B \xrightarrow{H^+} Q_A R_A Q_B^{\cdot -} R_B H^+$	(2) H^+ bound to R_B after pK shift due to formation of $Q_B^{\cdot -}$.
(3) $Q_A R_A Q_B^{\cdot -} R_B H^+ \xrightarrow{h\nu} Q_A^{\cdot -} R_A Q_B^{\cdot -} R_B H^+$ $\xrightarrow{H^+} Q_A^{\cdot -} R_A H^+ Q_B^{\cdot -} R_B H^+ \to Q_A R_A Q_B H_2 R_B$	(3) Second flash; direct binding of 2 H^+ to Q_B after transient formation of $Q_A^{\cdot -}$ and protonation of R_A.
(4) $Q_A - Q_B H_2 + Q_{pool} \to Q_A - Q_B + Q_{pool} H_2$	(4) QH_2 exchanged to pool for Q bound at Q_B site.

[a] The Fe atom that structurally bridges between Q_A and Q_B is not shown. Its E_m in chloroplasts ($E_{m7.5} = +370$ mV) may be lower than in bacteria, allowing the possibility of turnover in the $Q_A \to Q_B$ transition. The lower potential may result from bicarbonate instead of glutamic acid acting as a fifth ligand to the Fe, and also from the influence of the negative charge on Q_A^-, resulting in $Q_A Fe^{2+} Q_B \to Q_A^{\cdot -} Fe^{2+} Q_B \to Q_A Fe^{3+} Q_B^{2-}$ after one flash. The bicarbonate bound to Fe^{2+} may facilitate the H^+ uptake (Rutherford and Zimmerman, 1984; Renger et al, 1987; Petrouleas and Diner, 1987; Van Rensen et al., 1988).
From Wraight (1982), Maróti and Wraight (1988), McPherson et al. (1988).

The stoichiometry of the proton uptake associated with the quinone reduction shown in Fig. 5.14 is pH-dependent, with maximum values of 0.5 H^+/e^- for PQ_A^- and 0.8 H^+/e^- for $PQ_A Q_B^-$. The reason for values less than unity is that the oxidized states are already partly protonated. The proton uptake associated with semiquinone formation is not attributed to direct protonation of the semiquinone, but rather to the neighboring amino acids (the R groups in Table 5.9, e.g., Glu, Asp, His, Cys, or Tyr), whose pK values shift when the quinones are reduced (Fig. 5.15).

After formation of the quinol, $Q_B H_2$, at the reaction center, it is released into the quinone pool, and an oxidized quinone from the pool will then rebind to the reaction center protein. The quinol–quinone exchange at the Q_B site implies that this side of the reaction center structure must contain some kind of quinone channel. Several inhibitory herbicides [e.g., DCMU, atrazine (Fig. 5.1)] bind to the Q_B protein near the quinone-binding site and thereby block the binding of plastoquinone (Steinback et al., 1981; Velthuys, 1981).

The proton uptake by the quinone acceptor system initiates H^+ transloca-

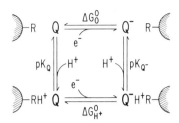

Figure 5.15. Protonation of reaction center amino acid residues linked to quinone reduction (McPherson et al., 1988).

tion across the membrane, and begins with reduction of Q_A and Q_B, which would then cause an upward pK shift of amino acids in the quinone neighborhood. This kind of pK shift in a membrane protein is called a "membrane Bohr effect" because of the conceptual analogy to the deprotonation–protonation reactions linked to the oxygenation–deoxygenation of hemoglobin (Perutz et al., 1980). After the $Q_B H_2$ has been released from the binding site and has entered the quinone pool, the role of the pool quinone is probably to facilitate electron and proton transfer across the hydrophobic membrane core. The redox proteins on either side of the pool provide the pathway for H^+ translocation across the rest of the membrane, and binding or docking sites for quinone on the oxidizing and reducing sides of the mobile quinone pool.

5.9 Quinone-Binding Proteins in Photosynthetic Reaction Centers

Two polypeptides of the bacterial photosynthetic reaction center, called L and M (light and medium size, respectively) are the quinone-binding proteins, defined initially by the quinone binding and photoaffinity probe experiments mentioned above, and then by x-ray diffraction analysis of crystals of the reaction center from the bacterium *Rps. viridis* (Deisenhofer et al., 1985; Michel et al., 1986a,b; Michel and Deisenhofer, 1988b) and *Rb. sphaeroides* (Chang et al., 1986; Allen et al., 1987a,b; Yeates et al., 1987), for which the quinone species are, respectively, MQ-9 (Q_A) and UQ-9 (Q_B) and UQ-10 (Q_A and Q_B). The other prosthetic groups in the reaction center are four bacteriochlorophylls, two bacteriopheophytins, and one iron atom. The L and M polypeptides each span the membrane through five α helices (Fig. 5.16A; Table 5.10). Both quinones are present in the *Rb. sphaeroides* crystals, but one quinone is bound tightly (Q_A) and the other much more loosely (Q_B) to the L-M polypeptides in *Rps. viridis*, because most of the Q_B is lost from crystals of the latter. The two quinones are separated by a distance of 18–19 Å between ring centers in *Rb. sphaeroides*, between which is located a high-spin Fe^{2+} atom that is not liganded directly to either Q_A or Q_B, but shares with them two pairs of histidine ligands. The distances between the bound quinone(s) and the pigment and redox prosthetic groups of the reaction center are tabulated for *Rb. sphaeroides* and *Rps. viridis* (Table 5.11), as is the relative orientation of the aromatic rings.

5.9 Quinone-Binding Proteins in Photosynthetic Reaction Centers 227

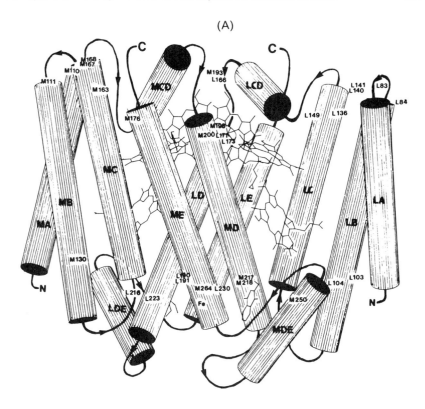

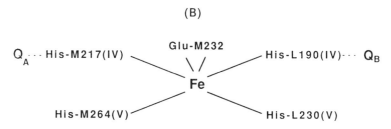

Figure 5.16. (A) Representation of the L and M subunit α-helices in the reaction center of the bacterium *Rps. viridis*. The view is parallel to the plane of the membrane. The special pair bacteriochlorophylls are at the interface of the L and M subunits between the D (IV) and E (V) α-helices, the accessory bacteriochlorophylls below α-helices LCD and MCD, and bacteriopheophytins near the C (III) α-helices. The binding site for Q_A (menaquinone in *Rps. viridis*) is between the MDC and MD α-helices and that for Q_B between the LDE and LD α-helices. The location of amino acids conserved between all L and M subunits and the D1 and D2 proteins is shown (Reprinted with permission from Michel and Deisenhofer, *Biochemistry*, 27, pp. 1–7. Copyright 1988 American Chemical Society.) (B) Amino acid ligands of the Fe^{2+}, Q_A, and Q_B in the bacterial reaction center. The bidentate ligand Glu-M232 is in the loop joining helices IV and V (Table 5.12).

Table 5.10. Transmembrane α-helices of the L and M polypeptides of the *Rps. viridis* reaction center

Subunit	Size[a]	Helix[b]	Spanning peptide
L	273 aa; MW = 30,571	I(A)	32–55
		II(B)	84–112
		III(C)	115–140
		IV(D)	170–199
		V(E)	225–251
M	323 aa; MW = 35,902	I(A)	46–75
		II(B)	110–139
		III(C)	142–167
		IV(D)	197–225
		V(E)	259–285

[a] The M_r values of the heavy (H), medium (M), and light (L) polypeptide subunits of the reaction center, determined by sodium dodecyl sulfate-polyacrylamide gel electrophoresis (SDS-PAGE) are 35,000, 28,000, and 24,000; as is often the case with hydrophobic polypeptides that do not bind SDS well, these M_r values are different from the true MW. The largest anomaly is for the H subunit which turned out to be the lightest, with 258 amino acids, and MW = 28,500.

[b] The notation A–E for the five membrane-spanning α-helices of the L and M polypeptides bears no relation to the notation Q_A and Q_B for the primary and secondary bound quinone acceptors.

From Deisenhofer et al. (1985b), Michel et al. (1986a).

Table 5.11. Center–center distance and relative orientation between Q_A, Q_B, and other pigment and redox prosthetic groups in reaction centers of *Rb. sphaeroides* and *Rps. viridis*

	Distance (Å)[a]		Angle between ring normals (degrees)[b]	
	Rb. sph.	Rps. vir.	Rb. sph.	Rps. vir
$Q_A:BPhe_A$	13.0	14.0	35	35
$Q_B:BPhe_B$	15.0	NA	35	NA
$Q_A:Q_B$	18.5	NA	20	NA
$Q_A:Fe$	11.0[c]	9.0	NA	NA
$Q_B:Fe$	8.0	NA	NA	NA

NA, not applicable; no Q_B in *Rps. viridis*.

[a] Estimated error of coordinates, ±0.4 Å.
[b] Estimated error for BPhe rings, ±6°, for quinones ±15°.
[c] Distance from Fe to center between two carbonyl oxygens.

From Allen et al. (1987b).

5.9 Quinone-Binding Proteins in Photosynthetic Reaction Centers

The nonheme iron between Q_A and Q_B is at an approximate center of symmetry between Q_A and Q_B in the *Rps. viridis* and *Rb. sphaeroides* structures, with a line connecting the center of the special pair and the Fe defining a twofold axis of rotational symmetry. The Fe atom is coordinated by a total of four histidine residues, L190, L230, M217, and M264 in *Rps. viridis*, and one glutamic acid, M232 (Fig. 5.16B). Of the four histidines, two belong to helices IV(D) and V(E) of the L-, and two are located on helices IV and V of the M-subunit. One histidine from the M polypeptide, M217 in *viridis* and M219 in *sphaeroides*, is involved in the binding of the quinone acceptor, Q_A, and one from the L polypeptide, L190, binds Q_B in *sphaeroides* and *viridis*, although Q_B is not seen in the latter crystals. Q_A and Q_B are positioned in the membrane with their isoprenoid chains kinked between $n = 1$ and 3 and most of the remaining chain lying parallel to the plane of the bilayer (Fig. 5.17).

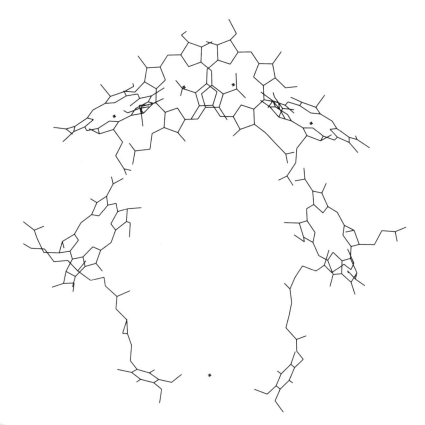

Figure 5.17. Arrangement of pigment and quinone groups in *Rps. sphaeroides* reaction center showing orientation of quinones and quinone isoprenoid chains. M or Q_B side is on left and L or Q_A side is on right. The phytyl tails of the bacteriochlorophylls and pheophytins are not shown. The atoms of the quinone isoprenoid chains were removed after C-24. (Drawing by C.H. Chang and M. Schiffer using partially refined coordinates, 1988; cf., Chang et al, 1986).

Q_BH_2 must be able to leave the Q_B site and move through the reaction center protein complex to the quinone pool near the center of the bilayer (Table 5.9). It is replaced at the Q_B site by a quinone from the pool. The 45–50 carbon ubiquinone isoprenoid chains require a channel in the protein for access and egress of the Q/QH_2. The channel formed inside the protein by the Q_B-isoprenoid is large enough for the Q_B quinone ring to pass back and forth without structural change in the protein, but a structure change would be required for dissociation of Q_A in *Rb. sphaeroides* (Allen et al., 1988).

The electron transport rate changes enormously, by a factor of $\sim 10^6$, as the electron passes through Q_A. The 200-ps transfer to Q_A succeeds in trapping the free energy of the reaction center excited state in the electron transport chain. Most of the positive free energy change ($\Delta E_m = 0.6$ V) in the electron transfer reaction from bacteriopheophytin to Q_A is necessary for a high quantum yield (≥ 0.9). However, removal of Q_A and substitution of quinone analogues with an E_m that is 0.15 V more negative than the native Q_A does not significantly affect the quantum yield (Gunner and Dutton, 1988). The transition in electron transfer rate constants at Q_A allows the E_m of Q_A/Q_A^- to be titrated through the fluorescence yield (Cramer and Butler, 1969). The oxidation of Q_A^- and transfer to Q_B, forming Q_B^-, is coupled to an amino acid pK shift and H^+ uptake, processes associated with local protein conformational changes that characteristically take place on a millisecond or submillisecond time scale. Thus, Q_A can be considered to be a dividing point in the redox reactions between the trapping of the electron from the excited state for photochemistry and the use of this energy for formation of the $\Delta\tilde{\mu}_{H^+}$. A major question at the present time with respect to structure and function of the bacterial reaction center is why electron transfer to BPhe and Q occurs predominantly on one side, the L- or Q_A-side (Figs. 5.16, 5.17, Chap. 6, section 6.6).

Significant sequence identity exists between the reaction center L and M subunits from the different bacterial sources (Williams et al., 1983, 1984; Youvan et al., 1984; Michel et al., 1986a), and between L–M and the chloroplast-encoded D1 and D2 proteins (L and M corresponding to D1 and D2), as shown in Table 5.12. The D1 polypeptide, whose translation is stimulated by light, has the highest turnover and synthesis rate of any thylakoid polypeptide. The frequency of sequence identity is not great in the first quarter of the sequences and the last 40 residues, but becomes significant starting at positions L83, M110, and D1,D2-109. His residues L173 and M200 of *Rps. viridis*, corresponding to D1-198 and D2-198, are the ligands to the Mg atom of the special pair BChl. His-L153 and His-M180 coordinate the Mg of the auxilary BChl, without obvious analogous residues in D1–D2. Five residue coordinate the iron. The (Fe–Q_A)- and (Fe–Q_B)-binding His residues in D2 and D1, respectively, corresponding to His-M217 and His-L190 in Table 5.12 are D2-215 and D1-215. The other two Fe-binding His residues in the O_2 evolving reaction center predicted by the sequence alignment of Table 5.1 would be D2-269 analogous to His-M264 in Fig. 5.16B, and D1-272 analogous

5.9 Quinone-Binding Proteins in Photosynthetic Reaction Centers

Table 5.12. Comparison of the amino acid sequences of the L and M subunits from *Rps. viridis* (LV, MV, Michel et al., 1986a), *Rb. capsulatus* (LC, MC; Youvan et al., 1984), and *Rb. sphaeroides* (LS, MS) reaction centers (Williams et al., 1983, 1984) and the D1 and D2 proteins from spinach chloroplasts ((Zurawski et al., 1982; Hirschberg and McIntosh, 1983; Rochaix et al., 1984; Alt et al., 1984)

```
(numbering: LV)     1                  21                 41
                    ↓                  ↓                  ↓    [I]
LV                  ALLSFER-KYRVRGGTLIGGDL-FDFWV--GPYFVGFFGVSAIFFIF    43
LC                  ALLSFER-KYRVPGGTLIGGSL-FDFWV--GPFYVGFFGVTTIFFAT    43
LS                  ALLSFER-KYRVPGGTLVGGNL-FDFWV--GPFYVGFFGVATFFFAA    43
D1                MTAILERRESESLWGRF-CNWITSTENRLYI-GWFGVLMIPTLLTATSVFIIAFIA   54
D2                MTIAVGKFTKD-EKDLFDSMDDWLRRDRFVFVGWSGLLLFPCAYFALGGWFTGTTF    56
MS     AEYQNIFSNVQVRGPADLGMTEDVNLANRSGVGPFSTL-LGWF-GNAQL--GPIYLGSLGVLSLFSGL  64
MC     AEYQNFFNQVQVAGAPEMGLKEDVDTFERTPAGMFNIL--GWM-GNAQI--GPIYLGIAGTVSLAFGA  63
MV     ADYQTIYTQIQARGPHITVSGEWGDNDRVGKPFYSYWL--GKI-GDAQI--GPIYLGASGIAAFAFGS  63
(numbering: MV)          21                 41                 61
                         ↓                  ↓                  ↓

        [I]     61                     81
        ____    ↓                      ↓         [II]
                                                 ____
LV      LGVSLIGYAASQGPTWDP-------FAISINPPDLKYGL-GAAP---------LLEGGFWQAITVCA   93
LC      LGFLLILWGAAMQGTWNP-------QLISIFPPPVENGL-NVAA---------LDKGGLWQVITVCA   93
LS      LGIILIAWSAVLQGTWNP-------QLISVYPPALEYGL-GGAP---------LAKGGLWQIITICA   93
D1      APPVDIDGI-REPVSGS-LLYGNNIISGAIIPTSAAIGLHFYPIWEA-ASVDEWLYNGGPYELIVLHF  119
D2      VTSWYTHGLASSYLEGCNFLTAAVSTP----ANSLAHSLLLLWGPEAQGDFTRWCQLGGLWAFVALHG  119
MS      MWFFTIGIWFWYQAGWNPAVFLRDLFFFSLEPPAPEYGLSFAAP---------LKEGGLWLIASFFM   122
MC      AWFFTIGVWYWYQAGFDPFIFMRDLFFFSLEPPPAEYGL-AIAP---------LKQGGVWQIASLFM   120
MV      TAILIILFNMAAEVHFDPLQFFRQFFWLGLYPPKAQYGM-GIPP---------LHDGGWWLMAGLFM   120
                          81                    101

        [II]  101                  121                  141
        ____  ↓                    ↓       [III]        ↓
                                           _____                    *ᵃ
LV      LGAFISWMLREVEISRKLGIGWHVPLAFCVPIFMFCVLQVFRPLLLGSWGHAFPYGILSHLDWVNNFG  161
LC      TGAFCSWALREVEICRKLGIGFHIPVAFSMAIFAYLTLVVIRPMMMGSWGYAFPYGIWTHLDWVSNTG  161
LS      TGAFVSWALREVEICRKLGIGYHIPFAFAFAILAYLTLVLFRPVMMGAWGYAFPYGIWTHLDWVSNTG  161
D1      LLGVACYMGREWELSFRLGMRPWIAVAYSAPVAAATAVFLIYPIGQGSFSDGMPLGISGTFNFMIVFQ  187
D2      AFALIGFMLRQFELARSVQLRPYNAIAFSGPIAVFVSVFLIYPLGQSGWFFAPSFGVAAIFRFILFFQ  187
MS      FVAVWSWWGRTYLRAQALGMGKHTAWAFLSAIWLWMVLGFIRPILMGSWSEAVPYGIFSHLDWTNNFS  190
MC      AISVIAWWVRVYTRADQLGMGKHMAWAFLSAIWLWSVLGFWRPILMGSWSVAPPYGIFSHLDWTNQFS  188
MV      TLSLGSWWIRVYSRARALGLGTHIAWNFAAAIFFVLCIGCIHPTLVGSWSEGVPFGIWPHIDWLTAFS  188
                                                                    BChl
        121                   141                  161

              181    [IV]       201
              ↓    *      *     ↓
LV      YQYLNWHYNPGHMSSVSFLFVNAMALGLHGGLILSVANPGDG-------DKVKTAEH---------EN  213
LC      YTYGNFHYNPFHMLGISLFFTTAWALAMHGALVLSAANPVKG-------KTMRTPDH---------ED  213
LS      YTYGNFHYNPAHMIAISFFFTNALALALHGALVLSAANPEKG-------KEMRTPDH---------ED  213
D1      AEH-NILMHPFHMLGVAGVFGGSLFSAMHGSLVTSSLIRETTENESA--NEGYRFGQEEEETYNIVAAH  252
D2      GFH-NWTLNPFHMMGVAGVLGAALLCAIHGATVENTLF-EDGDGANT--FRAFNPTQAEEETYSMVTAN  251
MS      LVHGNLFYNPFHGLSIAFLYGSALLFAMHGATILAVSRFGGERELEQIADRGTAAER--------AA  249
MC      LDHGNLFYNPFHGLSIAALYGSALLFAMHGATILAVTRPGGERELEQIVDRGTASER--------AA  247
MV      IRYGNFYYCPWHGFSIGFAYGCGLLFAAHGATILAVARFGGDREIEQITDRGTAVER--------AA  247
        ↑                  ↑              ↑    *
        sp.p.              Fe             Fe    241
```

Table 5.12 (continued)

```
           221              [V] 241                 261
            |                    |                   |
LV   QYFRD--VVGYS-IGALSIHRLGLFLASNIFLTGAFGTIASGPFWTRGWPEWWGWWLDIPFWS*     273
LC   TYFRD--LMGYS-VGTLGIHRLGLLLALNAVFWSACCMLVSGTIYFDLWSDWWYWWVNMPFWADMAGG 278
LS   TFFRD--LVGYS-IGTLGIHRLGLLLSLSAVFFSALCMIITGTIWFDQWVDWWQWWVKLPWWANIPGG 278
D1   GYFGRLIFQYASFNNSRSLHFFLAAWPVVGIWFTALGISTMAFNLNGFNFN-QSVVDSQGRVINTWAD 319
D2   RFWSQ-IFGVA-FSNKRWLHFFMLFVPVTGLWMSALGVVGLALNLRAYDFVSQEIRAAEDPEFETFYT 317
MS   LFWRW--TMGFN-ATMEGIHRWAIWMAVLVTLTGGIGILLSGTVV-DNWYVWGQNHGMAPLN*      307
MC   LFWRW--TMGFN-ATMEGIHRWAIWMAVMVTLTGGIGILLSGTVV-DNWYVWAQVHGYAPVTP*     306
MV   LFWRW--TIGFN-ATIESVHRWGWFFSLMVMVSASVGILLTGTFV-DNWYLWCVKHGAAPDYPAYLPA 311
                           ↑  ↑                                ↑
                          261 Fe                    281       301
```

```
LC   ING*                                        281
LS   ING*                                        281
D1   IINRANLGMEVMHE--RNAHNFPLDLAAIEAPSTNG*        353
D2   KNILLNEGIRAWMAAQDQPHEN-LIFPEEVLPRGNAL*       353
MS
MV   TPDPASLPGAPK*                               323
```

[a] His-M182 (*sphaeroides*) coordinates to the Mg of the auxiliary BChl on the Q_B side, but His-L-153 probably does not coordinate to the Mg of the BChl on the Q_A side (Yeates et al., 1988). The position of the transmembrane helices of the *Rps. viridis* reaction center is indicated by bars above the L subunits and below the M. The histidine ligands, H, to the Mg atoms of the two BChl of the reaction center in membrane-spanning peptide IV (special pair, sp.p.), to the auxiliary BChl in the bacterial sequences, and to the nonheme iron (Fe) are indicated (*), as is the Glu-M234 ligand to the Fe (*). (Adapted from Michel and Deisenhofer, 1988a, with addition of LS and MS from *Rps. sphaeroides*)

to His-L230 (Fig. 5.18A) Glu-M232 provides the fifth ligand to the Fe in the bacteria, again without an obvious corresponding residue in D2.

A five-helix model for the D1 and D2 polypeptides seems likely from (i) sequence homology with L and M polypeptides (Deisenhofer et al., 1985b; Trebst, 1986) and a similar distribution of hydrophobic residues, and (ii) experimental topographical analysis using antibodies to the peptide loops predicted to be exposed on the two sides of the membrane (Sayre et al., 1986). From these data, a five-helix model of the folding pattern of the D1 polypeptide in the membrane similar to those of L and M has been derived (Fig. 5.18B) and a hypothesis proposed for coordination of Q_A and Q_B by amino acids of the D2 and D1 polypeptides, respectively (Fig. 5.18A). As noted above, the D1 polypeptide, or "Q_B protein," is known to be the binding side of quinone analog inhibitors, such as DCMU and atrazine. The D1 polypeptide has also been called the "herbicide-binding protein." It is also intimately involved in binding the manganese needed for O_2 evolution on the lumen side of the membrane. The correct insertion of the D1 polypeptide into the membrane and binding of manganese for the oxygen-evolving complex is known to require prior cleavage of a peptide segment from the COOH-terminal end of a "pre-D1" protein (Diner et al., 1988).

The quinone and herbicide binding region in the D1 polypeptide, expanded in Fig. 5.18B, can be defined by amino acid mutation sites that conve

5.9 Quinone-Binding Proteins in Photosynthetic Reaction Centers   233

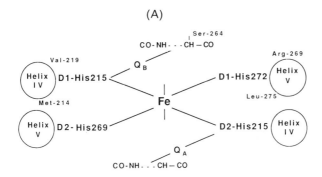

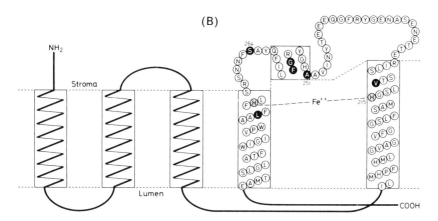

Figure 5.18. (A) Proposed binding of quinone-iron acceptor complex by D_1 and D_2 polypeptides of PS II based on analogy with the bacterial reaction centers. The amino acid changes in herbicide-resistant mutants are indicated. (Adapted from Trebst, 1986.) Amino acids Ser-264, Phe-255, Ala-251, and Val-219 are known sites of mutations conferring herbicide resistance that are grouped together in this model on one side of the membrane. Bicarbonate may replace Glu as the fifth ligand to the Fe. (B) Orientation of the D1 polypeptide in the thylakoid membrane showing amino acid residues (in black) whose mutation leads to herbicide resistance (Rochaix and Erickson, 1988). Polypeptide orientation based on studies of Trebst (1986), and Sayre et al. (1986) using antibodies made to peptides predicted to be in external aqueous phase. (C) Representation of quinone (a) and inhibitor (b, c) binding sites in *Rb. sphaeroides*. The sites are made of a pocket lined with residues from part of the D and E helices of the L subunit and the loop joining the helices. The suggested positions of UQ_0 and the inhibitors terbutryn and o-phenanthroline are shown in a, b, and c, respectively (Paddock et al., 1988. Reprinted by permission of Kluwer Academic Publishers).

(C)

(a)

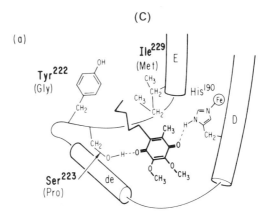

(b)

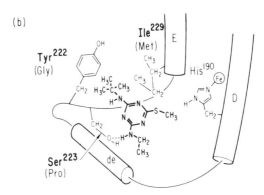

(c)

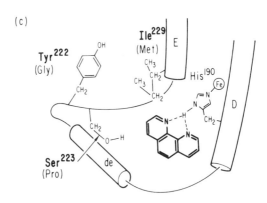

Figure 5.18 (*continued*)

5.10 Summary

Table 5.13. Properties of chloroplast herbicide-resistant mutants

Organism	Relative resistance[a]				Mutation	$Q_A \rightarrow Q_B$ electron transfer
	A	D	B	M		
Chlamydomonas reinhardtii	2	17	1	—	V219 → I	Normal
	25	5	—	1,000	A251 → V	Slow
	15	0.6	1	—	F255 → Y	Normal
	15	3	10	—	G256 → D	Slow[b]
	100	20	250	—	S264 → A	Slow
	1	5	4.5	—	L275 → F	Normal[b]
Amaranthus hybridus and *Solanum nigrum*	1,000				S264 → G	Slow
Anacystis nidulans	10	100			S264 → A	Slow

[a] A, atrazine; D, DCMU; B, bromacil; M, metribuzin. Resistance levels are relative to wild type.
[b] Unpublished results of Rochaix and Erickson.
From Rochaix and Erickson (1988).

herbicide resistance. The region of herbicide, and presumably quinone, binding, occurs in a large peripheral peptide loop linking transmembrane helices IV(D) and V(E) on the stromal side of the membrane. The binding of herbicides at the Q_B binding site can be probed through their perturbation of the EPR signal of Fe between Q_B, Q_A in the ferric (+3) state (Diner and Petrouleas, 1987). A representation of the quinone and herbicide binding site in *Rb. sphaeroides*, based in the crystal structure of the reaction center, is shown in Fig. 5.18C. The properties of the herbicide-resistant mutants are summarized in Table 5.13. The insertion of innocuous herbicide-resistant genes into the chloroplasts of crop plants is one of the long-range goals of plant molecular biology.

5.10 Summary

Long (isoprenoid)-chain quinones are present at high concentration in energy-transducing membranes where they act as mobile carriers of electrons and protons in the bilayer phase and at electron transfer branch points. The branched pathway to the alternate cytochrome oxidase in plant mitochondria does not generate a $\Delta\tilde{\mu}_{H^+}$ or ATP, but instead releases heat energy that protects flower-bearing tissue and volatilizes insect attractants. The quinone located near the center of the membrane bilayer has the common property in mitochondria, chloroplasts, and bacteria of mediating electron transport and H^+ translocation from the reducing end of the chain, dehydrogenases in respiratory chains and the chloroplast photosystem II, to the cytochrome b–c_1 and b_6–f complexes. The rate-limiting step is quinol oxidation and is approximately 10 ms in oxygenic photosynthesis. The quinone connection across membranes of physiological thickness does not involve transmembrane "flip-flop," but the long-chain quinone probably facilitates conduction across the central region of the membrane bilayer, on either side of which reside quinone-binding sites in the intrinsic redox proteins. The lateral mobility of quinone in the membrane bilayer is characterized by a diffusion coefficient, $D \simeq 10^{-9}-10^{-7}$ $cm^2 \cdot s^{-1}$, in fused or megamitochondria, that can be measured by fluorescence recovery after photobleaching (FRAP) of a fluorescent quinone analog, allowing diffusion over an rms distance of $\sim 1,100$ Å in 10 ms. The large cytochrome $b-c_1$ and oxidase complexes also have measureable diffusion coefficients ($D \simeq 10^{-11}-10^{-10}$ $cm^2 \cdot s^{-1}$).

Lateral segregation of PS II and I in the thylakoid membrane can be demonstrated by the different composition of inside-out and right-side-out membranes derived, respectively, from the appressed grana and stroma-grana endregions. The segregation implies that lateral mobility of plastoquinone is needed for redox communication between PS II and b_6-f complexes. Plastocyanin must also be able to diffuse over a large distance in order for the b_6-f complex to transfer electrons to the PS I reaction center. The diffusible quinone interacts on its reducing and oxidizing side with quinone-binding polypeptides that have been identified by covalent labeling of polypeptides using azido-labeled quinone analogs as photoaffinity probes. Two bound quinones, Q_A and Q_B, serve as electron acceptors in the photosynthetic reaction centers. In the bacterial reaction center, where structural data are more complete, the electron excited by a single flash is transferred from bacteriopheophytin to the bound quinone, Q_A, in ~ 200 ps. The electron transfer rate decreases by a factor of 10^6 as the electron is transferred from Q_A to Q_B ($\Delta E_m \simeq +85$ mV), forming the stable $Q_B^{\dot-}$ on a time scale of 0.1–0.5 ms, as a neighboring amino acid is protonated. The slower electron transport rate is probably associated with pK and conformation changes necessary for protonation–deprotonation events in the quinone binding protein complex. After a second flash, the fully reduced and protonated quinol, Q_BH_2, is released from the D1 polypeptide into the quinone pool, and is replaced at the Q_B binding site by another quinone. The two-flash periodicity in Q_B reduction leads to a description of this process as the "two-electron gate."

The structure of the quinone-binding sites in bacterial reaction centers is presently known at a resolution of 2.3–3.0 Å from x-ray diffraction and structure analysis of the reaction center complexes from the purple bacteria *Rps. viridis* and *Rb. sphaeroides*. The quinone species are MQ-9 (Q_A) and UQ-9 (Q_B) in *viridis* and UQ-10 (Q_A and Q_B) in *sphaeroides*. An Fe^{2+} atom between Q_A and Q_B, separated by ~ 18 Å, is on an axis of symmetry through the (BChl)$_2$ special pair dimer. The Fe^{2+} is not a direct ligand to Q_A-Q_B, but shares with them two pairs of histidine residues (L190, L230, M217, and M264 in *viridis*). His M-217 (*viridis*), or M-219 (*sphaeroides*), is also involved in binding Q_A, and His L-190 in binding Q_B (*sphaeroides*). Q_B is generally not retained in *viridis* crystals. Appreciable amino acid sequence identity exists between the bacterial L and chloroplast D1 subunits, and between M and D2. Histidine residues L-173 and M-200, corresponding to D1-198 and D2-198, are the amino acid ligands to the Mg atom of the reaction center BChl special pair. D2-215 and D1-215 are the (Q_A–Fe)- and (Q_B–Fe)-binding residues analogous to *viridis* M-217 and L-190, and D2-269 and D1-272 are analogous to M-264 and L-230. These sequence identities imply that the structure of the PS II reaction center will turn out to be similar to that of the bacteria, although PS II differs in containing additional small polypeptides including those of cytochrome b-559. It has been shown by protein topographical analysis that the D1 polypeptide, like L and M, has five membrane-spanning peptide helices. The Q_B-binding domain on the D1 polypeptide can be localized through mutations to quinone analog inhibitors and herbicides to a peripheral peptide loop linking the D and E trans-membrane helices on the stromal side of the membrane. One of the long-range goals of plant molecular biology is the insertion of innocuous herbicide-resistant genus coding for the D1 protein into the chloroplasts of crop plants.

Problems

58. Use Eq. 1 and the information on diffusion constants to check the characteristic distance of diffusion during each of the half-times listed in Table 5.6.
59. By what factor will the root-mean-square distance of diffusion of plastocyanin that occurs in 0.3 ms change if its diffusion constant were altered by a factor of two (e.g., by a change of internal pH or ionic strength)?
60. If the 60 mV increase in E_{m7} (150 mV compared to +90 mV in the unbound quinone pool) of the Q/QH_2 couple at the chromatophore Q_n binding site is attributed to the relative binding constants of the two redox species (Robertson et al., 1984), what is the relative binding constant of QH_2 compared to Q at this site?
61. Diffusion of spherical objects: The translational (D_T) and rotational (D_R) diffusion constants of a sphere are inversely proportional to the radius and volume, respectively, i.e., D_T (cm$^2 \cdot$s^{-1}) = $\mathbf{k}T/6\pi\eta a$ and D_R (radians$^2 \cdot$s^{-1}) = $\mathbf{k}T/8\pi\eta a^3$, where $\mathbf{k}$ is the Boltzmann constant, T the absolute temperature ($\mathbf{k}T = 4 \times 10^{-14}$ ergs at 300 K), and η = the viscosity in poise. (a) Calculate the expected diffusion constants of a spherical protein of 120,000 molecular weight ($a = 40$ Å) in a membrane with a viscosity = 1 poise (poise is unit of viscosity in cgs units, with dimension of g cm$^{-1} \cdot$s^{-1}; viscosity of H_2O at room temperature = 0.01 poise). (b) How does the value of D_T compare to the FRAP determination of the cytochrome b–c_1 complex? (c) What is the expected ratio of the value of D_T (or D_R) in the membrane vs. that measured in H_2O?

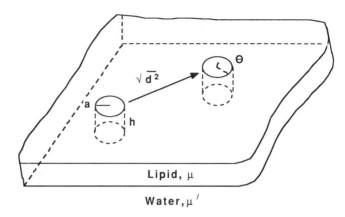

Figure 5.19. Model for cylindrical particle embedded in a lipid bilayer membrane bounded by aqueous phases on both sides. (Based on Saffman and Delbrück, 1975.)

62. Diffusion of cylinders (Fig. 5.19): Calculate (i) the translational diffusion constant and (ii) net distance moved in 10 ms of a "cylindrical" ($h/a = 3$; a = radius; $h = 50$ Å) protein (Saffman and Delbrück, 1975), partly submerged from the aqueous medium of viscosity $\mu' = 0.01$ poise into a membrane of viscosity $\mu = 1$ poise. (b) If the diffusion constant, $D_T = (\mathbf{k}T/4\pi\mu h)[\log \mu h/\mu' a - \gamma]$, where γ = Euler's constant = 0.5772, what is the rotational diffusion constant $[D_R \text{ (radians}^2 \cdot \text{s}^{-1})] = \mathbf{k}T/4\pi\eta a^2 h$ of the object completely submerged in the membrane? Note: for translational motion in a plane and rotation about an axis, $\overline{d}^2 = 4D_T t$; $\overline{\theta}^2 = 2D_R t$, where $\overline{d}^2$ and $\overline{\theta}^2$ are the mean square displacement and angular rotation in time t.
63. Calculate the activation energy for the electron transfer from BPh to Q_A according to Eq. 25 (Chap. 2) if $\Delta G° = -0.6F$ ($\Delta E_m \simeq +0.6$ V) for the transfer, and the reorganizational energy, $\lambda \simeq +0.6F$ ($\lambda = 0.6 \pm 0.1$ V, Gunner and Dutton, 1988; 0.64 V, electron transfer from the reaction center BChl dimer to Q_A, Feher et al., 1988). F is the Faraday constant.

Chapter 6
Photosynthesis: Photons to Protons

6.1 Light Energy Transfer

Most of the chlorophyll or bacteriochlorophyll molecules in the photosynthetic organelle serve as an antenna for light gathering. Approximately one chlorophyll molecule out of 500, and one bacteriochlorophyll molecule out of 50–100, is connected to the electron transport chain through special reaction center (bacterio)chlorophylls that trap excitation energy because they can transfer electrons with high efficiency on a picosecond time scale from the excited state to an acceptor. The theory of light energy transfer is central to photosynthesis, and also important for topographical studies of distance mapping between sites in macromolecules labelled with fluorescent probes.

Absorption of light energy, which occurs on a time scale of $\sim 10^{-15}$ s, can lead to coherent excitation of many (N) pigment molecules whose excited energy state ψ^* can be written as a linear combination of the excited states ϕ_j^* of the N individual (j) pigment molecules i.e.,

$$\psi^* = \sum_{j=1}^{N} c_j \phi_j^*, \tag{1}$$

where the c_j values are weighting constants determined by the extent to which the different ϕ_j^* share in the excitation. This coherent sharing of the excitation energy by two or more identical molecules is called an exciton interaction (Pearlstein, 1982). After a time ≤ 1 ps the excitation will tend to lose its coherence. The theory of light energy transfer in the latter case, and also between unlike molecules, can be described by the theory of resonance energy transfer (Förster, 1959). Because of its relevance to both photosynthesis and the mapping of distances in macromolecules, this chapter will concentrate on the theory of energy transfer by the resonance mechanism. Because the mechanism involves transfer of energy from the excited state of the pigment molecules, this state can also give rise to fluorescence. Following is a brief description of the mechanism of fluorescence.

Fluorescence and energy transfer occur on a time scale (10^{-10}–10^{-8} s) that is slow compared to the time of light absorption ($\sim 10^{-15}$ s) and also slow compared to the period of nuclear vibrations. The tendency of fluorescence to be shifted to longer wavelengths (lower energies) relative to the energy of the absorbed light is shown most readily for a two-atom (diatomic) molecule when the potential energy of the molecule is plotted as a function of the separation, r, between the two nuclei of the molecule [Fig. 6.1; note the resemblance to the nuclear potential energy diagrams utilized for discussion of electron transfer mechanisms (Figs. 2.10 and 2.11)]. The features of this model for fluorescence are: (i) the energy increases when the nuclei are very close together (small r) because of nuclear repulsion, and also increases at very large values of r because of the attraction of the intervening negatively charged electrons. The equilibrium distance, r_2, in the excited electronic state is generally larger than that, r_1, of the ground state. (ii) A series of vibrational energy levels are drawn in the well of the curves for the ground and first excited (*) electronic states. Substates associated with rotational levels are not shown. The distance between a vibrational level and the potential energy function is equal to the kinetic energy associated with the vibration, so that a vibrational level has zero kinetic energy associated with it at the position where it intersects the potential energy curve. The energy gap between vibrational levels is typically 2–3 kcal/mol, whereas that between electronic levels is on the order of 30–100 kcal/mol. (iii) At low temperatures the molecule in the ground state tends to be found at position r_1, in the lowest vibrational level. (iv) Upward and downward arrows represent absorptive and emissive transitions, respectively. (v) All transitions except those from the ground vibrational state tend to occur between the end points of the vibrational levels where the kinetic energy is zero and the molecule therefore spends the most time. (One can think of the oscillation of a pendulum about its equilibrium position. The two positions it occupies for the longest time are those at the extremes of its

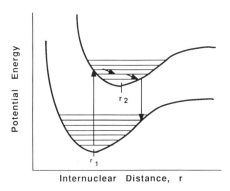

Figure 6.1. Energy of ground (bottom curve) and first excited (upper curve) states of a diatomic molecule as a function of internuclear distance. The equilibrium positions of these states are at r_1 and r_2. The characteristic times required for (i) the upward absorption transition and (ii) relaxation to the bottom vibrational level of the first excited state and fluorescence are 10^{-15} and 10^{-9} s, respectively (Herzberg, 1950; Turro, 1974).

6.1 Light Energy Transfer

motion.) (vi) After absorption of a quantum in $\sim 10^{-15}$ s, the first excited electronic state (*) is also in an excited vibrational state. Because the time for vibrational transitions is $10^{-11}-10^{-12}$ s, but the characteristic time for the transition between electronic states resulting in fluorescence is about 10^{-9} s, the molecule has time to relax to the lowest vibrational state before fluorescence occurs. The fluorescence will then occur from the lowest vibrational level of the excited electronic state. (vii) Thus, the energy of the emitted photon will be smaller, and the wavelength longer (red-shifted; "Stokes shift") because of: (a) energy loss to vibration and heat in the excited state, and (b) the shift of the equilibrium position of the excited state potential energy function to r_2 (Fig. 6.1).

6.1.1 Mechanism of Resonance Energy Transfer

Light energy transfer requires resonance or matching of the emissive transitions of a donor molecule and the absorptive transitions of an acceptor. This can be depicted as shown in Fig. 6.2 if the energy levels are drawn independently of internuclear distance.

The dipole–dipole interaction that is responsible for resonance energy transfer is analogous to the interaction responsible for the attractive Van der Waals forces between atoms and molecules, for which the electrostatic interaction between two neutral atoms is zero in the first approximation. However, the electron clouds fluctuate in their motion around the atoms so as to create

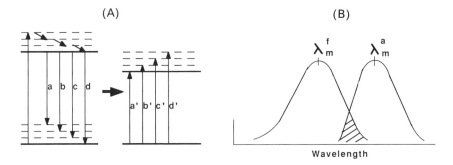

Figure 6.2. (A) Relative energy levels of donor (top) and acceptor (bottom) molecules participating in resonance energy transfer. The energy change "a" corresponding to fluorescence emission of the donor from the lowest vibrational state is drawn equal to the absorptive transition "a'," "b" to "b'," etc. The probability of light energy transfer is thus proportional to the overlap between the emission spectrum of the donor molecule and the absorption spectrum of the acceptor (hatched area below). (B) Overlap (hatched area) of fluorescence emission spectrum of the energy donor $[f(\lambda)]$ and absorption spectrum $[\varepsilon(\lambda)]$ of the acceptor that is necessary for resonance energy transfer from donor to acceptor. The wavelengths λ_m^f and λ_m^a correspond, respectively, to the fluorescence emission peak of the donor and the absorption peak of the acceptor.

electrical dipoles that fluctuate. Such an electrical dipole produces an electrical field that can polarize a neighboring atom, in which it creates a dipole moment by induction (Bohm, 1951). The energy of this dipole–dipole interaction results from the interaction of the electric field of the first atom with the dipole of the second, or that of the field of the second atom with the dipole of the first. In resonance energy transfer, absorption of light by the donor molecule (D) produces an excited state with a lifetime of $\sim 10^{-9}$ s having a dipole moment in the excited state that is usually much larger than that in the ground state. This results in an induced dipole in some properly oriented neighboring atoms that can act as energy acceptors. At the level of quantized or discrete energy states, the creation of the dipole moment and the excited state of the acceptor must be coupled to energetic decay of the donor from the excited state to the ground state (Fig. 6.2A). The presence of the energy acceptor adds another pathway for the decay of the excited state of the donor. The excited state of the donor molecule has a certain natural rate of decay, $k_F°$, due to fluorescence (and heat). If energy transfer also occurs with rate constant k_T from the donor excited state, then the rate of decay of the excited state is increased to $(k_F° + k_T)$, which can be monitored by a correspondingly faster rate of decay of fluorescence.

An expression for resonance energy transfer can be developed as follows: (i) The energy of interaction between donor and acceptor depends critically on the distance of separation. It varies as the inverse sixth power of the distance, R, of separation, i.e., as R^{-6}. This follows from the same kind of derivation for dipole–dipole interactions that is used for describing Van der Waals forces (Bohm, 1951). (ii) The rate constant for energy transfer, k_T, is proportional to (a) the energy of interaction, U_{DA}, between donor and acceptor, and (b) to the rate constant for fluorescence emission in the absence of energy transfer, $k_F°$, because both processes originate from the same excited state (Fig. 6.2). Therefore,

$$k_T \propto (U_{DA} \cdot k_F°);$$

since $U_{DA} \propto R^{-6}$,

$$k_T \propto (k_F°/R^6),$$

or

$$k_T = k_F° \frac{R_o^6}{R^6}, \qquad (2)$$

where the constant of proportionality is written as R_o^6. This constant of proportionality must, in fact, have the dimensions of distance to the sixth power in order that the dimensionality be the same on both sides of the equation. It can be seen from Eq. 2 that the rate constant for energy transfer equals the rate constant for fluorescence ($k_T = k_F°$) when $R = R_o$. (iii) The rate constant for energy transfer, k_T, can be written as the inverse of a characteristic time for transfer,

$$k_T = \frac{1}{\tau_T}. \tag{3}$$

Similarly, the rate constant for fluorescence in the absence of energy transfer is

$$k_F^\circ = 1/\tau_F^\circ. \tag{3a}$$

The rate constant, k_F, for the decay of the excited state, detected through decay of fluorescence, in the presence of energy transfer, is

$$k_F = (k_T + k_F^\circ) = 1/\tau_F. \tag{3b}$$

(iv) The energy from the excited state can decay through two branches, fluorescence and energy transfer. The yield, ϕ_F or ϕ_T, of fluorescence or energy transfer, the fraction of energy going through each branch or the branching ratio, is given by the ratio of the rate constants:

$$\phi_F = \frac{k_F^\circ}{k_F^\circ + k_T}, \tag{4}$$

and

$$\phi_T = \frac{k_T}{k_F^\circ + k_T}. \tag{5}$$

The fraction of energy going to all branches must add to unity, so:

$$\phi_F + \phi_T = 1. \tag{6}$$

The yield or efficiency of energy transfer, ϕ_T, can also be expressed in terms of the lifetimes, τ, of fluorescence:

$$\phi_T = \frac{k_T}{k_T + k_F^\circ} = \frac{(k_T + k_F^\circ) - k_F^\circ}{k_T + k_F^\circ},$$

and from Eqs. 3a and b

$$= 1 - \frac{k_F^\circ}{k_F^\circ + k_T} = 1 - \frac{\tau_F}{\tau_F^\circ}. \tag{7}$$

ϕ_T is thus equal to one minus the ratio of the lifetimes of the excited state, detected by fluorescence, in the presence and absence of energy transfer.

6.2 Use of Energy Transfer as a Spectroscopic Ruler

The distance, R, between energy donor and acceptor for which the yield or probability of energy transfer is 1/2 (efficiency of transfer = 50%) is a characteristic distance. $\phi_T = 1/2$ when $k_T = k_F^\circ$ and $\tau_F = \tau_F^\circ/2$ (Eqs. 5 and 7). From Eq. 2, $k_T = k_F^\circ$ when $R = R_o$. That is, when loss of excitation energy from the excited state proceeds with equal rate along the two branches, the rate constant for energy transfer is equal to that for fluorescence in the absence of energy transfer. Thus, the constant R_o is the characteristic distance between donor and acceptor for which the transfer efficiency = 50%.

R_o can be evaluated from spectroscopic quantities as (Förster, 1959):

$$R_o = (9.79 \times 10^3)(J\theta^2 \phi_F^\circ n^{-4})^{1/6}, \tag{8}$$

in Å, where ϕ_F° is the quantum yield of fluorescence in the absence of energy transfer, n is the refractive index of the medium, J is a normalized integral of the overlap between the spectra of the corrected donor fluorescence emission and acceptor absorption (Fig. 6.2B) so that:

$$J = \int_{\lambda_1}^{\lambda_2} f_D(\lambda)\varepsilon(\lambda)\lambda^4 \, d\lambda \bigg/ \int_{\lambda_1}^{\lambda_2} f(\lambda) \, d\lambda, \tag{8a}$$

θ^2 is an orientation factor that takes into account the relative alignment of the emission dipole of the donor and the absorption dipole of the acceptor, and $\theta^2 = 2/3$ for random orientation of donor and acceptor.

Example 6.1. Because of the inverse 6th power of distance (R^{-6}) dependence of the rate of energy transfer on distance, the efficiency decreases rapidly as R is increased beyond R_o. For example, what is the efficiency of energy transfer at a distance, $R = 2R_o$, between donor and acceptor?

Using $k_T = k_F^\circ(R_o/R)^6$, $k_T = k_F^\circ(R_o/2R_o)^6 = k_F^\circ(1/2)^6 = k_F^\circ/64$, and eqn. 6.5 for the efficiency of energy transfer, ϕ_T:

$$\phi_T = \frac{k_T}{k_F^\circ + k_T} = \frac{k_F^\circ/64}{k_F^\circ + k_F^\circ/64} = \frac{1}{65}$$

The $(1/R)^6$ dependence of resonance energy transfer has been checked by constructing chemical spectroscopic rulers (Fig. 6.3). Resonance energy transfer was measured from the α-naphthyl donor to a dansyl group, separated by oligomers of poly-L-proline (Stryer and Haugland, 1967). The distance between donor and acceptor was varied from 12 to 46 Å. The absorption and emission spectra of the donor and acceptor are shown in Fig. 6.4A and B. The uncertainty in the estimated distances between the centers of the donor and

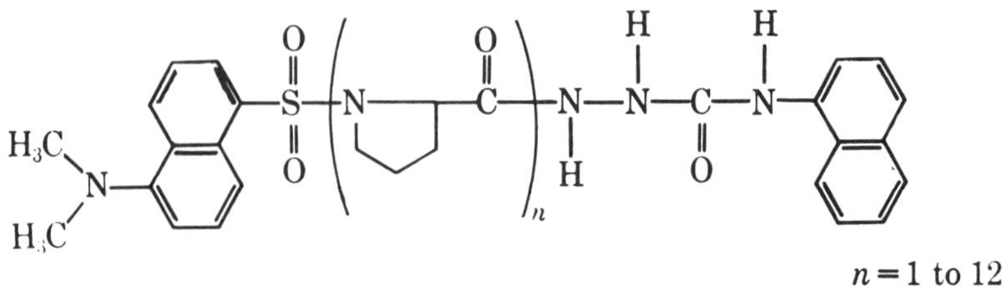

Figure 6.3. A model system [dansyl-(L-prolyl)$_n$-α-naphthyl] for measurement of the distance dependence of resonance energy transfer (Stryer and Haugland, 1967).

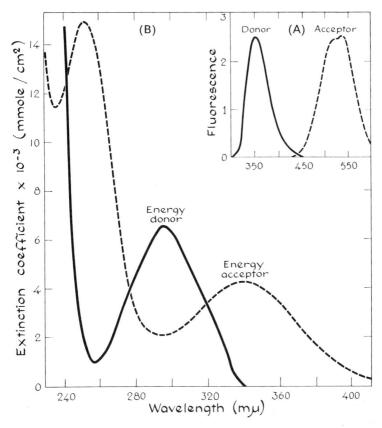

Figure 6.4. Absorption (A) and emission spectra (B) in ethanol of the energy donor [1-acetyl-4-(1-naphthyl)semicarbazide] and acceptor [dansyl-L-prolyl-hydrazide] (Stryer and Haugland, 1967).

acceptor chromophores is approximately ± 3 Å. The efficiency of energy transfer as a function of the center–center distance is plotted in Fig. 6.5 for dansyl-(L-prolyl)$_N$-α-naphthyl, with α-naphthyl and dansyl groups separated by defined distances ranging from 12 to 46 Å. The energy transfer is 50% efficient for a distance of 34.6 Å. The solid line drawn through the points corresponds to the transfer efficiency derived from $k_T = k_F°(34.6/R)^6$ and $\phi_T = k_T/(k_T + k_F°)$.

The characteristic R_o value for resonance energy transfer between tryptophan and NADH is approximately 25 Å. Tryptophan is the dominant natural fluorophore in proteins because its extinction coefficient at 280 nm $\varepsilon_{mM} \simeq 5 \times 10^3$) is approximately four times that of tyrosine. Energy transfer from tryptophan to bound NADH is convenient to study because their emission and absorption spectra have maxima that are typically 325–340 nm and 340 nm, respectively.

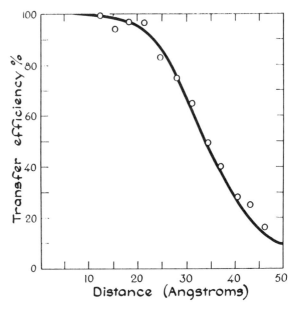

Figure 6.5. Efficiency of energy transfer in dansyl-(L-prolyl)$_n$-α-naphthyl as a function of distance between naphthyl donor and dansyl acceptor. The energy transfer is 50% efficient at a distance of separation of 34.6 Å ($R \equiv R_o$), and has an efficiency of 98.5% and 1.5%, respectively, when $R = R_o/2$, and $R = 2R_o$ (Stryer and Haugland, 1967).

An experimental method for detecting the existence of resonance energy transfer relies on the appearance of fluorescence from the energy acceptor which is quantitatively related to the disappearance of fluorescence from the donor. As the efficiency of energy transfer increases, the fluorescence from the donor will decrease and that from the acceptor will increase. Another variation on this situation is that fluorescence from the acceptor may be quenched so that fluorescence from the donor will be quenched as the efficiency of transfer to the acceptor increases. The quenching of fluorescence by the iron of heme proteins was used in this way to determine the distance 25–35 Å, between the center from the heme of cytochrome c adsorbed to cytochrome oxidase, and the nearest heme of the latter (Vanderkooi et al., 1978; Dockter, 1978; see Chap. 7, section 7.3).

6.2.1 Mapping ATPase

The chloroplast ATPase (CF_1) contains five different subunits ($\alpha, \beta, \lambda, \delta, \varepsilon$) in a stoichiometry of 3:3:1:1:1 and a molecular weight of about 400,000 (Chap. 8). Resonance energy transfer has been used as a spectroscopic rule to measure the distance between fluorescence probes bound to the nucleotide

6.2 Use of Energy Transfer as a Spectroscopic Ruler

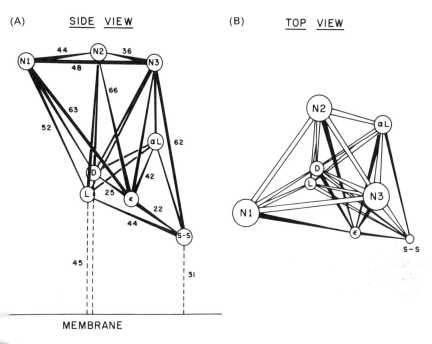

Figure 6.6. The spatial relationship in CF1 ATPase, measured by fluorescence energy transfer, between the three nucleotide binding sites (N1-N3), three sites on the γ subunit, (L), (D), and (S-S), and one on the epsilon (ε). D and L are γ (-SH) groups labeled in the dark and light, respectively, and (S-S) the γ subunit disulfide. (A) Side view and (B) top view, relative to the membrane surface, with distances in Å (McCarty and Hammes, 1987).

binding sites of CF_1 component using Eqs. 2–7 (Snyder and Hammes, 1985). More than 30 distances have been measured (Fig. 6.6). With present techniques, distances between ~10 and 90 Å can be measured with a precision of almost 10%. The specific sites labeled with fluorescent probes include: (i) three nucleotide binding sites located on the three β polypeptides; (ii) a single disulfide, a single sulfhydryl labeled in the dark, and a single sulfhydryl labeled in the light, all on the γ subunit; (iii) a single sulfhydryl on the ε subunit; (iv) an amino group on one α subunit; and (v) one Tyr on each of three β subunits. The energy transfer parameters between these donors and acceptors are summarized in Table 6.1.

The yield of energy transfer or fluorescence was determined from the lifetimes of fluorescence in the presence and absence of energy transfer, and ϕ_F° from the latter lifetime. R_o was determined (Eq. 8) from the yield of fluorescence, ϕ_F°, in the absence of energy transfer, the overlap integral, J, which is typically 10^{-13}–10^{-14} cm$^3 \cdot$ M^{-1}, the orientation factor $(\theta^2)_{avg}$ which 2/3 for a randomly oriented population of donors and acceptors, and the refractive index of the medium, $n = 1.4$ (Problem 65).

Table 6.1. Energy transfer parameters of fluorescence probes of CF_1

Donor	Location	Fluorescence max(λ_m^e, nm)	ϕ_F°	Acceptor	Location	Absorbance max(nm)	R_c
CPM	Dark-SH[a]	470	0.78	TNP-ATP	N_1-N_3	418/480	4
CPM	Dark-SH	470	0.78	FM	DiSH	495	5
PM	Light-SH	375/395	0.30	NBD	β-Tyr	390	3
PM	Dark-SH	375/395	0.30	NBD	β-Tyr	390	3

[a] Abbreviations: dark-SH and light-SH, dark- and light-site γ-sulfhydryls of CF_1, respectively; ϕ_F°, quantum yield of fluorescence in the absence of energy transfer; N_1-N_3, nucleotide binding sites 1–3 on CFl; D, γ-disulfide site; β-Tyr, NBD-reactive tyrosine residues on β subunit; NBD, N-(7-nitro-2,1,3-benzoxadia CPM, N-[4-[-(diethylamino)-4-methylcoumarin-3-yl]maleimide; PM, pyrenylmaleimide; FM, fluoresce maleimide. The value of $[\Theta^2]_{avg}$ is assumed to be 2/3, and the index of refraction, 1.4.
From Snyder and Hammes (1985).

Example 6.2. With this information, the distance between the known sites labeled with donor and acceptor fluorescence probes (Fig. 6.6) can be determined. From Eqs. 2 and 4:

$$R = R_o(\phi_T^{-1} - 1)^{1/6} \tag{9}$$

For energy transfer to TNP-ATP at nucleotide binding site I (Fig. 6.6) from the γ-sulfhydryl modified in the dark by CPM, $\phi_T = 0.41$ (Snyder and Hammes, 1985) and $R_o = 45.6$ Å (Table 6.1). Substituting these values into Eq. 9 $R = 48.1$ Å, for the distance between the N1 and D sites in Fig. 6.6.

6.3 Light Energy Transfer in Photosynthesis: The Phycobilisome

The transfer of light energy between the different antenna pigments that gather the light energy for photosynthesis is illustrated by the phycobilisome membrane system, which is one of three antenna complexes, in addition to photosystem I (PS I) and photosystem II (PS II), in cyanobacteria and the eukaryotic red algae. The phycobilisomes are regularly arranged on the surface of the thylakoid membrane as hemiellipsoidal or hemidiscoidal aggregates (Fig. 6.7. and B), containing 300–800 phycobilin pigments. The biliproteins in the phycobilisome are organized for a high efficiency of light energy transfer, so that the geometric proximity of pigment-protein molecules to the reaction center in the thylakoid membrane has the same order (Fig. 6.7B) as the hierarchy of absorption maxima (Fig. 6.8), i.e., phycoerythrins or phycoerythrocyanins at the periphery, phycocyanins in the middle, and allophycocyanins nearest to the reaction center in the thylakoid membrane (Gantt, 1980; Glaze 1985).

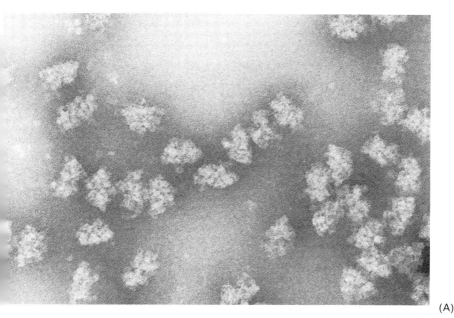

Figure 6.7. Electron micrograph (A) of phycobilisomes from the red alga, *Porphyridium cruentum*. (From E. Gantt.) (B) Three-dimensional arrangement of α-helices (X, Y, A–H) in the α-β monomer in C-phycocyanin from the cyanobacterium *Mastigocladus laminosum* derived from x-ray structure analysis of the C-phycocyanin from *M. laminosum* and *Agmenellum quadruplicatum* (Schirmer et al., 1987). The molecular weights of the PCα and PCβ subunits of the cyanobacterium, *Agmenellum quadruplicatum*, predicted from the gene sequence, are 17,620 and 18,335 (Lorimier et al., 1985). (C) Structure of the phycocyanin (PC) and phycoerythrin (PEC) hexamers in the red region of the phycobilisome and of allophycocyanin trimers (A, B, C) in the core. The core complexes A, B, C are made of allophycocyanin and linker peptide (between core complexes and from the core to the thylakoid membrane) complexes. The α and β PC genes in the cyanobacterium, Anabaena 7120, are part of an operon encoding the β and α subunits of PC, two linker polypeptides associated with PC, and a fifth reading frame that may encode another linker protein involved in attachment of the phycobilisome (Belknap and Haselkorn, 1987). [(B) and (C) from Zuber, 1986.]

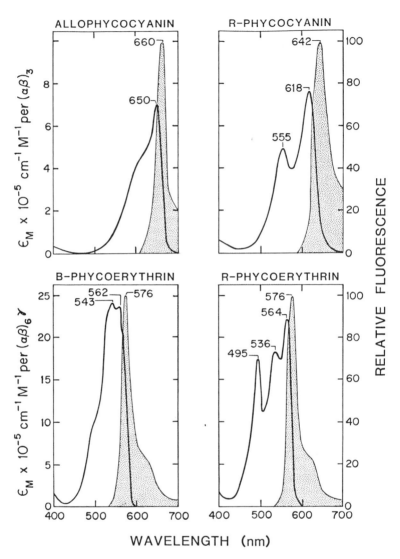

Figure 6.8. Absorption and emission spectra of biliproteins [From Glazer (1985). Reproduced with permission from the Annual Review of Biophysics and Biophysical Chemistry, vol. 14. © 1985 by Annual Reviews Inc.]

The absorption and fluorescence emission maxima (λ_m^e) of allophycocyanin ($\lambda_m^e = 660$ nm), R-phycocyanin ($\lambda_m^e = 642$ nm), and phycoerythrin ($\lambda_m^e = 576$ nm) are shown in Fig. 6.8.

The tendency of the light energy to be transferred to longer wavelength and lower energy according to the resonance mechanism described above can be seen in the time dependence of the fluorescence emission spectra

6.3 Light Energy Transfer in Photosynthesis

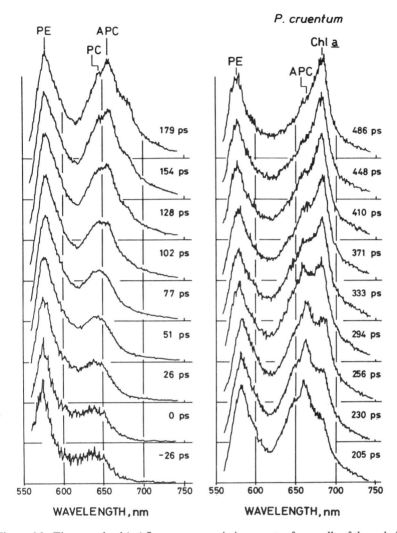

Figure 6.9. Time-resolved (ps) fluorescence emission spectra from cells of the red alga *P. cruentum* measured after excitation with a 540 nm, 6 ps pulse. PE, phycoerythrin; PC, phycocyanin; APC, allophycocyanin; chl *a*, chlorophyll *a* [From Glazer (1985). Reproduced, with permission, from the Annual Review of Biophysics and Biophysical Chemistry, vol. 14. © 1985 by Annual Reviews Inc.]

PE → PC → APC → Chl *a*, from cells of the red alga, *P. cruentum*, measured on a picosecond time scale (Fig. 6.9). The rise times of the fluorescence emission of β-phycoerythrin, R-phycocyanin, allophycocyanin, and chlorophyll *a* have been estimated to be 0, 12, 24, and 50 ps, respectively (Porter et al., 1978), supporting the PE → PC → APC → Chl *a* pathway for energy transfer inferred from the principles of resonance energy transfer.

6.4 Structures of Photosynthetic Antenna Pigment–Protein Complexes

The pigment molecules involved in antenna light-gathering function are bound to proteins of relatively small size, 6–30 kDa (Zuber, 1986). The chlorophylls and bacteriochlorophylls (Fig. 6.10) and carotenoids are noncovalently bound, and bilins are bound covalently. Four types of pigment-binding polypeptides have been distinguished in antenna complexes: (i) globular phycobiliproteins in the extramembrane phycobilisomes of cyanobacteria, red algae, and *cryptophyceae*, discussed above, and the bacteriochlorophyll *a*-binding protein crystallized from green bacteria, discussed below; (ii) fibrillary polypeptides in the extramembrane chlorosome structure of green photosynthetic bacteria; (iii) hydrophobic transmembrane polypeptides in bacteria; and (iv) polypeptides with both transmembrane domains and prominent polar regions in algae and higher plant chloroplasts.

The transition from the use of the extrinsic phycobiliproteins in cyanobacteria, which do not possess chlorophyll *b*, to the membrane-bound proteins that bind chlorophylls *a* and *b* in the green algae and higher plants that are their evolutionary descendants is a major change in the development of photosynthesis. The organism, *Prochloron didemni* (Lewin, 1976), which is phylogenetically closer to cyanobacteria than higher plants but contains a chlorophyll *a/b*-binding membrane protein, may be an evolutionary link.

6.4.1 Antenna of the Photosynthetic Bacteria

The bacteria contain the simplest antenna system. The structure of one of the antenna proteins, the bacteriochlorophyll *a* binding protein from a green bacterium, derived from x-ray diffraction analysis of the crystallized water-soluble complex, is shown in Fig. 6.11A. The polypeptide forms a 15-strand β-sheet that is wrapped around seven bacteriochlorophyll molecules. The seven magnesium atoms are each five-coordinated so that each pigment requires one amino acid to coordinate to the magnesium. It is important to note that only five of the seven pigment molecules, 1, 3, 4, 6, and 7 in Fig. 6.11A, can be liganded by His, and therefore other amino acids such as glutamine or asparagine must serve as ligands. This bacteriochlorophyll *a*–protein complex is organized as a trimer including 21 BChl *a* molecules in which at least the 7 BChl of one subunit, and probably the majority of the 21 molecules, act as a single unit in the delocalization of absorbed light energy.

In the purple photosynthetic bacteria, the *Rhodosprillaceae* and *Chromatiaceae*, two or three types of antenna complex have been found in the membrane, each with a different absorption maximum: the LH1 or B870 complex closest to the reaction center, and the LH2 or B800–850 and B800–820 complexes at increasing distances from the reaction center (Fig. 11B), where the numbers refer to the absorption maxima (nm) in the infra-red part of the spectrum. The "concentric" arrangement of antenna, focusing energy

6.4 Structures of Photosynthetic Antenna Pigment–Protein Complexes

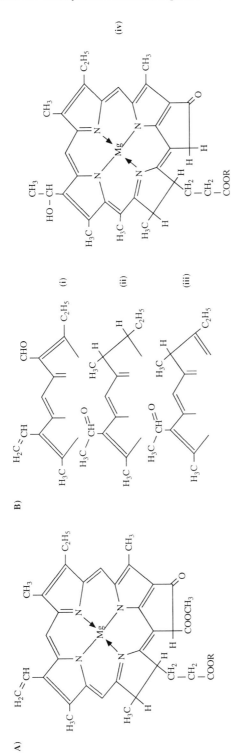

Figure 6.10. Structures of chlorophyll a (A), chlorophyll b (B, i) with the esterifying alcohol, R, phytol, $C_{20}H_{39}$, and bacteriochlorophylls a, b, and c (B, ii-iv). BChl a and b are found in the purple photosynthetic bacteria and can have geranylgeraniol instead of phytol. The BChl c from *C. aurianticus*, a 7,8-dihydroporphyrin without the 10-carboxymethyl group, with the esterifying alcohol mostly stearol, is usually the major chlorophyll component in green bacteria. BChl c can also have farnesyl as the esterifying group, and there are at least six different homologues with different side chains.

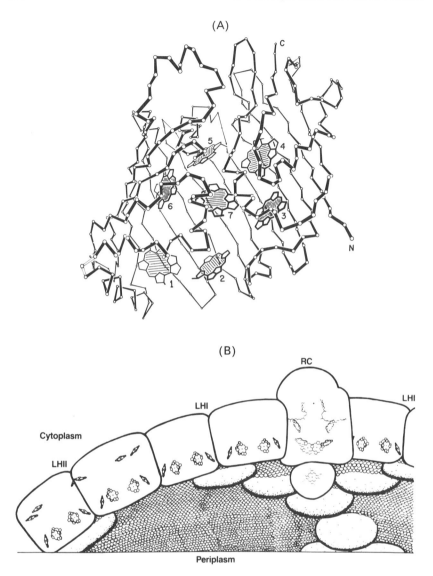

Figure 6.11A, B. (A) View of the bacteriochlorophyll *a* subunit from a green bacterium showing the seven BChl molecules. The side chains are not shown (Matthews et al., 1979; Tronrud et al., 1986). (B) Proposed arrangement of bacteriochlorophyll antenna molecules around the reaction center in the cytoplasmic membrane. LH1 complex has six BChl arranged in a ring, LH2 has the six in a ring and three more on the cytoplasmic side of the membrane (Hunter et al., 1989).

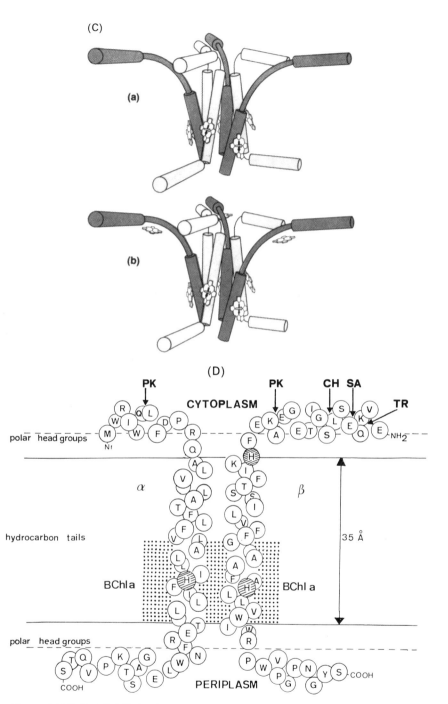

Figure 6.11C, D. (C) Minimal models for LH1(a) and LH2(b) BChl antenna complexes, showing α-helical regions as cylinders. The Q_x and Q_y dipole transitions are shown as dark and light arrows (Hunter et al., 1989). (D) Model of the bacterial α-β heterodimeric antenna complex. Histidine is the probable ligand for BChl *a*, and proteolysis sites on the cytoplasmic side are indicated (Brunisholz, 1985; Zuber, 1986).

(E)

```
                                                   X X              X        |   || X
                RS. RUBRUM B 870         MWRIWQLFDPRQALVGLATFLFVLALLIHFILLSTERFNWLEGASTKPVQTS
                RP. VIRIDIS B 1015       ATEYRTASWKLWLILDPRRVLTALFVVYLTVIALLIHFGLLSTDRLNWWEFQRGLPKAA
                RP. SPHAEROIDES B 870    MSKFYKIWMIFDPRRVFVAQGVFLFLLAVMIHLILLSTPSYNWLEISAAKYNRVAVAE
LHP-α           RP. CAPSULATA B 870      MSKFYKIWLVFDPRRVFVAQGVFLFLLAVLIHLILLSTPAFNWLTVATAKHGYVAAAQ
                RP. SPHAEROIDES B 800-850   MTNGKIWLVVKPTVGVPLFLSAAFIASVVIHAAVLTTTTWLPAYYQGSAAVAAE
                RP. CAPSULATA B 800-850  MNNAKIWTVVKPSTGIPLILGAVAVAALIVHAGLLTNTTWFANYWNGNPMATVVAVAPAQ
              *)CHLOROFLEXUS B 806-865   MQPRSPVRTNIVIFTILGFVVALLIHFIVLSSPEYNWLSNAEGG

                RS. RUBRUM B 870         EVKQESLSGITEGEAKEFHKIFTSSILVFFGVAAFAHLLVWIWRPWVPGPNGYS
                RP. VIRIDIS B 1015       ADLKPSLTGLTEEEAKEFHGIFVTSTVLYLATAVIVHYLVWTAKPWIAPIPKGWV
                RP. SPHAEROIDES B 870    ADKSDLGYTGLTDEQAQELHSVYMSGLWPFSAVAIVAHLAVYIWRPWF
LHP-β           RP. CAPSULATA B 870      ADKNDLSFTGLTDEQAQELHAVYMSGLSAFIAVAVLAHLAVMIWRPWF
                RP. SPHAEROIDES B 800-850   TDDLNKVWPSGLTVAEAEEVHKQLILGTRVFGVMALIAHFLAAAATPWLG
                RP. CAPSULATA B 800-850  MTDD--KAGPSGLSLKEAEEIHSYLIDGTRVFGAMALVAHILSAIATPWLG
              *)CHLOROFLEXUS B 806-865   MRDDDDLVPPKWRPLFNNQDWLLHDIVVKSFYGFGVIAAIAHLLVYLWKP
```

N-TERMINAL POLAR, CHARGED DOMAIN | CENTRAL HYDROPHOBIC DOMAIN | C-TERMINAL POLAR, CHARGED DOMAIN

(F)

Equilibration with (αβ)₃ LH2 unit — 1 ps

Transfer among LH2 units, then to LH1 — 10-50 ps

Collection of excitations at BchI 896, then transfer to RC — 50-100 ps

Photochemistry – charge separation in reaction centre — 100 ps

P870 / P870⁺

6.4 Structures of Photosynthetic Antenna Pigment–Protein Complexes

transfer toward longer wavelengths, is similar to that discussed above for the phycobilisomes. The minimal structural unit of these bacterial antenna complexes is a trimer of heterodimers of small (50–60 amino acids) α-and β-units (Fig. 6.11C). The similarity of the sequence of these bacterial antenna proteins, particularly with respect to a central hydrophobic domain being surrounded by NH_2 and COOH-terminal polar regions, is shown in Fig. 6.11D. Direct excitation of one of the BChl molecules of the ring of either LH1 or LH2 transfers energy to other members of the ring within 1 ps (Fig. 6.11F). Excitation transfer between different LH1 or LH2 trimers takes place in 10–50 ps. Within 50–100 ps the excitation is concentrated in the antenna pigment adjacent to the reaction center, with the characteristic time delay for charge separation in the reaction center $\simeq$ 100 ps after the initial photon absorption (Fig. 6.11F).

The infrared absorbing antenna complex B870α, β from the purple bacterium, *Rhodospirillim rubrum*, is a 1:1 dimer of the small (MW = 6,079 and 6,101) α and β polypeptides, as deduced from hydrophobicity analysis of the sequence (Chap. 1) and topographical analysis of the arrangement in the membrane (Fig. 6.11C). The COOH- and NH_2-termini of both polypeptides were found on the periplasmic and cytoplasmic sides of the membrane, respectively, and a BChl *a* molecule is proposed to be coordinated by a histidine residue on the periplasmic side. This arrangement resembles that of the hemes of the membrane-bound *b* cytochromes (see Chap. 7.4).

6.4.2 Chloroplast Antenna Complexes

The pigment proteins of PS I and PS II can be divided into two domains: (i) light-harvesting complex (LHC) that is not essential for the structure of the reaction center and core complex (CC) that is required for an active reaction center (Chitnis and Thornber, 1988; Green, 1988). There are at least four different chlorophyll-xanthophyll polypeptides in LHC of photosystem II (LHC II) (Fig. 6.12A), and approximately the same number in PS I. The LHC IIb component contains about one-half of the chlorophyll and one-third the protein in plant thylakoid membranes, and has been implicated in the regulation of the distribution of light energy between the two photosystems. The predicted folding in the membrane of the LHC IIb polypeptide is shown in Fig. 6.12B. This polypeptide, containing 9–13 chlorophyll (5–7 Chl *a*, 4–6 Chl *b*, and 2–3 xanthophyll) molecules, is one unit of a nine-membered, trimer of trimers, complex whose trans-membrane nature has been studied through image reconstruction and electron diffraction analysis of two dimensional ordered arrays (Fig. 6.12C). The presence of an Arg residue on the stromal

---

Figure 6.11E, F. (E) Sequences of bacterial α and β antenna proteins (Zuber, 1986). (F) Approximate time course of transfer of excitation energy through the antenna to the reaction center (Hunter et al., 1989).

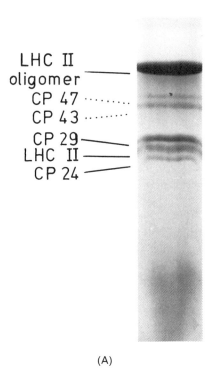

(A)

Figure 6.12. (A) SDS-polyacrylamide gel electrophoresis (SDS-PAGE) of photosystem II chlorophyll-protein complexes; most of the LHC II (the major Chl a/b complex of PS II) is preserved in its oligomeric form. CP 29 is thought to be a core-associated Chl a/b complex of PS II, and CP 24 a more peripheral complex; CP 47 and CP 43 (dotted lines) are chloroplast-encoded Chl a complexes that are closely associated with the reaction center (figure conburted by B.R. Green). (B) Model of the transmembrane folding and sidedness of the LHC IIb polypeptide (Peter and Thornber, 1988) inferred from sequence (Pichersky et al., 1985), hydrophobicity (Thornber, 1986), and topographical probe analysis (Bürgi et al., 1987). Nine to thirteen chlorophyll molecules (Chl a:Chl b = 1.0–1.4) are proposed to be coordinated to the polypeptide, which also contains 2–3 xanthophylls, through conserved His, Gln, and Asn residues (arrows) (Peter and Thornber, 1988). (C) trans-membrane model of 9 LHC II subunits forming this complex, inferred from image reconstruction of two dimensional crystals. The polypeptide chain protrudes 20 Å on the stromal side and 5–10 Å on the lumen side from the thylakoid membrane bilayer [From Kühlbrandt, 1984). Reprinted by permission from Nature, vol. 307, pp. 478–480. Copyright Macmillan Magazines Ltd.]

side of the putative transmembrane helix I in the model for LHC IIb (Fig. 6.12B), and of Glu and Arg, potentially salt-bridged, residues, on the lumen side of helices I and III, is important for accumulation of LHC in the membrane. The role of these residues was tested by in vitro expression and import into *Lemna* chloroplasts of mutant protein coded by an altered oligonucleotide (Kohorn and Tobin, 1987). This experiment provides an example of a protocol for testing the function of an intrinsic thylakoid membrane protein.

6.4 Structures of Photosynthetic Antenna Pigment–Protein Complexes 259

(B)

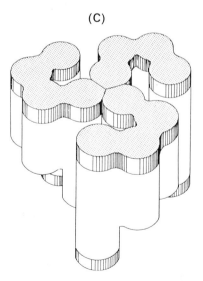

(C)

Figure 6.12 (continued)

The LHC IIb protein(s) are encoded by nuclear genes, often called *cab* genes. They are members of a multigene family, containing between 3 and 16 genes depending on the plant species with a high degree of identity at the amino acid level. Genes from at least 10 different plant species have been sequenced (Chitnis and Thornber, 1988). It is not known whether the multiple copies of LHC II-coding genes form the basis for the structural heterogeneity of the LHC II protein particles depicted above (Fig. 5.10D), which includes LHC II bound to the water-splitting enzyme complex (10–17 nm particles as seen by freeze-etch electron microscopy) and mobile LHC II complexes (seen as 8 nm particles), in a ratio of about 1:3 (Staehelin, 1986). LHC IIa (CP 29) is not part of the multigene family and is coded for by a single-copy gene.

Regulation of the Light Energy Distribution

Phosphorylation of LHC II polypeptides regulates the distribution of excitation energy between PS II and PS I. The phosphorylation occurs at the stroma-exposed NH_2-terminus (Fig. 6.13A). The phosphorylation is accomplished by a kinase that is activated when an excess of excitation absorbed by the PS II antennae causes an increase in the PQH_2/PQ ratio (Bennett et al., 1980). The E_m value at which the kinase is activated corresponds to the E_m of the PQ pool (Fig. 6.13B). In the field of kinase control and regulation (Hanks et al., 1988), this is the only example of a redox-controlled kinase. The kinase has been purified as an M_r 64,000 polypeptide that phosphorylates LHC II in a reconstituted system that does not, however, exhibit redox control (Coughlan and Hind, 1987). Mutants deficient in the cytochrome b_6–f complex (Chap. 7, section 7.3) that do not have kinase activity imply a connection between the complexes. (Gal et al., 1987; Wollman and Lemaire, 1988).

The mobility of some of the LHC II antenna regulates the distribution of excitation energy, so that when PS II receives an excess of excitation energy, antenna pigment moves from PS II to PS I. The increased negative charge on phosphorylated LHC II and the resulting intermolecular charge repulsion causes increased lateral movement of the LHC II from the center of the grana stacks to the end regions of the grana and to the stroma lamellae. At least two different LHC II populations can be distinguished, one containing an M_r 27,000 polypeptide that is slowly phosphorylated and tightly bound to the PS II reaction center, and an M_r 25,000 polypeptide that is peripherally bound, more rapidly phosphorylated and mobile (Fig. 6.13D). The LHC II migration to the stroma increases the antenna cross-section of PS I (Fig. 6.13C), which can be measured as an increase and a decrease, respectively, of $\sim 15\%$ in the rate of PS I and PS II electron transport under limiting light (Farchaus et al., 1982), and in vivo using photoacoustic spectroscopy (Canaani et al., 1984). The light energy distribution may also be environmentally regulated through (i) increases in the Chl a/b and Chl/Q_A ratios that occur in the presence of higher ambient light intensity, and (ii) changes in the pigment content dependent upon the light spectrum (Anderson, 1986).

6.4 Structures of Photosynthetic Antenna Pigment–Protein Complexes

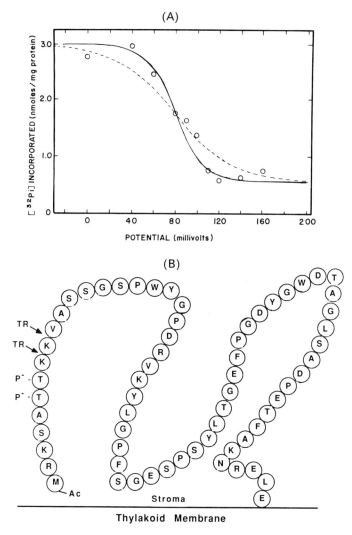

Figure 6.13. (A) Redox titration of LHC II kinase activity (Millner et al., 1982; compare with Horton and Black, 1980). (B) Possible threonine phosphorylation (-P) and lysine (K) trypsin (TR) sites on LHC II (see Mullet, 1983; Bennett et al., 1987). (C) Model relating change in phosphorylation of LHC-II, its lateral mobility, change in absorptive cross-section of PS II and PS I, and change from stacked (State I) to more unstacked (State II) grana [From Barber (1982). Reproduced, with permission, from the Annual Review of Plant Physiology, vol. 33. © 1982 by Annual Reviews Inc.] Phosphorylation and dephosphorylation are catalyzed by the kinase and a phosphatase that are inhibited, respectively, by sulfhydryl-directed reagents and fluoride. (D) Proposal for the dynamic organization of the heterogeneous LHC II. (Based on Larsson et al., 1987.)

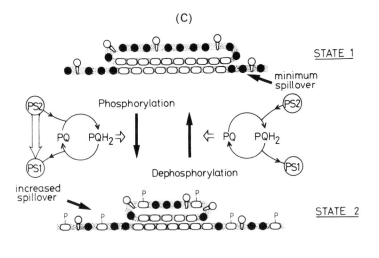

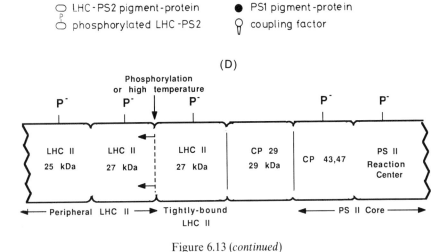

Figure 6.13 (*continued*)

Figure 6.14. (A) Model of a chlorosome attached to the cytoplasmic membrane of the green bacterium *C. aurianticus* (Reprinted with permission from Feick and Fuller, *Biochem.*, vol. 23, p. 3683. Copyright 1984 American Chemical Society.) (B) Model of 5.6 kDa polypeptide with seven possible BChl *c* binding sites. The BChl *c* molecules interact by H-bonded nucleophilic interactions between the C=O group of ring V and the –CH(CH$_3$)–OH group of ring I. Residues X2, X7, etc., are Ser, Thr residues in the back of the α-helix involved in polypeptide interactions. The chlorosome subunit composed of 12 α-helical 5.6-kDa units (Wechsler et al., 1985), possibly arranged to facilitate orientation and stacking of the antenna pigment (Brune et al., 1987). Areas for binding BChl *c* are hatched. (From Zuber, 1986.)

6.4 Structures of Photosynthetic Antenna Pigment–Protein Complexes 263

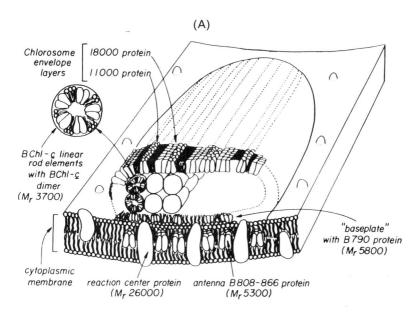

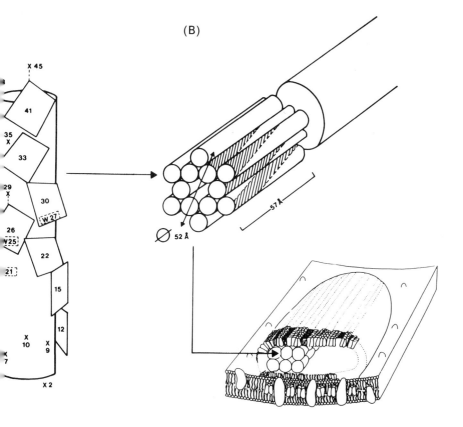

6.4.3 Energy Transfer from the Chlorosome

The photosynthetic apparatus of the thermophilic, facultative green photosynthetic bacterium *Chloroflexus aurantiacus* (Fig. 6.14) is located in two compartments: the chlorosome functioning as a light-harvesting organelle [~100 nm long, 30 nm wide, and 10 nm thick, Fig. 6.14A], containing BChl *c* as the major pigment, and the cytoplasmic membrane containing BChl *a* light harvesting pigment and the reaction center. A 5.6-kDa BChl *c* binding polypeptide has been sequenced that binds seven BChl *c* molecules at glutamine and asparagine residues (Fig. 6.14B), with the chlorosome 5.2 × 6 nm subunit containing 12 of these polypeptides (Fig. 6.14C).

6.5 Structure of Photosynthetic Reaction Centers

Light energy absorbed by the antenna chlorophyll or bacteriochlorophyll molecules migrates rapidly (10^{-11}–10^{-12} s) between pigment molecules by exciton and resonance energy transfer until it reaches a reaction center, where the electron is trapped with a quantum efficiency of approximately 1.0, because of the rapid (~10^{-12} s) transfer of the electron from the excited state, P*, of the reaction center to an acceptor, concomitant with formation of a cation radical, P$^+$. The absorption spectrum of the reduced reaction center BChl *a* dimer of *Rb. sphaeroides* has a room temperature maximum at 865 nm (Fig. 6.15). The absorption bands at 800–810 nm in the 5 K spectrum are associated

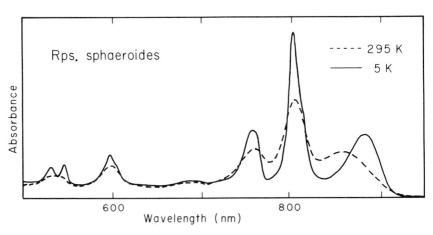

Figure 6.15. Absorption spectra of reduced *Rb. sphaeroides* reaction centers at room temperature and 5 K. The long wavelength band near 865 nm at 295 K arises from the special pair, (BChl)$_2$, the 800–810 nm band mostly from the auxiliary BChl, and bands at (532, 755 nm) and (542, 765 nm) with the two BPhe [From Kirmaier and Holten (1987). Reprinted by Permission of Kluwer Academic Publishers, Copyright © 1987.]

6.5 Structure of Photosynthetic Reaction Centers

Table 6.2. Typical properties of reaction centers of photosynthetic bacteria and of the two photosystems of plants, algae, and cyanobacteria

	Purple bacteria		Green bacteria	Oxygenic photosynthesis	
	Rb. sphaeroides	Rps. viridis	(Chloroflexaceae)[a]	PSII	PSI[b]
λ_{max} of P (nm)	870	940	865	680	700
No. of polypeptides	3	4	2	4(D1, D2, b-559)	2(psIA1, psIA2)[d]
Auxiliary or voyeur pigment	BChol a	BChol b	BChl, BPhe	nk	Chl a
E_m (P$^+$/P), V	0.45		0.36	+1.17	+0.5
Electron acceptor(s)	BPhe a	BPhe b	BPhe	Phe	Chl-Vit K$_1$-[4Fe-4S]
Secondary acceptor	[Q$_A$-Fe^{2+}-Q$_B$][c]		Menaquinone	PQ$_A$-Fe^{2+}-PQ$_B$	[4Fe-4S]
Electron donor	Cyt c_2	Membrane-bound Cyt c (4 heme)	Cyt c	Tyrosine[e] residue	Plastocyanin or Cyt c

nk, Not known.

[a] From Blankenship (1984); in addition, the sequence of the L- and M-subunits of the green thermophile, *Chloroflexus aurantiacus*, consisting of 310 and 306 residues, respectively, show a high degree of homology with those of purple bacteria (Ovchinnikov et al., 1988a, b).
[b] The reaction center complex in PS I differs from the conventional definition because many antenna chlorophyll molecules are bound to the psIA1 and psIA2 polypeptides that also bind the reaction center special pair.
[c] Q$_A$ is MQ-9 or UQ-10, and Q$_B$ is UQ-9 and UQ-10 in *Rps. viridis* and *Rb. sphaeroides*, respectively.
[d] The minimal purified PS I reaction center complex also contains a number of small polypeptides [M$_r$ (kDa) = 18, 16, 14, 9, and 4].
[e] See section 6.8.1.

mostly with the auxiliary BChl, and those at 532, 755, 542, and 765 nm mostly with the two BPhe (Fajer et al., 1975). The absorption maxima in reaction centers from the bacterium, *Rps. viridis*, containing BChl *b* and BPhe *b*, are shifted to longer wavelengths.

The original evidence for the existence of a BChl dimer as the reaction center was the observation that the electron paramagnetic resonance (EPR) line width of the P^+ radical was more narrow by a factor of $\sqrt{2}$ than that of BChl in solution (McElroy et al., 1969; Norris et al., 1971). The reaction center, P, is distinguished by (i) its long wavelength absorption band (e.g., P700, P680, P870, etc.), (ii) E_m values (~ 0.3–0.5 V) slightly more positive than the E_m of the immediate electron donor, (iii) an electron acceptor that is part of this redox system and has electronic overlap or contact with the reaction center excited state, and (iv) a unique polypeptide composition and structure in which the prosthetic groups of the reaction center and primary and secondary electron acceptors are arranged so as to accomplish transmembrane electron transfer. The difference in midpoint potential of P/P^* and P^+/P is approximately equal to the free energy, hv, of the absorbed photon (Chap. 2, section 2.9). A summary of the properties of some photosynthetic reaction centers is presented in Table 6.2. The evidence that led to the concept of two photosystems and two reaction centers in plant and algal oxygenic photosynthesis (quantum requirement, action spectra, first discovery of a reaction center, partial light reactions) is presented, for example, in Rabinowitch and Govindjee (1969) and Clayton (1980).

6.6 Structure of the Reaction Center Proteins: Transmembrane Charge Separation

The photosynthetic reaction centers from *Rps. viridis* and the *Rb. sphaeroides* are presently the only membrane proteins or protein complexes for which crystals have been obtained that diffract to high resolution, and the crystal structure has been solved (Deisenhofer et al., 1985a, b; Chang et al., 1986; Allen et al., 1987a,b; Yeates et al, 1987; Kühlbrandt, 1988b). The problem of crystallizing membrane proteins involves finding a detergent–amphiphile system that can compensate for the hydrophobicity of the surfaces of the protein normally exposed within the membrane, allowing ordered contacts to form between the polar surfaces of the protein (Fig. 6.16).

The *Rps. viridis* reaction center contains four polypeptides, L (273 amino acids), M (323aa), H (258aa), and a bound cytochrome containing four hemes. The reaction center of *Rb. sphaeroides* does not have the bound cytochrome, but only the three polypeptides, L (281 amino acids), M (307aa), and H (260aa).

After energy transfer to the reaction center dimer, an electron is transferred from the special pair in ~ 3 ps to bacteriopheophytin, and then in ~ 200 ps to the quinone acceptor, Q_A (Fig. 6.17C). The identity of the BPhe electron

6.6 Structure of the Reaction Center Proteins

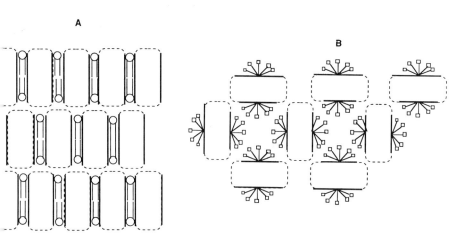

Figure 6.16. Basic types of protein crystals: (A) Schematic of two-dimensional membrane crystals and intercalated lipid (o–) ordered in a third dimension that tend to be small and useful mostly for electron microscopy. (B) Schematic of membrane proteins crystallized with detergent (□–) bound to the hydrophobic protein surfaces. The hydrophilic surfaces and contacts are indicated by dashed lines. (Adapted from Michel, 1983.) (C) Crystal of the reaction center of *Rb. sphaeroides*, 0.5 mm × 3 mm. (Contributed by C.-H. Chang and M. Schiffer.)

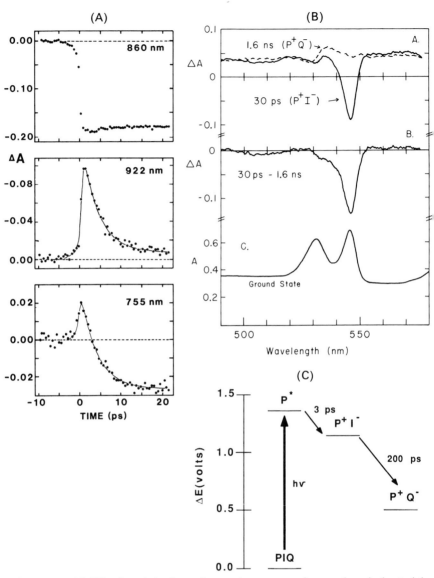

Figure 6.17. (A) Kinetics of the fast primary electron transfer reactions in bacterial chromatophores: oxidation of the reaction center at 860 nm ($t_{1/2} \leq 1$ ps), decay of stimulated light emission at 922 nm that reflects the formation and decay of the excited state, P*, and reduction of BPhe ($t_{1/2}$ of ~5 ps) at 755 nm after an oppositely directed ΔA due to oxidation of P (Woodbury et al., 1985. Reprinted with permission from *Biochemistry*, 24, 7516. Copyright 1985 American Chemical Society). (B) Laser flash-induced absorbance change in the pheophytin region of the spectrum showing that only one pheophytin band is reduced to BPhe⁻ in the flash, as compared to the two that can be chemically reduced in the ground state. Curve A: Separate spectra seen 30 ps and 1.2 ns after the flash. Curve B: Difference between 30 ps and 1.2 ns. Curve C: difference spectra for chemical reduction in the ground state showing two pheophytin components (Kirmaier et al., 1985). (C) Representation of primary electron transfer reactions of *Rb. sphaeroides*. The $P^+B^-\phi$ intermediate is not shown because it was not detected (Kirmaier and Holten, 1987). Analysis of the transition dipole moment of the excited state indicates that the reaction center special pair itself has substantial charge-transfer character (Lockhart and Boxer, 1988).

6.6 Structure of the Reaction Center Proteins

acceptor was inferred from the optical spectra of very fast ($\leq$ ps) absorbance changes (Fig. 6.17A), EPR signal changes, and the persistence of these changes at very negative ambient redox potentials as well as cryogenic temperatures. Comparison of the spectra of laster flash-induced absorbance changes with those of the ground state show that only the pheophytin absorbing at longer wavelengths, and the member of the pheophytin pair that is on the "L" side of the membrane, is reduced by the flash (Fig. 6.17B). The structural data for the position of the auxiliary BChl in the reaction centers indicates that the monomeric BChl (B) positioned between the special pair which is the electron donor (P) and the BPhe (ϕ) should facilitate electron transfer to the BPhe acceptor. A clearly defined reduced state, B^-, formed in less than 10 ps by very short 150 femtosec (1.5×10^{-13} s) laser flashes was not detected (Wasieliewski and Tiede, 1986; Fleming et al., 1988), indicating that the primary electron transfer occurs as:

$$PB\phi \xrightarrow{hv} P^*B\phi \xrightarrow[t_{1/2} = 3 \text{ ps}]{} P^+B\phi^-$$

If the BChl is reduced, the reaction would be: $PB\phi \rightarrow P^*B\phi \rightarrow P^+B^-\phi \rightarrow P^+B\phi^-$.

The high rate and quantum yield of the primary charge separation process in bacterial photosynthesis depend on the ability of the protein to stabilize the charge transfer states and to keep the reorganization energies small (Chap. 2, section 2.10) for the electron transfer reactions (Creighton et al., 1988). The electron transfer from Q_A to Q_B decelerates by a factor of $\sim 10^6$ from ~ 200 ps for Q_A reduction to 0.1 ms for Q_A oxidation, the latter time scale characteristic of those associated with H^+ uptake and protein conformational changes. Thus, energy conversion in the reaction center relies on a picosecond electron transfer from the excited state to trap the electron from the antenna, and millisecond events to convert the redox energy to a $\Delta\tilde{\mu}_{H^+}$.

The transfer of the electron from the special pair across the membrane dielectric to the quinone creates an electrical potential (an "electrogenic" reaction) that can be monitored by a shift of the absorption bands of the carotenoid in the membrane (Chap. 3, section 3.4.2). In *Rps. viridis*, the oxidized special pair, $(BChl)_2^+$, is reduced by the closest heme of the attached four heme *c* cytochrome, with the heme Fe-BChl *b* Mg distance approximately 21 Å, and the π clouds of these two ring structures separated by approximately 11 Å, another example of long-range in vivo electron transfer (Chap. 2, section 2.10.5).

6.6.1 Structure of the Bacterial Reaction Center

The crystal structure analysis indicates the presence of 11 trans (~ 40 Å)-membrane helices, spanning the membrane bilayer from the cytoplasmic to periplasmic side (Fig. 5.18A). Five helices are contributed by the L polypeptide, five by the M subunit, and one by H (Fig. 6.18), consistent with the prediction made from the primary sequences (Appendix III) of the distribution of hydro-

270   6: Photosynthesis: Photons to Protons

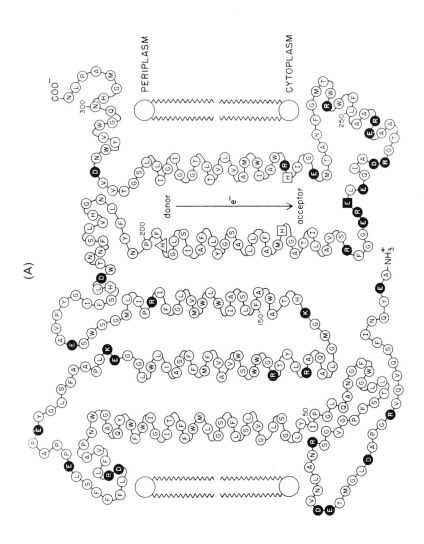

6.6 Structure of the Reaction Center Proteins

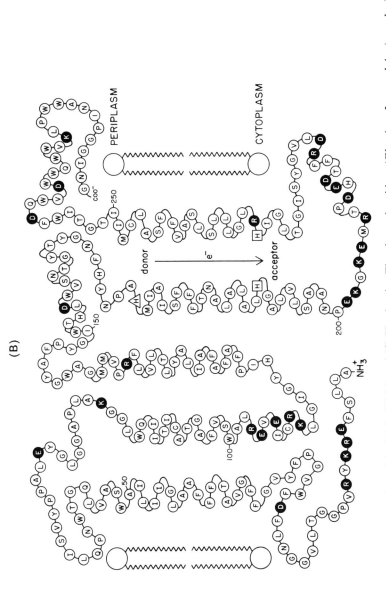

Figure 6.18. Schematic representation of the (A) L, (B) M, and (C) H subunits. The charged residues (Glu, Asp, Lys, and Arg) are depicted as filled circles. Ligands to the accessory (voyeur) BChl (hexagon), donor (BChl)$_2$ (△), and Fe (□) are marked (Williams et al., 1986). The notation L (light), M (medium), and H (heavy) referred to the M_r values on SDS gels. Hydrophobic proteins often run anomalously on these gels. Subsequent structure and sequence analysis showed H to be the smallest of the three polypeptides.

272　　6: Photosynthesis: Photons to Protons

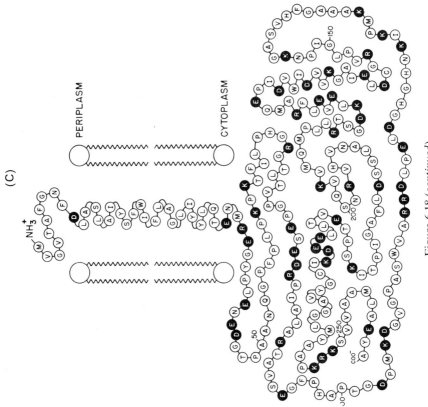

Figure 6.18 (continued)

6.6 Structure of the Reaction Center Proteins

phobic residues (Williams et al., 1984). The prosthetic groups are bound noncovalently to the L and M polypeptides. The thickness of the hydrophobic domain of the *Rb. sphaeroides* reaction center protein is approximately 30 Å with the polar regions of the membrane bilayer extending approximately 5 Å on either side of the nonpolar domain. The distance between the Fe atom at the center of symmetry between Q_A and Q_B at the cytoplasmic polar interface, and the center of the special pair which is approximately 2 Å into the nonpolar domain on the periplasmic side, is 28–30 Å.

The pigment molecules BChl and BPhe are arranged with an approximate twofold symmetry (section 5.9). The center of the special pair on one side of the bilayer is part of an apparent two-fold rotation symmetry axis going through the nonheme iron on the other side of the transmembrane structure (Fig. 6.19). The right and left branches of the schematic reaction center structure shown in Fig. 6.19A–C are often called "L" and "M" because most of the mass derived from the 11 transmembrane helices on those sides is from the L and M polypeptides, respectively (Fig. 5.16). Because L and M polypeptides are present on both sides, it seems simpler to instead use A and B for the Q_A and Q_B. The gross structural symmetry of the two sides is illustrated by the similarity of the distances and orientations on the A and B sides between the special pair, $(BChl)_2$, auxiliary BChl, BPhe, and Q_A or Q_B (Table 6.3; Fig. 6.19B).

The twofold symmetrical pigment arrangement would appear to create two branches for electron transfer from the special pair. However, as shown in Fig. 6.17B, only one pheophytin is reduced, and this pheophytin is known to be the BPhe on the "A" side from spectroscopic studies on oriented crystals (Knapp et al., 1985) and the lack of effect on the rate of primary electron transfer to removal of the BChl on the "B" side (Maróti et al., 1985).

The approximate twofold symmetry that relates the L and M polypeptides and the chromophore positions is a striking feature of the structure. Since the electron transport function is asymmetric, with electron transfer to the BPhe on the L side greatly preferred, by at least a factor of 12 (Michel-Byerle et al., 1988), it is important to examine the extent to which the symmetry is preserved. The amino acid residues on the L and M subunits that surround the special pair $(BChl)_2$, the accessory BChl, and the BPhe, for which the difference between the A and B side is conserved in *Rps. viridis*, *Rb. sphaeroides*, and *Rb. capsulata*, are tabulated in Table 6.4 (Tiede et al., 1988). There are many more differences between the A and B sides that are not conserved in the three bacteria. Clearly, the gross twofold symmetry is not absolute. The only charged residue in the region of the pigments is Glu-L104 (Table 6.4C) which, if protonated, could hydrogen bond to the ring V carbonyl of the A-side pheophytin. No such interaction would occur at the corresponding position M131, in *Rps. viridis* and *Rb. capsulata*, where the residue is Val. This difference in potential H-bond capability to the pheophytin on the A compared to the B side has been proposed to contribute to preferential electron transfer along

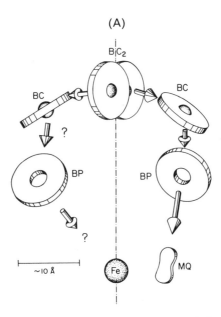

Figure 6.19. Spatial arrangement of prosthetic groups of the *Rps. viridis* reaction center in the membrane bilayer: (A) Schematic (From DeVault, 1986. Copyright © 1986 by Martinus Nijhoff Publishers. Reprinted by permission of Kluwer Academic Publishers; derived from Deisenhofer et al., 1985b). The existence of asymmetric electron transfer, using only the branch of the protein complex containing menaquinone-9 (Q_A), is shown. Next page: (B) Distances between the prosthetic groups of the *Rps viridis* reaction center (Michel-Byerle et al., 1988). The four-heme cytochrome of the *Rps. viridis* reaction center is not shown. (C) Orientation and arrangement of the prosthetic groups of the *Rps. viridis* reaction center in a stereo view including the four-heme cytochrome (HE) that would extend into the cell periplasm, the special pair (BC), bacteriopheophytin (BP), and Q_A [menaquinone-9 (MQ)] (Michel and Deisenhofer, 1988b).

the A side, although recent site-directed mutagenesis experiments in which the Glu-L104 was replaced by a Leu or Gln in *Rb. capsulata* did not indicate a critical role for the Glu residue (Kirmaier et al., 1988). Excess negative charge density on the B side BChl of the special pair $(BChl)_2$ has also been proposed as part of the explanation for the preferential electron transport along the A side of the reaction center (Michel-Beyerle et al., 1988).

Other conserved symmetry-breaking pairs are (Table 6.4): (i) Phe-L97 and Val- or Leu-M124 are close to BChl and BPhe on the L and M sides, respectively; (ii) Phe-L216 and Trp-M250 are located between BPhe and Q on the Q_B and Q_A sides. The larger ring of the Trp-M250 makes van der Waals contact with both $(BPhe)_A$ and Q_A, which could facilitate electron transfer between them.

6.6 Structure of the Reaction Center Proteins

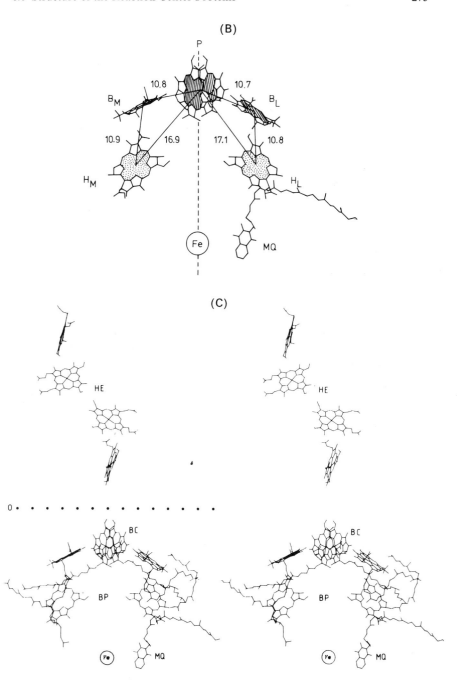

Figure 6.19 (*continued*)

Table 6.3. Distance and orientation between neighboring cofactors in reaction centers of *Rb. sphaeroides* and *Rps. viridis*

Cofactors	Distance between ring centers (Å)[a]		Angle (degrees) between ring normals	
	Rb. sph.	*Rps. vir.*	*Rb. sph.*	*Rps. vir.*
$(BChl_2)_A:(BChl_2)_B$	7.0	7.0	10	15
$(BChl_2)_A:BChl_A$	11.0	10.5	70	65
$(BChl_2)_B:BChl_B$	10.5	11.0	70	70
$BChl_A:BPhe_A$	10.5	10.0	60	70
$BChl_B:BPhe_B$	11.0	11.0	60	70
$BPhe_A:Q_A$	13.0	14.0	35	35
$BPhe_B:Q_B$	15.0	NA[b]	35	NA

[a] Error of coordinate, ±0.4 Å.
[b] NA, not applicable; population of bound Q_B is low in crystals of *Rps. viridis* (Fig. 6.19).
From Allen et al. (1987a).

Table 6.4. Amino acid residues in the neighborhood of (A) the special pair, $(BChl)_2$, (B) the accessory BChl, and (C) the bacteriopheophytin, for which polarity differences between A and B sides are approximately conserved

	A side				B side		
	sphaeroides	*viridis*	*capsulatus*		*sphaeroides*	*viridis*	*capsulatus*
A.							
L161	G	G	G	M188	S	S	S
L162[a]	Y	Y	Y	M189	L	I	L
B.							
L97	F	F	F	M124	V	L	V
C.							
L97	F	F	F	M124	V	L	V
L101	A	W	A	M128	W	W	W
L104	E	E	E	M131	T	V	V
L118	P	P	P	M145	A	A	A
M250	W	W	W	L216	F	F	F

[a] Tyr-L162 proposed to facilitate electron transfer from cytochrome heme to special pair in *Rps. viridis* (Michel et al., 1986b).
From Tiede et al. (1988).

6.6 Structure of the Reaction Center Proteins

6.6.2 Docking of Cytochrome c_2 to the Reaction Center

The model for the reaction center of *Rb. sphaeroides* has been examined (Allen et al., 1987a; Tiede et al., 1988) for charges that could complement the positively charged lysine residues of cytochrome c_2 (Chap. 4, section 4.4) and allow formation of the binary complex of complementary charged pairs (Table 6.5) through which the cytochrome can donate electrons to the reaction center (cf. Hall et al., 1987a, c; van der Waal et al., 1987). Differences between the charged residues that could be utilized by horse cytochrome c and cytochrome c_2 (mimicked by cytochrome c of *P. denitrificans*) are consistent with different orientations of these cytochromes that were detected by linear dichroism (Tiede, 1987). The distance of closest approach between the cytochrome heme and the tetrapyrrole rings of the special pair is 11 Å, and a bridging position of the aromatic Tyr-L162 might facilitate the electron transfer (Michel et al., 1986b; Allen et al., 1987a), as proposed earlier for electron transfer reactions involving cytochrome c (Chap. 2, section 10.5).

6.7 Reaction Centers of Plant and Algal PS I and II

6.7.1 PS I

For both PS I and II it is difficult to define the reaction center as precisely as in the purple bacteria because of the problem of defining the minimum photochemically active unit. These centers contain more polypeptides than do the bacteria, and the reaction center polypeptides also bind many antenna chlorophylls. There is no obvious homology between the sequences of the bacterial reaction center polypeptides and those of the plant and algal PS I reaction center. For the latter, two very homologous genes separated by 25 nucleotides have been identified from light-stimulated transcripts, with the reading frames corresponding to polypeptides of 83,200 (upstream gene, psaA1) and 82,500 (psaA2) (Fish et al., 1985). These genes appear to be highly conserved, and have 95% identity in the cyanobacterium, *Synechococcus sp.* PCC 7002 (Cantrell and Bryant, 1987). P700, like the bacterial reaction center, had been inferred to be a dimeric "special pair," (Chl $a)_2$, initially because the line width of the P700$^+$ free radical EPR signal was narrower by about a factor of $\sqrt{2}$ relative to the Chl a^+ monomer (Norris et al., 1971).

The electron acceptor system of PS I is the most complicated of all the reaction centers, with as many as five different prosthetic groups identified on the basis of distinct EPR spectra and E_m values. The electron acceptor system contains, in the order of more positive E_m, chlorophyll, possibly a phylloquinone (vitamin K_1), and three iron-sulfur clusters (Table 6.6), whose primary photochemistry is summarized in Fig. 6.20A.

A PS I complex has been isolated from spinach chloroplasts containing five small polypeptides in addition to the MW 83,000 and 82,000 psIA1 and

Table 6.5. Complementary charge pairs in the complex between the *Rb. sphaeroides* reaction center and cytochromes c and c_2

Cyt c	Cyt c_2	RC
K-10	K-10	D-M290
K-25	R-32	D-L155
(T-47)	K-55	D-L257 (R)[a]
E-50	E-59	K-L268 (D)
E-69	D-82	K-M108 (H)
—	D-93	R-M86
—	K-95	D-M87 (Q)
—	K-97	D-L261
K-79	K-99	D-M182
K-85	K-105	E-M99 (Q)
K-86	K-106	E-M94 (Y)

[a] Residues in parentheses are found in the *Rps. viridis* RC.
From Tiede et al. (1988).

Table 6.6. Redox and EPR properties of PS I reaction center and electron acceptors

Component notation	Identity and function	E_m	EPR properties
P700	Reaction center chlorophyll, probably a dimer of chlorophyll a	+500	Oxidized form: $g = 2.0026$, ~8 gauss linewidth
A_0	Intermediate electron acceptor, monomeric chlorophyll	~ −900 mV	Reduced form, Chl$^-$, $g = 2.0024$, 8 gauss linewidth
A_1	Electron acceptor, possibly phylloquinone (vitamin K_1)	Not known	Reduced form: $g = 2.0051$ ~10.5 gauss linewidth
Fe-S$_X$	Electron acceptor: most reducing [4Fe-4S] iron-sulfur center	−730 mV	Reduced form: g-values of 2.08, 1.88, 1.78
Fe-S$_B$	Electron acceptor; [4Fe-4S] iron sulfur center	−590 mV	Reduced form: g-values of 2.05, 1.92, 1.89
Fe-S$_A$	Terminal electron acceptor in PS I complex; [4Fe-4S] iron-sulfur center	−530 mV	Reduced form: g-values of 2.05, 1.94, 1.86

From Malkin (1987); Golbeck et al. (1987); Høj and Lindberg-Møller (1986); Høj et al. (1987).

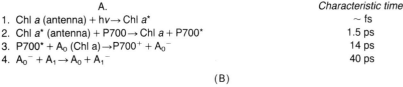

A.		Characteristic time
1. Chl a (antenna) + $h\nu \rightarrow$ Chl a^*		$\sim$ fs
2. Chl a^* (antenna) + P700 $\rightarrow$ Chl a + P700*		1.5 ps
3. P700* + A_0 (Chl a) $\rightarrow$ P700$^+$ + A_0^-		14 ps
4. A_0^- + $A_1 \rightarrow A_0 + A_1^-$		40 ps

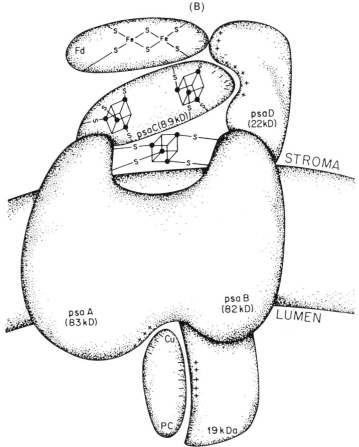

Figure 6.20. (A) Primary photochemistry of PS I (Wasielewski et al., 1987). The electrical charge separation in PS I occurs in a time ≤ 50 ps, setting this as an upper limit for light energy transfer and primary trapping (Trissl et al., 1987), with a more accurate assessment of 14 ps based on measurement of absorbance changes (Shuvalov et al., 1986; Wasielewski et al., 1987). The estimate of 40 ps for $A_0^- + A_1 \rightarrow A_0 + A_1^-$ is based on a transient absorbance change at 730 nm, the isobestic point of P700$^+$ (Fenton et al., 1979). (B) Proposed topography of the PS I complex showing five of the seven polypeptides isolated in the PS I reaction center particle from spinach chloroplasts. The MW 83,000 and 82,000 psaA1 and psaA2 gene products contain the two P700 special pair chlorophyll molecules and the [4Fe-4S] acceptor cluster "X" (Golbeck et al., 1987). Two [4Fe-4S] acceptor clusters, "A" and "B", are bound in the M_r 9,000 polypeptide. M_r 22,000 and 19,000 polypeptides interact with ferredoxin and plastocyanin, respectively, at the stroma and lumen surfaces. (From R. Malkin.) The function of the two other polypeptides in the complex is not known. This complex does not fit the rigorous definition of a reaction center since the large polypeptides that bind the special pair chlorophylls also bind many (ca., 40) antenna chlorophylls.

psIA2 gene products, for a total of seven. Five of these seven polypeptides have a known function in binding functional redox groups or binding peripheral polypeptides, plastocyanin on the lumen side as a donor to PS I and ferredoxin as an acceptor on the stromal side of the membrane (Fig. 6.20B).

6.7.2 PS II

The plant PS II reaction center clearly differs from those of bacteria and PS I on the electron donor side due to the presence of the water splitting system as the electron donor. The primary acceptor is pheophytin (Klimov and Krasnovskii, 1982), which accepts an electron in ~ 3 ps (Schatz et al., 1988). The reduction of Q_A then occurs in ~ 500 ps, and transfer between plastoquinones

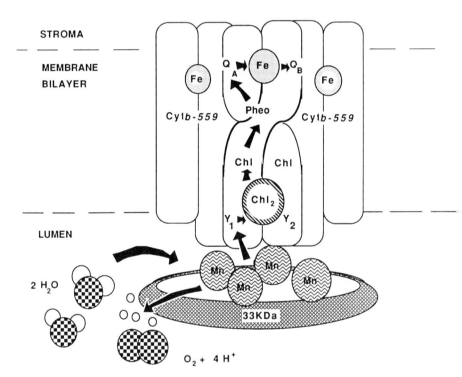

Figure 6.21. Representation of the minimal PS II reaction center with the extrinsic 33 kDa polypeptide. Cytochrome b-559 in the reaction center is depicted as a 9 kDa–4 kDa heterodimer (Widger et al., 1985) cross-linked by a heme (Fe) on the stromal side of the membrane (Tae et al., 1988), and present in two copies, on opposite sides of D1-D2. D1-D2 contain the binding sites for the P680 reaction center, the donors Y_1 and Y_2, pheophytin, and Q_A-Fe-Q_B. Y_1 and Y_2 are fast and slow electron donors to the P680 reaction center which are Tyr residues on the D1 and D2 polypeptides (Barry and Babcock, 1987; Debus et al., 1988; Vermaas et al., 1988; see section 6.8.1). (Drawing by G.S. Tae.)

Q_A and Q_B on the much slower 0.1–1 ms time scale. The pronounced sequence identities between the L and M polypeptides of the bacterial center, and two D_1 and D_2 polypeptides of plants and algae (Table 5.12), argue that the latter are part of the PS II reaction center. A reaction center core particle of PS II that contains four to six polypeptides and can carry out some photochemistry with high quantum efficiency, but does not evolve oxygen, consists of one copy of the D1 and D2 polypeptide, 4-5 Chl, 2 Phe, 2 plastoquinones, 1 Fe, and one or two copies of the heterodimeric cytochrome b_{559} (Nanba and Satoh, 1987). Except for the cytochrome b_{559}, the complement of pigments and prosthetic groups is analogous to that of the bacterial reaction center preparations. A proposed similar arrangement of ligands to the pigments, Fe, and quinone groups was noted previously (Table 5.12; Fig. 5.18). Some caveats to the analogy are: (i) it is not proven that P680 is a dimer (Rutherford and Acker, 1986); (ii) there is not a conserved His residue in the D1 and D2 sequences at the position corresponding to a His ligand for an auxiliary bridging Chl analogous to those (L153, M180, *viridis*; L153, M182, *sphaeroides*) in the bacterial reaction centers (Table 5.12). It appears, however, that histidine is not an obligatory ligand to the auxiliary BChl molecules, but that other residues, particularly Gln or Asn residues, could serve as well. Mutagenesis of a His ligand to the auxiliary BChl in wild-type *Rb. capsulatus* has shown that it can be replaced (Bylina and Youvan, 1988); (iii) the probable replacement of the Glu ligand to the PS II nonheme Fe by bicarbonate has been noted (footnote, Table 5.9). A representation of the PS II reaction center including the extrinsic 33-kDa polypeptide that participates in binding the manganese needed for O_2 evolution is shown (Fig. 6.21).

6.8 Photosynthetic Water Splitting, O_2 Evolution, and Proton Release by PS II

"The ability of plants to exert a salutary effect on the air," i.e., to produce O_2, was discovered by Priestley in 1771 (Rabinowitch, 1945). Although the possibility that bicarbonate might provide the source of the O_2 has been proposed most recently by Metzner et al. (1979), ^{18}O labeling experiments showed that the predominant source of O_2 must be H_2O (Ruben et al., 1941; Radmer and Ollinger, 1980), as does the flash periodicity of four for O_2 evolution (Fig. 6.22).

The photosynthetic unit for O_2 evolution in *Chlorella* is 2,500 chlorophylls per O_2, approximately 600 chlorophylls used to gather the light energy used to generate each of the 4 positive charges needed to oxidize H_2O to O_2,

$$2H_2O \xrightarrow{4h\nu} O_2 + 4H^+ + 4e^-$$

($E_{m7} = 0.82$ V; $E_{m5} = 0.93$ V near the pH of the thylakoid lumen).

This reaction shows a well-defined periodicity of four as a function of the number of single turnover (1e$^-$ transferred/flash) flashes applied to chloro-

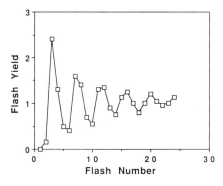

Figure 6.22. The yield of O_2 evolution per flash, normalized to the average steady-state yield, as a function of flash number (based on Joliot et al., 1969; Kok et al., 1970).

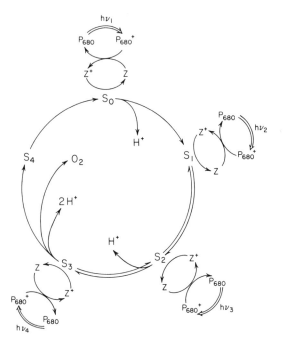

Figure 6.23. Kok's clock: S state cycle for electron transport and H^+ release in photosynthetic O_2 evolution. The pathway of electron transport is from the S states that are different Mn redox states of an Mn-protein complex, to Tyr-161 of the D_1 protein, the primary donor to P680, and then to the reaction center, P680 (Ort, 1986).

plasts (Fig. 6.22), showing that photosystem II units function independently to accumulate the four oxidizing equivalents needed to split one molecule of H_2O to O_2. The first maximum occurs on the third flash. Water splitting occurs in a concerted four electron transfer event and the water splitting enzyme complex accepts and stores four electrons for each O_2 formed and released. The release of protons inside the thylakoid lumen (Fig. 6.23) concomitant with water splitting occurs in the majority of the oxygen evolution complexes (OEC) with a stoichiometry of $(1, 0, 1, 2)$ as a function of flash number (Saphon and Crofts, 1977; Förster and Junge, 1985; Renger et al., 1987).

It can be seen in Fig. 6.22 that maxima for O_2 evolution occur on the 3rd,

6.8 Photosynthetic Water Splitting, O₂ Evolution, and Proton Release   283

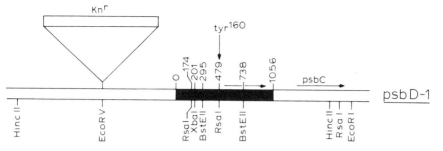

Figure 6.24. The 1056-bp *psb*D-1 gene coding for D2 protein from *Synechocystis* 6803 showing restriction sites, the site (479) of the base substitution in Tyr-160 and an upstream kanamycin (Kn$^+$) resistance site (Debus et al., 1988).

7th, 11th, etc., flashes until the oscillations die out. Successive light-induced electron transfer events in the PS II reaction center advance the OEC through five oxidation states, S_i, $i = 0$–4, with O_2 evolved and released as the highest oxidation state, S_4, reacts back to the lowest, S_0 (Kok et al., 1970). In the model, S_0 and S_1 are stable in the dark, and populated at a ratio, typically 1:3 (Joliot and Kok, 1975) that explains the maximum of the flash O_2 yield on the 3rd rather than the 4th flash (Fig. 6.22). The successive S_i states are higher valence states of a binuclear or tetranuclear manganese complex in the water oxidizing enzyme. The S-state cycle including the (1, 0, 1, 2) pattern of H$^+$ release is summarized by "Kok's clock" (Fig. 6.23). The following features of the model explain the oscillatory behavior of O_2 evolution. (i) The water-splitting centers act independently. (ii) The S_i state transitions are one quantum events in which P680 oxidizes the tyrosine [Tyr-OH → TyrO$^{\cdot}$ + H$^+$, probably Tyr-161 ("Y_1") on the D1 polypeptide (Fig. 6.21)], and the S_i Mn states are oxidized by this tyrosine. (iii) The oscillation is damped because a small fraction of the centers do not advance because of "misses" ($\sim 10\%$ in chloroplasts, 20% in algal cells), and another small fraction can advance through two states because of "double hits." (iv) S_0 and S_1 states are stable in the dark. S_2 and S_3 survive for several seconds before they deactivate to S_1, and the steady-state population of S_4 is very small because it rapidly reacts to form O_2 and the S_0 state.

6.8.1 Applications of Molecular Biology to Photosynthetic Electron Transport: Studies on Tyrosine as the Primary Electron Donor to P680$^+$

EPR studies had indicated that the electron carrier donating directly to P680$^+$ was a quinone cation radical. This concept was contradicted, and a tyrosine implicated as the relevant carrier, when it was found that deuteration of plastoquinone methyl groups did not affect the width of the primary donor EPR signal, whereas a significant narrowing of this signal occurred when the cyanobacteria were grown with deuterated tyrosine (Barry and Babcock, 1987). The PQ was deuterated via deuterated methionine fed to a methionine auxotroph of the cyanobacterium *A. variabilis*. From examination of the sequences (Table 5.12) and proposed structures of the D1 (Fig. 5.18B) and D2 polypeptides, it was hypothesized that one of the tyrosine pair (Y-161 of the D1 polypeptide and Y-160 of D2), which are close to the histidine ligand(s) proposed to coordinate P680 on the lumen side of the bilayer, is the direct electron donor to P680$^+$. The cyanobacterium, *Synechocystis* 6803, was used for site-directed mutagenesis because it is photoheterotrophic (i.e., grows without PS II) and transformable (takes up DNA). The mutagenesis of these genes is complicated by the presence of three copies of the gene coding for D1 (*psb*A) (Golden et al., 1986) and two copies of that coding for D2 (*psb*D). One copy of D2 was deleted, and a genomic fragment containing the other D1 gene (*psb*D-1) isolated using restriction enzymes (Fig. 6.24). The mutagenesis of *psb*D-1 was accomplished using a synthetic 15 base oligonucleotide that matched the *psb*D-1 gene at all 15 positions except at one base coding for Tyr-160 (codon, TAC; mutation, TAC → TTC, phenylalanine) The mutant could be generated by filling in the mutant strand to make double stranded DNA, selecting for the mutant versus the wild type DNA strand, and then finally transforming back into *Synechocystis*. The resultant mutant grew photosynthetically, but lacked an EPR signal that was known to be the slowly turning over companion of the primary donor to P680$^+$. It was concluded that the slow EPR signal arises from Tyr-160 of the D2 polypeptide (Y_2) on the slowly functioning branch of the PS II reaction center. It was then shown that Tyr-161 of the D1 polypeptide (Y_1) is the fast (50–250 ns) electron donor to P680$^+$ (Debus et al., 1988; Vermaas et al., 1988; Ikeuchi et al., 1988).

6.8.2 Active PS II-Enriched Preparations

Studies of PS II structure, redox reactions, and primary photochemistry (Fig. 6.25A), and polypeptide content (Tables 6.7 and 6.8), have been aided by the use of highly active (500–2000 μmol O_2/mg chl-h) PS II particle preparations (Stewart and Bendall, 1979; Berthold et al., 1981; Kuwabara and Murata, 1982; Dunahay et al., 1984). The water splitting redox components on the oxidizing side of PS II that have been studied in these particles are summarized (Table 6.7), along with the sequenced polypeptide components of PSII (Table 6.8).

A minimum PS II reaction center complex that can accomplish photochemistry with a high quantum efficiency has been prepared that consists of only the four polypeptides, D_1, D_2, and the two small cytochrome b_{559} subunits (Nanba and Satoh, 1987), and one to two small hydrophobic polypeptides. This five to six polypeptide complex does not bind Q_A or Q_B and is not sufficient for the water splitting reactions. The 7–8 proteins listed in Table 6.8 must be present (Fig. 6.25B) as part of the minimum structural unit

6.8 Photosynthetic Water Splitting, O₂ Evolution, and Proton Release

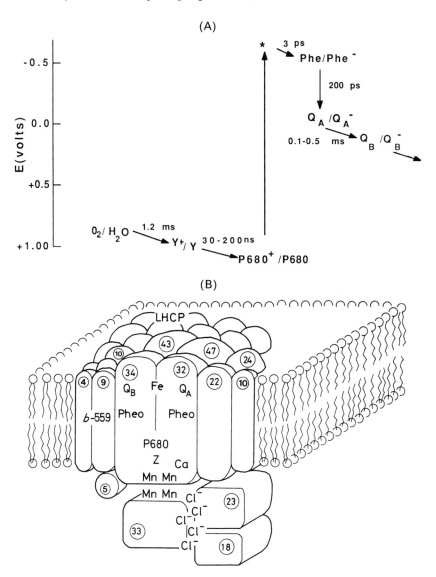

Figure 6.25. (A) Kinetics and E_m values of PS II components involved in a linear description of primary redox events. (B) Topography of photosystem II polypeptides (Murata and Miyao, 1987). Approximate molecular weights are shown, along with the presence of Ca^{++}, Cl^- and manganese; Y is a tyrosine (Y-161) in the D1 polypeptide that is the primary donor to P680, Q_A and Q_B the quinone secondary electron acceptors, and LHCP the light-harvesting pigment protein.

Table 6.7. Redox components with well-defined function associated with the photosystem II reaction center

Species	Detection	Function	Stoichiometry (per P680)	Binding Site, Polypeptide
P680	Optical, EPR	Reaction center	1	D_1, D_2[a]
Tyrosine	Optical, EPR	Primary donor to P680	1	D_1 (Tyr-161 on D_1 protein, Y
Cytochrome b_{559} heme	Optical, EPR	Probable protective function against stress (e.g., heat, light, O_2)	2	4, 9 kDa
Mn	Optical, EPR, EXAFS	Water oxidation	4	Interface: (D_1–I extrinsic 33
Phe a	Optical	Primary electron acceptor	2	D_1, D_2
PQ_A, PQ_B	Optical, EPR	Primary quinone-Fe^{++} electron acceptor	1	D_1, D_2
Fe	Optical, EPR, Mössbauer	Primary quinone-Fe^{++} acceptor complex	1	D_1, D_2

[a] By structural analogy with L, M polypeptide of bacterial reaction center.

Table 6.8. PS II polypeptides necessary for O_2 evolution[c]

$M_r \times 10^3$	Molecular weight	Nomenclature	Prosthetic Grou
47–51	56,246 (Morris and	CP47 (psbB)[a]	Antenna Chl
41–45	51,816 (Herrmann, 1984)	CP43 (psbC)	
33	39,465 (Rochaix et al., 1984; Alt et al., 1984)	D_2 (psbD)	
33	38,950 (Zurawski et al., 1982) 34,600 (Hirschberg and McIntosh, 1983)	D_1, HBP[b], Q_B-Protein (psbA)	P680, Chl, Phe, Q_B, Fe
33 (Extrinsic)	26,663 (Oh-oka et al., 1986)	33 kDa	Possibly Mn
9	9,162 (Herrmann et al., 1984;	Cyt b_{559} (psbE, F)	2 Heme b
4	4,268 (Widger et al., 1985)		

[a] psbA-F are chloroplast-encoded proteins; the extrinsic 33-kDa polypeptide, as well as the extrinsic 23- 17-kDa proteins are nuclear-encoded.
[b] Herbicide-binding protein.
[c] 4.8 kDa psb I gene product probably in reaction center and O_2-evolving preparations (Ikeuchi and Inc 1988).

6.8 Photosynthetic Water Splitting, O_2 Evolution, and Proton Release 287

in a complex that is competent for O_2 evolution (Ghanotakis et al, 1987). In addition to the proteins listed in Table 6.8, several other small hydrophobic polypeptides are present in PS II, and may be essential for O_2 evolution: (i) hydrophobic M_r 22,000 and 24,000 components that bind chlorophyll and the extrinsic M_r 23,000 protein, (ii) two M_r 10,000 proteins and possibly several small polypeptides ($M_r \leq 5,000$).

6.8.3 Role of Manganese

Redox transitions involving O_2 invariably involve catalysis by a rare earth metal or ion. Four manganese ions are known to be intrinsic to PS II and necessary for its function (Radmer and Cheniae, 1977; Yocum et al., 1981). Removal of two of four Mn present was found to totally inactivate O_2 evolution (Kuwabara et al., 1985). The other metalloprotein in the PS II reaction center is the cytochrome b_{559}, which contributes one or two hemes to the reaction center on the stromal side of the membrane (Fig. 6.21), but which does not seem to have a redox function in the primary photochemistry or in the pathway from water to P680. There is a structural requirement for the cytochrome in charge separation and water splitting. The actual redox function of this cytochrome may be related to the ability of photosystem II to respond to stress (Cramer et al., 1986). PSII is the most heat-labile of the photosynthetic membrane protein complexes. The lability relates to the loss or displacement of manganese, and PS II is readily damaged by heat (T $\geq$ 40°C), chilling, and water stress. It is also the site of photoinhibition, which may be related to the strong oxidant ($E_m \simeq +1.2$ V) generated by P680$^+$ (Thompson and Brudvig, 1988).

Although aqueous manganese(II) has a pronounced six-line EPR signal, this signal is not representative of most of the membrane-bound Mn, which is largely EPR-silent. The membrane-bound Mn probably has an average oxidation state greater than $+2$ since (i) complete reduction of the Mn ensemble to $+2$ causes increased extractability (Tamura and Cheniae, 1985), and light or oxidizing conditions are required for photoactivation of Mn-depleted chloroplasts on PS II particles in the presence of added Mn(II). The isolated Mn(III)/Mn(II) couple bound to an enzyme such as superoxide dismutase has an E_m value of 0.18–0.32 V (Lawrence and Sawyer, 1978), much below the $E_{m7} = 0.82$ V of the O_2/H_2O couple.

The manganese oxidation states associated with the S states have been studied through: (i) the pattern of absorbance changes in the near ultraviolet (300–380 nm), which is complicated by background absorbance changes and electrochromic band shifts due to changes in the net charge of the system (Dekker et al., 1984a,b; Renger and Weiss, 1986; Lavergne, 1987; Saygin and Witt, 1987; Kretschmann et al., 1988); (ii) EPR spectra, particularly of the S_2 state, and (iii) EXAFS (extended x-ray absorption fine structure) and x-ray K-edge spectra. The properties of the S states as they relate to the (1, 0, 1, 2)

Table 6.9. Pattern of electron transfer, H^+ release, net charge, Mn valence, and O_2/H_2O oxidation states associated with the S states of H_2O splitting

S-state transitions	$S_0 \dashrightarrow S_1$	$\dashrightarrow S_2$	$\dashrightarrow S_3$	$\dashrightarrow (S_4)$	$\dashrightarrow S_0$
Absorbed quanta	1	1	1	1	
Electron extraction $P680^+ \leftarrow e^- \ldots S_i$	1	1	1	1	
H^+ release	1	0	1	2	
Changes of positive surplus charge	0	1	0	-1	
Surplus charge	0 0	$+$	$+$		0
Possible states of water	OH^- OH^-	OH^- O^{2-}	OH^- O^{2-}	O^{2-} O^{2-}	$-O_2$ $+2H_2O$	OH^- OH^-
Possible states of Mn if two manganese are engaged in water oxidation	Mn^{2+} Mn^{3+}	Mn^{3+} Mn^{3+}	Mn^{3+} Mn^{4+}	Mn^{4+} Mn^{4+}	Mn^{4+} Mn^{4+Y_1}	Mn^{2+} Mn^{3+}
Possible states of Mn if four manganese are engaged in water oxidation	Mn^{2+} Mn^{3+} Mn^{2+} Mn^{2+}	Mn^{3+} Mn^{3+} Mn^{2+} Mn^{2+}	Mn^{3+} Mn^{4+} Mn^{2+} Mn^{2+}	Mn^{4+} Mn^{4+} Mn^{2+} Mn^{2+}	Mn^{4+} Mn^{4+} Mn^{3+} Mn^{2+}	Mn^{2+} Mn^{3+} Mn^{2+} Mn^{2+}

After Saygin and Witt (1987).

pattern of H^+ release and the oxidation states of a manganese dimer or tetramer are summarized in Table 6.9. The surplus charge in the S_2 and S_3 states is a consequence of unequal stoichiometries of e^- transfer and H^+ release in the preceeding steps. The unexpected variation in the H^+ stoichiometry may result from participation of titratable protein groups.

The pattern of H_2O and Mn states shown in Table 6.9 is a working hypothesis. Because there are four Mn per PS II reaction center; it is of interest that tetranuclear Mn analogues of the Mn OEC complex [e.g., $(Mn_4O_2(O_2CMe)(2,2'-bipyridine)]$ have been synthesized with distance parameters that are similar to those derived from EXAFS data for the OEC complex. In addition, the above compound has oxide bridges between all Mn centers, an average oxidation state of $+3$, and two pairs of inequivalent Mn atoms (Vincent et al., 1987; Vincent and Christou, 1987).

The question of whether the active Mn complex consists of one or two dimers, or a tetramer, is presently under debate. The S state clock of Kok and Joliot is redrawn in Fig. 6.26A using the different oxidation states of an Mn dimer partly for the purpose of simplicity and also because oxygen reduction in cytochrome oxidase of the respiratory chain uses a binuclear center consisting of heme a_3 and Cu_B (Chap. 7, section 7.3). Additional information on the Mn cluster should be attainable through x-ray, EPR, and visible spectroscopic analysis. The S_2 state has been characterized by (i) a 1.8 eV shift to higher energy, relative to the S_1 state, of the Mn x-ray absorption K-edge (Goodin et al., 1984), and (ii) a multiline (~ 18 line) EPR signal near $g = 1.98$

6.8 Photosynthetic Water Splitting, O₂ Evolution, and Proton Release

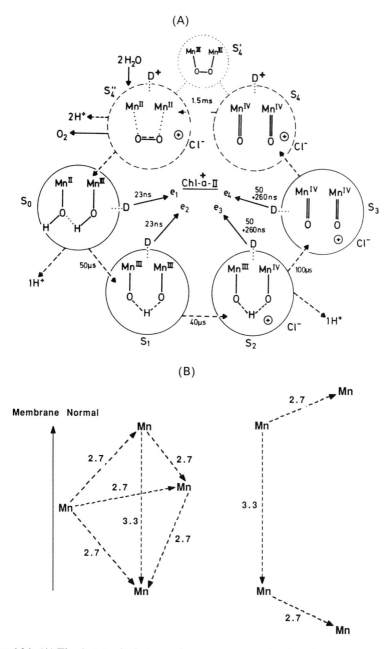

Figure 6.26. (A) The S state clock drawn for an Mn dimer showing the pattern of e^- transfer, H^+ release, and the S-state dependent half-time (50 or 250 ns) of P680$^+$ reduction by the immediate electron donor, Y_1. (After Witt, 1988.) (B) One possible geometrical arrangement (distances in angstroms) of a tetrameric (left) or two dimeric (right) Mn clusters, derived from EXAFS measurements using polarized synchrotron radiation and oriented chloroplasts. (Based on George et al., 1989.)

that could arise from a Mn(III)–Mn(IV) mixed valence dimer (Dismukes and Siderer, 1981) or a 3Mn(III)–Mn(IV) tetramer (de Paula et al., 1986; Brudvig and Crabtree, 1986). The broadening of the multiline signal in ^{17}O-enriched H_2O indicates binding of H_2O, OH^-, or H_2O_2 to the Mn cluster in the S_2 state (Hansson et al., 1986); there is a $g = 4.1$ EPR signal that arises from a subset, or different conformation or redox state of the Mn cluster also associated with the S_2 state, and possibly a precursor of the conformation associated with the multiline signal (de Paula et al., 1986; Zimmerman and Rutherford, 1986; Cole et al., 1987; Hansson et al., 1987). (iii) Fourier-transformed EXAFS spectra, which can probe the ligand environments around a particular metal with high resolution (ca. 10^{-2} Å), have shown that the manganese in the S_1 state exists in an asymmetric probably tetrameric cluster (Fig. 6.26B), with at least two Mn–Mn distance components of 2.7 Å (Yachandra et al., 1986) and 3.3 Å, the latter distance vector oriented perpendicular to the membrane plane, with probable bridging oxide or hydroxide ligands (George et al., 1989).

6.8.4 The Extrinsic Polypeptides and the Requirement for Ca^{2+} and Cl^-

The functions of the three extrinsic polypeptides, M_r 33, 23, and 17,000 bound to the lumen side of the OEC complex are to (i) stabilize or bind part of the Mn cluster and to shield it from exogenous reductants, and (ii) provide an optimum interaction between cation (Mn^{2+}, Ca^{2+}) and anion (Cl^-) cofactors and the PS II core polypeptides (Babcock, 1986). The 17- and 23-kDa polypeptides affect binding of Ca^{2+} and Cl^-, so that the binding constant of Ca^{2+}, approximately 0.3 mM in the presence of these two polypeptides, increases by about 10-fold in their absence. Solubilization of only the extrinsic 33-kDa polypeptide by (1 M NaCl + urea) can cause Mn release, whereas release of the extrinsic 17-kDa and 23-kDa polypeptides does not. Extraction with $CaCl_2$ can release all three extrinsic polypeptides, whereas the depleted particles retain almost all the Mn (Ono and Inoue, 1983) which is able to promote rebinding of the 33-kDa protein. The amino acid sequence of the 33-kDa polypeptide indicates a region with possible homology to Mn-superoxide dismutases (Oh-Oka et al., 1986). These data suggest that the Mn is bound jointly by the intrinsic D_1 and D_2 polypeptides, with two of the four Mn more accessible to the aqueous phase (Fig. 6.25A) and stabilized or bound by the extrinsic M_r 33,000 polypeptide (Table 6.10). When PS II membranes are depleted of the extrinsic polypeptides under conditions that allow retention of the functional Mn, substantial activity can be restored by addition of high concentrations of Ca^{2+} (~ 10 mM) and Cl^- (200 mM required if 33-kDa protein is absent; 10 mM if only the 17-kDa protein is removed) (Table 6.10).

The Ca^{2+} requirement is absolute as no other cation can substitute. However, the Cl^- can be replaced by Br^- or NO_3^-, or with lower specificity by I^-, and HCO_3^-, and the optimum response is associated with a stereochemical

Table 6.10. Properties of extrinsic PS II polypeptides

	Role in O_2 evolution	Mn binding	Salt effects
M_r 17,000[a]	Nonessential	No effect	10 mM Cl^- substitute for 17-kDa
M_r 23,000[a]	Helpful; depletion of both 18, 23-kDa blocks at S_2 or S_3	No effect	10 mM Ca^{2+} substitute for 23-kDa
M_r 33,000	essential in low Cl^- concentration	Responsible for stabilization or retention of 2 Mn	200 mM Cl^- can substitute for 33-kDa, restore 20% activity; 4 Mn needed for maximum binding of 33-kDa

[a] Not present in cyanobacteria.
From Homann (1986); Ghanotakis et al. (1987); Murata and Miyao (1987).

requirement for a particular ionic volume, ca. 0.025 nm^3 (Critchley, 1985). The Cl^- requirement can be demonstrated at neutral pH in chloroplasts from halo- or salt-tolerant (~ 0.25 M NaCl) plants such as mangrove that function optimally with a salt concentration of 0.25 M NaCl. The PS II activity of such halo-tolerant plants is inhibited in salt solutions whose concentration (~ 10 mM) would be close to optimum for chloroplasts from plants such as spinach or pea that cannot tolerate such high salt levels. The chloride requirement can be demonstrated in spinach or pea chloroplasts by depleting the Cl^- to a level < 1 mM (Critchley, 1985). Cl^- removal results in an inability to reach the higher S states, manifested in a lower yield of the S_2 multiline EPR signal (Damoder et al., 1986). The oxygen-evolving complex in Cl^--deficient thylakoids cannot store more than two oxidizing equivalents (Theg et al., 1984; Itoh et al., 1984; Zimmerman and Rutherford, 1986), perhaps because of a decrease in the E_m of the higher S states so that they can no longer serve as effective oxidants (Homann et al., 1986). Like Ca^{2+}, Cl^- may also have a role in facilitating protonation of H_2O ligands at the Mn centers during H_2O oxidation (Table 6.9), and maintaining net charge neutralization and protein conformation (Govindjee et al., 1985).

6.9 The Cyclic and Noncyclic Electron Transfer Chains

Cyclic electron transport in photosynthetic bacteria uses the transmembrane reaction center and cytochrome bc_1 complexes (Chap. 7, section 7.4) connected by the diffusive carriers ubiquinone and cytochrome c_2 to accomplish H^+ translocation, and synthesis of ATP and pyrophosphate (ca. 800–900 and 100–150 μmol/mg BChl-h, respectively, in chromatophores of *Rh. rubrum* under continuous illumination; Nyrén et al., 1986).

PS I-linked ferredoxin-dependent cyclic phosphorylation can also be demonstrated in chloroplasts (Tagawa et al., 1963) which must supply sufficient ATP for CO_2 fixation. ATP and/or the $\Delta\tilde{\mu}_{H^+}$ generated by this cyclic pathway can support transport of acetate, sugars, and phosphate in algae and cyanobacteria (Simonis and Urbach, 1973), and transport of bicarbonate used for CO_2 fixation in cyanobacteria (Ogawa et al., 1985). The requirement of ATP from the PS I cyclic pathway for CO_2 fixation may depend upon the yield of ATP ($ATP/2e^-$) from the noncyclic ($H_2O \rightarrow NADP^+$) pathway (Chap. 7, section 7.5), and from pseudocyclic phosphorylation in which O_2 ($O_2 + e^- \rightarrow O_2^{\dot{-}}$) rather than $NADP^+$ is the terminal electron acceptor (Egneus et al., 1975). The PS I-dependent cyclic electron transport could be especially important in generating a $\Delta\tilde{\mu}_{H^+}$ or ATP for CO_2 fixation or active transport when PS II, which is very labile, is inactivated under conditions of environmental stress. Ferredoxin is an important branch point in these reactions. It can support electron transfer reactions to many different acceptors in chloroplasts and algae (Chap. 4, section 4.9.3), including noncyclic electron transport to $NADP^+$ via the ferredoxin-$NADP^+$ oxidoreductase (FNR) (Chap. 4, section 4.9.3; Carrillo and Vallejos, 1987), or to O_2 in pseudocyclic electron flow, and a pathway of cyclic phosphorylation.

6.9.1 Noncyclic Electron Transport and H^+ Translocation

The linear noncyclic electron transport chain in chloroplasts shown in the Z format with the redox scale as the ordinate (Fig. 6.27A) deposits 2–3 H^+ into the thylakoid lumen for each electron transported from H_2O to $NADP^+$, i.e., $H^+/e = 2$–3. The diffusible plastoquinol (PQH_2) formed by PS II is oxidized to the semiquinone by the Rieske iron-sulfur protein of the cytochrome $b_6 f$ complex. Ultimately, each electron is transferred to the high potential donors to P700, the ISp, cytochrome f, plastocyanin [or cytochrome c in some algae or cyanobacteria (Chap. 4, section 4.8)], P700, and then to $NADP^+$ using the P700 photoreaction and ferredoxin. The role of the $b_6 f$ complex in quinol oxidation is discussed below at greater length (Chap. 7, section 7.4). The sites of H^+ deposition (Fig. 6.27B) are H_2O oxidation in PS II ($H^+/e = 1$) and quinol oxidation by the cytochrome $b_6 f$ complex, for which $H^+/e = 1$–2 (Chap. 7, section 7.4). A doubling of the H^+/e ratio associated with electron transport through the cytochrome $b_6 f$ complex can be observed in chloroplasts (Hangarter et al., 1987) and may occur through the mechanism of an H^+ pump or a Q cycle involving this complex (Chap. 7, section 7.4). If the H^+/e ratio for steady-state whole chain electron transport is (i) 2 or (ii) 3, and if in each case 3 H^+ are utilized per ATP synthesized (chapts. III, VIII), then the maximum vaue of the $ATP/2e^-$ ratio for noncyclic electron transport would be (i) 1.33 or (ii) 2.0, respectively. The maximum $ATP/2e^-$ ratio that has been observed experimentally for the noncyclic pathway was 1.26 (Table 6.11, A) close to the value predicted if $H^+/e = 2$ for the noncyclic chain (Chap. 7, section 7.5, Table 7.8).

6.9 The Cyclic and Noncyclic Electron Transfer Chains 293

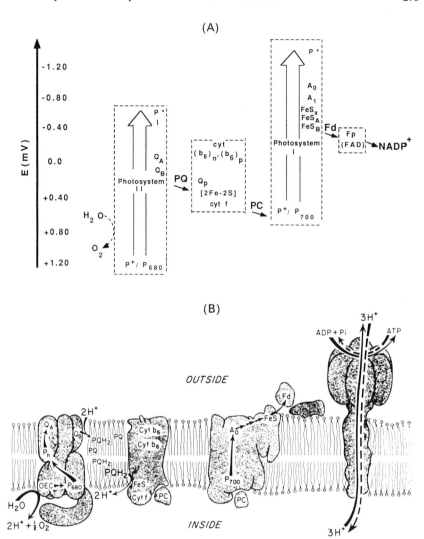

Figure 6.27. (A) "Z" scheme for pathway of oxygenic noncyclic electron transport chain to $NADP^+$ as terminal electron acceptor. Membrane-bound complexes and proteins discussed above in this chapter, and in Chap. 7, Section 7.4, are enclosed by dashed boxes. Q_p is a quinone binding site on the p (lumen) side of the membrane in $b_6 f$ complex. I, pheophytin (ϕ in section 6.6). (B) Sites of H^+ deposition in noncyclic electron transport (from Ort and Good, 1988). (C) Proposed ferredoxin-dependent pathway for cyclic phosphorylation around PS I involving the cytochrome $b_6 f$ complex. (i) Light-induced transfer of one electron in PS I reduces Fd and oxidizes PQH_2 to the anionic semiquinone ($PQ^{\overline{\cdot}}$). (ii) Reduction of heme b_p by $PQ^{\overline{\cdot}}$, completing the oxidant-induced reduction of heme b_p (see Chap. 7, section 7.4). (iii) Reduction of heme b_n by reduced ferredoxin. (iv) Cooperative two electron oxidation of the two hemes by PQ, forming PQH_2 and completing the cycle. An alternative pathway would involve ferredoxin reduction of PQ via a putative (?) ferredoxin-plastoquinone oxido-reductase (drawing by P.N. Furbacher).

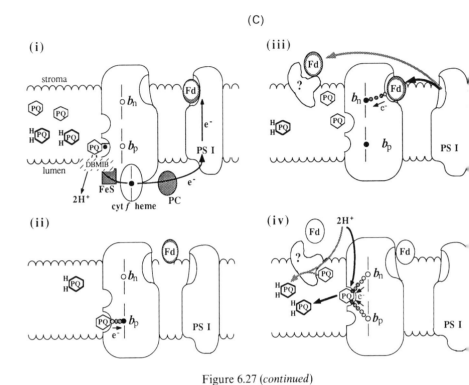

Figure 6.27 (continued)

Table 6.11. Chloroplast photophosphorylation under conditions of noncyclic and cyclic electron transport

Acceptor	Electron Transfer Rate (μeq/mg Chl-h)	Photophosphorylation Rate (μmol ATP/mg Chl-h)	P/2e^-
A. Methyl viologen (noncyclic)	200	505	1.26
B. Ferredoxin/O_2 (noncyclic + PS I cyclic)	96	308	1.60
C. Ferredoxin/$NADP^+$ (noncyclic + PS I cyclic)	159	501	1.58

From Hosler and Yocum (1985).

When ferredoxin was used as an acceptor in the presence of either $NADP^+$ or O_2, the P/2e ratio increased to 1.6 (Table 6.11B, C). The higher $P/2e^-$ ratio in the presence of ferredoxin indicates the operation of a PS I-dependent cyclic pathway, involving the plastoquinone pool and cytochrome $b_6 f$ complex. A hypothesis for the pathway of this PS I cycle is proposed in Fig. 6.27C. Oxidation of P700 would cause oxidation of the PS I donors and of PQH_2 to the semiquinone that could reduce the lumen-side heme of cytochrome b_6 (see Chap. 7, section 7.4). The concomitant reduction of the stromal side heme from the reducing side of PS I would result in both hemes of cytochrome b_6 being reduced. Their cooperative oxidation by plastoquinone located near the center of the bilayer, forming PQH_2, would complete the cycle.

6.10 Summary

Light energy is transferred by exciton and resonance mechanisms from the antenna pigment molecules of the photosynthetic membrane to the reaction center on a picosecond time scale. For the mechanism of resonance energy transfer, which is also relevant to mapping macromolecules such as CF_1 ATPase, the rate constant for energy transfer varies inversely as the sixth power of the distance of separation of donor and acceptor, and the transferred energy tends to be shifted ("red-shifted") toward lower energy. This can be demonstrated in the photosynthetic pigment system of the phycobilisomes of cyanobacteria or red algae, where the energy is transferred from phycoerythrin → phycocyanin → allophycocyanin → chlorophyll *a*. Besides the globular phycobiliproteins and the bacteriochlorophyll *a*-binding protein from green bacteria that have been crystallized, there are at least three other types of antenna polypeptides: fibrillary subunits in the chlorosome structure of green bacteria, small hydrophobic transmembrane polypeptide in bacteria, and larger proteins, with both transmembrane and prominent polar domains in algae and chloroplasts. The chlorophyll and bacteriochlorophyll in these pigment-proteins are 5-coordinated, with the magnesium coordinated by one amino acid that can be histidine, glutamine, or asparagine. The small hydrophobic bacteriochlorophyll-binding proteins span the membrane bilayer once as an $\alpha\beta$ dimer, with one histidine residue in each polypeptide involved in pigment coordination on the periplasmic side of the bilayer. The major light harvesting chlorophyll polypeptide of PS II (LHC II) is believed to span the membrane three times. Its NH_2-terminus, like that of almost all thylakoid membrane proteins, is on the stromal side of the membrane. Regulation of the distribution of excitation energy between the two photosystems can occur through phosphorylation of the LHC II polypeptides by a membrane-bound kinase near the NH_2-terminus. This is a unique example of a kinase under redox control. Phosphorylation is half-maximal and the kinase half-activated at an $E_{m7} \simeq +100$ mV that is close to the E_{m7} of plastoquinone. This regulatory mechanism is activated when excess excitation energy absorbed by PS

II results in reduction of the quinone pool to quinol, thereby activating the kinase and phosphorylating the LHC II. This results in lateral movement of the LHC II from the grana, that are enriched in PS II, to PS I in the peripheral region of the grana and the stroma, causing a redistribution of the pigment and the absorption of excitation energy from PS II to PS I.

Excitation energy is transferred to the absorption bands of the reaction center, centered at 680, 700, 865, 870, and 940 nm in PS II, PS I, green bacteria, and the purple bacteria *Rb. sphaeroides* and *Rps. viridis*, that lie on the long wavelength side of the antenna absorption peaks, as expected from resonance energy transfer theory. The reaction center traps excitation energy with a quantum efficiency close to 1.0 because of the rapid (~ 3 ps) transfer of the electron from its excited state to the first electron acceptor, bacteriopheophytin in the bacteria, pheophytin in PS II, and probably a chlorophyll in PS I. The reaction center from *Rps. viridis* containing four polypeptides and *Rb. sphaeroides* containing three are presently the only membrane proteins or protein complexes for which crystals have been obtained that diffract to high resolution, and the structure solved. The L, M, and H (light, medium, and heavy) polypeptides in *viridis* and *sphaeroides* contain, respectively, 273, 323, 258, and 281, 307, and 260 amino acids. The "H" notation turned out to be a misnomer based on anomalous mobility on gels. These three polypeptides result in a structure with 11 transmembrane helices, an approximate twofold axis of symmetry defined by the center of the special pair $(BChl)_2$ on one side of the bilayer and the Fe on the other. A pair of auxiliary BChl, a pair of BPhe, and a pair of bound quinones (Q_A and Q_B; Q_B not bound strongly in crystals of *viridis*), are distributed approximately symmetrically about this axis. The symmetry extends to the similarity of distances and orientations between the prosthetic groups on the two sides. However, electron transfer across the membrane bilayer occurs almost entirely through the Q_A side of the structure, the first detectable event being the 3 ps transfer to the A side bacteriopheophytin. The functional asymmetry may be associated with asymmetry of the amino acid and electron density distribution on the A and B sides.

The reaction center complex from the chloroplast PS I and II has not been defined as precisely as in the purple bacteria because these centers are structurally more complicated and also bind antenna chlorophyll. The major polypeptide products of the PS I center, which are believed to bind the special pair chlorophyll, $(Chl)_2$, have MW = 83,200 and 83,500. Of the five other polypeptides identified in the PS I complex, three have a known function, the M_r 9,000 binding two [4Fe-4S] clusters, the M_r 22,000 binding ferredoxin, and the M_r 19,000 binding plastocyanin on the lumen side of the membrane. The minimum PS II reaction center contains five different polypeptides, the M_r 33,000 D1 and D2 which have extensive sequence identity to the bacterial L and M subunits, the small α and β subunits of the dimeric cytochrome b_{559} dimer, two dimers binding two hemes, and at least one other small hydrophobic polypeptide. The electron acceptor system of PS I contains as many as five different prosthetic groups, chlorophyll, possibly vitamin K_1, and three

[4Fe-4S] centers identified from EPR spectra and redox titrations. Electron transfer to the first detectable acceptor has a characteristic time of 14 ps. Although the sequence identities between L-M and D1-D2 suggest appreciable functional analogy, proof is still lacking that the P680 special pair chlorophyll is dimeric. Besides charge separation across the membrane, the PS II reaction center functions on its donor side to oxidize H_2O to O_2. The four-electron transfer event needed for the reaction can be observed as a four-flash periodicity in O_2 evolution, described as 4 S states, with an H^+ release pattern of 1, 0, 1, 2. The pathway from H_2O to P680 ($E_m = +1.17$ V) contains four manganese ions that, in turn, donate to a tyrosine ($E_m \simeq +0.9$ V) which is the donor to the P680$^+$. Studies on the mechanism and structure of PS II have been aided by highly enriched PS II preparations. The minimum PS II particle that is active in O_2 evolution contains at least 7-8 distinct polypeptides, the five of the minimum reaction center preparation, and the chlorophyll binding polypeptides CP47 and CP43 (MW $\simeq$ 56,000 and 52,000), and one polypeptide bound extrinsically to the lumen side ("33 kD," MW = 26,663). The H_2O-oxidizing Mn complex probably has an average oxidation state $> +2$, increasing to $+3$ or $+3$ to $+4$ as the S states advance. Information on the Mn-Mn and Mn-ligand interactions, as well as distances, can be obtained from EPR and EXAFS spectra. Besides manganese, Ca^{2+} and Cl^- ions are essential for O_2 evolution.

The electron transport pathway in photosynthetic bacteria is called "cyclic" because the electrons are not removed from the transport system by reduction of a metabolite that is withdrawn, as in the $NADP^+$ reduction pathway of "noncyclic" oxygenic photosynthesis. The oxygenic system also supports a cyclic pathway linked to PS I that supplies extra ATP for CO_2 fixation and, at least in algae and cyanobacteria, utilizes ATP or $\Delta\tilde{\mu}_{H^+}$ to support transport of acetate, sugars, or phosphate. These pathways energize the membrane by H^+ translocation. In oxygenic photosynthesis, the maximum steady-state stoichiometry for ATP synthesis, ATP/2e$^-$ = 1.33, in noncyclic electron transport is associated with a stoichiometry, H^+/e^- = 2, for proton translocation.

Problems

64. If the value of the characteristic distance for resonance energy transfer from tryptophan to NADH is 25 Å, what is the yield of this transfer if the distance of separation is (a) 12.5 Å and (b) 50 Å?
65. (a) Write the differential equation for the decay of the fluorescent excited state (S^*) of a molecule that loses energy only because of fluorescence emission. (b) Write a similar differential equation if the energy loss is also to heat. (c) Define the lifetime of fluorescence for parts (a) and (b) in terms of the constants of the differential equations in (a) and (b). (d) Write the differential equation for the decay of the excited state of a fluorescent molecule if it loses energy only to fluorescence and energy transfer to an

acceptor molecule. (e) What is the yield or efficiency of energy transfer if the first-order rate constants for fluorescence in the absence of the acceptor, $k_F°$, and for energy transfer, k_T, are 3×10^9 s^{-1} and 10^9 s^{-1}, respectively. (f) What will the rate constant for energy transfer be in (e) when the distance between donor and acceptor is equal to R_o?; when it is equal to $2R_o$? (g) What is the distance between donor and acceptor if the fluorescence lifetime τ_D in the absence of an acceptor is 10.0 ns, if the fluorescence lifetime in the presence of the acceptor, $\tau_{D \to A}$, is 9.846 nsec, and $R_o = 30$ Å?

66. Graph the efficiency of resonance energy transfer as a function of distance (10–75 Å) between the fluorescence probes CPM and TNP-ATP bound to CF_1 if the characteristic distance, R_o, for energy transfer is 45.2 Å (Snyder and Hammes, 1985).

67. Calculate (a) the value of R_o and (b) the interprobe distance for energy transfer between a dansyl energy donor and fluorescein acceptor attached to the CF_1. $\tau_F°$ and $\phi_F°$ are 6 ns and 0.21, respectively, for the dansyl donor in the absence of energy transfer, and ϕ_F in the presence of energy transfer = 0.05. The overlap integral, J, = 7×10^{-14} cm$^3 \cdot$ M^{-1} (Baird et al., 1979).

68. Calculate the efficiency of light energy transfer *in vivo* from β-phycoerythrin to the rest of the phycobilisome pigment-proteins, according to Förster theory, if the fluorescence lifetimes (i) in vivo, and (ii) of the isolated β-PE pigment-protein are 70 ps and 2.4 ns, respectively.

69. (a) Calculate the contribution of ATP from PS I cyclic or pseudocyclic phosphorylation to total ATP synthesis in a plant that fixes CO_2 by the C-3 Calvin–Benson–Bassham pathway (3 ATP:2 NADPH) if the ATP:2e$^-$ ratio for noncyclic electron transport is 1.33.

Chapter 7
Light and Redox-Linked H^+ Translocation: Pumps, Cycles, and Stoichiometry

7.1 Introduction

Two mechanisms of H^+ translocation have been discussed thus far, the uptake and release of H^+ by ubi- and plastoquinone (Chap. 5), and the release of H^+ in the photosynthetic water splitting reaction (Chap. 6.8). The probable involvement of protein acid–base groups in the pathway of H^+ uptake to Q_A and Q_B in the photosynthetic reaction center was discussed in Chap. 5, section 5.8. Particular proteins are known to function in H^+ translocation. Many of the ideas on structure and function of proteinaceous H^+ pumps originated with bacteriorhodopsin (section 7.2). An H^+ pump is an energy-linked H^+ translocation system in a transmembrane protein or protein complex. Transmembrane H^+ pump function has also been documented for the electron transport protein complex cytochrome oxidase of respiratory membranes (section 7.3), and has been proposed for the cytochrome bc_1 ($b_6 f$) complex in respiratory and photosynthetic membranes (section 7.4). The bacteriorhodopsin and cytochrome oxidase pump systems, as well as the chloroplast CF_0-CF_1 ATPase pump, all have a characteristic maximum rate of H^+ translocation, 200–500 H^+/pump-s, corresponding to a common millisecond rate-limiting step and implying a common mechanism.

7.2 Bacteriorhodopsin, a Well-Characterized Light-Driven H^+ Pump

Bacteriorhopsin (bR) is an intrinsic membrane protein that is a primitive energy transducer and light-dependent proton pump. Its structure has been extensively characterized by electron diffraction studies, although it has not yet been possible to obtain crystals that diffract well enough for X-ray structural analysis. bR forms crystalline patches in the cytoplasmic membrane of the salt-tolerant *Halobacterium halobium*. It imparts the purple color to these membranes. These purple bacteria are found in briney waters such as the Dead Sea in Israel where the NaCl concentrations (>4 M) approach saturation. The bR pigment system is synthesized under conditions of limiting O_2 when oxidative phosphorylation is inefficient, and its physiological role is to transduce light energy into a $\Delta\tilde{\mu}_{H^+}$ that can be utilized for ATP synthesis and active transport (Stoeckenius, 1985).

Bacteriorhodopsin contains one molecule of all-*trans* retinal bound through a protonated Schiff base to the ε-amino group of Lys-216 in the 248 residue (MW = 26,653) protein. This contrasts with the 11-*cis* protonated Schiff base of the retinal of the dark-adapted visual pigment-protein, rhodopsin. This chromophore linkage, and nearby negative charges in the protein, determine the peak (λ_{max} = 568 nm in light-adapted bR) of a broad strong absorption band in the green region of the visible spectrum. Absorption of a photon causes isomerization of the all-*trans*-retinal to the 13-*cis* isomer (Fig. 7.1A), and a decrease from ~11 to 3-4 in the pK value of the Schiff base in the excited state (Chap. 2, section 2.11), allowing its deprotonation. The H^+ release occurs in the L → M step of the bR photochemical cycle that occurs on illumination, and which consists of discrete spectral intermediates linked to H^+ extrusion from the cell membrane to the outside medium (Figs. 7.1B, 3.16B). Protons are taken up from the cytoplasm during restoration of the original bR$_{568}$ state, thus generating a $\Delta\tilde{\mu}_{H^+}$ across the membrane. One (Lozier et al., 1976) or two (Ort and Parson, 1979; Govindjee et al., 1980) protons are extruded and taken up with each photochemical cycle, generating a membrane potential of ca. -200 mV, negative inside. The rate-limiting step of the bR photocycle and its coupled H^+ translocation is approximately 5 ms. One reason that the bR, in contrast to rhodopsin, can act as a continuous H^+ pump is that H^+ uptake from the cytoplasm regenerates the original bR$_{568}$ state, whereas the isomerization of rhodopsin leads to dissociation of the retinal. The two chromophore structures of bR are associated with two protein structures that accommodate the all-*trans* and 13-*cis* bR. These two protein conformations are shown as a rectangular → bent conformation that occurs in the L$_{550}$ → M$_{412}$ transition (Fig. 7.1B). This transition has been proposed to disconnect the Schiff base from a residue (A2, e.g., Asp-85, Asp-212) connected to the cell exterior, bringing it to a residue (A, e.g., D-96, section 7.2.1) that is closer to the cell interior.

The ordered arrangement of bR in the purple membrane allowed determination of its tertiary structure by electron diffraction analysis of the two-dimensional crystalline array to a resolution of 2.8 Å (Henderson and Unwin, 1975; Baldwin et al., 1988). This is the highest level of resolution thus far attained by electron diffraction and image reconstruction of two-dimensional crystalline protein arrays. The structure consists of seven transmembrane,

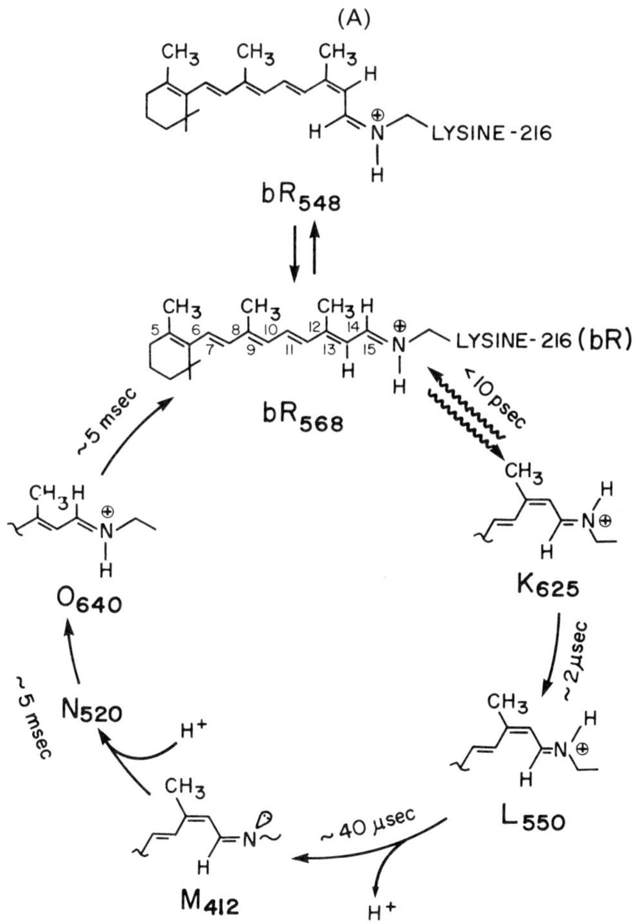

7.2 Bacteriorhodopsin, a Well-Characterized Light-Driven H^+ Pump 301

Figure 7.1. (A) Proton-pumping photocycle of bacteriorhodopsin showing light-induced 13-*trans* → *cis* transition of retinal (Khorana, 1988). Absorption maxima and room temperature decay times of the major spectral intermediates are indicated. Only the first step described by the wavy arrow is light dependent. Light adapted bR_{568} contains all-*trans* protonated Schiff base retinal, while intermediate M_{412} contains an unprotonated 13-*cis* Schiff base. The deprotonation and H^+ release is shown as being most closely related to the L → M transition, but may not be exactly correlated with these discrete spectroscopic intermediates. There may be two M intermediates with the slower M component correlating with H^+ pumping. The rate-limiting step for H^+ translocation in the photocycle is 5 ms for the M → O transition. (B) Schematic of bR photocycle showing relation of protein conformation change associated with 13-*cis*, 14-S-*trans*, 15-*anti* chromophore structure and H^+ transfer to bR spectral intermediates (Lugtenburg et al., 1988).

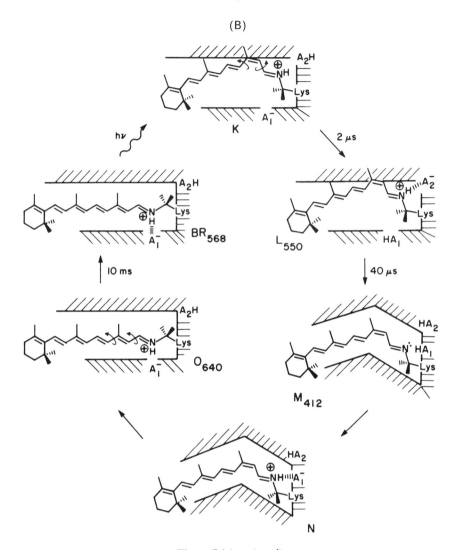

Figure 7.1 (continued)

35–40 Å, α-helical rods, packed 10–12 Å apart, and inclined to each other and the normal to the membrane plane at angles of 0–20° (Fig. 7.2A). The seven helices are arranged in a crescent-shaped cluster as determined by neutron (Fig. 7.2B) or electron (Fig. 7.2C) diffraction. Trimeric organization of the seven-helix crescent structure can be seen with both diffraction methods. Use of a photoactivable retinal analog showed that the all-*trans*-retinal bound by the Schiff base at Lys-216 in helix G (helix nomenclature, Fig. 7.3A) is in contact with Ser-193 and Gln-194 in helix F (Huang et al., 1982). Neutron diffraction studies indicate that the retinal is located within the protein, with

7.2 Bacteriorhodopsin, a Well-Characterized Light-Driven H$^+$ Pump

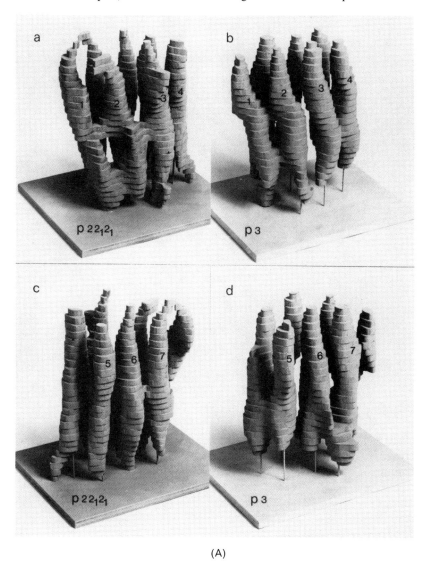

(A)

Figure 7.2A, B. (A) Models of the seven transmembrane α-helices of bacteriorhodopsin in orthorhombic [(a) and (c)] and trigonal [(b) and (d)] forms. Panels (a) and (b) show helices 1–4 (A–D) in the foreground, and panels (c) and (d) show helices 5–7 (E–G). The top and bottom of the helices are in contact with the aqueous solvent; the rest is in contact with the lipid (Leifer and Henderson, 1983). (B) Density map of the seven bacteriorhodopsin helices in a section parallel to the membrane plane, showing the position of the retinal pigment determined by neutron diffraction of bR with selectively deuterated retinal. (a) Superimposed on the bR structure are the center of deuteration of the cyclohexene ring, the middle of the polyene chain, and the center of deuteration of the Schiff base. The calculated position of the Schiff base nitrogen is marked by N. (b) Perspective with all-*trans* retinal in a plane perpendicular to the membrane with the polyene chain making a tilt angle at 20° (Heyn et al., 1988).

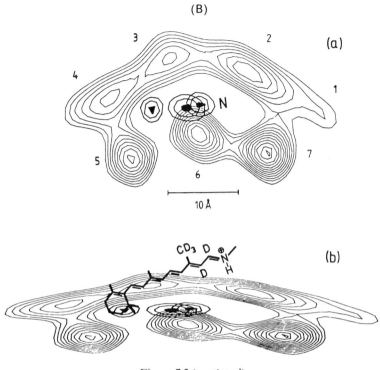

Figure 7.2 (continued)

the nitrogen of the Schiff base between helices B and F and the ring near helix D (Fig. 7.2C). Using the technique of resonance energy transfer to and from the retinal as a spectroscopic ruler (Chap. 6, section 6.1) with dye molecules [CoEDTA, Ru(bpy)$_3$] pressed against the membrane surface, and the assumption of uniformly and randomly oriented energy donors and acceptors, the center of the polyene chain was located at 10 ± 3 Å from a membrane surface (Kometani et al., 1987).

Electron diffraction analysis (Henderson and Unwin, 1975), neutron diffraction (Trewhella et al., 1983; Seiff et al., 1985; Heyn et al., 1988), determination of the amino acid sequence of the protein (Ovchinnikov et al., 1979) and the nucleotide sequence of the gene (Dunn et al., 1981), and the use of proteases and chemical modifiers as probes of bR topography in the membrane (Ovchinnikov, 1987), argue for the folding pattern of the bR molecule in the membrane bilayer shown in Fig. 7.3A and B. Relatively few charged groups are present in the membrane bilayer, some are thought to be neutralized by salt-bridging, and three seem to function in the bR H$^+$ pump (Table 7.1). The excess of negative carboxylic amino acids in the hydrophobic bilayer may also be partly neutralized by bound divalent cations (Katre et al., 1986) or by protonation of the carboxyl groups.

7.2 Bacteriorhodopsin, a Well-Characterized Light-Driven H^+ Pump

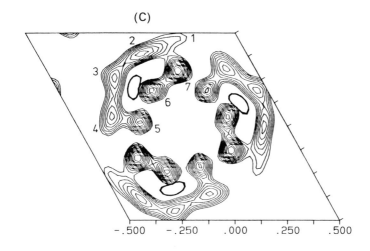

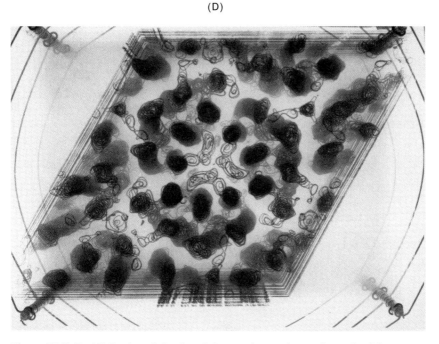

Figure 7.2C, D. (C) Projected density of the purple membrane determined from neutron diffraction intensities and electron microscopy phases at 8.7 Å resolution, showing trimeric unit. The side of the hexagonal unit cell corresponds to 63 Å (Seiff et al., 1985). (D) The electron density map of deoxycholate-extracted purple membranes showing a trimer of seven helix bR monomers arranged in a crescent. The view is from the cytoplasmic side of the membrane and the depth through the map is 54 Å (Tsygannik and Baldwin, 1987).

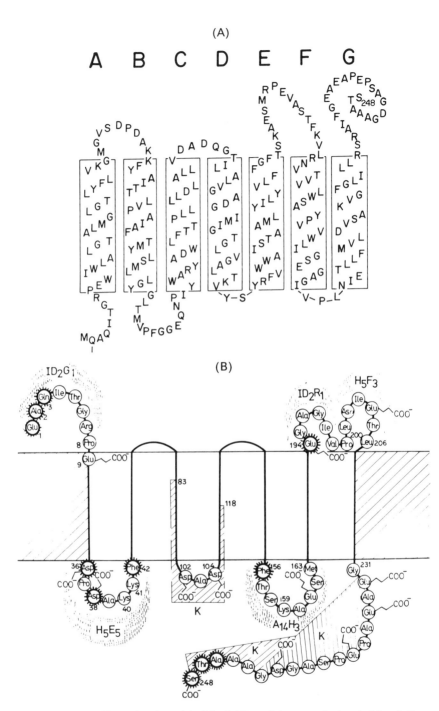

Figure 7.3. Two-dimensional model of the folding of the seven hydrophobic α-helices of the bacteriorhodopsin polypeptide in the membrane bilayer, showing (A) a secondary structure model. Tyr and Asp residues have been changed by mutation. NH$_2$- and COOH-termini are on the outside and cytoplasmic sides, respectively, of the bacterial cell. (From Khorana, 1988.) (B) Exposed helix-connecting loops that react with exogenous proteases and peptide-specific antibodies, and the retinal linked by a protonated Schiff base to Lys 216 (Ovchinnikov et al., 1985).

7.2.1 Mechanism of H$^+$ Pumping: Involvement of bR Spectral Intermediates

The bR proton pump is driven by the photochemical 13-*trans* → 13-*cis* isomerization of the retinal prosthetic group and deprotonation of the Schiff base nitrogen that occurs during the bR photocycle, involving at least four spectroscopically defined ground state intermediates, K (610 nm), L (550 nm), M (412 nm), and O (640 nm) (Fig. 7.1A). The torsional isomerization about the $C_{13} = C_{14}$ double bond (Smith et al., 1984) appears to occur as fast as 0.5 ps to a discrete intermediate, J (Mathies et al., 1988), formed even faster than the K intermediate shown in Fig. 7.1. In the absence of the Schiff base linkage to the bR lysine-216, the rotation around the double bond would be prevented by a large (~50 kcal/mol) activation barrier in the polyene retinal dye. In the presence of a protonated Schiff base, however, the rotational barrier of the 13–14 double bond is reduced to approximately 12 kcal/mol (Tavan et al., 1985), facilitating this conformational change. The *trans* → *cis* isomerization, which is linked to the p*K* change of the Schiff base, is followed by deprotonation of the Schiff base nitrogen in the L (550 nm) → M (412 nm) transition (Lewis et al., 1974), and formation of the transmembrane potential ($\Delta\psi$) in the time range of the L → M transition (Helgerson et al., 1985; Trissl, 1985). The decay rate of the M intermediate decreases with increasing $\Delta\psi$ (Helgerson et al., 1985). H$^+$ translocation across the entire bilayer and the resultant generation of a transmembrane $\Delta\psi$ must involve (Chap. 2, section 2.11) very rapid H$^+$ movement across the membrane bilayer either (a) by a "proton wire" mechanism in which discrete protonated intermediates would not be detected, or (b) several discrete proton transfer steps involving particular amino acids and perhaps including H$_2$O channels on each side of the Lys-216 Schiff base.

The proton wire mechanism appears to be excluded by the detection of several discrete amino acid intermediates using UV-visible and infrared spectroscopy (Engelhard et al., 1985). The possible amino acids with accessible p*K* values in the seven transmembrane helices of bR (Fig. 7.3) that might contribute to a transmembrane H$^+$ translocation pathway are summarized in Table 7.1.

In addition to spectroscopic studies, the role of particular amino acids in the structure and proton pumping function of bacteriorhodopsin has been studied by site-directed mutagenesis. The mutagenesis studies utilized a totally synthetic bR gene with closely placed unique restriction sites, development of an efficient expression system for bR, and a protocol for purification of denatured bR mutant proteins using the ability of denatured bR to bind retinal, refold, and regenerate the native chromophore (Khorana, 1988). The effect of individual mutation of all 11 Tyr (Mogi et al., 1987) and the Asp residues has been tested because of the proposed role of these groups in the translocation mechanism (Fig. 7.1B). The mutations were: Tyr → Phe, Asp → Asn and Glu; for other mutants, Arg → Gln, Glu → Gln, Pro (50, 91, 186) → Ala, Pro-186 → Leu, Val, and Gly, and Trp → Phe. Many of the altered residues resulted in an altered extent and rate of regeneration of the bR

Table 7.1. Possible H^+-translocating residues in transmembrane helices of bacteriorhodopsin

Protonatable groups	Possible residues, residue number (Fig. 7.3A)	Residues implicated in pump, experimental tests
A. Low pK		
1. Aspartic (D)	85, 96, 115, 212 (total of 8, incl. loop regions)	3 Asp implicated in bR cycle by light-dark FTIR[a] spectra (Engelhard et al., 1985). All eight Asp mutagenized. D-85, D-96, D-212 → Asn mutants have greatly reduced H^+ pumping (Mogi et al., 1988; Khorana, 1988).[b]
2. Glutamic (E)	9, 194, 204	
B. Middle pK		
1. Histidine	None	
C. High pK		
1. Tyrosine (Y)	26, 43, 57, 64, 79, 83, 147, 150, 185 (total of 11)	2 Tyr implicated in bR cycle by light–dark FTIR and UV spectra (Roepe et al., 1987a, b; Lin et al., 1987). All 11 Tyr changed to Phe (Mogi et al., 1987). Y-185 → F, significant inhibition of H^+ pump (Hackett et al., 1987; Khorana, 1988), and altered FTIR spectrum for bR → K and bR → M states (Braiman et al., 1988).
2. Threonine (T)	17, 24, 46, 47, 55, 89, 90, 107, 121, 142, 178, 205	
3. Serine (S)	59, 141, 183, 193, 214	

[a] Fourier-transform infrared spectroscopy.
[b] Control rates of H^+ pump: 3 H^+/bR-s (Problem 70).

chromophore spectrum, and these residues on helices B, C, G, and F (including D-85, D-96, D-212, Y-57, Y-185) were inferred to interact with the retinal.

H^+ pumping was most strongly impaired by mutation of three of the four Asp residues, D-85 and D-96 in helix C and D-212 in helix G. The Tyr-185 → Phe mutation also caused a marked inhibition of the H^+ pump. This mutation also resulted in the loss of difference absorbance bands in the infrared that are due to tyrosine protonation in the bR → K photoreaction, and deprotonation in the bR → M transition (Braiman et al., 1988). The residues whose mutation most affected the bR chromophore, and the Asp → Asn mutants that inhibited H^+ pumping are summarized in Fig. 7.4, where the helices are aligned along the contours determined by the electron diffrac-

7.2 Bacteriorhodopsin, a Well-Characterized Light-Driven H^+ Pump

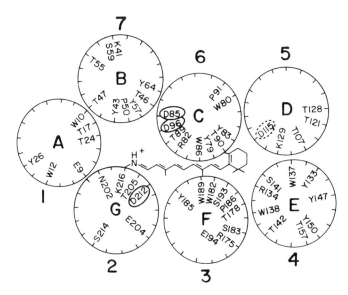

Figure 7.4. Helical wheel model of bR showing Asp residue (shaded) on helices C and G whose mutation inhibited H^+ pumping, and other amino acids, mostly on helices C and F, that are inferred to interact with the retinal (Khorana, 1988).

tion map of the bR. A possible role of H_2O channels in the H^+ pathway on each side of the Schiff base was suggested by the presence of bound H_2O in bR (Varo and Keszthelyi, 1985), and the ability of NH_2OH and $NaBH_4$, respectively, to bleach and reduce the protonated Schiff base (Mogi et al., 1988).

7.2.2 Comparison of Bacteriorhodopsin and Halorhodopsin

There are three other rhodopsin-like pigments in the membrane of *H. halobium*: halorhodopsin (hR) is an anion pump; there are two sensory rhodopsins (Spudich and Bogomolni, 1988) not discussed here, one of which discriminates color, and another whose photocycle components are determinants of attractant and repellent phototaxis. The halorhodopsin (hR) protein (274 amino acids, MW = 26,961) mediates a light-driven pump driving Cl^- import (Fig. 7.5) and has a significant primary sequence identity with bR (Fig. 7.6).

The most conserved regions of the hR and bR sequences are those in the transmembrane α-helical segments that are also proposed for hR, where 36% of the residues are conserved as opposed to 19% in the loops connecting the helices. Conspicuous among these conserved residues is the region around the retinal-binding residue Lys-242 in the seventh helix of hR that corresponds to Lys-216 in bR. There is no significant homology with the opsin proteins of eukaryotes. The seven hR helices also have an amphipathic character that can be quantitated by calculation of a hydrophobic movement across the helix

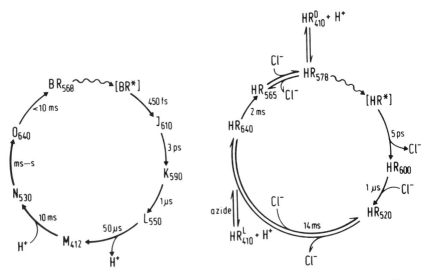

Figure 7.5. Comparison of photochemical cycles of bR and hR (Oesterhelt and Tittor, 1989).

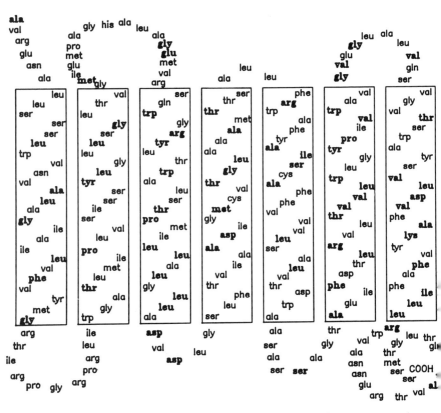

Figure 7.6. Amino acid sequence of halorhodopsin arranged into seven membrane spanning helices in analogy to bR. Residues in bold face are conserved between hR and bR. NH$_2$- and COOH-termini are on the extracellular and cytoplasmic sides of the membrane (Blanck and Oesterhelt, 1987; also discussed by Lanyi, 1988).

7.3 Cytochrome Oxidase (Mitochondrial Complex IV) as a Proton Pump

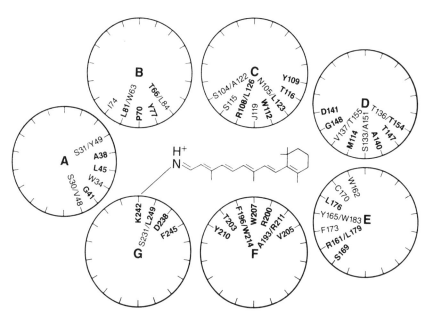

Figure 7.7. Helical wheel projection of seven α-helices of hR. Seven helices oriented to form an intrahelical or lumenal space. Bold residues are conserved between hR and bR. Amino acids with functional relevance are shaded (Oesterhelt and Tittor, 1989).

(Eisenberg et al., 1984), in which hydrophilic residues are segregated from hydrophobic in a helical wheel projection of the helices (Fig. 7.7). Except for helix C, a disproportionately large fraction of the conserved residues are concentrated in the hydrophilic segments of the helices. In both hR and bR, retinal acts as the light-triggered switch for ion translocation. The different ion pump functions of the two proteins must lie in the different specific ion [(H^+/OH^-) vs Cl^-] binding sites in bR and hR. Unlike bR, hR does not contain the two Asp residues in helix C that were shown to be important for the bR H^+ pump. A model for the Cl^- pump involves key roles of Arg residues (11/hR) in the mechanism: Cl^- may initially bind to the Arg-rich loop between helices A and B containing 4 Arg on the extracellular side, traverse the protein to the Schiff base, and then be translocated to an Arg-containing site on the cytoplasmic side from which it would be released (Lanyi, 1988). For both bR and hR, a major question is the mechanism of the switch of the protein from an ion acceptor to an ion donor.

7.3 Cytochrome Oxidase (Mitochondrial Complex IV) as a Proton Pump

Cytochrome oxidase in mitochondria and some aerobic bacteria has the dual role of catalyzing (i) the reduction of O_2 to water and (ii) generation of a $\Delta\tilde{\mu}_{H^+}$

at the terminal energy conservation site of the respiratory chain (Wikström et al., 1981). Its basic structure seems to have been conserved through evolution (Wikström et al., 1985). Cytochrome oxidase contains four redox centers, two a hemes (a and a_3) and two Cu ions (Cu_A and Cu_B), and 12–13 polypeptides in mammalian mitochondria. The three largest polypeptides are mitochondrially encoded and the two largest bind the two hemes and two coppers (Chap. 4, section 4.1.1). The three large polypeptides constitute the entire cytochrome oxidase of the aerobic bacterium, *Paracoccus denitrificans*. Two simpler oxidases have been found in the extreme thermophilic bacterium, *Thermus thermophilus*: (i) a $c_1 aa_3$ two-polypeptide complex in which the two hemes and two Cu are bound to a single M_r 55,000 polypeptide in redox environments essentially identical to those of the eukaryotic oxidase (Fee et al., 1986); (ii) an alternate ba_3 oxidase in the same *T. thermophilus* containing only one M_r 35,000 polypeptide, to which is bound one heme a_3, one heme b presumably replacing heme a, and two Cu (Zimmerman et al., 1988).

7.3.1 Mitochondrial Myopathies and Cytochrome Oxidase

Many of the nuclear encoded subunits of the mammalian oxidase are tissue-specific (Kuhn-Nentwig and Kadenbach, 1985; Capaldi, 1988), and there are tissue-specific dysfunctions of human cytochrome oxidase. An example that may involve developmentally different fetal and adult forms of the oxidase is that of a reversible infantile mitochondrial myopathy due to oxidase deficiency. Oxidase activity was only 10% of normal shortly after birth, but returned to a normal level when the baby was 18 months old (Capaldi, 1988).

7.3.2 Amino Acid Sequences and Hydropathy Plots of Subunits I–II; Electron Diffraction Studies; Location of Heme and Copper Binding Sites

The sequences of subunit I from many different mitochondrial sources (shown for *Paracoccus denitrificans*; Table 7.2A) can be readily aligned (Wikström et al., 1985), as can those of subunit II. The availability of a large family of sequences with a similar distribution of hydrophobic residues (Wikström et al., 1984, 1985; Raitio et al., 1987) allows hypotheses to be made regarding the folding of the protein in the membrane, the position of the heme and copper ligands, and their position in the membrane bilayer. The structure of cytochrome oxidase has also been analyzed by electron diffraction using two dimensional crystalline arrays (Fig. 7.8).

The diffraction and image reconstruction studies indicated that the oxidase complex spans the membrane as an asymmetric Y-shaped molecule that can crystallize as a monomer in detergent-rich sheets (Fuller et al., 1982) or as a dimer (Deatherage et al., 1982) (Fig. 7.8). The dimeric molecule, visualized

7.3 Cytochrome Oxidase (Mitochondrial Complex IV) as a Proton Pump

Table 7.2. (A) Amino acid sequence of subunit I from cytochrome oxidase of the aerobic bacterium, *Paracoccus denitrificans*

```
1        10        20        30        40        50        60        70        80
MSAQISDSIEEKRGFFTRWFMSTNHKDIGVLYLFTAGLAGLISVTLTVYMRMELQHPGVQYMCLEGMRLVADAAAECTPN

        90       100       110       120       130       140       150       160
AHLWNVVVTYHGILMMFFVVIPALFGGFGNYFMPLHIGAPDMAFPRLNNLSYWLYVCGVSLAIASLLSPGGSDQPGAGVG

       170       180       190       200       210       220       230       240
WVLYPPLSTTEAGYAMDLAIFAVHVSGATSILGAINIITTFLNMRAPGMTLFKVPLFAWAVFITAWMILLSLPVLAGGIT

       250       260       270       280       290       300       310       320
MLLMDRNFGTQFFDPAGGGDPVLYQHILWFFGHPEVYMLILPGFGIISHVISTFARKPIFGYLPMVLAMAAIAFLGFIVW

       330       340       350       360       370       380       390       400
AHHMYTAGMSLTQQTYFQMATMTIAVPTGIKVFSWIATMWGGSIEFKTPMLWALAFLFTVGGVTGVVIAQGSLDRVYHDT

       410       420       430       440       450       460       470       480
YYIVAHFHYVMSLGALFAIFAGTYYWIGKMSGRQYPEWAGQLHFWMMFIGSNLIFFPQHFLGRQGMPRRYIDYPVEFSYW

       490       500       510       520       530       540       550
NNISSIGAYISFASFLFFIGIVFYTLFAGKPVNVPNYWNEHADTLEWTLPSPPPEHTFETLPKPEDWDRAQAHR
```

Nine histidine residues found to be identical in many oxidase sequences are underlined.
From Raitio et al. (1987).

Table 7.2. (B) Amino acid sequence of subunit II from the *P. denitrificans* Oxidase

```
1        10        20        30        40        50        60        70        80
MAIATKRRGVAAVMSLGVATMTAVPALAQDVLGDLPVIGKPVNGGMNFQPASSPLAHDQQWLDHFVLYIITAVTIFVCLL

        90       100       110       120       130       140       150       160
LLICIVRFNRRANPVPARFTHNTPIEVIWTLVPVLILVAIGAFSLPILFRSQEMPNDPDLVIKAIGHQWYWSYEYPNDGV

       170       180       190       200       210       220       230       240
AFDALMLEKEALADAGYSEDEYLLATDNPVVVPVGKKVLVQVTATDVIHAWTIPAFAVKQDAVPGRIAQLWFSVDQEGVY
                                                    ↑    *
       250       260       270       280       290
FGQCSELCGINHAYMPIVVKAVSQEKYEAWLAGAKEEFAADASDYLPASPVKLASAE
 *  ↑ *    *
```

Subunit is made in a precursor form with an N-terminal extension. Arrows indicate invariant carboxylic acids that may be involved in binding of cytochrome *c*. Asterisks mark proposed ligands for copper.
From Raitio et al. (1987).

by electron microscope studies of two-dimensional crystalline assays of the oxidase and image reconstruction, protrudes 50 Å from the membrane from the C (cytoplasmic)-side and 10–15 Å on the M (matrix) side. A dimeric form can be prepared from purified bovine oxidase which has a higher activity than the monomer and shows biphasic Eadie–Hofstee plots for kinetic parameters that an be interpreted in terms of two interacting catalytic sites (Bolli et al., 1985), and a dimer of the oxidase has higher activity for H^+ translocation activity (section 7.3.4). The large protuberance of the oxidase on the cyto-

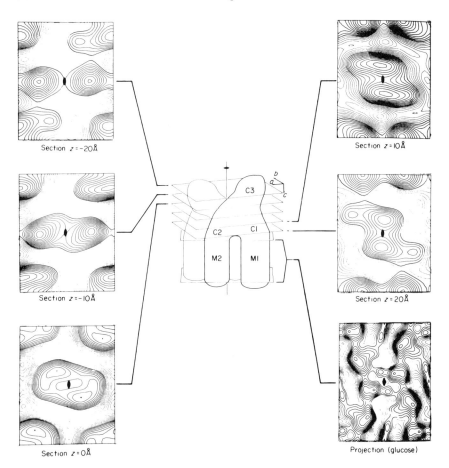

Figure 7.8. Sections through the three-dimensional map of cytochrome oxidase parallel to the membrane. Bold contours correspond to protein. On the sections, the a axis (100 Å) is horizontal, the b axis (124 Å) vertical. The projection of the vesicle crystal embedded in glucose in which bold contours correspond to protein, and weak contours to lipid, is in correct alignment with the uranyl acetate map. The upper five rectangles passing through the crystal show the planes of the sections and the directions of the crystallographic axes, with a spacing of 10 Å between the sections. The lower two rectangles show the upper and lower membrane surfaces (Deatherage et al., 1982).

plasmic side indicates tht the active domain of a membrane protein can extend far from the imagined planar boundary of the lipid bilayer.

EPR and fluorescence energy transfer studies have shown that both heme irons and the two coppers are found in the cytoplasmic half of the membrane (Fig. 7.9), liganded to subunits I and II. Cytochrome c binds ($K_D \simeq 10^{-8}$ M) to a specific site on the protruding cytoplasmic side (Fig. 7.9), transferring electrons to Cu_A and heme a that are subsequently transferred to the binuclear heme a_3/Cu_B reaction site where oxygen binding and reduction take place.

7.3 Cytochrome Oxidase (Mitochondrial Complex IV) as a Proton Pump

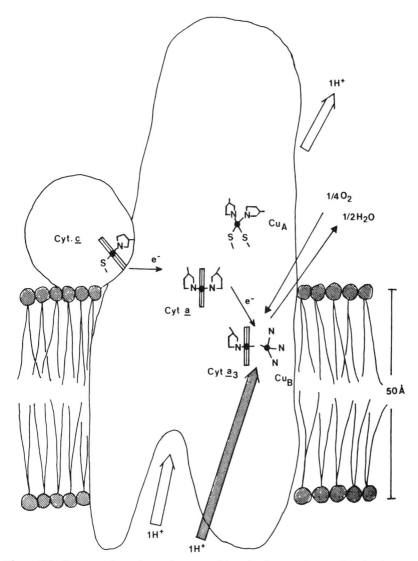

Figure 7.9. Cross-section of cytochrome oxidase in the membrane showing heme *a* and Cu binding sites. (From Wikström, 1987. With permission from *Chemica Scripta*, Cambridge University Press.)

7.3.3 Structural Models of the Redox Centers

The Cu_A Site on Subunit II

The underlined region of the subunit II sequence (Table 7.2B) containing conserved two Cys and one His residues is similar to that of Cu-binding sites in blue copper proteins, implying that three of the four ligands of a Cu on subunit II are two Cys and one His residues (Fig. 7.10). This Cu is in a polar

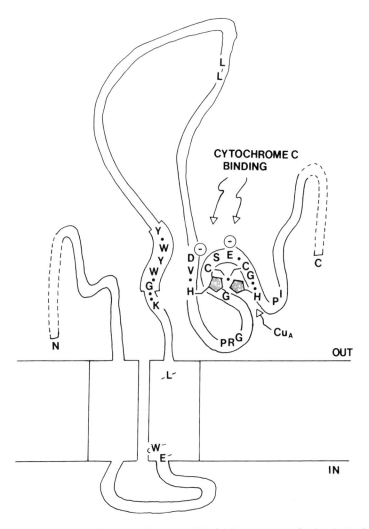

Figure 7.10. Schematic diagram of the possible folding pattern of subunit II of cytochrome c oxidase showing ligands of the Cu_A (Holm et al., 1987). The residues fully conserved in subunit II in a large number of cytochrome c oxidase sequences are noted. The invariant acidic residues that dock cytochrome c (Capaldi et al., 1983; Millett et al., 1983) are marked by a minus sign.

domain on the subunit II polypeptide (Wikström and Saraste, 1984), close to conserved acidic residues D-112, D-158, and E-198 of subunit II, two of which are strictly conserved and are believed to pair with complementary charges in cytochrome c (Chap. 4). The identity of the fourth Cu ligand is thought to be His-209 (Table 7.2), which is the only other conserved His residue. Because the Cu_A-heme a pair are the initial acceptors for the electron from cytochrome c, the electron transfer pathway within the oxidase is cyt $c \rightarrow (Cu_A \rightarrow$ heme $a) \rightarrow (Cu_B$-heme $a_3)$ (Gelles et al., 1986; Holm et al., 1987).

7.3 Cytochrome Oxidase (Mitochondrial Complex IV) as a Proton Pump   317

Binding of Heme a, Heme a_3, and Cu_B to Subunit I

If subunit II binds Cu_A, and all of the metal groups are bound in subunits I–II, then heme a, heme a_3, and Cu_B must be ligated by subunit I (Holm et al., 1987). These redox groups require five or six His residues for coordination (two for heme a, one for heme a_3 because it must also bind O_2, and two or three for the Cu_B), which should be found in the nine conserved His residues in the sequence. The distances between the redox centers [12–20 Å between the heme irons (Ohnishi et al., 1982) and < 5 Å between heme a_3 and Cu_B (Chance and Powers, 1985)], and analogy with cytochrome b of the bc_1 complex (see below, section 7.4) suggested a model for a compact structure using the trans-membrane helices VI, VII, and X, that contain two His residues, as predicted from hydropathy plots. Heme a was proposed to cross-link helices VI and X through His-266 and His-398 (Table 7.2A) near the cytoplasmic (p) surface of the cytochrome. The second His residue in helix VI near the center of the bilayer, His-273, would provide the proximal His ligand to heme a_3, and the second His on helix X, His-406, together with His-322 and His-323 on helix VII, would provide the three needed for coordination of the Cu_B (Holm et al., 1987; Fig. 7.11). The intramolecular cross-linking of subunit I by heme a is an example of a structural motif that is prevalent in membrane-bound cytochromes (see Figs. 6.21 and 7.17D).

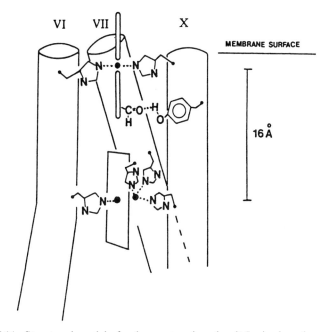

Figure 7.11. Structural model of redox centers in subunit I, viewing along the membrane plane. Transmembrane domains VI, VII, and X are used to bind two hemes and a copper, heme a through His-266 (helix VI) and His-398 (helix X), and heme a_3 (through His-273 on helix VI)-Cu_B (through His-322 and His-323 on helix VII and His-406 on helix X) near the center of the bilayer (Finel et al., 1987).

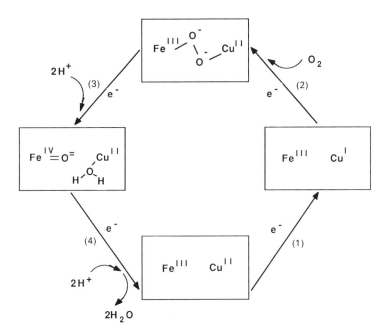

Figure 7.12. Proposed mechanism of O_2 reduction by heme a_3-Cu_B as described in the text (from Wikström, 1987).

Pathway of O_2 Reduction (Fig. 7.12)

Starting with oxidized (Fe^{III}) heme a_3 and Cu_B (Cu^{II}), one-electron reduction of the Fe^{III}-Cu^{II} complex from initially reduced cytochrome c through Cu_A and heme a forms Cu_B^+ (step 1). Reduction by a second electron concomitant with O_2 binding and electron transfer to bound dioxygen by reduced heme a_3–Cu_B forms a peroxy structure (step 2), transfer of a third electron from the respiratory chain to the binuclear complex, together with another internal oxidation and uptake of $2H^+$ forms a ferryl intermediate, ($Fe^{4+} = O^{2-}$, Cu^{2+}-H_2O) (step 3), and transfer of the fourth electron together with uptake of the last $2H^+$ would complete the reduction of O_2 to H_2O and leave heme a_3 and Cu_B in the oxidized initial state (Wikström, 1987). Steps 2 and 3 would involve net transfer of two electrons, which is more favorable thermodynamically (Fig. 2.3).

7.3.4 Proton Translocation Across the Membrane by Cytochrome Oxidase

The reduction of O_2 to H_2O should require one H^+ for each electron transferred from cytochrome c, i.e., $O_2 + 4e^- + 4H^+ \to 2H_2O$. Depending on whether the proton comes from an internal or external mitochondrial compartment,

7.3 Cytochrome Oxidase (Mitochondrial Complex IV) as a Proton Pump

this will result in an internal or external alkalinization. An acidification of the medium can be observed when ferrocyanide or ferrocytochrome c was added as reductant to rat liver mitochondria, demonstrating that the reduction of O_2 by the oxidase is associated with H^+ release from the outer surface of the membrane or H^+ translocation across the membrane. The acidification of the medium linked to redox function of the cytochrome oxidase can also be observed with cytochrome oxidase reconstituted into artificial membrane vesicles. When reduced cytochrome c is added to the vesicles as a source of electrons, H^+ efflux into the medium is observed in parallel with oxidation of cytochrome c (Fig. 7.13A).

The stoichiometry of the acidification caused by the oxidation of cytochrome c can be seen in Fig. 7.13A in the absence of FCCP to be approximately $H^+/e = 1$, because the acidification caused by oxidation of two equivalents of cytochrome c and by addition of two equivalents of HCl are equal. The observation that the alkalinization of the medium expected for the proton uptake accompanying the O_2 reduction, $O_2 + 4H^+ + 4e^- \rightarrow 2H_2O$, is observed only in the presence of the protonophoric uncoupler, FCCP (Fig. 7.13A and C) argued that the latter proton originates from the inside of the vesicle. The proton disappearance from the inner matrix space of the mitochondrion sensed by a fluorescein-conjugated phospholipid, acting as a pH indicator, together with H^+ efflux to the eternal medium accompanying the oxidation of cytochrome c (Thelen et al., 1985), argues for transmembrane H^+ movement associated with redox turnover of the cytochrome oxidase acting as a redox-linked electrogenic proton pump (Fig. 7.14A). The $\Delta\tilde{\mu}_{H^+}$ generated by the oxidase pump is approximately -240 mV (Wikström and Saraste, 1984).

Much of the discussion on the mechanism of the oxidase H^+ pump (Malmström, 1985; Krab and Wikström, 1987; Prince, 1988) has centered on the pH dependence of the E_m values of the heme a redox centers (Wilson et al., 1976a; Blair et al., 1986). Redox-dependent binding of H^+ and/or OH^- could be direct, involving OH^- ligand exchange at the redox center or H^+ binding to heme side chains. It could also be indirect, involving pK changes at distant residues on the protein, e.g., the effect of O_2 on protonation/deprotonation of distant residues in hemoglobin ("Bohr effect"; Perutz, 1978; Ho and Russu, 1987). A $\Delta E_m/\Delta pH = -59$ mV dependence (Chap. 2, section 2.5) is not a necessary criterion for thermodynamic coupling of e^- and H^+ at a redox center, although its presence is highly suggestive. A smaller absolute value for $\Delta E_m/\Delta pH$ could result, for example, from a small separation of the pK values of the oxidized and reduced states, pK_o and pK_r (Fig. 2.2). The mechanism of such a proton pump may also involve a succession of redox-dependent changes of hydrogen bond strength (pK changes; membrane "Bohr" effect) of amino acids (e.g., His, Tyr) that span the membrane, as in the case of bacteriorhodopsin discussed above, perhaps initially involving the heme a formyl or propionate groups (Fig. 7.14B). The role of the amino acids in such a translocation pathway has not been discussed for cytochrome oxidase to the extent that it has been for bR and hR. However, a ligation change of Cu_A involving

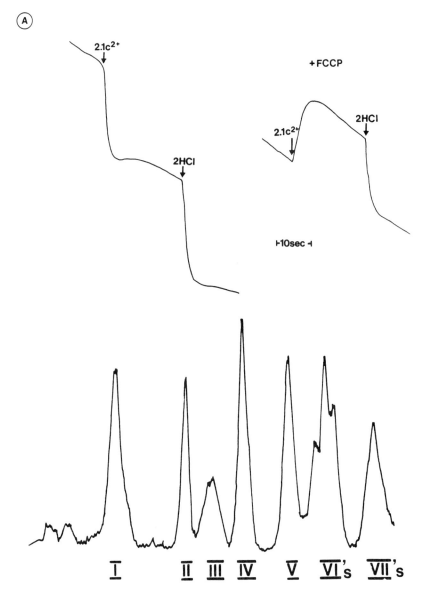

Figure 7.13. Proton efflux observed in different reconstituted cytochrome oxidase preparations characterized by gels (densitometric tracings shown below the recorder tracings of pH changes). (A) Control oxidase with respiratory control ratio = 5.5. (B) Subunit III-deficient oxidase from control treated with detergent and separated by anion-exchange fast performance liquid chromatography. (C) Subunit III—containing oxidase fraction from same prep as (B). The pH change of the medium in response to the pulse of cytochrome c was measured (Finel and Wikström, 1986) in the presence and absence of the protonophoric uncoupler FCCP, and with valinomycin to prevent inhibition of H^+ movement by a membrane potential (Chap. 3, section 3.12.1).

7.3 Cytochrome Oxidase (Mitochondrial Complex IV) as a Proton Pump 321

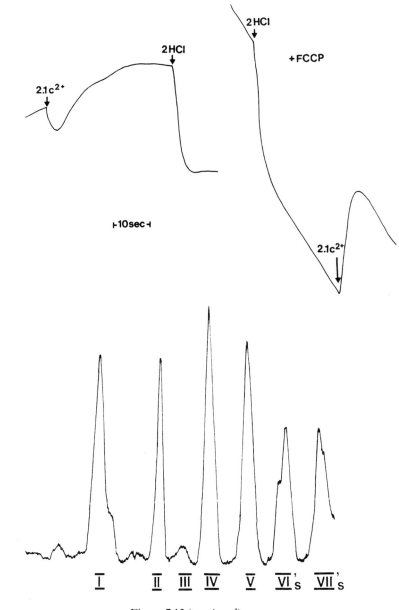

Figure 7.13 (*continued*)

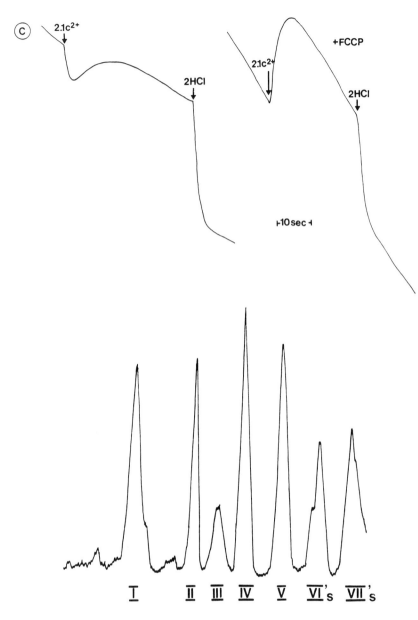

Figure 7.13 (*continued*)

7.3 Cytochrome Oxidase (Mitochondrial Complex IV) as a Proton Pump 323

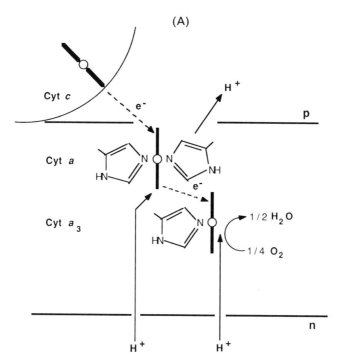

Figure 7.14A, B. (A) Schematic view of proton translocation by the cytochrome oxidase redox-linked proton pump. The location of O_2 reduction has been drawn near the center of the bilayer. The electrogenic nature of the reaction would be the same if O_2 reduction occurred at the C (cytoplasmic) side. (B) Possible mechanism for redox driven proton pump in cytochrome oxidase. Proposed mechanism based on finding that the strength of the heme a formyl hydrogen bond in situ increases by 2–2.5 kcal/mol upon reduction of heme a in situ. (I) Proposed structure of cytochrome oxidase in which formyl group is hydrogen bonded to proton donor, X-H, associated with protein. (II) Proposed mechanism: The heme a formyl group is indicated by $>C=0$, and the iron valence is shown. H-bonded chains with donors R_r and R_l occur to the right (r) and left (l) of the formyl group. Reduction is shown as a 2 step process that results in (c). The hydrogens involved in the H^+ translocation are labeled a–d to identify their motions. (Reprinted with permission from Babcock and Callahan, *Biochemistry*, vol. 22, pp. 2314–2319. Copyright 1983 American Chemical Society).

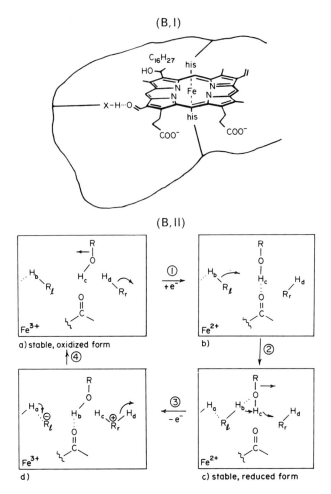

Figure 7.14 (continued)

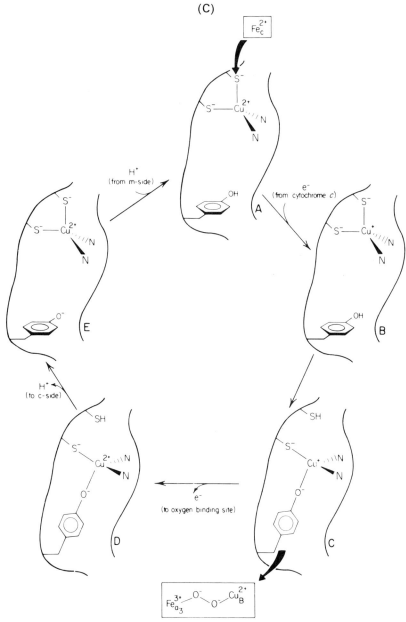

Figure 7.14C. Alternative proton pump model based on a calculation indicating that ligation of Cu_A^+ by 2 Cys thiolate ions and 2 His imidazole is unstable. Reduction of Cu_A^{2+} by cyt c and H^+ uptake will lead to a Tyr residue exchanging with the thiolate, which is protonated by the Tyr residue, as a ligand of Cu_A. When the Cu_A is reoxidized by the binuclear a_3–Cu_B oxygen binding center, the Cys residue is deprotonated, displaces the Tyr to reform the original ligand configuration, and H^+ uptake reprotonates the tyrosine. Thus, the H^+ pathway at the Cu_A site is H^+ (n side) → Tyr^- → S^- → SH → H^+ (p side) (Gelles et al., 1986.)

a Tyr residue that is dependent on the redox state of Cu_A has been proposed as the basis of the oxidase H^+ pump (Fig. 7.14C). Because the critical requirement of the H^+ pump is that membrane siddedness and H^+ direction be preserved, the pump can be uncoupled if Tyr^- and S^- are protonated or deprotonated by groups on the wrong side of the Cu_A site (Gelles et al., 1986).

The mechanism of the cytochrome oxidase proton pump may turn out to be similar in mechanism to that of bacteriorhodopsin except that the former is redox-linked and the latter driven by light. A unique aspect of H^+ translocation in the oxidase as opposed to bacteriorhodopsin is that it involves the H^+-dependent oxygen reactions. This has led to the proposal that oxygen itself is involved in the translocation mechanism (Mitchell et al., 1985).

Subunits I and II that bind the metal centers are certainly involved in the H^+ translocation mechanism. The third mitochondrial-encoded subunit, subunit III, has also been implicated in this function. Removal of SU III can result in a significant decrease in the H^+/e^- ratio for H^+ translocation mediated by the oxidase (Fig. 7.13B; also, Nalecz et al., 1985; Prochaska and Reynolds, 1986). However, even though the H^+ translocation is reduced by depletion of subunit III, there is a residual H^+ movement that can be significant, as in the case of a two-subunit (M_r 45,000 and 28,000) oxidase from the bacterium *Paracoccus denitrificans* from which subunit III is missing ($H^+/e \simeq 0.6$). The depletion of SU III that resulted in loss of H^+ extrusion in the experiment of Fig. 7.13B also caused monomerization of the oxidase, and it has been proposed that subunit III stabilizes a dimeric form of cytochrome oxidase that is necessary for H^+ translocation (Nalecz et al., 1985; Finel and Wikström, 1986).

7.4 The Q Cycle and H^+ Translocation in Complex III and Chloroplast $b_6 f$ Complexes

The ubiquinol:cytochrome c oxidoreductase (complex III, bc_1 complex) and its analog, the plastoquinol:plastocyanin oxidoreductase ($b_6 f$ complex), are ubiquitous in energy transducing membranes that utilize light or an obligatory aerobic respiratory chain. They provide the pathway for transferring electron equivalents from the reducing side of the electron transport chain through the ubiquinol or plastoquinol pool to cytochrome c or plastocyanin (Fig. 5.5). The rate-limiting step for electron transfer through this complex is a few milliseconds, so that like the bR, cytochrome oxidase, and ATPase H^+ pumps, the specific rate of H^+ pumping in the cytochrome bc_1 complex should be approximately several hundred H^+/complex-s.

The polypeptide composition of the photosynthetic complexes and the bacterium *P. denitrificans* is simpler than those of mitochondria (Table 7.3). For example, the chloroplast cytochrome $b_6 f$ complex has four polypeptides, with M_r values of 33,000 (cytochrome f), 23,000 (cytochrome b_6), 20,000 (ISp), and 17,000 ("subunit IV") (Fig. 7.15). The two cytochromes and subunit IV

7.4 The Q Cycle and H$^+$ Translocation

Table 7.3. Composition of complex III and cytochrome $b_6 f$

Source	Redox centers (nmol/mg protein)			Turnover no. (s^{-1})	Polypeptides ($M_r \times 10^{-3}$)
	Cyt c_1 or f	Cyt b	ISp		
Mitochondria, Bovine heart (Hauska et al., 1983; Wikström and Saraste, 1984; Rieske and Ho, 1985)	4	8	4	100–4,000	46–53 (Core protein I), 43–45 (Core II), 30–31 (III, cyt b), 28–31 (IV, cyt c_1), 24–25 (V, RFeS), 12–14 (VI), 8–12 (VII), 6–9 (VIII). [eleven in Gonzalez-Halphen, 1988]
Yeast (Siedow et al., 1978)	4.6	9.3	4.6	70–200	As above, perhaps missing VIII
P. denitrificans (Berry and Trumpower, 1985; Yang and Trumpower, 1986)	13.2	19.4	—	500	62(c_1), 39(b), 20; from genes, 50(b), 45(c_1), 20(ISp) (Kurowski and Ludwig, 1987)
Chloroplasts, spinach[a]					
(Hurt and Hauska, 1981)	7.3	13.1	8.3	3; 60–100[b]	34, 33, 23.5, 20, 17.5[c]
(Clark and Hind, 1983a)	8.2	15.7	Present	3	37, 33.5, 22, 19, 16.5
(Doyle and Yu, 1985)	10.8	20.6	—	2–3	Same as Hurt and Hauska (1981)
(Black et al., 1987)	9.0	18.0	10	20–35	33, 23.5, 19.5, 17
Cyanobacteria, *A. variabilis* (Krinner et al., 1982)	4.5	9.0	1–1.5	5	38 or 31, 22.5(2), 16.5, <10(2)
Chromatophores, *Rb. sphaeroides*					
(Gabellini et al., 1982)	5.0	10.0	Present	2	40, 34, 25, <10
(Yu et al., 1984)	8.3	8.3	15.0	25	48, 30, 24, 12
(Wilson et al., 1985)	—	—	—	—	30, 24.5, 12.5, 11
(Andrews et al., 1988)	Ratio of centers is 1:2:0.7			300	40, 37, 27, 12

[a] chloroplast rates measured with plastocyanin except Doyle and Yu (1985).
[b] ton-X-100 used in original sucrose gradient was later found to be inhibitory (Hauska, 1986).
[c] ifth polypeptide, $M_r \simeq 20,000$, has been reported in the green alga *C. reinhardtii* (Lemaire et al., 1986).

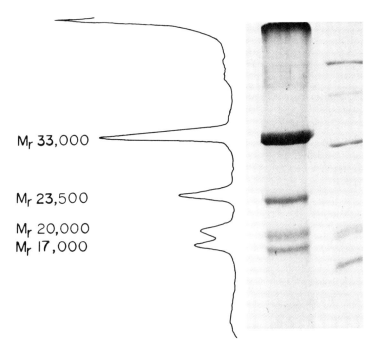

Figure 7.15. Sodium dodecyl sulfate-polyacrylamide gel electrophoresis (SDS-PAGE) gel of the cytochrome b_6–f complex in spinach chloroplasts. M_r 33,000 (cyt f), 23,000 (cyt b), 20,000 (Rieske ISp), 17,000 (subunit IV) (Black et al., 1987).

are encoded by the chloroplast genome, and the ISp is nuclear-encoded. In the mitochondria, cytochrome c_1 is also coded by a nuclear gene. The assembly of the complex in both organelles requires coordination of synthesis from the two sources, and protein import from the cytoplasm, and little is known at present about the details of such assembly processes. A key role of the chloroplast ISp is suggested by the large increase of its mRNA on illumination. A unique aspect of the assembly of cytochrome b_6 in the complex is its requirement for ATP in illuminated chloroplasts (Willey and Gray, 1988). The position of the chloroplast genes in the plastid chromosome is shown in Fig. 7.16A, where the bicistronic nature of the transcript for cytochrome b_6 and subunit IV can be noted. The order of the genes in the operon coding for the ubiquinol-cytochrome c reductase in the *Rb. capsulatus* genome is b, c_1, and ISp (Fig. 7.16B).

Cytochrome b is the only organelle-encoded polypeptide of the mitochondrial complex III. The stoichiometry of heme b:heme c (cytochrome c_1 or f):[2Fe-2S] center is 2:1:1, in all complexes, with the two b hemes coordinated within a single polypeptide (Nobrega and Tzagoloff, 1980; Widger et al., 1984; Wikström and Saraste, 1984). The monomeric molecular weight of the

7.4 The Q Cycle and H⁺ Translocation

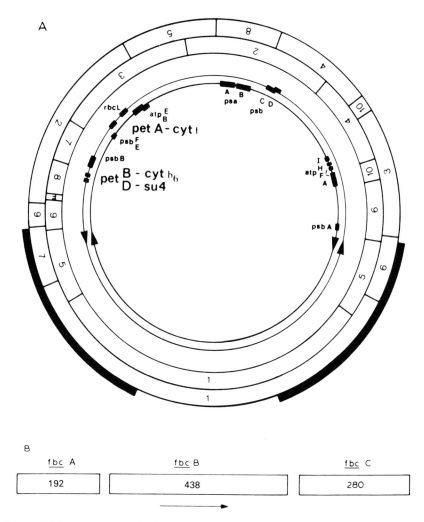

Figure 7.16. Depiction of (A) the spinach plastid chromosome emphasizing the location of the genes for cytochromes b_6 and f (figure contributed by R.G. Herrmann), and (B) the 3,874-base pair fbc operon of *Rb. capsulatus* (Redrawn from data in Gabellini and Sebald, 1986; also, Davidson and Daldal, 1987b.) (A) The location and transcription polarity of the genes on the two strands are indicated by the inner circles and arrowheads. psbA-F denote genes for the D1 herbicide-binding protein, 51- and 44-kDa chlorophyll *a* apoproteins, D2-32-kDa protein, and cytochrome b_{559} M_r 9,000 and 4,000 subunits (psbE and F). (B) The fbc operon consists of the genes for the Rieske ISp, cytochrome *b* and cytochrome c_1 proteins (fbc A, fbc B, and fbc C, whose length in amino acids are shown). [(A), From Cramer et al., 1987].

bovine mitochondrial complex is about 200,000 and it constitutes ~10% of the inner membrane protein. The reason for the variation in polypeptide number for the complex from different sources is not known. One explanation is that the smaller subunits may have specialized or regulatory functions as in mammalian mitochondrial cytochrome oxidase (Chap. 4, section 4.1.1) and ATPase (Chap. 8, sections 8.1 and 8.3.2).

Mitochondrial Myopathies and Complex III; Possibility of Redox Therapy

A 17-year-old girl had suffered from lactic acidosis and muscle weakness since she was 9 years old. Morphological findings indicated bizarre mitochondria indicative of mitochondrial myopathy. The bioenergetic capacity of the resting muscle was found from ^{31}P-nuclear magnetic resonance assay (Chap. 1, section 1.14.3) of phosphocreatine to inorganic phosphate to be 15% of normal, and decreased further with exercise. The activity of succinate dehydrogenase, cytochrome oxidase, and ATPase in biopsy samples was normal, but succinatecytochrome c reductase activity was decreased by about 40-fold. The cytochrome c_1 level was normal, but cytochrome b and several other subunits of complex III were missing. After signed permission and informed consent, artificial redox compounds [ascorbate and menadione, $E_{m7} = +10$ mV (Chap. 2, section 2.2)] that would bypass complex III and reduce cytochrome c directly were administered. This treatment led to a marked improvement in exercise capacity and increased levels of muscle phosphocreatine (Kennaway et al., 1984; Capaldi, 1988).

7.4.1 Structure and Function of Components of the Complex: Cytochrome b

The amino acid sequences of mitochondrial and chromatophore cytochrome c_1, and chloroplast cytochrome f were compared and discussed in Chap. 4 (Tables 4.8, 4.9; Fig. 4.15), as were the sequences of the Rieske ISp from mitochondria and chromatophores (Table 4.12). Although there is substantial local identity in particular regions (heme, nonheme iron coordination sites), overall the conservation of sequence identity is not high. The amino acid sequences of the heme-binding domain of cytochrome b, as well as the location of membrane-spanning hydrophobic regions, have been conserved in evolution to a high degree (Widger et al., 1984; Saraste, 1984; Hauska et al., 1988). The chloroplast, mitochondrial, and Rb. capsulatus cytochrome polypeptides contain 214, 380–385 (typically) and 438 residues. The chromatophore and mitochondrial sequences are more similar in length and sequence identity (Table 7.4), but their homology is high only over these first 214 residues, which constitute the heme binding domain (Widger et al., 1984).

The information contained in the COOH-terminal domain of the mitochondrial b cytochromes that is missing in the shorter chloroplast cytochrome b_6 may reside in subunit IV, a polypeptide of the chloroplast b_6–f complex

7.4 The Q Cycle and H$^+$ Translocation

Table 7.4. Amino acid sequences of the b cytochromes of mitochondrial complex III, photosynthetic bacteria, and the chloroplast b_6–f complex

```
          1         10        20        30        40        50        60        70        80
MY:    MAFRKSNVYLSLVNSYIIDSPQPSSINYWWNMGSLLGLCLVIQIVTGIFMAMHYSSNIELAFSSVEHIIRDVHNGYILRY
MA:    MRILKSHPLLKIVNSYIIDSPQPANLSYLWNFGSLLALCLGIQIVTGVTLAMHYTPSVSEAFNSVEHIMRDVNNGWLVRY
RbC:   ...KWLHDKLPIVGLVYDTIM-IPTPKNLNWWWIWGIVLAFTLVLQIVTGIVLAIDYTPHVDLAFASVEHIMRDVNGGWAMRY
Cp:    SKVYDWFEERLEIQAIADDITSKYVPPHVNIFYCLGGITLTCFLV-QVATGFAMTFYYRPTVTDAFASVQYIMTEVNFGWLIRS

          81        90       100       110       120       130       140       150       160
           ↓          ↓
MY:    LHANGASFFFMVMFMHMAKGLYYGSYRSPRVTLWNVGVIIFILTIATAFLGYCCVYGQMSHWGATVITNLFSAIPFVGND
MA:    LHSNTASAFFFLVYLHIGRGLYYGSYKTPRTLTWAIGTVILIVMMATAFLGYVLPYGQMSLWGATVITNLMSAIPWIGQD
RbC:   IHANGASLFFLAVYIHIFRGLYYGSYKAPREITWIVGMVIYLLMMGTAFMGYVLPWGQMSFWGATVITGLFGAIPGIGPS
Cp:    VHRWSASMMVLMMILHVFRVYLTGGFKKPRELTWVTGVVLGVLTASFGVTGYSLPWDQIGYWAVKIVTGVPDAIPVIGSM

          161       170       180       190       200       210       220       230       240
           ↓                             ↓
MY:    IVSWLWGGFSVSNPTIQRFFALH-YLVPFIIAAMVIMHLMALHIHGSSNPLGITGNLDRIPMHSYFIFKDLVTVFLFMLI
MA:    IVEFIWGGFSVNNATLNRFFALH-FLLPFVLAALALMHLIAMHDTGSGNPLGISANYDRLPFAPYFIFKDLITIFIFFIV
RbC:   IQAWLLGGPAVDNATLNRFFSLH-YLLPFVIAALVAIHIWAFHTTGNNNPTGV...DTLPFWPYFVIKDLFALALVLLG
Cp:    LVELLRGSASVGQSTLTRFYSLHTFVLPLLTAVFMLMHFLMIRKQGISGPL..........GHNYYWPNDLLYIFPVVIL
                                            *

          241       250       260       270       280       290       300       310       320
MY:    LALFVFYSPNTLGHPDNYIPCNPLVTPASIDPEWYLLPFYAILRSIPDKLLGVITMFAAILVLLVLPFTDASVVRGNTFK
MA:    LSIFVFFMPNALGDSENYVMANPMQTPPAIVPEWYLLPFYAILRSIPNKLLGVIAMFAAILALMVMPITDLSKLRGVQFR
RbC:   FFAVVAYMPNYLGHPDNYIQANPLSTPAHIVPEWYFLPFYAILR...KFFGVIAMFGAIAVMALAPWLDTSKVRSGAYR
Cp:    GTIACNVGLAVLEPSMIGEPADPFATPLEILPEWYFFPVFQILRTVPNKLLGVLLMASVPAGLLTVPFLENNKFQNPFRR

          321       330       340       350       360       370       380
MY:    VLSKFFFFIFVFNFVLLGQIGACHVEVPYVLMGQIATFIYFAYFLIIVPVISTIENVLFYIGRVNK
MA:    PLSKVVFYIFVANFLILMQIGAKHVETPFIEFGQISTIIYFAYFFVIVPVVSLIENTLVELGTKKNF
RbC:   PKFRMWFWFLVLDFVVLTWVGAMPTEYPYDWISLIASTYWFAYFLVILPLIGATEKPEPIPASIEEDFNSHIG
Cp:    PVATTVFLVGTVVAL-WLGIGATLPIDKSLTLGLF
```

MY and MA, mitochondria from yeast (Nobrega and Tzagoloff, 1980) and the fungus *A. nidulans* (Waring et al., 1981); RbC, *Rb. capsulatus* (Gabellini and Sebald, 1986; Daldal et al., 1987; Davidson and Daldal, 1987a, b); Cp, chloroplasts of spinach (Heinemeyer et al., 1984), with numbering as in the yeast sequence.

(*) End of cytochrome b_6 sequence; start of bicistronic downstream reading frame for subunit IV is MGVTKKPOLNDPVLRAKLAKGM. Cp, residues not shown: GEPA after Tyr-226, V after N-311; RbC, residues not shown: 16 residues before the first residue (K) shown, 12 residues between V-213 and D-218, and 22 residues after R-284. The two membrane-spanning helices that are underlined are cross-linked by two hemes coordinated to the two pairs of conserved His (↓) in each helix (see Fig. 7.17D).

coded by a reading frame for 155 residues (Table 7.4). This is inferred from (i) a bicistronic messenger RNA for cytochrome b_6 and subunit IV read from cytochrome b_6 to subunit IV (Heinemeyer et al., 1984), and (ii) a 30% homology between residues 260–350 of the mitochondrial cytochromes and subunit IV as shown in Table 7.4 (Widger et al., 1984). The single cytochrome b polypeptide encoded by the mitochondrial genome then corresponds to a split gene product in the chloroplast.

The hydropathy functions of the heme binding domain of these b cytochromes are even more highly conserved. The cross-correlation (Appendix III) of the hydropathy for residues 1–211 (end of cyt b_6 sequence) is high between all of the cytochromes, and particularly between the mitochondrial-chromatophore sequences, but not between the b cytochromes and other intrinsic membrane proteins of a similar size and hydrophobicity. This implies that the hydropathy function for the heme binding domain of the b cytochromes is a meaningful structural parameter (Widger et al., 1984). Sequences of 20–25 uncharged and hydrophobic residues in intrinsic membrane proteins are generally assumed to span the hydrophobic membrane bilayer in an α-helical conformation because (i) the solved crystal structures for the photosynthetic reaction centers and the electron diffraction data for bacteriorhodopsin define ample precedent for such structures; (ii) 20–25 residues in an α-helix with a pitch of 1.5 Å/residue are predicted to span a membrane bilayer of 30–40 Å in width. Hydropathy analysis indicated that the mitochondrial and chromatophore b cytochrome contain as many as eight membrane-spanning α-helices, and the chloroplast cytochrome b_6 would contain four such helices in its 214 residues (Fig. 7.17B). Two of the transmembrane helices would be cross-linked by the a pair of conserved His residues on the n and a pair on the p side of these helices (Fig. 7.17C), showing the structural motif of heme-cross-linking of intramembrane peptides in the b cytochromes (compare with Fig. 7.11, for cytochrome oxidase).

The hemes of the mitochondrial cytochrome b are known to be coordinated by histidines in both proximal and distal positions (Carter et al., 1981). With two hemes/cytochrome, four His per cytochrome polypeptide are involved in heme coordination. The four conserved His residues are found as two pairs in hydrophobic membrane spanning peptides II (His 82, 96) and IV [183 and 197 (198 in the chloroplast b_6)], with a His pair on each side of the membrane bilayer.

Sites of Inhibitor Binding; Protease Accessibility; Implications for Sidedness

The details of the cytochrome b folding model have been analyzed using inhibitor-resistant mutants of yeast (di Rago and Colson, 1988) and mouse mitochondrial cytochrome b (Howell and Gilbert, 1987), and of the *Rb. capsulatus* cytochrome (Daldal, 1987). Most of the inhibitors of the b–c_1 and b_6–f complexes affecting cytochrome b act with a preferential sidedness, either on the n or p side (Table 7.5A). Because the inhibitors are quinone analogues, this suggests the presence of specific quinone binding sites, Q_n and Q_p, on each side of the membrane. A caveat to the concept of well-defined Q_n and Q_p sites is that the majority of the well-characterized inhibitors act on the p side of the membrane (Table 7.5A), so that the p site is better characterized than the n. Chloroplasts do not have an n side inhibitor, and photoaffinity azidoquoinone analogues bind at a p side on subunit IV of the b_6–f complex (Doyle et al., 1989), so that the Q_n–Q_p concept seems best established in chromatophores and mitochondria. Sites of inhibitor action have been located in the yeast mitochondrial cytochrome b by nucleic acid sequencing of mutants resistant to the p side inhibitors mucidin, myxathiazol, and stigmatellin, and to the *n* side inhibitors DCMU and antimycin A (Table 7.5B). The mutations

7.4 The Q Cycle and H+ Translocation

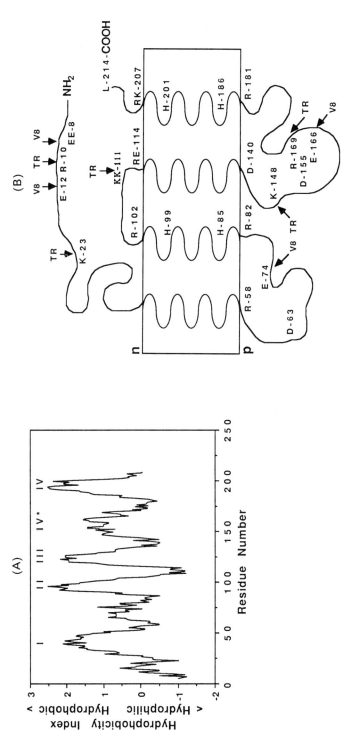

Figure 7.17A, B. (A) Hydropathy plot of the spinach chloroplast cytochrome b_6, showing the four main peaks of hydrophobicity that are thought to correspond to membrane spanning helices of ~20 residues. Peak III is probably a peripheral peptide domain. (B) Four helix model of monomeric chloroplast cytochrome b_6 in the membrane bilayer inferred from comparison of sequences (Table 7.4), homology of hydropathy plots, the assumption that the conformation of the membrane spanning peptides is α-helical, and determination of protein topography using proteases and peptide-directed antibodies. Distance between heme edges, ca. 12 Å. The sidedness of the membrane is indicated by sides n (stroma) and p (lumen), for the sign of the $\tilde{\mu}_{H^+}$ generated by the H+ translocation.

334 7: Light and Redox-Linked H⁺ Translocation

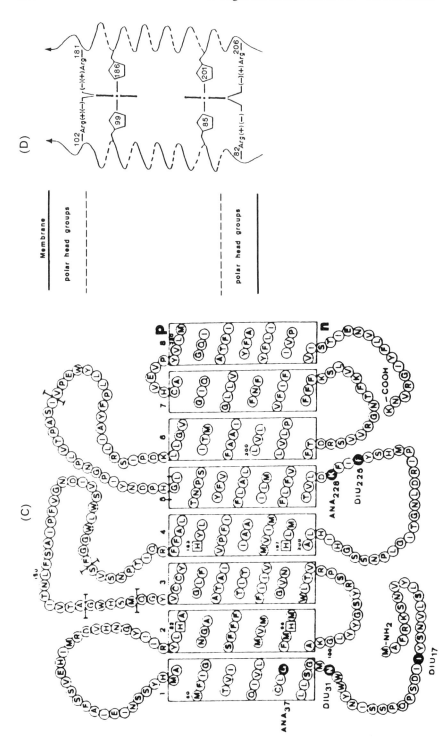

7.4 The Q Cycle and H⁺ Translocation

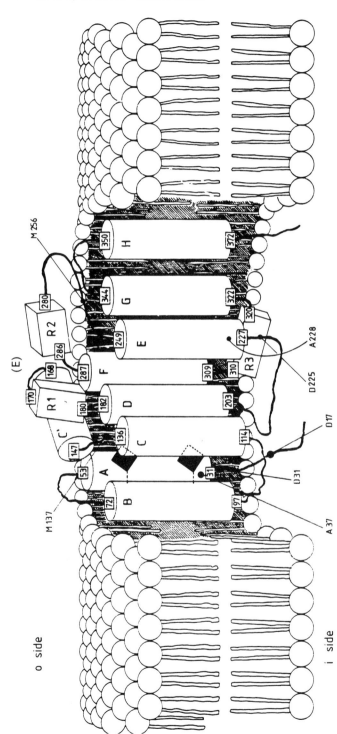

Figure 7.17C, D, E. (C) Eight helix model of monomeric yeast cytochrome b obtained from hydropathy plot (Appendix III) showing the sites of mutation to n (DCMU, antimycin A) and p (myxathiazol) side inhibitors of cytochrome b in mitochondria and chromatophores (di Rago et al., 1988). (D) Intramolecular cross-linking of two transmembrane α-helical segments of cytochrome b of bc_1 or $b_6 f$ (numbering for b_6–f) complex by two hemes spanning the bilayer, one on each side. Cross-linked segments each contain two conserved His residues, separated by 13 residues in chromatophores and mitochondria, and 13 and 14 residues in the two helices of chloroplasts (cf., Widger et al., 1984; Hauska et al., 1988). (E) Three-dimensional representation of the organization of the yeast cytochrome b polypeptide (Brasseur, 1988).

Table 7.5A. Properties of some common inhibitors of the cytochrome bc_1 or $b_6 f$ complex[a]

Compound	Side of Inhibition	Comment	References
Antimycin A	n	Mit: Tight binding to b_{ox} and b_{red} but $K_{dox} < K_{dred}$; red shift of b_{562} spectrum; causes "crossover" of b and c_1; elim. $Q_n^{\doteq}$ EPR signal	Berden and Slater (1972); Bowyer and Trumpower (1981); de Vries et al. (1981)
		Chr: Inhibits b_{561} oxid., c_1 reduction $Q_n^{\doteq}$ EPR signal, Cyt b_{561} red shift. Stoichiometry, 1:1 monomeric complex. Chl: inhib. cyclic phosphorylation	van den Berg et al., (1979); Robertson et al. (1984); Moss and Bendall (1984)
DBMIB	p	Chl: Shifts EPR signal of ISp from 1.89 → 1.94; stoichiometry, 0.5 with b_6–f.	Malkin (1982); Graan and Ort (1983)
DNP-INT	p	Chl: Inhibits oxid. of PQH_2; cyt b azido-labeled inhibitor binds to cyt b, ISp.	Trebst et al., (1978); Oettmeier et al., (1983); O'Keefe (1983)
Halogenated hydroxypyridines	n	Chl: Inhibits cyclic phosphorylation, $P/2e^-$ of non-cyclic, but not electron transport.	Hartung and Trebst (1985)
NQNO(HQNO)	n	Chl: Increase of amplitude of b_6 photoreduction; shifts cyt b_6 E_m in vitro.	Selak and Whitmarsh (1982); Clark and Hind (1983b); Jones and Whitmarsh (1985)
Myxothiazol	p	Mit: Binds to ISp or ISp-cyt. b; redshift of b_{566}; g_x shift of ISp; does not inhibit succinate reduction, fumarate oxidation of cyt b. Mucidin is a qualitatively similar inhibitor.	Von Jagow et al. (1984)

7.4 The Q Cycle and H⁺ Translocation

Table 7.5A (continued)

Compound	Side of Inhibition	Comment	References
Myxothiazol	p	Chr: Blocks reduction of ISp, UQH$_2$ oxid., but not oxid. of ISp, at 3/complex. Blue shift, E_m shift ($-90 \to -40$ mV) of b_{566}. Binding independent of anti A; displaces UHDBT.	Meinhardt and Crofts (1982)
Stigmatellin	p	most effective inhibitor of PQH$_2 \to$ PC in isolated b_6–f complex.	Oettmeier et al. (1985)
UHDBT	p	Chr: Shifts E_{m7} (280 $\to$ 350 mV) of ISp; blocks oxid of ISp; competes with UQ for site on ISp; Increased oxid of cyt c (Chr), cyt f (Chl).	Whitmarsh et al. (1982); Crofts (1985)

[a] The inhibitory sites are on the p or n side of the b cytochrome, the sides toward or away from which protons are translocated. The notation n and p is used to provide a single notation for chromatophores, chloroplasts, and mitochondria for the side of the membrane on which positive (p) and negative (n) $\tilde{\mu}_H$. values are established as a result of electron transport (see Chap. III). Inhibitors on the n side, close to cytochrome b_{561} in chromatophores, prevent oxidation of cytochrome b and stabilize oxidant-induced reduction of cytochrome b, but do not affect reduction of the ISp by ubiquinol. Inhibitors on the p side are closely associated with the ISp and with cytochrome b_{566} in chromatophores, prevent reduction of ISp by quinol and oxidant-induced reduction of cytochrome b, but do not affect the reduction of cytochrome b that may still occur in b-c_1 via the site on the n side. Reduction of cytochrome b is completely inhibited only by blocking sites on both n and p sides. Chr, chromatophores; Chl, chloroplasts, Mit, mitochondria.
DBMIB, 2,5-dibromo-3-methyl-6-isopropyl-benzoquinone.
DNP-INT, dinitrophenyl ether of 2-iodo-4-nitrothymol.
HQNO, 2-n-heptyl-4-hydroxyquinoline-N-oxide.
NQNO, 2-n-nonyl-4-hydroxyquinoline-N-oxide.
UHDBT, 5-n-undecyl-6-hydroxy-4,7-dioxobenzothiazol.

Table 7.5B. Sites of inhibitor-resistant Mutations and Quinone Binding in Mitochondrial Cytochrome b

Inhibitor	Side	Residue
DCMU	n	17-Ile → Phe
		31-Asn → Lys
		225-Phe → Leu or Ser
		226-Ile → Phe
Antimycin A	n	37-Gly → Val
		38-Gly → Val[a]
		228-Lys → Met
Myxothiazol	P	142-Gly → Ala[a]

[a] Mouse mitochondrial cytochrome b; all others in yeast.
From di Rago and Colson (1988); Howell and Gilbert (1987).

to DCMU resistance in the yeast mitochondrial cytochrome at positions 17 (Ile → Phe) and 31 (Asn → Lys) near the polar NH_2-terminus define the orientation of the NH_2-terminus on the cytoplasmic or n side (Fig. 7.17C). Yeast cytochrome mutations at Gly-37 → Val that yield resistance to antimycin A are five residues into the bilayer on the n side (Fig. 7.17C). Mutations to DCMU at positions 225 (Phe → Leu or Ser) and 226 (Ileu → Phe) and to antimycin at position 228 (Lys → Met) would also be on the n side of the mitochondrial or chromatophore cytochromes in the model of Fig. 7.17C.

7.4.2 Topography of the Other Redox Components of the Complex

Cytochrome f

As discussed (Chap. 4, section 4.7), the heme of cytochromes f and c_1 is on the p side of the membrane, most likely in the aqueous phase or at the membrane interface.

Rieske ISp

The amino acid sequences derived from DNA sequence analysis of the Rieske ISp from *Neurospora crassa* mitochondria (Harnisch et al., 1985) and *Rb. capsulata* (Gabellini and Sebald, 1986) have been compared (Table 4.13). A location of the ISp at the p side interface is implied by its accessibility to membrane-impermeant redox reagents from the cytoplasmic side of the membrane (Trumpower, 1981b), its presence in an M_r 16,000 water-soluble chymo-

7.4 The Q Cycle and H^+ Translocation

tryptic peptide made from the detergent-solubilized *N. crassa* protein (Harnisch et al., 1985), and the presence of its immediate oxidant, the heme of cytochrome c_1 or f, on the p side. Thus, the prosthetic groups of cytochromes f, c_1, the [2Fe-2S] center, plastocyanin, and the c cytochromes are all in contact with the polar phase, on the p side of the membrane. Selective extraction of the mitochondrial ISp by high salt concentrations caused a loss of its characteristic electron paramagnetic resonance (EPR) signal, and resulted in loss of the cytochrome c reductase activity as well as the oxidant-induced reduction of cytochrome b (Fig. 7.18). The activities were restored upon reconstitution of the depleted system with purified ISp.

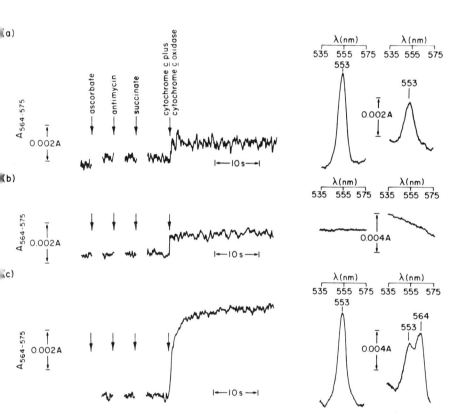

Figure 7.18. Demonstration of oxidant-induced reduction of mitochondrial cytochrome b involved in complex III and requirement of Rieske [2Fe-2S] ISp. Cytochrome c and oxidase were added to oxidize cytochrome c_1 initially reduced by succinate and ascorbate. Antimycin A was present to prevent reoxidation of reduced cytochrome b. Spectra on the right were obtained before and after addition of cyt c and cyt oxidase. (a) Complex III depleted of Rieske ISp. (b) Complex III omitted from the suspension to show that background absorbance change is nonspecific due to a scattering artifact. (c) Complex III as in (a), and reconstituted with Rieske ISp (Trumpower, 1981b).

7.4.3 Oxidant-Induced Reduction of Cytochrome b

The observation of oxidant-induced reduction (Wikström and Berden, 1972) in the cytochrome bc_1 and $b_6 f$ complexes, together with the measurement of H^+/e values >1 for these complexes, was important in subsequent studies of cytochrome b function, and in the formulation of the Q cycle mechanism (Mitchell, 1976; Garland et al., 1975). Addition of oxygen to the respiratory chain or ferricyanide to the complex resulted in the expected oxidation of cytochrome c_1, but also caused reduction of cytochrome b (as shown in Fig. 7.18A) in the presence of the inhibitor antimycin A, which blocks its reoxidation (Table 7.5). The explanation is that the semiquinone ($Q^{\dot-}$) generated by oxidation of the quinol (QH_2) is a stronger reductant than the quinol (Chap. 2, section 2.8), capable of reducing the cytochrome b (Fig. 7.18B). From Chap. 2, section 2.8, one can calculate the semiquinone formation constant (K_s) if the E_m of $Q/Q^{\dot-} = E_m$ (cytochrome b heme) $= -90$ mV, and E_m (Q/QH$_2$) $= +70$ mV (Problem 73).

Transmembrane Electron Transport Pathways of Cytochrome b

In contrast to the proton pump models discussed for the cytochrome oxidase and bacteriorhodopsin complexes, the model for electrogenic H^+ translocation that is favored by most workers in the field is a "Q cycle" (Mitchell, 1976; Matsuura et al., 1983; Wikström and Saraste, 1984; Crofts, 1985; Rich, 1985) that can utilize the transmembrane heme arrangement shown in Fig. 7.17D. The experimental observations, in addition to the structural concepts, on which the Q cycle is based are summarized in Table 7.6: There is general agreement that $H^+/e^- = 2$ in the mitochondrial b–c_1 complex. H^+ translocation across the chloroplast $b_6 f$ complex is also electrogenic, but the exact value of the H^+/e^- stiochiometry is less certain. For the chloroplast complex, $H^+/e^- = 2$ and 1 under conditions of low and high $\Delta\tilde{\mu}_{H^+}$, respectively, and the reaction is not electrogenic under the latter conditions (Graan and Ort, 1983).

The kinetics and redox potential dependence are similar for the reduction of cytochrome b_{561} (b_n) and the formation of the slow electrochromic phase (slow $\Delta\psi$) in the presence of antimycin A (Glaser and Crofts, 1984). The reduction of cytochrome b_{561} by a single flash in the presence of antimycin, and of b_{566} by a second flash, or by a first flash when b_{561} is chemically reduced (Crofts et al., 1983; Meinhardt and Crofts, 1983), is as expected from the Q cycle model (Fig. 20).

Mechanism of the Q cycle

The oxidation of the quinol by one electron transferred to the high potential oxidant, at an ambient pH above the pK of the semiquinone, would result in

7.4 The Q Cycle and H⁺ Translocation

Table 7.6. Data underlying the Q cycle formulation of electron transfer in the bc_1 or b_6f complex[a]

1. Structural data on transmembrane arrangement of two hemes of cytochrome b (Fig. 7.17D).
2. $H^+/e^- = 2$ for quinol oxidation in mitochondria, and for chloroplasts in a single turnover flush experiment.
3. Oxidant-induced reduction of cytochrome b (Fig. 7.18C); dependence of oxidant-induced reduction on presence of oxidized ISp (Fig. 7.18; compare with Prince et al., 1982).
4. Millisecond electrochromic carotenoid band shift (Fig. 7.19A and B) dependent on quinol oxidation (Fig. 7.19C) in chromatophores and chloroplasts (Chap. 3, section 3.4.2) indicative of a millisecond electrogenic charge separation (Jackson and Dutton, 1973; Joliot and Delosme, 1974).
5. Inhibitors blocking reactions of bc_1 and b_6f complexes at n or p sides; define Q binding sites (Chap. 2, section 2.8, Chap. 5, section 5.7) on n, p sides (Table 7.5A).
6. Flash-induced reduction of two transmembrane hemes in chromatophores; b_{561} (b_n) and b_{566} (b_p) hemes identified by spectrum and E_m value (Crofts, 1985).
7. Demonstration of $\Delta\psi$-driven electron conduction from reduced heme b_n to heme b_p of the mitochondrial cytochrome b (West et al., 1988).

[a] See Fig. 7.20.

---

release of $2H^+$, formation of the semiquinone anion, Q^-, at pH values greater than its pK, and an H^+/e^- value of 2, according to the following steps (Fig. 7.20).

(i) Oxidation of the high potential part of the e^- transport chain by O_2 or light results in a one-electron oxidation by the ISp of quinol (QH_2) to the semiquinone (Q^-), with $2H^+$ deposited on the p side. Q^- has a negative enough E_{m7} (Chap. 2), to reduce b_p whose E_{m7} (-100 to 0 mV in the different membranes, Table 7.7), makes it difficult to reduce with QH_2 (Fig. 7.20A and B).

The quinol reacts with the oxidized ISp (FeS^+) $b-c_1$ complex in a second-order process (Crofts et al., 1983),

$$QH_2 + (FeS^+ \cdot b_p) \to Q + (FeS \cdot b_p^-) + 2H^+, \quad (1)$$

with the equilibrium constant,

$$K_{eq} = \exp\left\{(E_m(FeS) + E_m(b_p) - 2E_m(Q)) \cdot \frac{F}{RT}\right\} \quad (2)$$

The E_{m7} values of the ISP and the quinone pool are 280–290 mV and ~65–90 mV, respectively, in mitochondria, chromatophores, and chloroplasts. Then, with E_m values in mV,

$$K_{eq} = \exp\{(100 + E_m(b_p)) \div 25.7\}, \quad \text{at pH} = 7 \text{ and } 25°C, \quad (3)$$

7: Light and Redox-Linked H⁺ Translocation

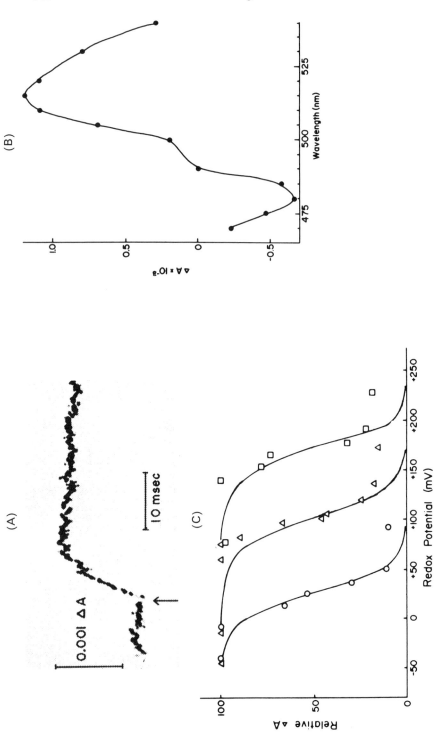

Figure 7.19. Time course (A), spectrum (B) and redox titration as a function of pH (C) of the millisecond electrochromic band shift arising

7.4 The Q Cycle and H⁺ Translocation

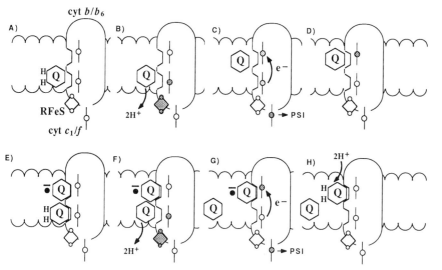

Figure 7.20. Q cycle model for the transfer of two electrons (first electron, top row A–D) from plastoquinol (ubiquinol) to the cytochrome b_6f or bc_1 complex. (A, B) The oxidation of the Rieske iron-sulfur (RFeS) center by cyt f, formation of the semiquinone ($Q^{\dot{-}}$, not shown) and transfer of $2H^+$ to the lumen, and reduction of heme b_p by $Q^{\dot{-}}$. (C, D) Transfer of electron from heme b_p to b_n. Oxidation of cyt f is also shown. (E) Transfer of electron from b_n to Q forming semiquinone, and binding of quinol at p-side. (F–H) Transfer of second electron and deposition of $2H^+$, as in (B–E) except that QH_2 is formed at n-side by one electron reduction of $Q^{\dot{-}}$ (drawing by P.N. Furbacher).

and
$$K_{eq} = 1.5, \quad \text{when } E_m(b_p) = -90 \text{ mV}. \tag{4}$$

(ii) The next event in the Q cycle would be quinone diffusion from the p to the n side and interheme electron transfer across an ~ 12 Å dielectric between the heme edges to the higher potential b_{562} or b_{561} in mitochondria and chromatophores ($E_{m7} = +50$ mV), or the other b_{563} heme in chloroplasts (Table 7.7), a step that would be electrogenic (i.e., the electron crossing the membrane dielectric would generate a membrane potential) (Fig. 7.20C). The reactions would be:

$$QH_2 + (FeS^+ \cdot b_p \cdot b_n) \to Q + (FeS\, b_p^- \cdot b_n) + 2H_p^+, \tag{5}$$

and
$$FeS \cdot b_p^- \cdot b_n \to FeS \cdot b_p \cdot b_n^-.$$

Then,
$$K_{eq} = \exp\{(E_m(FeS) + E_m(b_n) - 2E_m(Q)) \div 25.7\}. \tag{6}$$

If $E_m(b_n) = +50$ mV, as in Rb. sphaeroides chromatophores, then $K_{eq} = 10^{2.5} = 350$.

Table 7.7. E_m values and reduced α-band maxima of the two hemes of cytochrome b in mitochondria, chloroplasts, and chromatophores

	E_m of b_p, b_n	α-Band maxima of b_p, b_n (nm)
Mitochondria	$-50, +50$	566, 561
Chromatophores	$-90, +50$	566, 562
Chloroplasts	$-50, -50$[a]	563(564)

[a] Two hemes are isopotential ± 50 mV.

Thus, a Q cycle reaction in chromatophores or mitochondria that proceeds only as far as to b_p does not go forward to an appreciable extent, but is pulled forward by the presence of the coupled reduction of b_n when there is a large positive ΔE_m between b_p and b_n.

The question of how the quinone moves from the p to the n side in the Q cycle model has led to a variation on the model (Wikström and Saraste, 1984; Rich and Wikström, 1986), in which the motion of the semiquinone is restricted to the central region of the bilayer between the two hemes. Quinol is oxidized to a semiquinone that can reduce the b_p, as in the Q cycle. The resulting quinone is displaced by the semiquinone $Q^{\bar{\cdot}}$ formed in a second cycle of quinol oxidation at the p side. The second $Q^{\bar{\cdot}}$ can then accept an electron from the n side b heme forming quinol and completing the cycle. In this cycle, the semiquinone alternately reduces and oxidizes the b_p and b_n during successive turnovers of the complex, providing the mechanism for shuttling electron equivalents from the p to the n side.

(iii) The cycle would be completed at the n side by two successive one-electron reductions of quinone to semiquinone (Fig. 7.20E) and semiquinone to quinol (Fig. 7.20F–H). Thus, the Q_p site turns over twice for each turnover of Q_n. One problem with this aspect of the model is that the reduction of Q to $Q^{\bar{\cdot}}$ on the n side is not clearly thermodynamically facile. In chromatophores, where the E_{m7} of $b_n = +50$ mV, it depends critically on the E_{m7} of the $(Q \rightarrow Q^{\bar{\cdot}})$ reaction, which has been estimated to be $+30$ mV (Robertson et al., 1984). In chloroplasts, the $Q \rightarrow Q^{\bar{\cdot}}$ reaction could be supplied by ferredoxin as part of the photosystem I cyclic pathway (Fig. 6.27C). The second electron needed to reduce $Q^{\bar{\cdot}}$ to QH_2 would then be supplied by b_n.

(iv) The Q cycle would be completed by diffusion of the QH_2 formed at the n-site to the p side.

The Q cycle model developed for chromatophores is slightly different because of a 1:2, or even smaller, stoichiometry of bc_1 complex to reaction center as compared to 1:1 in chloroplasts. Two reaction centers per b–c_1 complex are excited per flash, the diffusible carrier, cytochrome c_2, turns over twice per flash, and in the steady state two quinols (QH_2) are formed per flash, one by the reaction center and one at the n side of the b–c_1 complex (Fig. 7.21) [Crofts, 1985].

7.4 The Q Cycle and H^+ Translocation

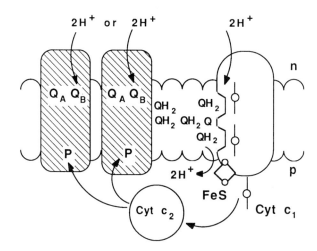

Figure 7.21. Cyclic electron transport pathway in photosynthetic bacteria utilizing the cytochrome bc_1 complex and the reaction center at a 1:2 stoichiometry. The bc_1 complex functions by the Q cycle mechanism, with two bc_1 complexes turning over per flash, each translocating $2H^+$ per turnover (Crofts, 1985; drawing by P.N. Furbacher).

Unique Aspects of the Chloroplast b_6–f Complex; Other Explanations for Electrogenic H^+ Translocation

The similarities of the sequences and hydropathy pattern of the $b_6 f$ complex and the heme-binding domain of the bc_1 complexes give ample reason to expect that if the mechanism for electrogenic H^+ translocation is a Q cycle in the cytochrome bc_1 complex, then the same mechanism must apply to the cytochrome b_6–f complex. However, the chloroplast cytochrome b_6 is different in its smaller size (Table 7.4), the extra residue (threonine) between the two heme-binding histidines in helix IV (Table 7.4; Fig. 7.17A), a single quinone binding site on subunit IV with none detected on the cytochrome b polypeptide (Doyle et al., 1988), and the presence of ferredoxin and ferredoxin: $NADP^+$ reductase on the n-side of the membrane. It differs also in the similar redox and spectral properties of the two b_6 hemes, which may be a consequence of the extra residue in helix IV resulting in a $100°$ change in the relative orientation of the two His rings. In any case, it has not been possible to demonstrate heme $b_p \rightarrow b_n$ electron transfer associated with the Q cycle mechanism and electronic communication between the heme b_p and b_n appears to be inefficient (Furbacher et al., 1989). The Q cycle would not be needed in any case in steady-state electron transport in chloroplasts because the H^+/e^- ratio = 1 for plastoquinol oxidation (Graan and Ort, 1983). Therefore, another mechanism is needed to explain the $H^+/e^- = 2$ ratio observed under conditions of low $\Delta\tilde{\mu}_{H^+}$ present in single turnover flash experiments in the laboratory. The precedent of the light- and redox-driven H^+ pumps in the bacteriorhodopsin and cytochrome oxidase systems discussed above suggest the possibility of such a pump in the $b_6 f$ complex, that could be gated by turnover of the iron-sulfur center (Girvin and Cramer, 1984) or the $QH^{\cdot}/QH_2$ couple (Lorusso et al., 1985; Joliot and Joliot, 1986). In addition, the transmembrane arrangement of hemes b_n and b_p in cytochrome b_6 can be utilized in PS I cyclic phosphorylation (Fig. 6.27C).

7.5 H^+ Translocation or Deposition Sites in the Mitochondrial, Chromatophore, and Chloroplast Electron Transport Chains; Stoichiometries of H^+ Translocation and ATP Synthesis

The mitochondrial respiratory chain contains three sites of H^+ translocation, at complex I, III, and IV (Fig. 3.5A). At the present time, the mechanism of a Q cycle for the cytochrome bc_1 complex of mitochondria and chromatophores is favored by most workers, whereas the mitochondrial cytochrome oxidase, as discussed, is thought to operate through a proton pump. The chloroplast noncyclic chain contains two sites of H^+ deposition or translocation, the H_2O-splitting reaction and the quinol oxidation reaction involving the cytochrome $b_6 f$ complex (Fig. 3.5C, 6.27B). The latter is the only site for H^+ translocation in the cyclic pathway of the photosynthetic bacteria.

The maximum number of protons that can be translocated by the respiratory and noncyclic photosynthetic electron transport chains can be calculated: (a) For the respiratory chain, the free energy, ΔG_{et}, available to support the $\Delta \tilde{\mu}_{H^+}$ obtained from NAPH oxidation is derived from the redox span NADH $\to O_2$, approximately equal to 1.2 V. The $\Delta \tilde{\mu}_{H^+}$ in units of volts ["proton-motive force" $= \Delta \tilde{\mu}_{H^+}/F$ (Chap. 1)] is ~ 200-240 mV. (b) For noncyclic photosynthetic electron transport, the comparable $\Delta \tilde{\mu}_{H^+}$ (predominantly a ΔpH) is supported by the free energy change, ΔG_{et}, derived from electron flow from $PQ_A^{\bar{\cdot}}$ to P700 ($\Delta E_m \simeq 0.6$ V), in the intermediate electron transport chain joining the two photosystems (Chap. 6), perhaps with a contribution from the donor side of photosystem II (PS II). Then the maximum number of protons translocated, $n_{H^+} = H^+/e$, per electron transported through these redox spans can be calculated from:

$$\Delta G_{et} + \Delta \tilde{G}_{H^+} = 0$$
$$-(1)(F)(\Delta E) + n_{H^+} \cdot \Delta \tilde{\mu}_{H^+} = 0 \tag{7}$$

For the respiratory chain,

$$-(F)(1.2) + n_{H^+} \cdot \Delta \tilde{\mu}_{H^+} = 0$$
$$-(F)(1.2) + n_{H^+}(0.24F) = 0 \tag{8}$$

7.5 H⁺ Translocation or Deposition Sites in the Mitochondrial

$$n_{H^+} = \frac{1.2}{0.24} = 5, \quad \text{or 10 per pair of electrons transferred.}$$

For the noncyclic photosynthetic chain,

$$-(F)(0.6) + n_{H^+} \cdot \Delta\tilde{\mu}_{H^+} = 0$$

$$-(F)(0.6) + n_{H^+} \cdot (0.24F) = 0 \tag{9}$$

$$n_{H^+} = 2.5.$$

The experimental values obtained for the H^+/e^- stoichiometry in the different partial reactions of the mitochondrial respiratory chain are summarized in Table 7.8.

The determination of H^+/e^- stoichiometries has been a controversial area. The range of values for NADH → O_2 in the respiratory chain is 5–6 for the H^+/e^-, or 10–12 for $H^+/2e^-$. Because the consensus value for H^+/ATP in chloroplasts, mitochondria, and chromatophores is 3 (Table 3.10), and one H^+ is needed for phosphate transport into the matrix (Fig. 3.27), the total $H^+/ATP = 4$ in mitochondria, as discussed (Chap. 3, section 3.9). If the $ATP/2e^-$ stoichiometry for oxidation of NADH-linked substrate is 3, then the H^+/e^- for the respiratory chain must be 6 (Table 7.8), and the $H^+/2e^- = 12$. If this $ATP/2e^-$ ratio is 3 1/4 (Table 7.9B), then the $H^+/2e^-$ ratio would have to be even greater, i.e., 13 (Lemasters, 1984), indicating that the assumed value of $\Delta\tilde{\mu}_{H^+}$ would have been slightly too large in the estimate made in Eq. 8.

Table 7.8. Stoichiometry of proton translocation in the respiratory and noncyclic photosynthetic chain

Reaction	H^+/e^-
A. *Photosynthesis* (Saphon and Crofts, 1977; Förster and Junge, 1985; Hangarter et al., 1987a)	
H_2O → P680 (PS II)	1 (Average of 4 S states)
PQH_2 → P700 (PS I)	1 (Steady-state)
	2 (Weak light or threshold of steady-state)
H_2O → P700 (PS I + PS II; noncyclic electron transport)	2 (Steady-state)
B. *Respiration* (Wikström and Saraste, 1984; Moody et al., 1987)	
NADH → ubiquinone (site 1)	1–2
Ubiquinol → cytochrome c (site 2)	2
Cytochrome c → O_2 (site 3)	2[a]
Sites (1 + 2 + 3)	5–6[b]

[a] If the measured stoichiometry of H^+ ejection from the energized oxidase is $H^+/e = 1$, then $H^+/e^- = 2$ when the translocation of H^+ for O_2 reduction is included (Fig. 7.14A).
[b] A value of H^+/e^- for respiratory electron transport at least equal to 6, and $H^+/2e^-$ at least 12, would be required if $H^+/ATP = 4$ for ATP synthesis incl. P_i transport (Chap. 3, section 3.9), and $ATP/2e^- = 3$ or 3 1/4 for NADH-linked substrates (Table 7.9B).

Table 7.9. Stoichiometry of ATP synthesis in partial and complete electron transport reactions of respiration and plant photosynthesis

Reaction	ATP/2e$^-$
A. Photosynthesis: chloroplasts (Flores and Ort, 1984; Hosler and Yocum, 1985)	
$H_2O \rightarrow$ ferredoxin/O_2 (PS II + PS I)	1.60
$H_2O \rightarrow$ methyl viologen (PS II + PS I)	1.2–1.3
Duroquinol $\rightarrow$ methyl viologen (PS I)	0.6
$H_2O \rightarrow$ dimethyl benzoquinone (PS II)	0.3
$H_2O \rightarrow$ dimethyl benzoquinone with preillumination (PS II)	0.5
B. Respiration (Chance and Williams, 1956[a]; Hinkle and Yu, 1979[b]; Lemasters et al., 1984[c]).	
Site I: NADH $\rightarrow$ UQ; Succinate $\rightarrow$ acetoacetate[c]	$\leq$ 1[a]; 1.25[c]
Sites I + II: NADH $\rightarrow$ Cyt c	$\leq$ 2[a]; 1.75[c]
Site III: TMPD-Cyt $c \rightarrow O_2$	$\leq$ 1[a]; 1.50[c]
Sites II + III: Succinate $\rightarrow O_2$	$\leq$ 2[a]; 1.3[b]; 2.0[c]
Sites I + II + III: NADH $\rightarrow O_2$	$\leq$ 3[a]; 2.0[b]; 3.25[c]

7.5.1 Stoichiometry of ATP Synthesis

The ATP/2e$^-$ or P/O stoichiometry has often been discussed in terms of an integral number of phosphorylation sites in the respiratory and photosynthetic chains, 3, 2, and 1 for electrons in the respiratory chain donated, respectively, by NADH, succinate, or cytochrome c. However, using the concepts of the chemiosmotic hypothesis (Chap. 3), one can see that the ATP/2e$^-$ stoichiometry will be determined by the stoichiometries of (i) proton translocation by the electron transport chain (H$^+$/2e$^-$), and (ii) the proton requirement (H$^+$/ATP) for ATP synthesis [ATP/2e$^-$ = (ATP/H$^+ \cdot$ H$^+$/2e$^-$)], and therefore need not be an integral number (Ferguson, 1986).

Values of n_{H^+} = 2–3 (H$^+$/2e$^-$ = 4–6) for chloroplasts, together with H$^+$/ATP = 3 for the H$^+$ requirement for ATP synthesis, imply that the maximum ATP/2e$^-$ ratio for noncyclic electron transport, $H_2O \rightarrow$ PS I, would be 1.33 (n_{H^+} = 2) or 2.0 (n_{H^+} = 3). The maximum ATP/2e$^-$ values measured for noncyclic electron transport in the absence of ferredoxin (Table 7.9A) are more consistent with n_{H^+} = 2. The steady-state ATP/2e$^-$ value of 1.2–1.3 for noncyclic electron transport, $H_2O \rightarrow$ methyl viologen would imply that ATP/2e$^- \simeq$ 0.6 for the PS II noncyclic reaction, because the ATP/2e$^-$ value determined for PS I = 0.6 (i.e., whole chain reaction = PS I + PS II). The explanation of the low ATP/2e$^-$ ratio for PS II associated with proton release from H_2O (Table 7.9A) may arise from the use of lipophilic quinone electron acceptors. The values > 1.3 measured with ferredoxin are attributed to the simultaneous operation of the PS I cyclic pathway (Fig. 6.27C).

As sumarized in Table 7.9B, the ATP/2e$^-$ values of 3 for NAD-linked (e.g., pyruvate, 3-hydroxybutyrate, malate, glutamate; Fig. 4.23) and 2 for FAD-linked respiratory substrates are the standard textbook values that defined the sites of phorphorylation in the respiratory chain (Tzagoloff, 1982). The sites are also defined by proton or charge-translocating protein complexes located in the mitochondrial inner membrane (Fig. 3.5A): site 1, the NADH–ubiquinone oxidoreductase—complex 1 (Chap. 4); sites 2 and 3, the ubiquinol–cytochrome c oxidoreductase and cytochrome oxidase.

7.6 Summary

An H^+ pump is an energy-linked H^+ translocation system in a transmembrane protein or protein complex. H^+ pump function has been documented in bacteriorhodopsin (bR), cytochrome oxidase, and H^+-ATPase. The rate-limiting step in all cases is a few milliseconds. bR forms purple crystalline patches in the cytoplasmic membrane of the salt-tolerant bacterium, $H.$ $halobium$. All-$trans$-retinal is bound through a Schiff base to Lys 216 of the 248 residue protein. Light absorption causes isomerization of the retinal and a large decrease in the pK of the Schiff base that allows subsequent H^+ release in the L → M step of the bR photochemical cycle. H^+ extrusion to the external medium and H^+ uptake from the cytoplasm result in formation of a $\Delta\tilde{\mu}_{H^+}$ across the membrane bilayer. This H^+ translocation pathway involves protonation and deprotonation of a chain of amino acids, in which Asp-85, Asp-96, Asp-212, and Tyr-185 have been implicated from analysis of site-directed mutants. Bound H_2O may also contribute to the pathway. The natural order of bR in the membrane has allowed a tertiary structure determination to ~3 Å resolution by electron diffraction and image reconstruction. The membrane structure of bR consists of a trimeric organization of seven transmembrane, 35–40 Å helical rods arrayed in a crescent-shaped structure. There are three other rhodopsin-like pigments in the $H.$ $halobium$ membrane, two sensory rhodopsins, and halorhodopsin (hR) which has significant primary sequence identity to bR and mediates a light-driven Cl^- pump.

Cytochrome oxidase in mitochondria and some aerobic bacteria has the dual role of generating a $\Delta\tilde{\mu}_{H^+}$ as it reduces O_2 to H_2O. The oxidase contains four redox centers, two a hemes (a and a_3), and two Cu ions (Cu_A and Cu_B), and 9–13 polypeptides in eukaryotes of which the three largest are encoded by mitochondria. The two largest of the latter bind the two hemes and two coppers. More simple oxidase preparations with only one or two subunits containing the metal centers have been isolated from aerobic bacteria. Deficiency in oxidase, or cytochrome bc_1 complex, has been recognized as a source of mitochondrial myopathy. The Cu_A is believed to be coordinated by two Cys and two His residues in eukaryotic subunit II, close to three acidic residues that are believed to complex with basic residues of cytochrome c. Cu_A-heme a are the initial electron acceptors for the electron from cytochrome

c, so that the electron transfer pathway within the oxidase is cyt $c \to (Cu_A \to$ heme $a) \to$ (heme a_3-Cu_B). Three redox groups are proposed to cross-link or coordinate residues of helices VI, VII, and X of subunit I: heme a cross-links His-266 (helix VI) and His-398 (helix X) on the p side of the membrane; heme a_3 coordinates to His-273 (helix VI), Cu_B to His-322 and His-323 (helix VII), and His-406 (helix X) in the center of the bilayer. A model for O_2 evolution involves transfer of two electrons from reduced heme a_3-Cu_B to bound O_2 forming a bridging peroxy intermediate, and complete reduction after transfer of two more electrons from the respiratory chain and binding of $4H^+$. In addition to the H^+ bound as part of the $O_2 \to H_2O$ reaction, 1 H^+ per e^- transferred is pumped by the oxidase complex from the matrix to the cytoplasmic side of the membrane, thereby generating a $\Delta\tilde{\mu}_{H^+}$ as large as -240 mV. The proposed mechanisms for H^+ translocation have focused on redox-linked H^+ binding changes of the redox groups, and not as in bR on the role of amino acids. A role of heme a the H^+ pump was suggested by an increase in the strength of its formyl group hydrogen bond in the reduced oxidase. The third subunit in the eukaryotic oxidase may be involved in H^+ translocation, perhaps indirectly through stabilization of a dimeric oxidase.

The cytochrome bc_1 and b_6f complexes are ubiquitous in energy transducing membranes that utilize light or an obligatory aerobic respiratory chain. The simpler complexes are in bacteria and chloroplasts, three and four polypeptides, respectively, including cytochrome b (two hemes), cytochrome c_1 or f (1 heme) and the high-potential iron-sulfur protein (1[2Fe-2S] center). The amino acid sequences in the heme binding domain (first 215 residues) of the b cytochromes and the hydropathy functions for these cytochromes have been extensively conserved in evolution, allowing a hypothesis for folding of the cytochrome b polypeptide in the membrane bilayer. Two of the transmembrane peptides contain a His residues on each side of the bilayer through which they are cross-linked on each side by a bis-histidine coordinated heme, placing one heme on the n side and one on the p side of the membrane. The function of the b cytochrome is in all cases associated with oxidant-induced reduction. In this reaction, the quinone (QH_2) is oxidized to the semiquinone, $Q^{\cdot -}$, by the Rieske [2Fe-2S] center on the p side of the membrane. The unstable $Q^{\cdot -}$ or $QH^{\cdot}$ is a stronger reductant than QH_2 and can reduce heme b_p. This reaction initiates a "Q cycle" pathway in chromatophores and mitochondria, in which the electron is next transferred across the bilayer from heme b_p to b_n. This pathway is believed to be responsible for electrogenic translocation of $2H^+$ across the bilayer for each electron transferred from QH_2 to the [2Fe-2S] center. Other H^+ pump mechanisms are also considered because it has not been possible to document electron transfer from b_p to b_n in chloroplasts, where it is likely that the b_6f complex is used in PS I-dependent cyclic electron transport.

The H^+/e^- values from all sources of H^+ translocation are: 5-6 for NADH $\to O_2$ in the respiratory chain, and 2 for steady-state noncyclic electron transport in oxygenic photosynthesis.

Problems

70. (a) Using the data shown, calculate the specific H^+ pump rate from lipid vesicles for (A) bR, and (B) bR fusion protein purified from *E. coli*, if both assays contain 71 pmol bR, and the vertical bar corresponds to 3 nmol H^+ and the horizontal bar to 1 min (Dunn et al., 1987). (b) How does the rate of H^+ efflux correspond with the characteristic time for H^+ release in the bR photocycle (Fig. 7.1)? (c) If the number of H^+ pumped per bR = 50 in the steady state, and if this H^+ movement were not compensated by movement of other charges, how many bR molecules must pump to generate a $\Delta\psi = -100$ mV across a bR-containing vesicle of 1 μm diameter (cf. Chap. 3, section 3.5)?

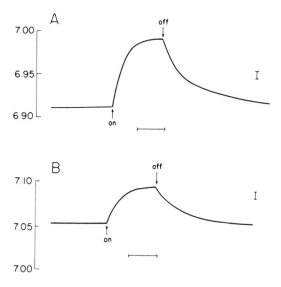

71. The enthalpy and energy changes of bR can be resolved into phases during the photocycle (Ort and Parson, 1979). The input energy of the photon is 49 kcal/mol. At 100 μs after the flash, the bR enthalpy is ~20 kcal greater than before illumination. After H^+ release and before rebinding, a large amount of heat (~45 kcal/mol) is released to the surroundings. At this time the total energy of the system is 35 kcal/mol less than it was before the flash. However, after this large energy release, the free energy of the system must be greater than before illumination, because relaxation to the initial state occurs spontaneously. Calculate the minimum decrease of entropy associated with the large increase of molecular order that must accompany the heat release.
72. Calculate the maximum H^+/e^- ratio that could be generated in the denitrifying electron transport chain of *P. detrificans* if the free energy released by electron transport would be used entirely to maintain a steady

proton motive force of 180 mV, and the total redox span of the electron transport chain is +750 mV, from NADH to the NO_3/NO_2^- couple ($E_{m7} = +430$ mV).

73. Calculate the equilibrium constant in chloroplasts for a Q cycle reaction in which one electron is transferred from PQH_2 [E_{m7} (PQ/PQH_2) = +100 mV] to heme (b_n) if the E_{m7} of heme $b_n = -50$ mV. Assume the E_{m7} (ISp) = +290 mV.

PART III   Utilization of Electrochemical Ion Gradients

Chapter 8
Transduction of Electrochemical Ion Gradients to ATP Synthesis

8.1 Introduction to the Structure and Function of the ATP Synthase

Proton pumping ATPases can be divided into three classes (Pedersen and Carafoli, 1987a,b; Nelson, 1988): (1) The eubacterial "F-type" that is present in bacteria such as *E. coli*, mitochondria, and chloroplasts, does not use a phosphorylated high-energy intermediate of the enzyme, and can function in both oxidative- and photo-phosphorylation to reversibly transduce $\Delta \tilde{\mu}_{H^+}$ and ATP. (2) vacuolar "V-type" ATPases, found in the vacuolar systems of eukaryotic cells, and perhaps descended more directly from the archaebacteria, functions in acidification of the organelle interior. (3) The "P-type" is commonly found in the plasma membrane, utilizes a phosphorylated high-energy intermediate, and is also involved in the translocation of Na^+, K^+ and Ca^{2+}. Most of the following discussion concerns the F-type used in organelle ATP synthesis. It is characterized by its multisubunit polypeptide composition. The notation EF_0F_1, CF_0F_1, MF_0F_1, and TF_0F_1 is used for the enzyme consisting of the membrane-bound (F_0) and peripheral (F_1) components from *E. coli*, chloroplasts, mitochondria, and thermophilic bacteria, respectively.

The F_1 sector, containing the catalytic site(s) for ATP synthesis and hydrolysis, can be seen as a ~ 90 Å diameter structure by electron microscopy in negatively stained organelle membranes or reconstituted liposomes (Fig. 8.1). The large distance (~ 40 Å) between catalytic sites on CF_1 was discussed in Chap. 6 in the context of the use of resonance energy transfer to measure intra-protein distances. The intrinsic hydrophobic F_0 complex acts as an H^+ channel or conduit of conformational changes necessary for ATP synthesis in the F_1 domain.

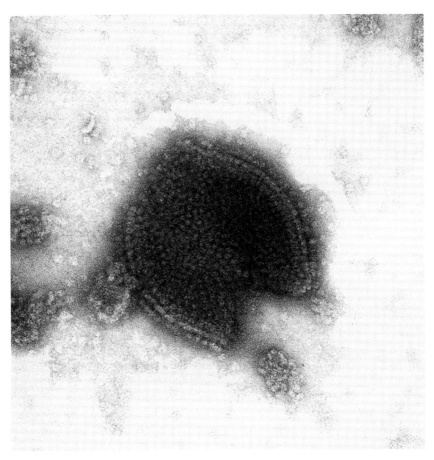

Figure 8.1. Electron micrograph of proteoliposome reconstituted with CF_0F_1 ATPase. The CF_0 sector has an approximately cylindrical shape with a height and diameter of 8.3 nm and 6.2 nm, respectively. CF_1 sector is a disc-like structure with a diameter and height of 11.5 and 8.5 nm, respectively. The stalk connecting CF_0 to CF_1 is 3.7 nm. (E.J. Boekema and P. Gräber, unpublished results; compare with Boekema et al., 1987.)

Sidedness

The F_0F_1 enzyme is located in the cytoplasmic membrane of photosynthetic bacteria and *E. coli* and in the inner membrane of mitochondria, with the F_1 unit located on the inner or matrix side of the membrane (Fig. 8.2A). The enzyme is oppositely oriented in the membranes of chloroplast thylakoids, chromatophores, and submitochondrial particles (SMP), where it is located on the outside or stromal side (Fig. 8.2B). The orientation of the ATPase correlates with the direction of electron transport-linked H^+ movement, which is outwardly directed for mitochondria and prokaryotes, and inwardly directed for thylakoids, chromatophores, and SMP.

8.1 Introduction to the Structure and Function of the ATP Synthase

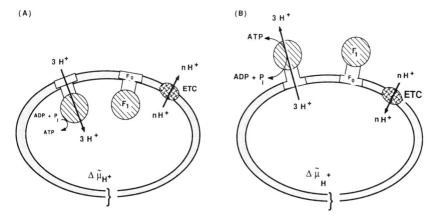

Figure 8.2. Orientation of the reversible H^+-ATP synthase (A) in mitochondria and prokaryotes where $\Delta\psi$ is the dominant component of the $\Delta\tilde{\mu}_{H^+}$ under physiological conditions, and (B) in chromatophores, submitochondrial membrane vesicles, and thylakoids where ΔpH is the dominant component of the $\Delta\tilde{\mu}_{H^+}$ because $\Delta\psi$ is dissipated by permeability to counter ions. In thylakoids, the H^+ flux has been tracked from the lumen across the CF_0F_1, and into the external medium with a half-time of about 40 ms (cf., Junge, 1987).

Reversibility

The F_0F_1 enzyme synthesizes ATP by using the H^+ flux through a favorable electrochemical gradient (large negative $\Delta\tilde{\mu}_{H^+}$) across the F_0 to the F_1 side of the membrane. Thus, the sign of the $\Delta\tilde{\mu}_{H^+}$ favorable for ATP synthesis would be negative for H^+ moving from the outside to inside (cytoplasm or matrix) in bacterial cells and mitochondria, and from the lumen (inside) to the outside (stroma or intermembrane space) in chloroplasts, chromatophores, or SMP. Utilizing ATP, the enzyme can also operate in the reverse direction to translocate H^+ from the F_1 side across F_0 to generate a $\Delta\tilde{\mu}_{H^+}$ that can be used for cell or organelle work such as transport (Chap. 9). Thus,

$$nH^+ (F_0 \text{ side}) + P_i + ADP \leftrightarrow ATP + H_2O + nH^+ (F_1 \text{ side}), \qquad (1)$$

with a net free energy change, $\Delta G = \Delta G_{ATP} + n\Delta\tilde{\mu}_{H^+}$ (Chap. 1).

Charge Balance, the ATP–ATP Translocator, and Electrogenicity in Mitochondria

The direction of the reversible H^+ movement in mitochondria through the ATPase was shown in Fig. 3.27, along with the flow of H^+ showing utilization by the ATP synthase, P_i transport, and ATP–ADP translocase enzymes, of the H^+ flux through the $\Delta\tilde{\mu}_{H^+}$. The number of H^+ that move electrogenically across the membrane generating the $\Delta\tilde{\mu}_{H^+}$ must be balanced by the number that return through the synthase and phosphate transport enzymes. Otherwise, the resulting hyperpolarization will cause inhibition of electron transport and H^+ translocation. Thus, if $4H^+/2e^-$ are translocated across one site in the

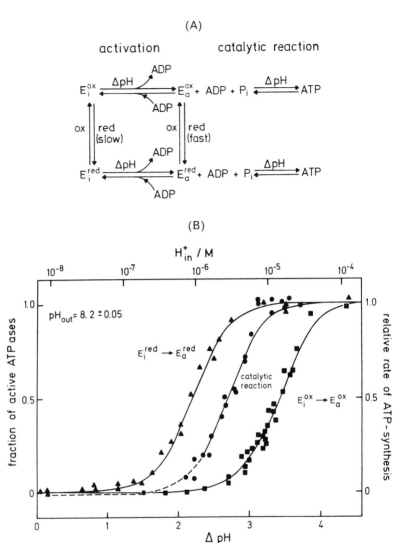

Figure 8.3. (A) Model for regulation of the H^+-ATPase in chloroplasts. In the dark, the ATPase, E_i^{ox}, is oxidized and inactive, it contains a tightly bound ADP, a disulfide (S—S) in the γ subunit, and a group with $pK_a = 5.9$ that must be protonated for the enzyme to be activated for ATP synthesis. Upon reduction of the disulfide, $E_i^{ox} \rightarrow E_i^{red}$, the pK of the activable group increases (>5.9) so that a smaller ΔpH is needed for activation. E_i^{ox} or E_i^{red} is transformed to the active form, E_a^{ox} or E_a^{red}, by energization (ΔpH) accompanied by the release of a tightly bound ADP. (Adapted from Junesch and Gräber, 1985.) (B) Activation of ATP synthase in oxidized and reduced states by ΔpH. (i) Catalytic reaction: illumination of reduced enzyme activates CF_1 to high rates of ATP synthesis (400 ATP/CF_0F_1-S, half-maximal activity at $\Delta pH = 2.7$) or hydrolysis (-90 ATP/CF_0F_1-S at $\Delta pH = 0$). (ii) The oxidized enzyme can catalyze the same maximum rate of synthesis, but the half-maximum rate requires $\Delta pH = 3.4$. (iii) ATP hydrolysis by activation of reduced, inactive enzyme ($E_i^{red} \rightarrow E_a^{red}$) is also shown. ΔpH generated by dark acid–base transition (Chap. 3.6) (Junesch and Gräber, 1987).

8.1 Introduction to the Structure and Function of the ATP Synthase

electron transport complex (e.g., complex III), $3H^+$ are returned per ATP synthesized through the ATP synthase (Chap. 3), and $1H^+$ with $H_2PO_4^-$ is returned through the phosphate transport protein. Tabulated in this way, an electrogenic reaction results from the ATP–ADP exchange that is linked to net outward transport of one extra negative charge on ATP relative to ADP (Chap. 3, section 3.9.1).

Regulation of the Enzyme; Oxidized and Reduced States of CF_1

The direction of the ATP synthase-ATPase reaction is determined by the $\Delta\tilde{\mu}_{H^+}$, the synthetic and hydrolytic reactions respectively favored by large and small values of the $(-\Delta\tilde{\mu}_{H^+})$ driving the H^+ current through F_0 to F_1. In chloroplasts, the threshold value of the ΔpH for ATP synthesis and the level of the ATP synthase activity is also regulated by the reduction of the enzyme (Fig. 8.3), involving a disulfide group (—S—S—) on one (γ) of the F_1 subunits (Fig. 8.3) (Shahak, 1982; Mills and Mitchell, 1982). Because the $\Delta\tilde{\mu}_{H^+}$ is small when chloroplasts are kept in the dark, it is necessary for the chloroplast enzyme to possess a regulatory mechanism other than the $\Delta\tilde{\mu}_{H^+}$, so that cellular ATP will not be dissipated at night through hydrolysis by the chloroplast ATPase. This mechanism involves deactivation in the dark and consequent activation of the CF_1 in the light via reduction of the γ subunit via photosystem I (PS I), thioredoxin, and ferredoxin. This results in two compo-

le 8.1. Molecular weights of F_0F_1 subunits

trinsic nponent	EF_0F_1	PBF_0F^a	TF_0F_1	$MF_0F_1^b$	CF_0F_1
	55,264 (uncA)	55,026	54,590	55,164	55,457
	50,316 (uncD)	50,852	51,938	51,595	53,831
	31,387 (uncG)	32,437	31,778	30,141	35,857
)SCP)	19,558 (δ) (uncH)	19,543	19,657	20,967 (OSCP)	20,512
)	14,914 (uncC)	14,307	14,333	15,065 (δ)	14,701
nimal)	—	—	—	5,652	—
	—	—	—	8,000	—
Pase hibitor	—	—	—	7,383	—
	—	—	—	13,000	—
mbrane ponent, F_0					
	30,258 (uncB)	12,460	26,589	25,000 (ATPase-6)	27,060
	17,233 (uncF)	10–11,000	17,268	10,000	20,900
	8,246 (uncE)	9,400	7,334	8,000	7,968
urth F_0 ubunit	—	—	14,595	~7,000 (A6L)	16,000

. *rubrum*, photosynthetic bacterium.
ef heart mitochondria.
here there is a dual notation for the subunit, the primary notation corresponds to that of the *E. coli* and roplast subunits; OSCP, oligomycin sensitivity conferral protein.

nents of energetic control over initiation of ATP synthesis by thylakoid membranes: (i) energetic transformation of the CF_1 to a catalytically active state; (ii) an energetic threshold for synthesis of ATP from particular concentrations of ADP and orthophosphate. The activation process dominates the thermodynamic requirement in illuminated chloroplasts below a threshold value of ΔG_{ATP} that depends on whether the CF_1 is reduced (threshold = 45 kJ/mol) or oxidized (threshold = 51 kJ/mol), corresponding to a difference in the threshold ΔpH of ~ 0.3 if the $H^+/ATP = 3$ (Hangarter et al., 1987b). The threshold difference in ΔpH between oxidized and reduced states was found to be somewhat larger when ATP synthesis is driven by a dark acid–base transition (Fig. 8.3B).

The complicated demands on the activity of the mammalian mitochondrial ATPase necessitate regulation by several interrelated factors including (i) the $\Delta \tilde{\mu}_{H^+}$ across the inner mitochondrial membrane, (ii) an ATPase inhibitor protein subunit (Table 8.1), (iii) bound nucleotide and/or phosphate, (iv) divalent cations, and (v) the reduction level of disulfide groups on the F_1 part of the molecule (Chernyak and Kozlov, 1986; Schwerzmann and Pedersen, 1986). The maximum rate of ATP synthesis, ca. 400 ATP/F_0F_1-s, is similar in chloroplasts (Junesch and Gräber, 1987) and mitochondria relieved of the restraint in $\Delta \tilde{\mu}_{H^+}$ (Matsumo-Yagi and Hatefi, 1988).

8.2 Preparation of H^+-ATPase

The F_0F_1 enzyme can be solubilized from membranes with a detergent such as cholate, deoxycholate, octylglucoside, or Triton X-100 (Penefsky et al., 1960). Denaturation can be prevented by minimizing the detergent concentration and time of treatment, and adding phospholipid, glycerol, or ATP which has a protective effect. The F_1 alone can be removed from the membrane and purified as a water-soluble protein by washing with distilled water or EDTA, extraction in chloroform, sonication, with all procedures carried out at room temperature because F_1 dissociates into subunits and loses its ATPase activity much more rapidly when incubated on ice ($\sim 4°C$) than at room temperature (see Problem 73).

8.3 Structure of F_0F_1 ATP Synthase

8.3.1 Subunit Composition

F_0F_1 ATP synthase complexes contain as many as 13 subunits, with most complexes containing 8–10 subunits (Pedersen and Carafoli, 1987a). As with other oligomeric protein complexes of mitochondria and chloroplasts, the subunits are synthesized by both organelle and cytoplasmic ribosomes. The five F_1 polypeptide subunits in bacteria and plants, and the five largest in mammalian mitochondria are $\alpha, \beta, \gamma, \delta, \varepsilon$ with a stoichiometry, in order of descending molecular weight, of $3\alpha:3\beta:1\gamma:1\delta:1\varepsilon$ corresponding to a molecular weight of approximately 385,000 (Fig. 8.4; Table 8.1). The three largest

8.3 Structure of F_0F_1 ATP Synthase

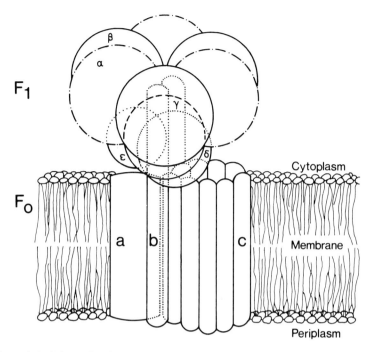

Figure 8.4. Schematic of EF_0F_1 *E. coli* complex (Schneider and Altendorf, 1987).

subunits, α, β, and γ, appear to be common to all prokaryotic and eukaroytic F_1's but the subunit analogies of other subunits with the eukaryotic mitochondrial MF_1 are more complicated. The bovine MF_0F_1 has perhaps as many as 13 different polypeptides, of which five, α, β, γ, δ, and ε (Notation for EF_0F_1; Table 8.1) form the external F_1 domain (Walker et al., 1985). The eukaryotic yeast and rat liver enzymes also have a larger number of subunits than those of bacteria and chloroplasts. A 7.4-kDa ATPase inhibitor protein is also associated with MF_1. The δ subunit of EF_1 is homologous to the MF_1 oligomycin sensitivity conferral protein (OSCP), which along with factor F_6 is involved in the attachment of F_1 to F_0. The OSCP is a 90 × 30 × 30 Å prolate ellipsoidal protein (Dupuis et al., 1983) that is often seen in electron microscope pictures as a stalk-like structure. The δ subunit is thought to be the connecting structure in chloroplasts and *E. coli*.

Regarding the membrane sector, F_0, the EF_0 contains three polypeptides, *a*, *b*, *c*, sometimes denoted χ, Ψ, and Ω, present at an approximate stoichiometry of 1:2:10 (Foster and Fillingame, 1982). The stoichiometry of the F_0 subunits in the membrane sector has not been clearly established in the other complexes. The eukaryotic F_0 is again more complicated, with four subunits in chloroplasts and bovine mitochondria. Of the eukaryotic F_0F_1 subunits encoded by organelle and nuclear DNA, only two of the MF_1 subunits, those for proteins A6L and ATPase-6 are encoded by mitochondrial DNA, the remainder by the nucleus. Six of the nine CF_0F_1 polypeptides, α, β, ε, *a*, *b*, and *c*, are encoded by chloroplast DNA.

8.3.2 Function of F_1 Subunits

α and β

The central importance of the β subunit is implied by (i) the high degree of primary sequence conservation between β subunits of *E. coli*, chloroplasts, and mitochondria. Antibody to the β subunit in fact cross-reacts with antibody from the corresponding subunit of all H^+-ATPases tested so far (Nelson, 1988). The β subunit sequence from rat liver mitochondria is shown in Table 8.2A. Sequences of the β subunit involved in nucleotide binding that show

Table 8.2. (A) Predicted amino acid sequence of most of the MF_1 β subunit from rat liver mitochondria,[a] (B) sequence identity with P-ATPases, and (C) sequence identity with adenylate kinase

1   50
NH_2...TATGQIVAVIGAVVDVQFEDGLPPILNALEVQGRESRLVLEVAQHLGEST
  51   100
VRTIAMDGTEGLVRGQKVLDSGAPIKIPVGPETLGRIMNVIGEPIDERGP
                                                  150
IKTKQFAPIHAEAPEFIENSVEQEILVTGIKVVDLLAPYAKGGKIGLFGG
                                               200
AGVGKTVLYMELINNVAKKHGGYSVFAGVGERTREGNDLYHEMIESGVIN
                                               250
LKDATSKVALVYGQMNEPPGARARVALTGLTVAEYFRDQEGQDVLLFIDN
                                             300
IFRFTQAGSEVSALLGRIPSAVGYQPTLATDMGTMQERITTTKKGSITSV
                                              350
QAIYVPADDLTDPAPATTFAHLDATTVLSRAIAELGIYPAVDPLDSTSRI
                                             400
MDPNIVGSEHYDVARGVQKILQDYKSLQDIIAILGMDELSEEDKLTVSRA
                                             450
RKIQRFLSQPFQVAEVFTQHMGKLVPLKETIKGFQQILAGDYDHLPEQAF
                           472
YMVGPIEEAVAKADKLAEEHGS-COOH

B.

β Subunit (298–305):	NH_2 ... T - T - K - K - G - S - I - T ... COOH
Na^+,K^+-ATPase (368–375):	... S - D[e]- K - T - G - T - L - T ...
Ca^{2+}-ATPase:	... S - D[e]- K - T - G - T - L - T ...

C.

β Subunit (156–170):	... G G A G V G K T V L I M E L I ...
Adenylate kinase (15–29):	... G G P G S G K G T Q C E K I V ...
β Subunit (248–256):	... G Q D V L L - F I D ...
Adenylate kinase (110–119):	... A Q P T L L L Y I D ...
β Subunit (311–333):	...Y V P A D D L T D P A P A T T F A H L D A T T ..
Adenylate kinase (172–194):	...K V N A E G S V D N V F S Q V C T H L D A L K...

[a] Predicted from cDNA sequence; from comparison with NH_2-terminal sequence of the protein, first residue (T) is not the NH_2 terminus.
From Garboczi et al. (1988).

8.3 Structure of F_0F_1 ATP Synthase

Table 8.3. Specific chemical modification and inhibition of ATPase in the β subunit of bovine mitochondria MF_1

Reagent	Residue(s) modified	No. β's modified for loss of capacity for rapid catalysis
DCCD	Glu-199	1–2
NBD-Cl[a]	Tyr-311	1
FSBI[b]	Tyr-345	1
2-azido-ATP	Tyr-345	1
	Tyr-368 (noncatalytic)	1
FSBA[c]	Tyr-368; His-427	3

[a] NBD-, 4-chloro-(7-nitrobenzo)-2-oxa-1,3-benzoxadiazole.
[b] p-Fluorosulfonylbenzoylinosine.
[c] p-Fluorosulfonylbenzoyladenosine.

identity to a phosphate binding sequence of P-type ATPases are shown in Table 8.2B, and three sequences in adenylate kinase with identical segments in Table 8.2C. (ii) In addition, the β subunit, possibly along with α, has been identified as the nucleotide-binding subunit in F_1 using photoreactive nucleotide analogs such as 2-azido-ATP and 5'-p-fluorosulfonylbenzoyladenosine (Abbott et al., 1984; Boulay et al., 1985; Bullough and Allison, 1986a). (iii) Specific chemical labeling in the bovine mitochondrial β subunit (Table 8.3) at a stoichiometry of one/MF_1 can result in complete inactivation of the ATPase activity (Vignais and Lunardi, 1985; Bullough and Allison, 1986b). The latter experiments also have implications for the question of site–site interactions between the three β and three α subunits (section 8.7.1).

The nucleotide photoaffinity probe labeling of α in CF_1 (Abbott et al., 1984), MF_1 (Boulay et al., 1985), ATP binding to EF_1 α subunit (Rao et al., 1987), and the inhibitory effect of a mutation (S373 → F) in the α subunit on steady-state catalysis (Noumi et al., 1984a) indicate that α is also involved in nucleotide binding and the catalytic site (Table 8.4). A search for the nucleotide binding sequence through sequence homology has indicated some related sequences in α and β, and homology of residues 150–156 in β to other known nucleotide binding sequences (Walker et al., 1982), such as the nucleotide binding pocket in adenylate kinase and in Ca^{2+}- or Na^+, K^+-ATPase (Table 8.2B, C), *ras* p21 transforming protein, and elongation factor Tu (Garboczi et al., 1988).

The Small F_1 Subunits, γ, δ, and ε

In spite of the similarity of the molecular weight heirarchy and subunit stoichiometry in the F_1 from different sources, the role of these subunits, particularly δ and ε, appears to vary between the different membrane systems, as summarized in Table 8.4. ATPase activity of EF_1, TF_1, and CF_1 seem to require a minimum α–β–γ unit. γ was found to be part of the H^+ gate in TF_0F_1 from reconstitution studies of H^+ translocation in liposomes (Fig. 8.5). Studies of γ mutants of EF_1 show that its COOH-terminal domain is important for ATPase activity and for normal interaction of EF_1 and EF_0 (Miki et al., 1986).

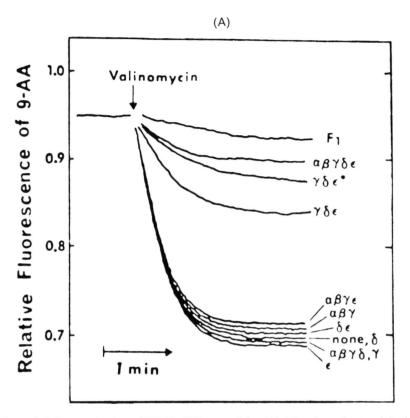

Figure 8.5. Reconstitution of TF_0F_1 ATPase activity. (A) $\Delta\tilde{\mu}_{H^+}$-dependent activity in F_0 (17.5 μg)-containing liposomes measured by self-quenching of the fluorescence probe, 9-aminoacridine (see Chap. 3), that occurs when protons move into the vesicle through reconstituted TF_0F_1 ATPase in response to valinomycin-initiated K^+ efflux. More quenching occurs when the H^+ channel is not blocked. The following amounts (μg) of different subunits were added to the F_0-liposomes: α, 31; β, 45; γ, 15; δ, 11; ε, 11, and addition of double aliquot is indicated by (*) (Kagawa, 1978). (B) Generation of electric current in planar bilayer membranes containing TF_0F_1. TF_0F_1 was incorporated into the bilayer membranes and current measured at picoampere resolution. After formation of the bilayer, one chamber (cis) was perfused with buffer to remove free TF_0F_1-liposomes. Perfusing medium containing ATP was added at the first arrow, and removed at the second. The ATP-induced current was suppressed by NaN_3, a specific inhibitor of TF_1, or a transmembrane potential of −180 mV, negative on the cis side (Hirata et al., 1986). (C) pH change caused by valinomycin-induced H^+ uptake in proteoliposomes made with different EF_0 subunits (Schneider and Altendorf, 1987).

8.3 Structure of F_0F_1 ATP Synthase

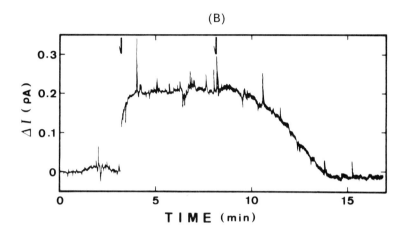

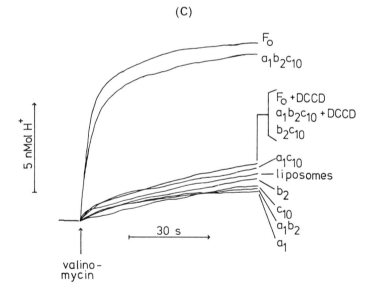

Table 8.4. Summary of consensus functions of F_0F_1 subunits

A. α, Involved in nucleotide binding; probably contains the catalytic site, or shares it with β, at the $\alpha-\beta$ interface.

β, Highly conserved in evolution; readily labeled with nucleotide analogs; contains catalytic site(s); some local sequences involved in nucleotide binding have been identified.

γ, Probably part of the minimum $\alpha-\beta-\gamma$ unit needed for ATPase activity and the H^+ gate. An "organizer" polypeptide, possibly involved in a "turnstile" mechanism for alternating sites; cross-links to β and ε in EF_1.

δ, Interacts with NH_2-terminal regions of α; δ (OSCP in mammalian mitochondria) needed for binding to EF_0; part of proton gate (TF_0F_1); prevents nonproductive H^+ leak; used for binding to F_0, and needed for maximum rates of photophosphorylation.

ε, Needed for binding to EF_0; ATPase inhibitor, as well in MF_0F_1; part of proton gate (TF_0F_1); inhibitor of CF_1 ATPase.

B. a, b, c, Essential components of proton channel; c especially well studied in all the membrane systems because of its specific reactivity with dicylohexylcarbodiimide (DCCD) involving one acidic amino acid residue in the hydrophobic domain and contributing two membrane spanning helices per subunit to the channel structure, with ~ 10 c/channel; a may contribute five membrane-spanning helices to the channel structure; b, involved in binding EF_1 to EF_0 and the membrane.

δ is required for binding CF_1 to CF_0, δ and ε for EF_1 binding to EF_0 and as an inhibitor of CF_1 ATPase (Table 8.4). EF_1 δ may interact with a large protruding segment of the EF_0 b subunit (Fig. 8.9) for successful EF_0F_1 binding. The MF_1 subunits analogous to EF_1 δ and ε in terms of function and amino acid homology are the OSCP and the δ subunits, respectively.

8.3.3 Reconstitution from Subunits

F_1 activities can be reconstituted from individual subunits. This is illustrated for TF_1 which can be reconstituted from its subunits mixed in the presence of ATP and Mg^{2+} into F_0-containing artificial liposomes (Fig. 8.5A; Table 8.5; Yoshida et al., 1977; Futai, 1977; Kagawa, 1978), for TF_0F_1 reconstituted into planar bilayer membranes (Fig. 8.5B), and for CF_0F_1 (Pick and Racker, 1979) and EF_0 incorporated into artificial liposomes (Fig. 8.5C). TF_0-proteoliposomes loaded with K^+ and the fluorescent pH indicator, 9-aminoacridine (Chap. 3, section 3.4.1) were used to determine which subunit of the TF_1 can gate or plug the H^+ channel. Addition of valinomycin to generate a K^+-diffusion potential (Chap. 3, section 3.4.2) drives exchange of internal K^+ for external H^+ through the TF_0 H^+ channels, resulting in

8.3 Structure of F_0F_1 ATP Synthase

Table 8.5. Reconstituted activities and subunit requirements of TF_1

Activities	Subunits
1. ATP hydrolysis	$\alpha\beta\gamma$
2. Binding to TF_0	δ and ε
3. Blockage or gating of H^+ permeability of TF_0	$\gamma\delta\varepsilon$
4. H^+ translocation, ATP synthesis, P_i–ATP exchange	All five subunits added to TF_0 in liposomes

From Kagawa (1978).

quenching of fluorescence unless TF_1 subunits ($\gamma\delta\varepsilon$) are added that can plug the CF_0 channel (Fig. 8.5A). This gating was more complete if the nucleotide binding subunits α and β were also present. Using proteoliposomes, made with different EF_0 subunits, measurement of the external pH change with a pH electrode showed that the H^+ uptake in response to the K^+-diffusion potential required the complete complement of EF_0 subunits, a, b, and c, and is blocked (Fig. 8.5C) by dicyclohexylcarbodiimide (DCCD) that binds specifically to the c subunit (section 8.5.3).

A hybrid ATPase could be formed from TF_1 and EF_1, with $T\alpha T\beta E\gamma$ resembling TF_1, $E\alpha E\beta T\gamma$ being relatively thermophilic, and $E\alpha T\beta E\gamma$ having properties similar to those of EF_1 (Takeda et al., 1982), showing (i) that α, β, and γ are the subunits essential for ATPase activity, (ii) the properties of thermophilicity cannot be ascribed to any single subunit, and (iii) not all permutations are allowed in this system because the combination $T\alpha E\beta T\gamma$ did not work. All possible combinations of the three subunits from the closely related *Salmonella typhimurium* and *E. coli* showed ATPase activity (Hsu et al., 1984), and the subunit β from *E. coli* could be used to reconstitute β-less chromatophores from *Rhodospirillum rubrum* (Gromet-Elhanon et al., 1985).

8.3.4 Is the Ion Translocated by $\Delta\tilde{\mu}_{H^+}$ H^+ or OH^-?

When one speaks of a direction for H^+ movement, it is, in fact, difficult to distinguish it from the alternative possibility of an electrically equivalent but oppositely directed OH^- movement. However, for the reconstituted TF_0 complex it has been shown from the external pH dependence of the H^+ conductivity of TF_0 inserted into liposomes, measured with a pH electrode as in Fig. 8.5C, that the translocated ion is probably H^+ and not OH^- (Sone et al., 1981). Considering H^+ as the substrate, the rate of H^+ uptake increased as the external pH decreased, generating a linear double-reciprocal graph. This graph could be characterized, for $\Delta\psi = -94$ mV, by Michaelis–Menten parameters, $V_{max}/mg = 31$ µg ion/min-mg for TF_0 and a $K_m = 9.5 \times 10^{-8}$ M for H^+. The average specific H^+ conductance was 47 $H^+/s \cdot TF_0$ [approximately 1/6 the maximum pump rate measured for the CF_0F_1 hydrolyzing 90 ATP/s (Fig. 8.3)]. The alternative OH^- movement is less likely because it would be outwardly directed from the vesicle lumen that has a limited OH^- capacity.

8.3.5 Direct Observation of Electrical Current Generated by TF_0F_1 ATPase

The H^+ ATPase from the PS3 thermophilic bacterium was incorporated in planar bilayers, where current can be measured on a subpicoampere scale. A current of 0.1–0.2 pA was observed when ATP was added to one side of the bilayer (Fig. 8.5B). The H^+ nature of the current was inferred from the flow of positive current from the F_1 to the F_0 side. Suppression of the current by a cis-negative potential of -180 mV [conditions for ATP hydrolysis, $\Delta G° = -7.5$ kcal/mol, $ATP/(ADP)(P_i) = 10^4$] showed that the stoichiometry of H^+ translocation in this experiment was $3H^+/ATP$ (Problem 76).

8.3.6 Crystallographic Studies of F_1

Rat liver MF_1 has been crystallized and at 9 Å resolution suggested to be a dimer with an asymmetric unit of 180,000 molecular weight, and approximate three- and twofold axes of symmetry (Amzel et al., 1982; 1984; Pedersen and Amzel, 1985). The overall structure was found to be asymmetric with two equal masses, presumably the two α–β subunits, and a third unequal and larger mass, thought to be an α and β subunit associated with one or more of the smaller single-copy F_1 subunits, γ, δ, or ε. The ability of ligands such as Mg^{2+} to bind to one (α) subunit, and inhibitors such as NBD-Cl and FSBI, to completely block ATPase activity when bound at a stoichiometry of $1:MF_1$ (Table 8.2), can be interpreted to mean that one β or α–β subunit is unique in its ligand binding and possibly catalytic properties. Electron micrsocopic studies of MF_1 negatively stained (Fig. 8.6), or decorated with antibody, shows the F_1-ATPase as an $80 \times 110 \times 120$ Å structure with an approximately symmetric hexagonal arrangement of six globules (α, β subunits) surrounding a central mass ($\gamma\delta\varepsilon$ subunits) that seems more closely aligned to one pair (α–β) of the six globules (Tiedge et al., 1985; Boekema et al., 1986). The question that will arise below (section 8.7.1) concerning the mechanism of the F_1-ATPase is whether this asymmetry has functional significance.

8.4 DNA Sequence of *Unc* Operon

The sequencing of the gene in *E. coli* followed isolation and mapping of uncoupled (*unc*) mutants that could not grow on succinate as a carbon source (Butlin et al., 1971). The genetic complementation groups were defined from mutants defective in EF_0F_1 that failed to grow on nonfermentable carbon sources such as succinate. Approximately 40% of these mutants respire normally but cannot carry out oxidative phosphorylation with the necessary efficiency. These uncoupled (*unc*) mutants map near 83 min on the *E. coli* chromosome. The gene–product relationships for EF_0F_1 now established

8.4 DNA Sequence of *Unc* Operon

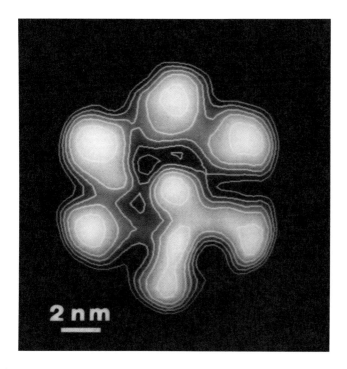

Figure 8.6. Image reconstruction analysis of electron micrograph of isolated MF_1-ATPase (top view). The asymmetric hexagonal image represents the largest (40% of the images) homogeneous subpopulation present in the data set. A seventh asymmetric mass near the center is thought likely to contain the γ, δ, and ε subunits (Boekema et al., 1986).

from the mutations and comparison of DNA and protein sequences, in the order of transcription, are *unc* I, B(a), E(c), F(b), H(δ), A(α), G(γ), D(β), and C(ε) arranged in a single transcriptional unit (Fig. 8.7A). The first 130 amino acid reading frame of the *unc* operon, before the *a* gene, codes for the 14 kDa *Unc*I protein which is not part of the ATPase structure but may function in assembly of the EF_0F_1. No polypeptide analogous to the *Unc*I is known for other F_0F_1 complexes. The F_1 subunits have the same order in the purple nonsulfur photosynthetic bacteria although the F_0 genes are missing from this operon. Similarly, the six CF_0F_1 genes coded by the chloroplast DNA are in the same order as in the *E. coli* operon (Fig. 8.7B). The gene clusters are probably related to the evolution of the F_0F_1 enzyme complex. The matching order with the *E. coli unc* operon of the six ATP synthase genes in chloroplasts and the nine genes of the cyanobacterium *Synechococcus* 6301 (Fig. 8.7B) provide strong evidence for an endosymbiotic origin of chloroplasts. Comparison of the amino acid sequences of the F_1 ATP synthase subunits indicates that they have been subjected to different evolutionary constraints.

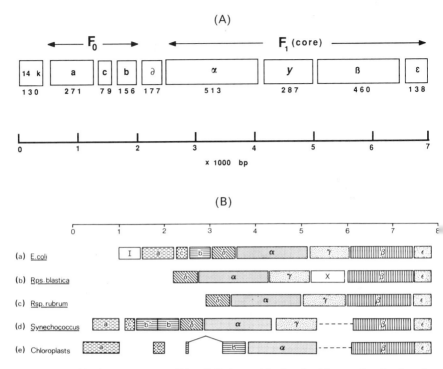

Figure 8.7. (A) The *unc* operon of *E. coli*. Polypeptide sizes (residues and codon length (bp) are shown (Adapted from Walker et al., 1984.) (B) Arrangement of *atp* genes encoding ATP synthase subunits in bacteria and chloroplasts. Gene X in *Rps. blastica* and I in the *E. coli* operon encode proteins of unknown function, and gene b' in *Synechococcus* is a duplicated and diverged form of *b* (Falk and Walker, 1988; *cf.*, Cozens et al., 1986, 1987).

The sequences of the catalytic α and β subunits are well conserved in contrast to those of the smaller subunits, presumably because of the central role of α and β in catalytic function.

The biogenesis of the EF_0F_1 enzyme complex in vivo provides an interesting problem in macromolecular complex assembly. Without the F_1, it might be predicted that assembly of F_0 would result in a depolarized, deenergized membrane and poor cell growth because of the uncapped passive F_0 proton channel. However, it is possible for F_0 and F_1 subunits to assemble independently, although the F_1 may protect the F_0 subunits from degradation in vivo (Aris et al., 1985).

The phenotypes of the mutants in the *unc* operon, some generated spontaneously and some site-directed, that have been confirmed by sequencing are summarized in Table 8.6. The functions altered by these mutations help to define the functions of the F_1 and F_0 subunits (Table 8.4).

8.4 DNA Sequence of *Unc* Operon

Table 8.6. Phenotypes of some sequenced EF_0F_1 *Unc* mutants

Mutants	Phenotype/Defect	Reference
*unc*A401, α (S373 → F)	Defective in steady-state ATP synthesis, although competent in single turnover; implies loss of α–β interaction or α function	Noumi et al. (1984a)
*unc*B402 *a* (chain term., am)	*a* Not present in membrane but *b* and *c* present; ~50% of normal F_1 binding.	Fillingame (1984)
*unc*B2019, *a* (S206 → L)	Impaired H^+ conductance of F_0	Cain and Simoni (1986)
*unc*B2001, *a* (H245 → Y)	Impaired H^+ conductance of F_0	Cain and Simoni (1986)
*unc*B, *a* (E219 → D, H, L, Q; H245 → E)	E219–H245 interaction in H^+ translocation	Cain and Simoni (1988)
*unc*B, *a* (R210 → Q; H245 → L)	Membranes H^+ impermeable; ATPase DCCD-resistant	Lightowlers et al. (1987)
*unc*B, *a* (E219 → Q)	Membranes H^+ impermeable	Lightowlers et al. (1988)
*unc*D11, β (S174 → F)	Loss of Mg^{2+} binding site; decreased inhibitor sensitivity.	Noumi et al. (1984b)
*unc*D, β (E41 → K; E185 → K; G223 → D; S292 → F)	Defective assembly of EF_0F_1	Noumi et al. (1986a)
*unc*D, β (R246 → H)	Active in unisite, but defective in multisite ATP hydrolysis	Noumi et al. (1986b)
*unc*E114, *c* (E42 → Q)	Defective coupling between EF_0, EF_1	Mosher et al. (1985)
*unc*E513, *c* (A25 → T)	Reduced H^+ permeability of F_0	Fimmel et al. (1985)
*unc*E408 *c* (L31 → F), *unc*E463	Slower H^+ translocation	Jans et al. (1983); Senior (1983)
*unc*E513 *c* (A25 → T)	F_0 has lower H^+ permeability; F_1 lost in presence of DCCD	Fimmel et al. (1983)
DG 27/10, *c* (A21 → V)	H^+ Impermeable	Hoppe and Sebald (1984)
*unc*E410, *c* (P64 → L)	Loss of energy-linked functions	Fimmel et al. (1983)
*unc*E501, *c* (A20 → P) E502	Revertants of *unc* E410	Fimmel et al. (1983)

Table 8.6 (continued)

Mutants	Phenotype/Defect	Reference
uncE106, c (G58 → D)	Negligible integration of c into membrane; a and b present in normal amounts. F_1 bound in proportion to c in membrane.	Fillingame (1984)
uncE105, c (D61 → G)	F_0 appears to assemble properly; F_1 binds with normal affinity; has normal activity. H^+ translocation of F_0 sector is blocked, but ATPase of F_1 bound to F_0 is normal	Fillingame et al. (1984)
uncE107, c (D61 → N)	As for uncE105 except that ATPase of F_1 bound to F_0 is inhibited by 50%	Fillingame et al. (1984)
uncE114 c (Q42 → E)	Normal F_1 binding to F_0; F_0 H^+ translocase uncoupled from F_1 ATPase.	Fillingame (1984)
uncE c (I28 → V or T)	DCCD-resistant	Senior (1985)
uncF515, b (G131 → A)	Uncoupled; low ATPase; prevents assembly; complements uncF476 (G9 → D).	Jans et al. (1985)
uncF476, b (G9 → D)	Reduced incorporation of b into membrane; complements uncF515; suppressed by a (P240 → L or A) or c (A62 → S) mutations.	Jans et al. (1984a); Kumamoto and Simoni (1987)
uncF469, b (W26 → stop)	Reduced incorporation of b into membrane.	Jans et al. (1984b)
uncF, b (G9 → D)	Prevents formation of H^+ pore.	Porter et al. (1985)
uncF, b (G131 → D)	Prevents formation of H^+ pore and binding of F_1 to F_0.	Porter et al. (1985)

8.5 Function of the Membrane-Bound Subunits *a*, *b*, and *c*

8.5.1 The Largest EF_0 Subunit, *a*

This subunit shows significant homology in a COOH-terminal region of ~80 residues with the MF_0 subunit ATPase-6 (Table 8.1). Subunit *a* is a hydrophobic protein and has five to seven hydrophobic segments that are sufficiently long to span the membrane bilayer, as inferred from a hydropathy plot (Appendix III) with the NH_2-terminal region and several loops exposed to the water phase (Fig. 8.8). The highest level of EF_1 binding to mutant membranes was found when subunit *a* was present together with *b* or *c*, indicating that

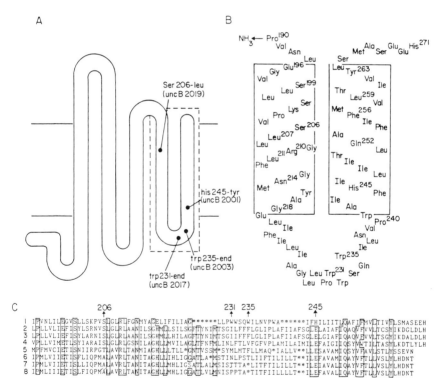

Figure 8.8. (A) Model for membrane folding of the EF_0 *a* polypeptide, which shows substantial sequence homology in the COOH-terminal region to the mitochondrial ATPase-6 subunit. The predicted position of some mutations in the *a* subunit is shown. (B) Sequence and proposed structure of two COOH-terminal transmembrane helices of the *a* subunit, and (C) aligned sequences of COOH-terminal region of EF_0 *a*, and ATPase-6 subunits of *A. nidulans*, *N. crassa*, *S. cerevisiae*, *D. melanogaster*, bovine, human, and mouse (Cain and Simoni, 1986; see Cox et al., 1986).

subunit a is involved in binding EF_1 to EF_0 (Table 8.4). The effect on EF_0 H^+ conductance of mutations at residues 206, 210, 219, and 245 and of a mutant at position 240 that suppresses b subunit mutant uncF476 at Gly-9 (Table 8.6), implies that subunit a has a function in the EF_0 proton channel and interacts with subunits b and c (Cain and Simoni, 1986; Kumamoto and Simoni, 1987; Lightowlers et al., 1988).

8.5.2 Subunit b

Subunit b has a hydrophobic segment (residues 1–30) at the NH_2-terminus while the remaining 80% of the molecule is hydrophilic (Hoppe and Sebald, 1984). The major hydrophilic segment, containing the COOH-terminal end, extends into the cytoplasmic water phase. The NH_2-terminal region has also been shown, using radiolabeled lipid-soluble probes, to be in contact with the membrane phospholipid. This suggests a working model for the b subunit in which an NH_2-terminal domain is inserted within the membrane, interacting with subunits a and c, and the rest of the b polypeptide extrinsic to the membrane forms the contact with EF_1 (Fig. 8.9). Consistent with this model, subunit b in EF_1-depleted membranes is sensitive to proteolysis, is protected from degradation by EF_1, and EF_1 does not bind to membranes in which the hydrophilic part of b has been removed by trypsin or from mutants missing subunit b. The subunits used for binding in the EF_1 would be δ and ε (Table 8.4). Structural models for F_0 sector b must take into account that the stoichiometry of subunit b in the membrane is $2b/F_0$ (Foster and Fillingame, 1982), and that cross-linking studies show that it exists in the membrane as a dimer (Hoppe and Sebald, 1984).

8.5.3 Molecular Properties of the Proteolipid Subunit c

Subunit c in $E.\ coli$ is present in large excess relative to a and b, with the approximate stoichiometry of $c:EF_0F_1 = 10 \pm 1$ and $c:b:a = 10:2:1$ (Foster and Fillingame, 1982). The subunit c proteolipid was first purified (Cattell et al., 1971) using binding of DCCD which forms a stable adduct with carboxyl groups, Asp or Glu residues, in a nonpolar environment. DCCD was shown to be an inhibitor of the membrane-bound ATPase of $S.\ faecalis$ membranes with no effect on soluble ATPase (Harold et al., 1969). Other experimental observations that indicated the DCCD-reactive subunit c to have a direct role in energy transduction were: (i) DCCD is a potent inhibitor of oxidative- and photophosphorylation; reaction with only one of the multiple (6–12) copies of the c subunit seems sufficient for inhibition (Sigrist-Nelson and Azzi, 1980). (ii) Chloroplasts, mitochondria, or $E.\ coli$ missing F_1 are leaky to protons. Addition of DCCD tends to repair the leak. The deficiency in EF_1 may occur either because the cells are genetically deficient or because the F_1 has been removed. An oxygen pulse added to anaerobic mitochondria or membranes

8.5 Function of the Membrane-Bound Subunits *a*, *b*, and *c*   375

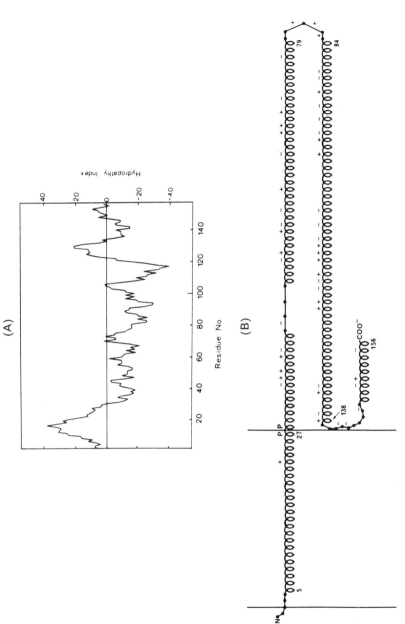

Figure 8.9. (A) Hydropathy plot (Walker et al., 1984.) and (B) model for folding in the membrane bilayer of EF_0 subunit *b* (Senior and Wise, 1983).

Table 8.7. Amino acid sequences of F_0 c subunits

Source	
	10 20 30 40 50 60 70
Bov MF_0	DIDTAAKFIGAGAAT-VGVAGSGAGIG--TVFGSLIIGYARNPSLKQQLFSYAILGFALSEAMGLFCLMVAFLILFAM
Spi CF_0	MNPLIAAASVIAAGLAVGLASIGPGVGQGTAAGQAVEGIARQPEAEGKIRGTLLLSLAFMEALTIYGLVVALALLFANPFV
EF_0	MENLNMDLLYMAAAVMMGLAAIGAAIGIGILGGKFLEGAARQPDLIPLLRTQFFIVMGLVDAIPMIAVGLGLYVMFAVA

[a] Bov, bovine mitochondria; spi, spinach chloroplasts; EF_0, $E.$ $coli$ bacteria. Also conserved in $N.$ $crassa$ and $S.$ $cerevisiae$ mitochondria and P53 thermophilic bacteria are residues 17–19, 24–25, 28, 30–32, 38, 41, 4 45–48, 54, 60, 63, 66–67, 70, 73, and 76.
From Walker et al. (1984).

of $E.$ $coli$ F_1^- strains elicits very little proton efflux, presumably because the protons return to the membrane as fast as they leave. Addition of DCCD can restore the proton efflux and can also restore active transport to the same H^+-leaky ATPase$^-$ mutants (Altendorf et al., 1974). (iii) A role of the DCCD-reactive residue of the c subunit in energy coupling is also implied by H^+-translocation, but not ATPase activity, being abolished in mutants where the DCCD-reactive Asp-61 is replaced by a glycine or asparagine (Table 8.6).

The number of residues/subunit c in mitochondria, chloroplasts, $E.$ $coli$, and thermophilic bacteria varies between 72 and 81 (Table 8.7). The sequences indicate the presence of two long hydrophobic peptide segments on each side of a polar domain, so that in $N.$ $crassa$ mitochondria, for example, the NH_2-terminal region is hydrophilic, residues 16–40 are hydrophobic, 41–52 are hydrophilic, and 53–80 are mostly hydrophobic. The residues in the hydrophobic domains seem the most conserved. The DCCD-binding residue is conserved near position 65, an acidic residue in the middle of an otherwise hydrophobic segment, which is a glutamic acid in all the sequences shown in Table 8.7 except that of $E.$ $coli$ where it is Asp-61.

The amino-terminal sequence is generally hydrophilic. Most of the acidic and basic residues of the proteolipid are clustered in the middle of the polypeptide chain. Extending approximately from residues 38–56 in mitochondrial, bacterial, and chloroplast proteins, this region is predicted to contain a turn domain connecting the two hydrophobic α-helices of approximately 25 residues in length (Fig. 8.10). This structural model is supported by (i) far ultraviolet circular dichroism determination of a high α-helical content ($\sim 80\%$) of subunit c of yeast MF_0 in detergent or incorporated into liposomes (Mao et al., 1982), (ii) discrete labeling of residues predicted to be in the nonpolar region by a hydrophobic photoaffinity probe (Hoppe and Sebald, 1984), (iii) impaired binding of EF_1 to EF_0 in the Q-42 $\rightarrow$ E mutant of subunit c (Mosher et al., 1985); and (iv) reactivity of the two Tyr residues predicted to be near the periplasmic surface (Fig. 8.10).

The acidic residue Asp-61 would be in the center of a hydrophobic α-helical domain. The insertion of this residue into the lipid bilayer requires energy. Addition of one more such residue by directed mutation into the nonpolar

8.5 Function of the Membrane-Bound Subunits a, b, and c

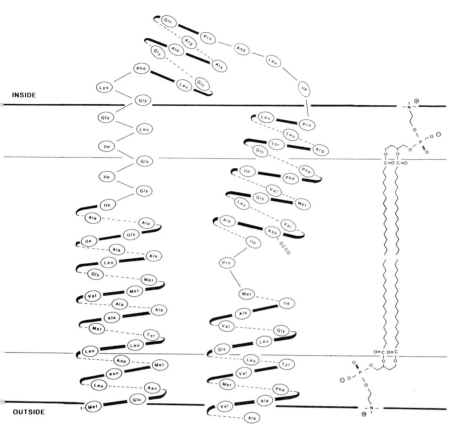

Figure 8.10. Predicted folding pattern in the membrane of EF_0 proteolipid subunit c polypeptide. (Reprinted with permission from Deckers-Hebestreit et al., *Biochemistry*, vol. 26, pp. 5482–5486. Copyright 1987 American Chemical Society.)

part of the protein, Gly-23 → Asp, apparently creates too large an energy barrier since this c mutant subunit is not incorporated into membranes (Jans et al., 1983). Many of the known mutations tend to cluster near Asp-61 and toward the side of the membrane where F_1 is thought to bind. The effect of amino acid substitutions such as Ile-28 → Thr or Val which cause resistance to normal DCCD concentrations, or Pro-64 → Leu which can block proton conduction Table 8.6, could be interpreted in terms of sensitivity of Asp-61 to a changed amino acid environment.

The proposed folding pattern of subunits a, b, and c of EF_0 is shown in Figs. 8.8–8.10. The extended loop of b and the hairpin loop of c are believed to bind or be directed to EF_1. The proposed folding of all three subunits in the membrane bilayer is shown in Fig. 8.11, along with particular residues that are implicated in channel function by mutation studies. A structure for the H^+ channel involving the a, b, and c subunits is discussed in section 8.8.

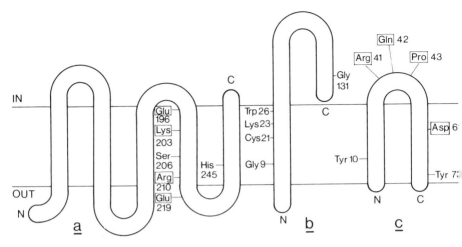

Figure 8.11. Proposed folding of EF_0 *a*, *b*, and *c* subunits in the membrane bilayer showing amino acids whose phenotypes have been studied by mutation (Table 8.6). Conserved amino acids are boxed. Residues 210, 219, and 245 of the *a* subunit have been proposed to have a function in H^+ translocation. Interactions between subunits are inferred by suppressor mutations at position 240 of *a* and position 62 of *c* that partly restore deficient function in the G9 → D mutant of the *b* subunit (Schneider and Altendorf, 1987; compare with Kumamoto and Simoni, 1987).

8.6 Mechanism of ATP Synthesis

8.6.1 Bound Nucleotides

The F_1 of chloroplast, mitochondrial, and bacterial membranes bind three to six ADP or ATP molecules in at least two different kinds of sites, "exchangeable" and "nonexchangeable" (Shavit, 1980; Cross and Nalin, 1982; Wise et al., 1984; Abbott et al., 1984; Yoshida and Allison, 1986). The nonexchanging or slowly exchangeable nucleotides are bound very tightly ($K_D \leq 5 \times 10^{-8}$ M; Kironde and Cross, 1987) and are not released or replaced by thorough washing, binding of nucleotides or hydrolysis of ATP. Nucleotides at exchangeable sites ($K_D = 1$–10 μM) can be released by binding of nucleotides at other sites, energization of the membrane, or hydrolysis of ATP by the isolated enzyme. The nonexchangeable and exchangeable nucleotides function, respectively, (i) in a noncatalytic structural or regulatory role and (ii) at sites of catalysis in ATP synthesis or hydrolysis. CF_1 contains at least three defined nucleotide binding sites (McCarty and Hammes, 1987; Xue et al., 1987): Site 1 contains exchangeable ADP that remains bound after extensive dialysis and gel filtration, but can exchange with ADP or ATP in the medium. MgATP binds to site 2 with very high affinity and is nonexchangeable, because it remains associated when nucleotides are bound at other sites and has a dissociation half-time of days. Binding at site 3 is readily exchangeable (McCarty and Hammes, 1987). The extent and pattern of the binding of nucleotides, nucleotide analogs, and other specific inhibitors of ATPase activity (Table 8.3) have been used to deduce the number and subunit locaton of catalytic and noncatalytic sites. The labeling pattern by nucleotide analogs indicates that all β subunits contain one or two nucleotide binding sites, with the second site noncatalytic if present. The identity of the amino acids near a nucleotide binding site on the β subunit has been obtained using the radiolabeled photoaffinity nucleotide analog, 2-azido $[\alpha\text{-}^{32}P]$ ADP (Fig. 8.12A). After photolabeling of the beef heart enzyme, the labeled β subunit was cleaved with the methionine-specific reagent, cyanogen bromide, showing that the label was localized in peptide A293-M358. Further analysis showed that residues L-342, I-344, Y-345, and P-346 (Table 8.3) were labeled (Garin et al., 1986). A similar labeling pattern in the CF_1 β subunit was found using a benzoyl-ATP analog, for which label was found in residues corresponding to Y-345 and D-352 (Admon and Hammes, 1987). A model for the nucleotide binding site in the β subunit (Figs. 8.12B, C) can be based on involvement of these residues, labeling of Y-345 with nucleotide analogs (Table 8.3), and Lys-162 near the binding site that is labeled by NBD-Cl. Sequence analogies of the β subunit with adenylate kinase were noted in Table 8.2, and can form the basis for a model of the β subunit based on analogy with the x-ray crystal structure of adenylate kinase and other adenylate-binding enzymes (Fig. 8.12D).

For the mitochondrial and bacterial enzymes, it has been proposed that β and α subunits are the sites of catalytic and noncatalytic sites (Bullough and Allison, 1986b) because: (i) a total of six nucleotides are bound per F_1; (ii) two nucleotide binding sites, one catalytic, have been defined per β subunit in the photosynthetic bacterium *R. rubrum* (Khananshvilli and Gromet-Elhanon, 1985). The catalytic sites may reside on the β subunits and the noncatalytic sites at the $\alpha-\beta$ interface. The properties of mutant *unc*A401 (Table 8.6) of EF_1 are of interest in this respect. In this mutant, Ser-373 of the α subunit is changed to a phenylalanine, and that mutant is defective in steady-state ATP synthesis, although competent in a single turnover. Because the catalytic site is thought to be on the β subunit, and the Ser $\rightarrow$ Phe mutation causes an already hydrophobic domain to be more hydrophobic, it was inferred that this mutant is defective in the $\alpha-\beta$ intersubunit interaction.

8.6.2 Energy Requirement for Nucleotide Release from ATP Synthase; Exchange Reactions

The energy requirement for ATP synthesis is usually discussed as being manifest in the ΔG for covalent bond formation of P_i and ADP (e.g., Chap. 1, section 1.14). However, it was found that a prominent effect of the ΔG input

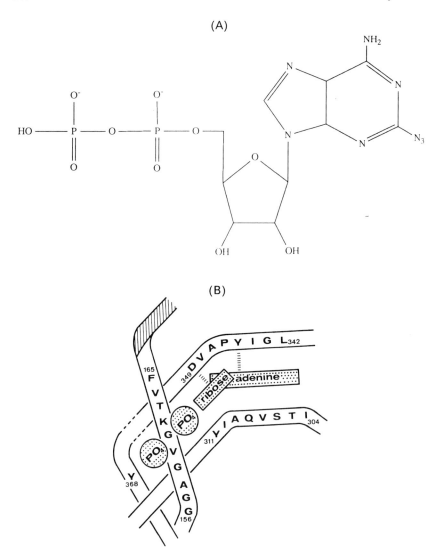

Figure 8.12. (A) Structure of nucleotide photoaffinity label, 2-azido-ADP. (B) Representation of nucleotide binding site on β subunit (Reprinted with permission from Garin et al., *Biochemistry*, vol. 25, pp. 4431–4437. Copyright 1986 American Chemical Society; Cross et al., 1987). (C) Representation of the β subunit of rat liver mitochondrial MF_1. The segment 141–333 is folded into a five-stranded β sheet structure connected by α-helices analogous to the structure (D) of adenylate kinase (Fry et al., 1986). The $\beta-\alpha-\beta-\alpha-\beta$ structure in the COOH-terminal half of the molecule is similar to the established nucleotide-binding domain motif (Rossmann et al., 1974). Stippled amino acids are analogous to adenylate kinase, and blackened sections to Na^+, K^+- and Ca^{2+}-ATPases. Adenine is seen edge-on as an open rectangle next to the phosphates. Lys-162 and Tyr-311 labeled by NBD-Cl are noted ((C) and (D) from Garboczi et al., 1988).

8.6 Mechanism of ATP Synthesis

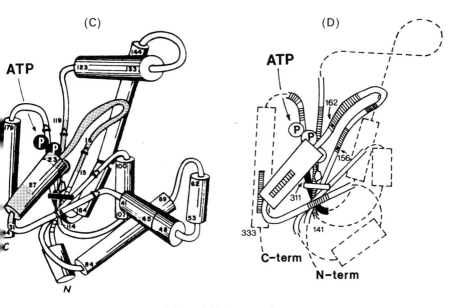

Figure 8.12 (continued)

in oxidative phosphorylation was to bring about a release of preformed ATP bound to MF_1 (Boyer et al., 1973; Cross ad Boyer, 1975). This was discovered by measuring the dependence of different exchange reactions in mitochondria and submitochondrial particles on the presence of energy-dissipating uncouplers. Exchange reactions provide a measure of the reversibility of overall reactions, or of an enzyme (E)-bound reaction that is part of an overall sequence. For example, the exchange of $^{32}P_i$ from medium orthophosphate to ATP ($P_i \rightleftarrows$ ATP) exchange,

$$^{32}P_i + ADP + E \rightleftarrows {}^{32}P_i \cdot E \cdot ADP \rightleftarrows E \cdot AT^{32}P \rightleftarrows AT^{32}P, \qquad (1)$$

provides a measure of the reversibility of the overall reaction of oxidative phosphorylation (Hackney and Boyer, 1978; see Fig. 8.13).

Measurement of phosphate–oxygen exchange provides another probe for reversal of the formation of enzyme-bound ATP. Oxygen exchange occurs because the bound P_i, without being released from the enzyme, can tumble or rotate at the active site allowing all four phosphate oxygens to form H_2O with equal probability as bound ATP is formed. For example, bound ATP formed from P_i containing four ^{18}O atoms would retain three ^{18}O in the terminal phosphoryl group, the β–γ bridge oxygen would be contributed by ADP, and one ^{18}O will be lost to water. If the ATP synthesis reaction is reversed before ATP is released from the enzyme, and the ATP takes up water from the medium to form bound ADP and P_i, the bound P_i will contain one ^{16}O derived from water while retaining its three ^{18}O atoms. If ATP is formed by another reversal on the enzyme, the probability is $\frac{1}{4}$ that it will contain an

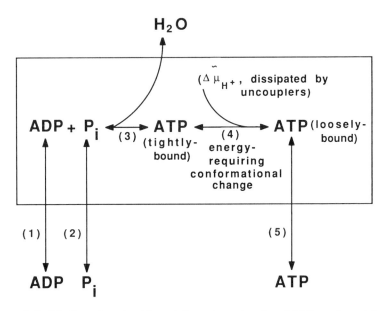

Figure 8.13. Binding of nucleotide and P_i at the active site of ATP synthase inferred from exchange experiments assuming a single site mechanism. Reaction steps occurring at the catalytic site are enclosed in the rectangle. (Based on Boyer et al., 1973; Cross and Boyer, 1975; Boyer, 1983)

Steps required for medium exchange reactions assuming a single site mechanism:

$$\text{Medium } P_i \rightleftarrows \text{HOH: 2, 3 (uncoupler-insensitive)}$$
$$\text{ATP} \rightleftarrows \text{HOH: 3, 4, 5 (uncoupler-sensitive)}$$
$$P_i \rightleftarrows \text{ATP: 2, 3, 4, 5 (uncoupler-sensitive)}$$

^{16}O from water. A large number of reversals is required to exchange all of the phosphate oxygens with those from water. The ATP formed by oxidative or photo-phosphorylation always contains some water oxygen, proving that there is always some reversal of formation of the bound ATP. These exchange reactions are classified as "medium-" and "intermediate-exchange," depending on whether the exchange occurs into the same component that was added from the medium, or into another intermediate formed on the enzyme that is subsequently released. Other examples of exchange reactions are:

2. Medium $P_i \rightleftarrows H_2O$ exchange, in which $[^{18}O]P_i$ binds, undergoes exchange, and P_i is released to the medium.
3. Intermediate $P_i \rightleftarrows H_2O$ exchange, in which $[^{18}O]$ATP binds, and the bound P_i that is formed undergoes exchange and is released to the medium.
4. Medium ATP $\rightleftarrows H_2O$ exchange, in which ATP binds, undergoes exchange, and is released to the medium.

Intermediate $P_i \rightleftarrows H_2O$ exchange measured as a functon of the ATP concentration shows that the number of water oxygens appearing in each P_i

8.7 Thermodynamic and Kinetic Constants for ATP Hydrolysis

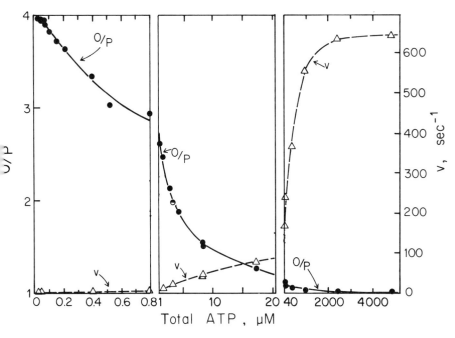

Figure 8.14. Dependence on ATP concentration of rate of ATP hydrolysis (v) and water oxygens (^{16}O) in released P_i. (After O'Neal and Boyer, 1984.)

released increases and approaches the theoretical maximum of four as the ATP concentration is decreased and approaches zero (Fig. 8.14). Therefore, the ATP is bound to the enzyme for a longer time, allowing more exchange, when the ATP concentration is low. At very low ATP concentrations the number of exchange events was calculated to be 400.

An energy requirement for release of ATP from MF_1 was inferred from the observation that the $P_i \rightleftarrows ATP$ and medium $ATP \rightleftarrows H_2O$ exchange reactions that require release of bound ATP in mitochondria and submitochondrial particles were sensitive to the presence of uncouplers that dissipated the energized state (Fig. 8.13), but the intermediate $P_i \rightleftarrows H_2O$ exchange was relatively insensitive (Figs. 8.13 and 8.15).

The concept that the free energy of enzyme-bound ATP is stored not in the high-energy bond, but in the energy of binding of ATP to the enzyme, was supported by measurement of the binding properties of ATP to MF_1.

8.7 Thermodynamic and Kinetic Constants for ATP Hydrolysis

ATP hydrolysis could be studied at a single catalytic site because MF_1 has a very high affinity, $K_a = 10^{12}$ M^{-1}, for ATP, and the rate of hydrolysis,

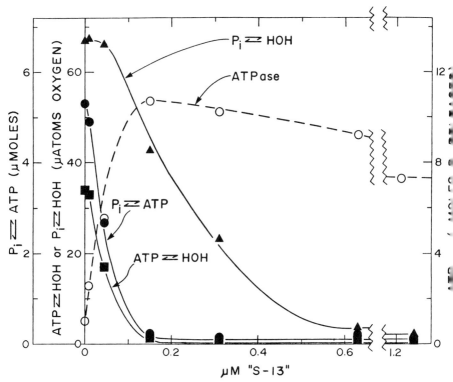

Figure 8.15. Dependence of rate of medium $P_i \leftrightarrow H_2O$ exchange, medium ATP $\leftrightarrow$ H_2O exchange, $P_i \rightleftarrows$ ATP exchange, and ATPase activity, as a function of concentration of the uncoupler, S-13 (Table 3.9). Rat-liver mitochondria were incubated for 5 min at 37°C with 5 mM $^{32}P_i$ (equivalent to 10^6 cpm), 5 mM ATP, with 0.98 atom percent of excess ^{18}O, and concentrations of S-13 as indicated (Boyer et al., 1973).

10^{-4} s^{-1}, at a single site is small. Using the following pathway for ATP binding, dissociation, and hydrolysis, the different forward and reverse rate constants have been determined for MF_1.

Table 8.8. Kinetically definable steps in ATP hydrolysis and synthesis

$$F_1 + ATP \underset{k_{-1}}{\overset{k_1}{\rightleftarrows}} F_1 \cdot ATP + H_2O \underset{k_{-2}}{\overset{k_2}{\rightleftarrows}} F_1 \cdot ADP \cdot Pi \underset{k_{-3}}{\overset{k_3}{\rightleftarrows}} P_i + F_1 \cdot ADP \underset{k_{-4}}{\overset{k_4}{\rightleftarrows}} ADP + F_1$$

The affinity of the "first site" of the F_1 for ATP was measured through determination of (i) the forward rate constant, k_1, 6.4×10^6 M^{-1} s^{-1}, similar to values previously found for ATP binding to myosin ATPase, and (ii) the rate of dissociation of bound ATP,

$$k_{-1} = 7 \times 10^{-6} \text{ s}^{-1}.$$

The association equilibrium constant for ATP binding to the F_1 enzyme, K_a, is then

8.7 Thermodynamic and Kinetic Constants for ATP Hydrolysis

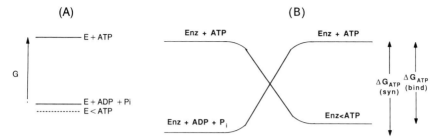

Figure 8.16. (A) Energy level diagram comparing states with free enzyme and free ADP + P_i (E + ADP + P_i), free enzyme, free ATP (E + ATP), ATP tightly bound to enzyme (E < ATP) (dashed level), showing that the energy required for synthesis of bound ATP is small. (B) Free-energy-reaction coordinate diagram showing that release of free energy stored in bound ATP can be equivalent to conversion of energy to bond energy of anhydride ATP. (Based on Jencks, 1980.)

$$K_a = k_1/k_{-1} = 10^{12} \text{ M}^{-1}$$

(Grubmeyer et al., 1982), with a standard free energy of binding, $\Delta G° = -16.3$ kcal/mol at 25°C. The free energy required to create a loose binding site is the 12–14 kcal/mol needed to change K_a to 10^2–10^4 M^{-1}. This ΔG requirement is similar to the amount previously calculated to be available from the steady-state phosphate potential in mitochondria and chloroplasts (Chap. 2, Problem 32). All three catalytic sites of the enzyme may have a high affinity for ATP, but this is not observed for more than one site, perhaps because of subunit interactions. The precedent in kinase enzymes for conversion of bond energy of ATP to binding energy of ATP has been noted above (Chap. 1, section 1.15.2), and can be described in a free energy level diagram (Fig. 8.16).

Step two in Table 8.8 is the catalytic reaction, and its equilibrium constant, $K_2 = 0.5$, was determined by measurement of bound ATP, ADP, and Pi. Since

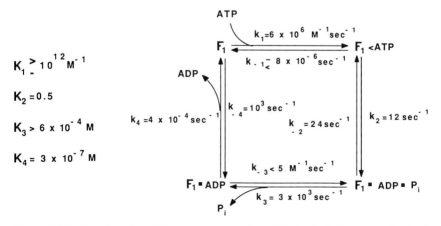

Figure 8.17. Kinetic and equilibrium constants for unisite hydrolysis and synthesis of ATP on the F_1 ATPase. (Adapted from Grubmeyer et al., 1982.)

step 2 involves a dynamic equilibrium between (ADP, P_i) and (ATP, H_2O), and the β–γ bridge oxygen of ATP is derived from ADP, the oxygen atoms of the γ-phosphoryl group of ATP should exchange O atoms with H_2O at the active site, as discussed above. The kinetic and thermodynamic parameters for "unisite" ATP synthesis and hydrolysis are summarized (Fig. 8.17).

8.7.1 Studies of ATP Hydrolysis with Isolated MF_1 and Membrane-Bound MF_1; Evidence for Site–Site Interactions

Interactions between F_1 catalytic sites and subunits can be inferred from the dependence of kinetic and thermodynamic parameters of nucleotide binding on concentration. When a 10-fold molar excess of F_1 was mixed with [α-^{32}P]ATP (unisite case) in a rapid flow and mixing apparatus (Fig. 8.18), the enzyme–substrate complex rapidly forms, and the nucleotide binds predominantly to the highest affinity site with a relatively slow rate ($\sim 10^{-4}$ s^{-1}). The $^{32}P_i$ orthophosphate formed at zero time (20% of the added [^{32}P]ATP) represents the equilibrium distribution of bound substrate and bound hydrolysis products resulting from single site catalysis. After rapid addition of a large excess (10 mM) of nonradioactive ATP to populate the second site (bisite case), about half of the bound [^{32}P]ATP was hydrolyzed and the $^{32}P_i$ was released in the first 5 ms after mixing with the second syringe (Fig. 8.18). One can calculate an approximate turnover number of $1 \div (3 \times 10^{-3}) = 300$ s^{-1} for this bisite situation, an increase of a factor of $\sim 10^6$ relative to the unisite case, with a concomitant decrease in the binding constant from 10^{12} to about 10^4. Thus, the binding and kinetic constants for release in the "unisite" and

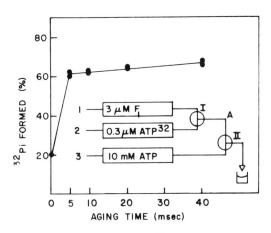

Figure 8.18. ATP-promoted hydrolysis of [γ-^{32}P]ATP bound to a single F_1 site as described in text (Cross et al., 1982).

"bisite" situations are:

	K_a	$k\ (s^{-1})$
Unisite	10^{12}	10^{-4}
Bisite	10^4	300

The *E. coli* F_1 ATPase showed a similar enhancement of the ATP hydrolysis rate as the ATP concentration was increased, which was attributed to positive cooperativity between different sites (Wise et al., 1984; see Hiller and Carmeli, 1985, evidence for CF_1 subunit cooperativity). The lack of this enhancement in *unc*A α-subunit mutants was attributed to the loss of such cooperativity. The properties of the membrane-bound MF_1 in sub-mitochondrial particles have been found to be similar to those of the solubilized F_1 (Table 8.9).

Table 8.9. Properties of membrane-bound MF_1 associated with the mechanism of ATP hydrolysis

---
1. High affinity ($K_a = 10^{12}$ M^{-1}) binding site for ATP
2. Hydrolytic step with $K_{eq} \simeq 1$
3. Increase of $\sim 10^6$ in the rate of ATP hydrolysis, and a decrease of ~ 8 orders of magnitude in the K_a, when ATP is bound to additional catalytic sites
4. Energy-dependent succinate or NADH-supported decrease in affinity of MF_1 for ATP. Dissociation of 1/3 of bound ATP in presence of NADH is kinetically competent, i.e., commensurate with initial rate of oxidative phosphorylation
5. Inhibition by DCCD or oligomycin of ATPase activity and high-affinity binding of ATP (Penefsky, 1985a)

From Penefsky (1985b, c).

The data on the concentration dependence of ATP hydrolysis could be explained by a model for reversible ATP synthesis and hydrolysis involving two or three alternating and interacting sites (Fig. 8.19A). The scheme is drawn so that steps 1, 2, 3, 4 correspond to the direction of ATP synthesis, with ATP bound tightly (<) in step 1 to the enzyme, E, to which ADP and P_i bind loosely (·). Input energy from the $\Delta\tilde{\mu}_{H^+}$ causes a conformational and binding change in step 2 so that the tightly bound ATP is raised in energy to a loosely bound state to allow release in step 3. The alternating site cycle is completed in step 4 when ATP is synthesized at the active site from ADP and P_i. The lack of an energy requirement for this step ($K_{eq} \simeq 2$) is a distinctive aspect of the model and is a consequence of the coupled tight binding ($\Delta G \ll 0$) of the ATP occurring in the same step. An alternating site model involving energy-dependent sequential and symmetric use of catalytic sites on all three β subunits or α–β subunit pairs is shown (Fig. 8.19B). The ability of one bound molecule of the inhibitor, FSBI, to completely block MF_1 ATPase activity (Table 8.3) is consistent with action of this inhibitor at any one of three equivalent alternating catalytic sites that are necessary for the reaction. The requirement of three molecules of the analog FSBA to fully inhibit the activity (Table 8.3) would imply that one or more of the sites to which FSBA binds

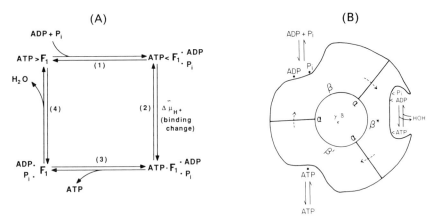

Figure 8.19. (A) Scheme for two alternating catalytic sites, in which tightly and weakly bound ligands are denoted (<) and (·), respectively. (Adapted from Kayalar et al., 1977; Hutton and Boyer, 1979.) Reaction sequence is clockwise for ATP synthesis and counterclockwise for hydrolysis. (B) The energy-dependent symmetric alternating three-site mechanism for F_1 ATPases. The scheme shown in one-third of the enzyme cycle assumes three interacting α–β catalytic subunits per ATP synthase. Upon energization, the catalytic sites are interconverted between three conformations involving loose (·) and tight (<) binding states for ligands. (Adapted from Gresser et al., 1982; compare with Adolfsen and Moudrianakis, 1973, 1976.) There is at present no evidence for rotational motion associated with activation of catalytic subunits relative to a core. Some evidence for conformational changes of the enzyme was obtained by measurement of the rotational correlation time of CF_1 in solution, which increased by a factor of three upon activation (Wagner et al., 1985).

are not completely essential for catalysis (Bullough and Allison, 1986a), or that the three sites for FSBA binding do not contribute equally to catalysis in any model (Wang 1986). The "positive cooperativity" of the bisite enzyme might arise from one β subunit hydrolyzing ATP at a low rate while activating the other two β subunits (Ysern et al., 1988). The dissociation of one-third of the bound ATP at a rate equal to the initial rate of ATP synthesis (Table 8.9) is most simply explained by catalysis occurring only on one of the β subunits which is unique because of its high affinity. In addition, the data on MF_1 structure discussed above (section 8.3.6) indicate that the time-average structure is asymmetric, with one of the three α–β pairs different from the other two. All current views of F_1 mechanism agree on an important function of intersubunit interactions, although not on (i) a requirement for symmetry and (ii) relative rotation of α–β subunits around the enzyme core.

8.8 Mechanism of Transduction of $\Delta\tilde{\mu}_{H^+}$ to ATP

The mechanism by which the $\Delta\tilde{\mu}_{H^+}$ is transduced to effect a change in nucleotide affinity at the active site could be (i) direct protonation of the F_1, perhaps at

8.8 Mechanism of Transduction of $\Delta\tilde{\mu}_{H^+}$ to ATP

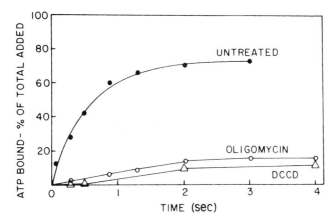

Figure 8.20. Inhibition by DCCD or oligomycin of ATP binding to submitochondrial particles (Penefsky, 1985a).

a conserved acid–base cluster of residues 411–417 and 423–429 of the (EF_1) β subunit (Kagawa, 1984). A second possibility (ii) is conformational change of the F_1 brought about by protonation of F_0, perhaps at the DCCD-binding Glu or Asp residue of the c subunit (section 8.5.3). A role for the latter mechanism was indicated by the decrease in high-affinity ATP binding of membrane-bound MF_1 caused by DCCD or oligomycin (Fig. 8.20).

The conformational change could be delivered through the b subunit to the catalytic F_1 section of the enzyme, as emphasized in one version of the model built mainly on information from the E. coli system (Cox et al., 1986). A ring of 6–10 c subunits would be arranged around a central core consisting of the membrane domain of two b subunits and the five membrane-spanning domains of one a subunit, so that the two c helices are equidistant from the center (Fig. 8.21A). The conserved acidic residue (Asp 61 in EF_0) in the c subunit could hydrogen bond to subunit a through Ser-206 or Arg-210 in membrane-spanning amphipathic helix IV or His-245 in helix V; residues that have been implicated in the EF_0 H^+ channel by mutagenesis studies (Table 8.6) of subunit a (Fig. 8.21B). If the b subunit extends from the membrane as an 80–90 Å helical hairpin (Fig. 8.9), it could interact with the F_1 δ and ε subunits, as well as the other b subunit in the dimer. Interaction of all 6–10 c subunits with the single copy of a, indicated by inhibitor data, could be accomplished by the rotational catalysis mechanism and sequential interaction with all c subunits (Fig. 8.21B) in the alternating site model. In this model, such a rotational motion of c relative to a and b would be driven by sequential protonation–deprotonation of the Asp-61 of the c subunit, utilizing the $\Delta\tilde{\mu}_{H^+}$, that would modulate its interaction with the positive residues of the a subunit helix IV. This would drive the sequential use of α–β catalytic subunits. Reaction of DCCD with Asp-61 would, of course, block this reaction, and Asp-61 → Asn or Gly mutants could be defective in oxidative phosphorylation and ATPase pumping (Table 8.6).

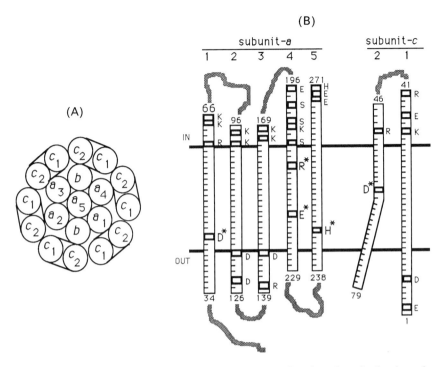

Figure 8.21. (A) Proposed packing of membrane spanning domains of subunits a, b, and c of EF_0, allowing sequential interactions between helix IV of subunit a and subunit c. (B) Proposed formation of a proton conducting pathway across the membrane bilayer caused by H bonding of Asp-61 of c with Ser-206, Arg-210, and Glu-219 residues of conserved amphipathic helix IV and His-245 of helix V of F_0 subunit a. Residues that are proposed to be involved in the proton pore are marked by (*) (Figure contributed by F. Gibson, based on Howitt et al., 1988; compare with Mitchell, 1985; Cox et al., 1986; Kumamoto and Simoni, 1987; Lightowlers et al., 1988.)

Some uncE and uncF mutants affecting the c and b subunits (Table 8.6) are of interest with respect to this model. F_1-stripped membranes of these mutants are not H^+-leaky and do phosphorylate, although with reduced efficiency. Compared to the wild-type, these mutants have substituted amino acids of larger size in the hydrophobic core of the c (Leu-31 → Phe; Ala-25 → Thr), or the b subunit (Gly-9 → Asp), which may then sterically inhibit $\Delta\tilde{\mu}_{H^+}$-induced rotation of the F_0 subunit assembly.

8.9 Other Classes of H^+-Translocating ATPases

The F_0F_1 (F) or eubacterial ATP synthase-ATPase is one of three known classes of ion translocating ATPase (Fig. 8.22). The other two are classified as plasma membrane- or phospho- (P), and vacuolar (V)-ATPases (Pedersen and Carafoli, 1987a; Nelson, 1988). All of the H^+-ATPases are electrogenic pumps. The ATPase classes are primarily distinguished by their subunit composition,

8.9 Other Classes of H$^+$-Translocating ATPases

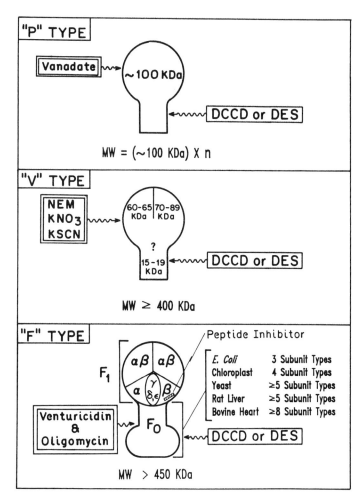

Figure 8.22. Comparison of F, V-, and P-ATPases showing approximate subunit composition, molecular weight, and action of some inhibitors (DES, diethylstibestrol; NEM, N-ethylmaleimide) (Pedersen and Carafoli, 1987a).

the presence of a phosphoenzyme intermediate in the case of the P class, the resulting sensitivity to characteristic inhibitors, and by the H$^+$/ATP ratio and the resulting tendency to synthesize (F) or hydrolyze (P and V) ATP (Table 8.10). The F_0F_1 and V-ATPases translocate only H$^+$, whereas the class of P-ATPases can also drive the translocation of Na$^+$, K$^+$, or Ca^{2+}.

8.9.1 The Vacuolar-Type ATPases

The name of this class of ATPase reflects its presence as a ubiquitous endomembrane H$^+$-ATPase in organelles other than mitochondria, chloroplasts,

Table 8.10. Properties of H^+-ATPases

	ATPase		
	Eubacterial (F)	Vacuolar (V)	Plasma membrane or phospho- (P)
A. Mass (kDa) and subunits	500; 8–12	200–500; 3–8	100; 1–2
B. Pump properties			
Electrogenic	+	+	+
Phosphorylated intermediate, vanadate sensitive	–	–	+
Anion dependent	–	+ (Cl^-)	–
Cation dependent	–	–	+ (K^+)
C. Inhibitors			
Vanadate	–	–	+
Nitrate	–	+	–
NEM	–	+	–
DCCD	+	+	–, +
NBD-Cl	+	+	–
Oligomycin (mitochondria)	+	–	–

From Nelson (1988).

and endoplasmic or sarcoplasmic reticulum. These ATPases are found in vacuoles of fungi and yeast, plant tonoplasts, lysosomes, the Golgi complex, synaptic vesicles, chromaffin granules, and clathrin-coated vesicles, and may be the most diversified group of the three ATPase classes (Nelson, 1988). Like the F_0F_1 eubacterial complex, it is multisubunit (Fig. 8.23A) with a molecular weight of $2-5 \times 10^5$. The vacuolar H^+-ATPases drive an electrogenic H^+ pump, do not utilize a phosphoenzyme intermediate, and are stimulated by anions (e.g., Cl^-). The function of the V-ATPases is to pump protons to the internal space to generate a $\Delta\tilde{\mu}_{H^+}$ that can be used for active transport. In chromaffin granules (Dean et al., 1986; Nelson and Moriyama, 1987), the $\Delta\tilde{\mu}_{H^+}$ generated by the ATPase provides the energy for transport of catecholamines such as epinephrine and serotonin (Chap. 9, section 9.14). The acidification of endocytic vesicles may also be important in ligand–receptor dissociation and consequent receptor recycling.

The vacuolar ATPase has a peripheral or extrinsic sector consisting of more than one copy of MW = 60,000 and 70,000 polypeptides and more than one peptide with MW = 30–40,000 (Nelson, 1988). It also has an intrinsic membrane bound sector containing a DCCD-binding proteolipid subunit from chromaffin granules whose molecular weight from the DNA sequence is 15,849 (Mandel et al., 1988). The existence of appreciable sequence identity with the 8 kDa proteolipid subunit from mitochondria and chloroplasts implies that the vacuolar and organelle (eubacterial) H^+-ATPases evolved

8.9 Other Classes of H⁺-Translocating ATPases 393

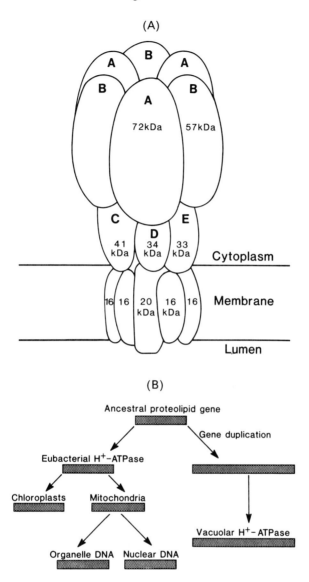

Figure 8.23. (A) Proposal for arrangement of the extrinsic and intrinsic polypeptides of the chromaffin granule H$^+$-ATPase. (Contributed by N. Nelson.) (B) Proposed model for evolution of eubacterial and vacuolar proteolipid subunits (Mandel et al., 1988).

from a common ancestral gene. A proposal for the polypeptide arrangement of the H^+-ATPase from chromaffin granules, and for the evolution of the proteolipid subunit, is shown (Fig. 8.23A and B).

8.9.2 H^+-Plasma Membrane or Phospho(P)-ATPase

The ATPase in the plasma membranes of the fungus *Neurospora crassa* and yeast (Serrano et al., 1986) translocates H^+ electrogenically, and that in the gastric mucosa exchanges H^+ for K^+ (Rabon et al., 1985). The function of these P-ATPases is to pump/secrete H^+ to the extracellular environment, and thereby to exert control of the intracellular pH, and as well on the $\Delta\psi$ and $\Delta\tilde{\mu}_{H^+}$ that are used for secondary transport of ions, amino acids, and sugars. The ATPase H^+ pump from *N. crassa*, with an H^+/ATP stoichiometry $= 1$ (Gouffeau and Slayman, 1981), can generate a $\Delta\psi$ of -350 mV, a $\Delta pH = 1.4$, and a resulting $\Delta\tilde{\mu}_{H^+} = -430$ mV. This H^+/ATP stoichiometry implies that the ATP hydrolysis reaction is difficult to reverse (Problem 79) and physiologically the enzyme will function in the direction of ATP hydrolysis. Although there is not yet agreement on the H^+/ATP ratio for the vacuolar H^+-ATPases, it appears that these also operate mostly in the direction of ATP hydrolysis, so that in the cell the F_0F_1 ATP synthases in mitochondria or chloroplasts produce ATP, and the phospho- and vacuolar ATPases consume it (Fig. 8.24).

The complete amino acid sequences of the plasma membrane H^+-ATPase from *N. crassa* (Hager et al., 1986; Addison, 1986) and yeast (Serrano et al.,

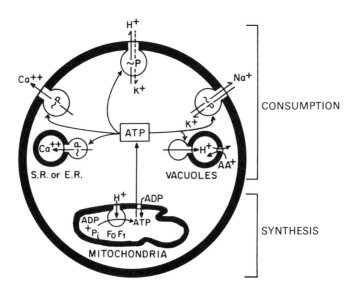

Figure 8.24. Intracellular relation between ATP supply from F_0F_1 ATP synthases and consumption by vacuolar and phospho-ATPases (Pedersen and Carafoli, 1987a).

8.9 Other Classes of H⁺-Translocating ATPases

```
H     191   TEAPEVVPGDILQVEEGTIIPADGRIVTDD.AFLQVDQSALTGESLAVDKH   240
Ca    140   IKARDIVPGDIVEVAVGDKVPADIRILSITTLRVDQSILTGESVSVIKH     190
Na,K  178   INAEEVVGDLVEVKGGDRIPADLRIISAN..GCKKNSSLTGESEPQTRS     226
Kdp   121   VPADQLRKGDIVLVEAGDIIPCDGEVIEGG...ASVDESAITGESAPVIRE   168

H     245   VFASSAVKRGEAFVVITATGDNTFVGRAAALVNAASGGISGH             285
Ca    208   LFSGTNIAAGKALGIVATTGVSTEIGKIRDQMAATEQDKTP             248
Na,K  241   AFFSTNCVEGTARGIVVYTGDRTVMGRIATLASGLEGGQTP             281
                                                  *
H     327   LAITIIGPVGLPAVVTTMAVGAAYLAKKAIVQKLSAIESLAGVEILCSDKTGTLTKN   386
Ca    300   VALAVAAIPEGLPAVITTCLALGTRRMAKKNAIVRSLPSVETLGCTSVICSDKTGTLTTN   359
Na,K  323   IGHIVANVPEGLLATVTVCLTLTAKRMARKNCLVKNLEAVETLGSTSTICSDKTGTLTQN   382
Kdp   256   VALLVCLIPTTIGGLLSASAVAGMSRMLGANVIATSGRAVEAAGDVDVLLLDKTGTTTLG   315

H     471   TCVKGAPLFVLKTVEEDHPIPEEVDQAYKNKVAEF     505
Ca    512   MFVKGAPEGVIDRCNYVRVGTTRVPMTGPVKEKIL     546
Na,K  503   LVMKGAPERILDRCSSILIHGKEQPLDEELKDAFQ     537
Kdp   392   MIRKGSVDAIRRHVEANGGHFPTDVDQKVDQVARQ     426

H     534   DPPRHDTYKTVCEAKTLGLSIKMLTGDAVGIAR        566
Ca    601   DPPRKEVMGSIQLCRDAGIRVIMITGDNKGTAI        633
Na,K  591   DPPRAAVPDAVGKCRSAGIKVIMVTGDHPITAK        623
Kdp   447   DIVKGGIKEAFAQLRKMGIKTVMITGDNRLTAA        479

H     604   ADGFAEVFPQHKYNVEILQQRGYLVAMTGDGVNDAPSLKKADTGIAVEG.SSDAARSAAD   663
Ca    673   ACCFARVEPSHKSKIVEYLQSYDEITAMTGDGVNDAPALKKAEIGIAMGS.GTAVAKTASE   732
Na,K  685   EIVEARTSPQQKLIIVEGCQRQGAIVAVTGDGVNDSPALKKADIGVAMGI.AGSDVSKQAAD   745
Kdp   488   DDFLAEATPEAKLALIRQYQAEGRLVAMTGDGTNDAPALAQAQVAVAMNS.GTQAAKEAGN   547
```

Figure 8.25. Amino acid identities between the *N. crassa* plasma membrane H⁺-ATPase, the plasma membrane Ca²⁺-, Na⁺, K⁺-ATPases, and the K⁺-transporting ATPase (product of *KdpB* gene) from *E. coli*. The aspartic acid phosphorylation site and a reactive lysine protected by ATP are indicated by an asterisk (*) and arrow (↓), respectively (Hager et al., 1986).

1986) have been obtained and are highly (74%) homologous. Both enzymes have several regions in which there is a high degree of identity (Fig. 8.25) with the amino acid sequences of the plasma membrane Na^+, K^+-ATPase from sheep kidney (Shull et al., 1985), the Ca^{2+}-ATPase from rabbit sarcoplasmic reticulum (Brandl et al., 1986), to a smaller extent with the K^+-ATPases of *E. coli* and *S. faecalis* (Hesse et al., 1984; Solioz et al., 1987). Some identity between the sequence around the aspartic acid phosphorylation site and a residues 298–305 of the β subunit of the F_0F_1 ATPase was noted above (Table 8.2).

A hydropathy plot of the *N. crassa* enzyme and a suggested folding pattern in the membrane are shown (Fig. 8.26).

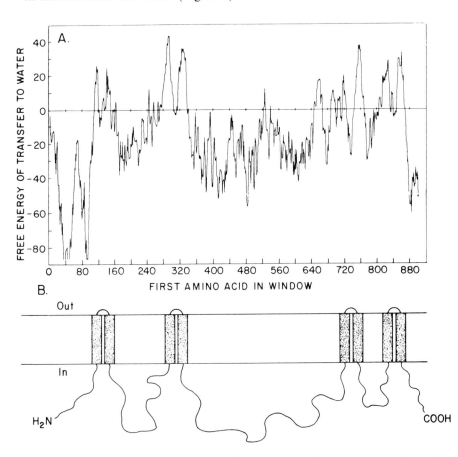

Figure 8.26. (A) Hydropathy plot of the *N. crassa* H^+-ATPase generated with a 20 residue window (Hager et al., 1986), and the resulting suggestions for folding of the protein involving (B) eight (Hager et al., 1986) or (not shown) 10 (Addison, 1986; Sussman et al., 1987) membrane-spanning helices. A single DCCD-reactive residue, Glu-129, that is conserved in the *N. crassa* and yeast sequences, would reside in the first proposed membrane-spanning segment.

8.9.3 Phospho- or (E_1-E_2)-ATPase

The H^+-plasma membrane or phospho-ATPase is a member of the class of E_1-E_2 ion pumps or ATPases, where E_1 and E_2 refer to two conformational states of the phosphoenzyme in the pathway of ion transport (Fig. 8.27A). The E_1 state of the enzyme is formed as an aspartyl phosphate intermediate and is inhibited by the phosphate analog vanadate. The phosphorylated aspartic acid, $D \sim P$, is on the cytoplasmic side of the membrane in the Na^+, K^+-ATPase in a sequence C-S/T-D $\sim$ P-K (Walderhaug et al., 1985). Except for the Na^+, K^+-ATPase, which also has a smaller M_r 55,000 glycoprotein subunit, these enzymes seem to consist of a single subunit of approximately 100,000 molecular weight.

8.9.4 Na^+,K^+-Phospho-ATPase

The sodium/potassium-dependent ATPase Na^+,K^+-ATPase is the integral membrane protein (Fig. 8.27B) that couples ATP hydrolysis to vectorial transport of Na^+ and K^+ across the plasma membrane. The electrogenic uptake of $2K^+$ per ATP hydrolyzed in exchange for $3Na^+$ extruded from the cytoplasm constitutes an electrogenic pump that generates a membrane potential negative on the cytoplasmic side. The pump action maintains the internal K^+ level. The resulting Na^+ electrochemical gradient drives Na^+-linked transport of sugars and amino acids, extrusion of Ca^{2+} and H^+ from the cell, secretion of salt from salt glands, kidney, and skin cells, regulates cell volume and intracellular pH through antiport activity, and restores ionic balance in excitable tissues. The enzyme consists of a catalytic subunit (α) with $M_r \simeq 110,000$ and at least one smaller subunit (β) with $M_r \simeq 55,000$. The ATP binding site and the phosphorylation site are located on the cytoplasmic side of the membrane, whereas the binding site for the cardiac glycoside inhibitor, ouabain, is located on the extracellular side (Cantley, 1981; Shull et al., 1985). The catalytic subunit can exist in the two major conformational states, E_1 (preferential binding of cytoplasmic Na^+ and/or ATP) and E_2 (preferred binding of extracellular K^+ and/or covalent phosphate). The transport of Na^+ and K^+ is linked to the conformational transition between these two states (Fig. 8.27C), perhaps involving a dimeric complex (Fig. 8.28). There is a conformation-dependent trypsin cleavage site ~ 20 amino acids from the NH_2-terminus that is rapidly cleaved in NaCl, but is inaccessible in KCl, a different cleavage site appearing under the latter conditions (Cantley, 1981). Information on the structure (Fig. 8.27B) and function of the enzyme has been obtained from analysis of the sequences of α (1,016 amino acids) and β (302 residues) subunits of the Na^+,K^+-ATPase (Shull et al., 1985), comparison with that of the Ca^{2+}-ATPase, topographical probing of the membrane-bound enzyme using exogenous proteases and peptide-directed antibodies (Ovchinnikov et al., 1987), and electron microscope-image reconstruction

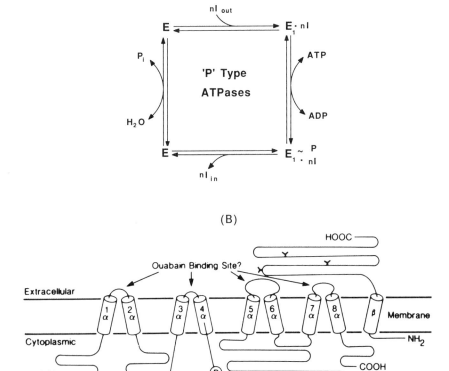

Figure 8.27. (A) General reaction scheme for ion transport by phosphoATPases. $E_1 E_2$ ion pumps are ATP-driven ion transporters. Ion (I) translocation is linked to phosphorylation of the carrier, E, in state E_1 to form $E_1 \sim P$. After operation of the pump to accomplish the translocation of n ions (I), the carrier is converted to the state, E_2, while the ion is released, completing the cycle. (Adapted from Pedersen and Carafoli, 1987a.) (B) Proposed model of the folding pattern in the membrane bilayer of the Na^+, K^+-ATPase (Reprinted with permission from Price and Lingrel, *Biochemistry*, vol. 27, pp. 8400–8408. Copyright 1988 American Chemical Society.) P and A designate the location of phosphorylation and ATP binding sites on the cytoplasmic side of the membrane. (C) Application of model in (A) to Na^+, K^+-ATPase: Na^+ binds to the E_1 state in the cytoplasm. Phosphorylation ($E_1 \sim P$) is associated with Na^+ translocation to the outside where release of Na^+ and binding of K^+ is associated with the E_2 phosphoenzyme. Dephosphorylation is linked to uptake and release of K^+ in the cytoplasm, reforming the E_1 state.

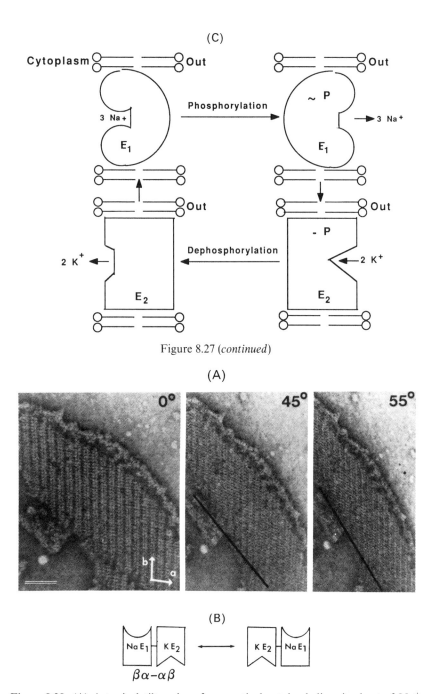

Figure 8.28. (A) A typical tilt series of a negatively stained dimeric sheet of Na^+, K^+-ATPase. Crystallization was induced by phospholipase A_2. The diagonal lines in the 45- and 55-degree tilt images show the direction of the tilt axis with respect to the specimen. The white arrows indicate the directions of the a and b lattice vectors. Scale bar, 1,000 Å. (B) Mixed dimer model for function of Na^+, K^+-ATPase. (Mohraz and Smith, 1988.)

studies of two-dimensional crystals of the Na^+,K^+-ATPase which, after overnight dialysis in the presence of phospholipase A_2, were exclusively dimeric (Fig. 8.28A). The dimeric structure suggested that the Na^+,K^+-ATPase may function as mixed $(\alpha\beta)_2$ dimer in which the two units of the dimer function 180° out of phase. That is, while one unit is in the E_1 conformation and pumps Na^+ from the cell, the other in the E_2 state transports K^+ into the cell (Fig. 8.28B). At the end of the cycle, the two units change conformation (Mohraz and Smith, 1988).

The similarity in size, primary sequence, position of hydrophobic domains, and phosphorylation and ATP binding sites implies a common evolutionary origin for the Na^+,K^+- and Ca^{2+}-ATPase enzymes. The region of greatest amino acid homology is near the phosphorylation site at Asp-369. The NH_2-terminal amino acid sequences located on the cytoplasmic side of the membrane show a high degree of homology between different Na^+,K^+-ATPases, but not with the Ca^{2+}-ATPase (Shull et al., 1985). Along with the Na^+-dependent tryptic cleavage site near the NH_2-terminus, this suggests that the NH_2-terminal hydrophilic lysine-rich domain may control the binding and passage of Na^+ and K^+ on the cytoplasmic side of the membrane.

8.9.5 Ca^{2+}-Transporting ATPase

Intracellular Ca^{2+} concentrations of resting state muscle cells are typically about 10^{-6} M, several orders of magnitude smaller than that of the extracellular fluid. The sarcoplasmic reticulum Ca^{2+}-ATPase, transporting two Ca^{2+} per ATP hydrolyzed, has a central role in the regulation of intracellular Ca^{2+} and in the contraction–relaxation cycle of muscle. Genes for the Ca^{2+}-ATPase of fast and slow twitch rabbit sarcoplasmic reticulum, coding for proteins of 997 and 993 amino acids (Brandl et al., 1986), have been sequenced. The gene products are very homologous in regions that are proposed to be an extensive cytoplasmic catalytic domain, sectors involved in the attachment of the cytoplasmic to the transmembrane domain, and putative transmembrane α-helices (Fig. 8.29).

The model of 10 transmembrane α-helices resembles the multi-helical model shown above (Figs. 8.26B and 8.27B) for the H^+- and Na^+,K^+-ATPases. A proposal for separate cytoplasmic transduction, phosphorylation, and nucleotide binding domains was based on assignment of functional sites to tryptic peptides (MacLennan et al., 1985), as well as measurement of a long distance (35–50 Å) between the Ca^{2+} and nucleotide binding sites (Scott, 1985). The stalk region made of amphipathic helices is proposed to connect the globular cytoplasmic domain to the membrane. The first three of these helices contain a large number, 16, of glutamic acids that may contribute to the initial (E_1 state) Ca^{2+} binding sites. The transport process involving the transition from the initial E_1–(Ca^{2+}) external state to the final E_2–(Ca_2^{2+})

8.9 Other Classes of H⁺-Translocating ATPases

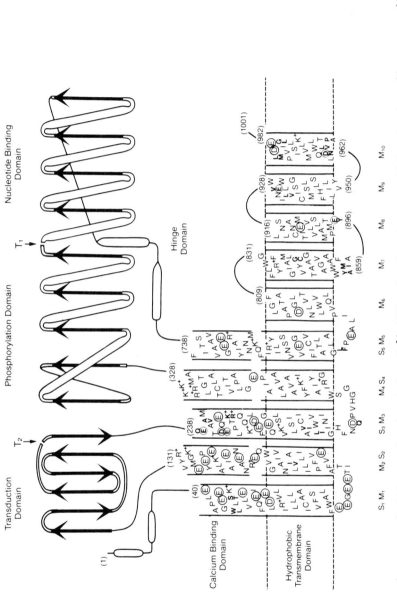

Figure 8.29. Proposed structure of rabbit sarcoplasmic reticulum Ca^{2+}-ATPase based on the gene and inferred amino acid sequences of fast and slow twitch rabbit muscle sarcoplasmic reticulum (Brandl et al., 1986; compare with MacLennan et al., 1985).

internal requires Ca^{2+}-dependent phosphorylation of the Asp-351 residue. The two Ca^{2+} are thought to be translocated through the helices of the stalk and into the hydrophobic channel by a resulting conformational change. Large-scale conformational changes of the sarcoplasmic reticulum ATPase occurring in the first few hundred msec after triggering of a single turnover of the enzyme have been measured by flash hydrolysis of caged ATP (Chap. 1, section 1.14) and measurement of the resultant structural change by time-resolved x-ray diffraction (Blasie et al., 1985). The structural change, transfer of about 8% of the enzyme mass from the cytoplasmic side into the membrane bilayer, occurs within one synchronous turnover of the ensemble of Ca^{2+}-ATPase molecules, and is associated with phosphorylation of the enzyme and extrusion of $2Ca^{2+}$ per ATPase. Crystal structure data also suggest the possibility of an $E_1 \rightarrow E_2$ monomer $\rightarrow$ dimer structural transition (Dux et al., 1985).

8.10 Summary

H^+ pumping ATPases can be divided into three groups: the eubacterial "F-type" present in chloroplasts, *E. coli*, and mitochondria that does not have a phospho-intermediate and can reversibly transduce $\Delta\tilde{\mu}_{H^+}$ and ATP, the vacuolar "V-type" that hydrolyzes ATP to acidify the organelle interior, and the plasma membrane "P-type" that uses a phospho-intermediate and is also involved in translocation of Na^+, K^+ and Ca^{2+}. Most of the discussion in this chapter centers on the F class that has definable membrane (F_0) and extrinsic (F_1) sectors, both of which are characterized by multisubunit structure. The F_1 contains the catalytic site(s) for ATP synthesis and hydrolysis, and the F_0 an H^+ channel or conduit of conformational change arising from the $\Delta\tilde{\mu}_{H^+}$. F_0F_1 is located in the cytoplasmic membrane of prokaryotes such as photosynthetic bacteria and *E. coli*, in the inner mitochondrial membrane, and in thylakoids where it is oppositely oriented. The maximum rate of ATP synthesis and of H^+ translocation in ATP hydrolysis is $\sim 400/F_0F_1$-s. The five F_1 subunits in bacteria and plants, and the five largest in the mammalian enzyme are α, β, γ, δ, ε, in a ratio 3:3:1:1:1 corresponding to a molecular weight of 385,000. The F_0 sector contains three or four subunits, of which the *a*, *b*, and *c* present in *E. coli* at a stoichiometry of 1:2:10 have been studied most thoroughly. Function of the F_1 subunits: α, nucleotide binding, catalytic site or shared site with β; β, conserved in evolution, contains catalytic, nucleotide binding site(s), sequence identities with other nucleotide binding proteins; γ, part of minimum $\alpha-\beta-\gamma$ unit needed for ATPase activity and H^+ gate; δ, function organelle-dependent; generally prevents a nonproductive H^+ leak; ε, also organelle-dependent; often an ATPase inhibitor and part of the H^+ gate. Subunits *a*, *b*, and *c* of the F_0 sector are all essential for reconstitution of H^+ channel activity, with five predicted transmembrane channel helices/*a*, two helices/*c*, and one/*b* which has a large polar extrinsic component that interacts with F_1. Genes for the subunits of the *E. coli* F_0F_1

8.10 Summary

are arranged as an operon, $acb\delta a\gamma\beta\varepsilon$. Evolutionary relatedness and an endosymbiotic origin of chloroplasts is implied by an identical order in the nine genes from the cyanobacterium *Synechococcus* 6301, the five F_1 genes in purple nonsulfur bacteria, and six of the chloroplast F_0F_1 genes.

Crystallographic studies of mitochondrial F_1 have proceeded to a level of 9 Å resolution, where the overall structure is seen to be asymmetric with two equal masses presumed to be two $\alpha-\beta$ pairs, and a third larger mass thought to be the other $\alpha-\beta$ associated with one or more of γ, δ, or ε. An approximately symmetric hexagonal structure, 80 × 110 × 120 Å, with six units surrounding a central mass, was discerned using electron microscopic analysis of two-dimensional arrays of F_1 negatively stained or decorated with antibody. Nucleotide binding sites, $3-6/F_1$, can be classified as exchangeable, probably functioning at or near the active sites, and non- or slowly exchangeable. Amino acids near these sites on the β subunit have been localized and identified by photolabeling with radioactive nucleotide analogs, with L-342, I-344, Y-345, and P-346 prominently labeled.

The $\Delta\tilde{\mu}_{H^+}$ for net synthesis of F_1-bound ATP is required to cause a conformational change in the enzyme and a binding change of the tightly bound ATP to allow its release. This was inferred from the sensitivity to uncouplers of exchange reactions ($P_i \rightleftarrows ATP$ and medium $ATP \rightleftarrows H_2O$) that require release of bound ATP, and from the measurement for the F_1 of a $K_a = 10^{12}$ M^{-1} for ATP when only one nucleotide binding site is occupied. Site–site interactions are implied by a decrease in the K_a of a factor of $\sim 10^8$, and a corresponding increase in the enzyme turnover time of a factor of $\sim 10^6$, when nucleotide is bound to more than one $\alpha-\beta$ subunit. This interaction can be explained by a symmetric alternating site model, or a model with nonequivalent subunits and positive cooperativity, in both cases involving three interacting $\alpha-\beta$ sites for ATP synthesis and hydrolysis. The synthesis of ATP from ADP and P_i can occur in one step of an alternating site cycle without an energy requirement because of coupling to tight binding of ATP in the same step.

The multisubunit vacuolar ("V") ATPase found in many organelle and membrane vesicle structures may be the most diversified class of ATPases. They can pump H^+ to an internal space to (i) generate a $\Delta\tilde{\mu}_{H^+}$ that can be used for active transport in chromaffin granules of the neurotransmitters epinephrine and serotonin, and (ii) to facilitate ligand–receptor dissociation. The P-type plasma membrane phospho-ATPase in *N. crassa* and yeast translocates H^+ electrogenically to generate a $\Delta\psi$ and $\Delta\tilde{\mu}_{H^+}$ that can be used for active transport. Its H^+/ATPase ratio = 1 implies that the enzyme functions in the direction of ATP hydrolysis. This enzyme is a member of a class of E_1-E_2 phospho-ATPases, where E_1 and E_2 designate two conformational states in the pathway of ion transport. The E_1 state involves an aspartyl-phosphate intermediate, inhibited by the analogue vanadate, that is linked to ion translocation. Conversion of the enzyme to the E_2 state is linked to ion release, completing the translocation. For the Na^+,K^+-ATPase that couples

ATP hydrolysis to uptake of $2K^+$ and extrusion of $3Na^+$ per ATP, cytoplasmic Na^+ is bound in the E_1 state, and extracellular K^+ in E_2. α and β subunits of the enzyme consisting of 1,016 and 302 residues, respectively, have been sequenced. The sarcoplasmic reticulum Ca^{2+}-ATPase that transports $2Ca^{2+}$/ATP has an important role in the regulation of intracellular Ca^{2+} at ~ 1 μM and in the contraction–relaxation cycle of muscle.

Problems

74. What is the sign of the enthalpy change associated with the association of F_1 subunits in vitro if the binding is much more stable at room temperature than at $4°C$?
75. Calculate the value of $\Delta\psi$ needed to drive ATP synthesis at $25°C$ if $\Delta G_{ATP} = +12$ kcal/mole in (i) chromatophores, (ii) mitochondria, and (iii) E. coli membrane vesicles ($pH_o = 7.5$), for which ($pH_i - pH_o$) would equal -3, $+0.5$, and 0. The direction of the transmembrane proton movement in chromatophores is the same as that in chloroplasts, and opposite to the direction of the H^+ movement in mitochondria and E. coli vesicles; $n_{H^+} = 3$ in all three systems.
76. As noted above in problem 75, the value of ΔpH in E. coli membrane vesicles is $\sim$ zero at $pH = 7.5$. Thus, the source of all free energy available from the electrochemical potential for driving ATP synthesis or active transport at pH 7.5 in these membranes resides in the $\Delta\psi$. It is known, however, that ΔpH can be solely responsible for the uptake at lower pH ($\sim pH$ 5.5) of glucose-6-phosphate. This is one indication that ΔpH and $\Delta\psi$ are both thermodynamically and enzymatically interchangeable. From problem 75(iii), determine the vaue of ΔpH that would completely substitute for $\Delta\psi$.
77. One of the ways in which the number of protons, n_{H^+}, needed to synthesize one molecule of ATP was obtained was by measuring in state 4-mitochondria all experimentally determinable quantities that appear in the thermodynamic equation linking ΔpH, $\Delta\psi$, ATP, ADP, and P_i. Assume (ATP) = 1 mM; (ADP) = 10 μM; (P_i) = 10 mM; $\Delta pH = 0.5$; $\Delta\psi = -147.5$ mV; $\Delta G° = 7$ kcal/mol. Calculate n_{H^+}.
78. What is the H^+/ATP ratio if the flow of current generated by ATP hydrolysis by TF_0F_1 incorporated in planar bilayer membranes is suppressed by a potential of -180 mV, cis-negative (see Fig. 8.5B), and the ΔG for ATP synthesis is $+12$ kcal/mol.
79. Calculate the ATP/ADP ratio that can be obtained from a $\Delta\tilde{\mu}_{H^+} = 5$ kcal/mol per mole of H^+ translocated, if the H^+/ATP ratio required for ATP synthesis is (a) three (F-ATPase) or (b) one (N. crassa ATPase). Assume the $\Delta G°$ for ATP synthesis is 8 kcal/mol and (P_i) = 10 mM. What is the implication for the reversibility of the ATPase activity in each case?

80. (a) What is the average H^+ conductance (H^+/CF_0F_1-s) through the CF_0F_1 corresponding to the maximum rates of ATP synthesis of 400 ATP/CF_0F_1-s, if H^+/ATP = 3? (b) The maximum conductance of an individual CF_0F_1 has been found to be 1 picosiemen (10^{-12} amp/V) (Lill et al., 1987). To what H^+ flux would this correspond in the presence of $\Delta\psi = 0.1$ V? 1 amp = 1 coulomb/s; 1 coulomb = 6.2×10^{18} electronic charges.
81. Calculate the approximate H^+ flux (number H^+/second) through the ATP synthetase channel of mitochondria, if
    (i) The rate of respiration or O_2 uptake is 1 nmol/mg membrane protein per second;
    (ii) 10% of the membrane protein is ATP synthase. Therefore, the rate of O_2 uptake is $\simeq$ 10 nmol/mg synthase/s;
    (iii) The molecular weight of the synthase is approximately 400,000. Calculate the number of O_2 molecules taken up per molecule of synthase per second;
    (iv) The ATP/$2e^-$ or ATP/O ratio in mitochondria is 3 for NADH-linked substrates;
    (v) The number of H^+ translocated through the reversible ATPase-ATP synthase per molecule of ATP synthesized is also 3.
82. (a) Calculate the number of ATP and GTP molecules that an *E. coli* cell must synthesize per second to support protein synthesis if its total weight is 10^{-12} g, its dry weight 25% of the total weight, its protein content 60% of its dry weight, the protein average molecular weight is 30,000, approximately 0.1% of the protein molecules are newly synthesized per second, and the synthesis of each peptide bond requires 2 ATP and 2 GTP. (b) If the $\Delta G°$ for ATP and GTP synthesis is 7.5 kcal/mol, and $\sim$90% of the cellular energy is used for protein synthesis, what is the rate of energy production per cell?

Chapter 9
Active Transport

9.1 Introduction

All living organisms must exchange material with the surrounding environment across their cell membranes, exporting waste materials and importing useful metabolites. The transport of such materials is catalyzed by proteins specific for particular molecules or groups of molecules. Emphasis will be placed on those processes that can be strictly defined as active transport: *The transport and accumulation of substrate against a concentration gradient without chemical modification of the substrate.* Other uptake processes such as group translocation (transport accompanied by chemical modification of the transported substrate) will be treated less extensively. This emphasis highlights the many similarities between the energetics of transport and of oxidative- and photo-phosphorylation. For a variety of reasons (e.g., the availability of useful mutants), the majority of transport systems to be discussed will be from the bacterial kingdom, but a number of important eukaryotic transport systems will also be analyzed.

Bacterial transport systems can be classified in terms of the location of the proteins involved in the transport process. Some transport systems appear to involve specific receptors in the outer membrane. Other systems, such as those involved in transport of arabinose, galactose, and histidine in *Salmonella*, contain a soluble protein located in the periplasmic space between the inner and outer membranes in addition to membrane-bound components. Transport catalyzed by these "shockable" systems (so called because the initial substrate binding proteins can be easily released from the cells by osmotic shock procedures) use ATP as the energy source for substrate accumulation. The third major class of transport systems, for which the well-studied *E. coli* lactose transport system is a good example (Kaback, 1986) and on which this chapter will place greatest emphasis, involves only membrane-bound proteins intrinsic to the bacterial inner membrane. These are often referred to as secondary transport systems because they utilize preexisting ion gradients as the energy source for substrate accumulation. The "y" gene coding for the lactose transport protein, or lactose permease, is one of three genes coding for proteins involved in lactose transport and metabolism. The three contiguous structural genes are adjacent to and downstream from a regulatory section of the DNA known as the *lac* operator-promoter, which together are known as the *lac* operon. In the absence of an inducer, a specific protein (the "repressor") binds to the *lac* operator. The repressor protein interferes with the proper binding of RNA polymerase to the operator and thus blocks production of messenger RNA for all three structural genes of the operon. When the inducer, a derivative of the lactose molecule, is present, it binds to the repressor protein and causes dissociation of the repressor from the DNA regulatory region, allowing RNA polymerase to begin transcription of the structural genes. This control in the *lac* operon at the level of transcription is of value to the cell because energy is not consumed for the biosynthesis of proteins involved in lactose metabolism unless lactose is actually available in the environment. A considerable number of other transport proteins exhibit similar patterns of induction by the transport substrate.

Bacterial cells can accumulate metabolites at internal concentrations as high as 10^5-fold greater than the external concentration. For molecules with charge z, the electrochemical potential of the accumulated solute, $\Delta\tilde{\mu}_S$, is given by (Chap. 1, section 1.11):

$$\Delta\tilde{\mu}_S = 2.3RT \log_{10}\frac{[S_i^{+z}]}{[S_o^{+z}]} + zF \cdot \Delta\psi \tag{1}$$

9.2 Evidence for Protein Carrier-Mediated Transport

Evidence for a carrier-mediated, rather than a nonspecific diffusion mechanism for transport, comes from: (i) The kinetics of transport processes. For simple diffusion, the rate of transport would be proportional to the difference in concentration of the transportable substrate on the two sides of the membrane. When the internal concentration of the substrate approaches zero, a simple diffusion mechanism would result in an initial transport rate that increases linearly with increasing external concentration (Fig. 9.1a). In fact, a plot of the initial rate of transport vs. substrate concentration, [S], generally has the shape of a rectangular hyperbola (Fig. 9.1b). The shape of this curve is identical to that derived from the Michaelis–Menten formulation for enzyme-catalyzed reactions and K_m values are routinely reported for carrier-mediated transport reactions. The fact that the rate of transport asymptotically reaches a maximum value implies that transport is catalyzed by a

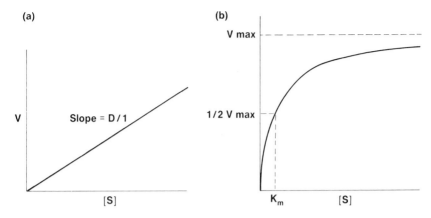

Figure 9.1. Initial kinetics of simple diffusion (a) and carrier-mediated transport (b). Plots of velocity versus the difference in substrate concentrations on the two sides of the membrane give a straight line (a) with a slope slope of $D/1$ and (b) a rectangular hyperbola approaching a maximum velocity (V_{max}) at high $[S]$. D is the diffusion constant, and 1 the membrane thickness (Adapted from Zubay, 1983, *Biochemistry*, © The Benjamin/Cummings Publishing Company.)

limited number of carriers that become saturated with substrate at concentration $\gg K_m$. (ii) The specificity of transport processes is a second characteristic of enzyme-catalyzed reactions, often involving transport of a single enantiomer of a chiral substrate (e.g., D- but not L-glucose). (iii) Important evidence for the protein nature of transport carriers came from analysis of bacterial transport mutants. For example, analysis of lactose transport activity indicated that loss of transport activity was associated with a mutation in a single gene, *lac y*, coding for the permease (Hobson et al., 1977). The *y* gene has been sequenced, and the *lac* permease synthesized in vitro (Wright et al., 1986). (iv) The lactose transport system has been successfully reconstituted using the purified *E. coli* lactose carrier protein incorporated into liposomes (Kaback, 1986).

9.3 Techniques for Studying Transport in Bacteria

Radioactively labeled substrates are available for virtually all transport systems of interest, so that transport can be conveniently monitored by following the uptake of radioactivity into bacterial cells. The cells can be trapped on membrane filters with pore sizes (e.g., 0.45 μm) smaller than the bacteria, untransported substrate removed by washing the filters, and the trapped bacteria then assayed for radioactivity. Alternatively, the flow dialysis technique has proven useful for measuring uptake (Kaback, 1976).

The problem of substrate metabolism subsequent to transport can be minimized by limiting the uptake experiment to short times, by using specific inhibitors of substrate metabolism, or by utilizing nonmetabolizable substrate analogs such as the glucose analog α-methylglucoside, the alanine/glycine analog α-aminoisobutyric acid (AIB), or the lactose analog thiomethyl-β-D-galactopyranoside (TMG). In addition, right-side-out cytoplasmic (inner) membrane vesicles devoid of the cytoplasmic enzymes that catalyze substrate metabilism can be employed for transport studies (Owen and Kaback, 1979). Because active transport of many substrates (e.g., those not utilizing periplasmic proteins, such as lactose or proline) in membrane vesicles at rates comparable to those observed with intact cells, it can be concluded that the outer membrane of the cell envelope is not involved in the transport of such substrates, and that all components of these transport systems are located in the cytoplasmic or inner membrane.

9.4 Structure of the Cell Envelope of Gram-Negative Bacteria

The envelope of Gram-negative bacteria consists of three layers that stain in the electron microscope and are chemically definable: (i) The outer membrane consists of protein, phospholipid, and lipopolysaccharide (LPS). It contains pores for small (MW $\leq$ 600) hydrophilic molecules and is a barrier for hydrophobic and amphiphilic molecules. (ii) The peptidoglycan monolayer is responsible for the cell shape and rigidity. (iii) The cytoplasmic or inner membrane contains most of the membrane-bound biological activity of the envelope. The envelope contains approximately 15% of the total cell protein and has a protein:lipid ratio of 2–3:1. The major phospholipids are: phosphatidylethanolamine (70–80%), phosphatidylglycerol (5–15%), and cardiolipin (5–15%). The fatty acid chains are mainly saturated (16:0) and unsaturated (16:1 or 18:1) fatty acid, respectively. The relative amounts of saturated and unsaturated chains are dependent on growth temperature, probably to regulate membrane fluidity.

The LPS, found exclusively in the outer half of the outer membrane of Gram-negative bacteria, is made of a core lipid, lipid A, containing several fatty acids linked to glucosamine, a core oligosaccharide containing three 2-keto-3-deoxyoctonyl (KDO) and two heptosyl, glucosyl, and galactosyl residues. Attached to the core is the O-antigen polysaccharide chain, approximately 300 residues long. Since phospholipid comprises the inner half of the outer membrane, it is an asymmetric bilayer membrane. The LPS appears to be involved in the barrier that Gram-negative bacteria possess toward hydrophobic and amphiphilic molecules such as antibiotics, ionophores (e.g., carbonylcyanide-p-trifluoromethoxy phenylhydrazone and valinomycin), and dyes such as methylene blue, crystal violet, or fluorescence probes.

The outer membrane has only a few major ($>10^4$ molecules/cell) protein components, as detected by sodium dodecyl sulfate-polyacrylamide gel electrophoresis (SDS-PAGE). There are also a number of minor (10^2–10^3 molecules/cell) outer membrane proteins (*omp*) that function in cell division-specific DNA binding, as receptors for bacteriophages and colicins, and in the transport of vitamins and metals.

9.5 $\Delta\tilde{\mu}_{H^+}$ Formation in Bacteria

A distinction can be made between primary and secondary active transport. In the former, the transport process is coupled directly to an energy-generating reaction, as in the Ca^{2+}-ATPase of the mammalian sarcoplasmic reticulum, where the energy released during ATP hydrolysis is utilized directly to drive Ca^{2+} movement across the membrane (Chap. 8, section 8.9). Secondary active transport utilizes ion gradients preformed by electron transport or ATP hydrolysis as the energy source for transport. Electron flow during bacterial respiration is coupled to proton translocation across the cytoplasmic membrane, as discussed above, with proton movement directed outwardly from the cytoplasm to the extracellular medium.

Both ΔpH and $\Delta\psi$ components of the proton-motive force have been measured in respiring bacteria. Using the lipophilic cation uptake methods described previously (Chap. 3, section 3.4), the $\Delta\psi$ of respiring *E. coli* cells was found to be -100 to -150 mV (Fig. 9.2), depending on the pH and cation

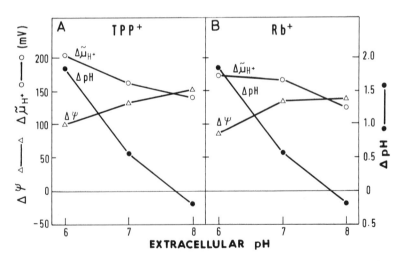

Figure 9.2. Dependence of ΔpH, $\Delta\psi$, and $\Delta\tilde{\mu}_{H^+}$ on extracellular pH in respiring *E. coli* cells. Succinate was the respiratory donor and ΔpH was measured by the distribution of [^{14}C]5,5-dimethyl-oxazolidine-2,4-dione (DMO). $\Delta\psi$ was measured by the distribution of (A) [^3H]tetraphenylphosphonium ion (TPP$^+$), or (B) the distribution of $^{86}Rb^+$ in the presence of valinomycin. $\Delta\tilde{\mu}_{H^+}$ (in mV, actually $\Delta\tilde{\mu}_{H^+}/F$) was calculated from ($\Delta\psi = -59 \cdot \Delta pH$) (Reprinted with permission from Zilberstein et al., *Biochemistry*, vol. 18, pp. 669–673. Copyright 1979 American Chemical Society.)

9.5 $\Delta\tilde{\mu}_{H^+}$ Formation in Bacteria

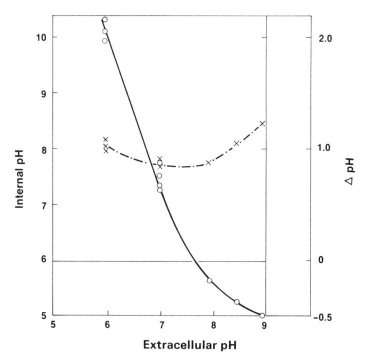

Figure 9.3. Intracellular pH and ΔpH in *E. coli* cells as a function of external pH. Succinate was the respiratory donor, ΔpH (o) was measured by the distribution of [^{14}C]DMO, and internal pH (x) calculated (Padan et al., 1976).

concentration of the external medium (Padan et al., 1976). Similar values have been obtained with a number of different respiratory electron donors, with electron acceptors other than O_2 (e.g., fumarate), with membrane vesicles instead of cells, and with aerobic bacteria other than *E. coli*. Parallel measurements of $\Delta\psi$ in "giant" *E. coli* cells using both lipophilic cation distribution and microelectrode techniques yielded similar values for $\Delta\psi$ (Felle et al., 1980). The ΔpH component of the $\Delta\tilde{\mu}_{H^+}$ has been found to have a similar value when measured by permeant weak acid distribution and by ^{31}P-nuclear magnetic resonance (^{31}P-NMR) in respiring cell and vesicles of *E. coli*, as well as in cells of the photosynthetic bacterium *Rb. sphaeroides* (Shulman et al., 1979; Nicolay et al., 1981).

One conclusion from these measurements is that the internal pH is maintained at an essentially constant value over a wide range (>3 pH units) of external pH values (Fig. 9.3), implying that pH homeostasis is of paramount importance to bacteria. Thus the value of ΔpH in respiring bacterial cells depends almost entirely on the external pH. For many bacteria, ΔpH is virtually zero at external pH values near or slightly above neutrality. It is only at acidic pH values that a ΔpH thermodynamically favorable for driving active transport or ATP synthesis is maintained by respiring bacteria. The decrease in the ΔpH component of $\Delta\tilde{\mu}_{H^+}$ with increasing external pH is balanced to some extent by an increase in $\Delta\psi$. The total proton-motive force near -200 mV maintained by *E. coli* during steady-state respiration is similar to values observed with other bacteria.

9.6 Active Transport of Sugars Coupled to H^+ Cotransport

9.6.1 Transport of Neutral Sugars: Evidence for H^+ Symport

Active transport of a number of sugars (e.g., lactose and other β-galactosides) is coupled to the cotransport or symport of one or more protons. The basic ideas for symport transport systems are illustrated in Fig. 9.4 (compare with Fig. 1.5) which shows how $\Delta\tilde{\mu}_{H^+}$, maintained either by respiration or ATP hydrolysis, can provide the energy for accumulation of sugars. Because sugars such as lactose are electrically neutral molecules, the $\Delta\psi$ (inside negative) maintained by bacteria could not contribute to the net driving force for lactose uptake unless lactose transport were coupled to the import of positive charge into the cells. Because ΔpH is often close to zero, the energy supply for H^+ symport is often dominated by the $\Delta\psi$ term.

Experimental support for the hypothesis that $\Delta\tilde{\mu}_{H^+}$ can provide the driving force for lactose uptake in *E. coli* comes from the observations that: (i) Protonophoric uncouplers, which collapse both ΔpH and $\Delta\psi$ (Chap. 3, section 3.6.7), are effective inhibitors of lactose transport (Kaback, 1976), and (ii) from the demonstration that transport can be driven by either $\Delta\psi$, or

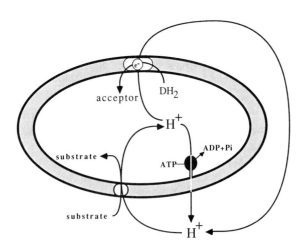

Figure 9.4. Suggested H^+ symport mechanism of sugar uptake in bacteria. Proton efflux arises from respiration and ATP hydrolysis. This diagram shows the mutual dependence of the flows, but not their stoichiometry.

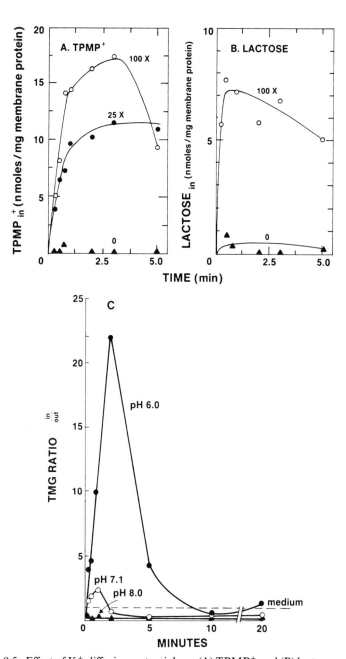

Figure 9.5. Effect of K$^+$ diffusion potentials on (A) TPMP$^+$ and (B) lactose uptake by *E. coli* vesicles. The vesicles, containing 0.1 M potassium phosphate buffer (pH 6.6) and valinomycin, were diluted into 0.1 M sodium phosphate buffer (pH 6.6) (–○–, –●–) or 0.1 M potassium phosphate buffer (pH 6.6) (–▲–), containing either [^{3}H]TPMP$^+$ (A) or [^{14}C]lactose (B). The numbers next to the lines represent changes in [K$^+$] (Reprinted with permission from Schuldiner and Kaback, *Biochemistry*, vol. 14, pp. 5451–5461. Copyright 1975 American chemical Society.) (C) ΔpH-driven TMG accumulation by *S. lactis* cells. Cells were washed with distilled water and then transferred into 0.1 M sodium phosphate buffer either at pH 6.0, 7.1, or 8.0) containing [^{14}C]TMG (Kashket and Wilson, 1973).

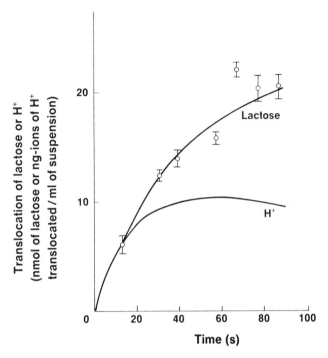

Figure 9.6. Time course of H^+ and lactose uptake by an anaerobic suspension of *E. coli* cells. H^+ uptake was followed using a sensitive pH electrode and lactose uptake using [^{14}C]lactose. The experiment was initiated by the addition of lactose (From West and Mitchell, 1972. Reprinted by permission from *Biochemical Journal*, vol. 132, pp. 587–592. Copyright © 1972 The Biochemical Society, London.)

ΔpH, in the absence of another energy source. The former is illustrated by the use of a K^+ valinomycin diffusion potential (monitored by triphenylmethylphosphonium ($TPMP^+$) accumulation, Fig. 9.5A) to support lactose uptake by *E. coli* vesicles. (Fig. 9.5B). The latter is illustrated by TMG uptake by the Gram-positive bacterium *S. lactis* after rapid addition of HCl to acidify the external medium (Fig. 9.6C). (iii) Lactose uptake by weakly buffered suspensions of *E. coli* cells results in a small pH increase in the external medium (Fig. 9.6), with an initial lactose:H^+ uptake stoichiometry of 1. Control experiments using *E. coli* cells lacking the lactose transport system showed no increase in external pH that was dependent on lactose addition.

9.6.2 Stoichiometry of H^+: Sugar Symport

Using the lactose transport system, and assuming a steady-state in which all of the free energy available in $\Delta\tilde{\mu}_{H^+}$ is utilized by the lactose transport system,

9.6 Active Transport of Sugars Coupled to H^+ Cotransport

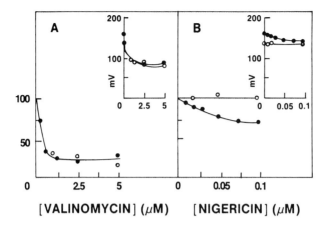

Figure 9.7. Effects of valinomycin (A) and nigericin (B) on the steady-state accumulation of lactose by respiring *E. coli* at pH 5.5 (●-●) and pH 7.5 (o-o). $\Delta\tilde{\mu}_{lactose}$ is shown as a percentage of the value obtained in the absence of ionophores. The insets show values for $\Delta\tilde{\mu}_{lactose}$ (in mV), calculated from $\Delta\tilde{\mu}_{lactose} = 59 \times \log[\text{lactose}]_{in}/[\text{lactose}]_{out}$ (Reprinted with permission from Ramos and Kaback, *Biochemistry*, Vol. 16, pp. 854–859. Copyright 1977 American Chemical Society.)

the H^+:sugar soichiometry, n, can be calculated (compare with Chap. 1, section 11.)

$$\Delta\tilde{\mu}_{lactose} + n\Delta\tilde{\mu}_{H^+} = 0 \qquad (2)$$

Substituting for $\Delta\tilde{\mu}_{H^+}$,

$$\log\frac{[\text{lactose}]_{in}}{[\text{lactose}]_{out}} = \frac{n}{Z}\Delta\tilde{\mu}_{H^+} = n\cdot\Delta\text{pH} - \frac{n\Delta\psi}{Z} \qquad (3)$$

This prediction can be experimentally tested by measuring $[\text{lactose}]_{in}/[\text{lactose}]_{out}$ as a function of $\Delta\tilde{\mu}_{H^+}$, ΔpH and $\Delta\psi$ (Fig. 9.7). At pH 5.5, where the $\Delta\tilde{\mu}_{H^+}$ maintained by *E. coli* vesicles consists of both ΔpH and $\Delta\psi$, both the ionophores, valinomycin and nigericin, that eliminate $\Delta\psi$ and ΔpH, respectively (Chap. 3, section 3.12.1), inhibit respiration-driven lactose uptake. At an external pH of 7.5, where $\Delta\text{pH} \simeq 0$, the uptake is inhibited by valinomycin, but not nigericin. These results are consistent with the hypothesis that lactose uptake occurs via an electrogenic lactose/H^+ symport, with a lactose:H^+ stoichiometry equal to 1. However, $\Delta\tilde{\mu}_{lactose}$ can be significantly less than $\Delta\tilde{\mu}_{H^+}$ so that the H^+:lactose stoichiometry may exceed 1 at some values of the external pH (Ramos and Kaback, 1977b; Booth et al., 1979). For example, the ratio of $\Delta\tilde{\mu}_{lactose}:\Delta\tilde{\mu}_{H^+}$ at pH 7.5 in *E. coli* vesicles is 2 (Fig. 9.8). In contrast, results obtained with intact *E. coli* cells, shown in Table 9.1, are more consistent with a H^+:lactose stoichiometry = 1, independent of external pH (Zilberstein et al., 1979). While the reason for this discrepancy is not clear, such discrepancies provide an example of the necessity of checking results obtained with vesicles against those obtained with cells.

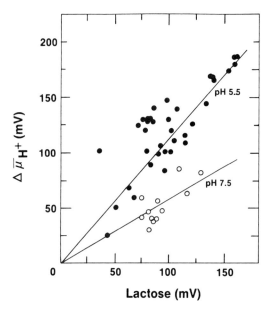

Figure 9.8. Relationship between $\Delta\tilde{\mu}_{H^+}$ and the steady-state level of lactose accumulation by respiring *E. coli* vesicles. Ascorbate + phenazine methosulfate was used as the respiratory electron donor ΔpH and $\Delta\psi$ were measured by the distribution of [^{14}C]acetate or [^{14}C]DMO and the [^{3}H]TPMP$^+$ cation, respectively. $\Delta\tilde{\mu}_{H^+}$ was varied by adding small amounts of uncoupler. $\Delta\tilde{\mu}_{lactose}$ was calculated as in Fig. 9.7 (Reprinted with permission from Ramos and Kaback, *Biochemistry*, vol. 16, pp. 4271-4275. Copyright 1977 American Chemical Society.)

Table 9.1. $\Delta\tilde{\mu}_{H^+}$ and $\Delta\tilde{\mu}_{lac}$ at different external pH values (Reprinted with permission from Zilberstein et al., *Biochemistry*, vol. 18, pp. 669-673. Copyright 1979 American Chemical Society.)

External pH	$\Delta\psi$ (mV)	ΔpH (mV)	$\Delta\tilde{\mu}_{H^+}$ (mV)	$\Delta\tilde{\mu}_{lac}$ (mV)	$\Delta\tilde{\mu}_{lac} : \Delta\tilde{\mu}_{H^+}$
6.0	102 ± 7.0	105 ± 1.1	207 ± 6.6	162 ± 5.8	0.78
8.0	152 ± 4.0	−12 ± 1.6	138 ± 4.6	142 ± 6.2	1.03

9.6.3 Distinguishing an H$^+$ Symport from an OH$^-$ Antiport

The data that support the operation of sugar/H$^+$ symport in bacteria are equally consistent with a sugar/OH$^-$ antiport because it is not possible to discriminate on thermodynamic grounds between these two mechanisms. An attempt to distinguish between these two possibilities utilized uptake of a positively charged hexose analog, 2-amino-2-deoxyglucose, by the hexose transport system of the green alga *Chlorella vulgaris* (Komar et al, 1983). At pH values greater than the pK of the 2-amino group where the neutral.

9.6 Active Transport of Sugars Coupled to H^+ Cotransport

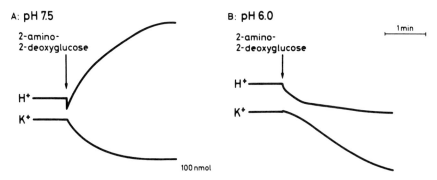

Figure 9.9. H^+ and K^+ movements caused by the addition of 2-amino-2-deoxyglucose to a suspension of *C. vulgaris* cells. An upward deflection indicates uptake of the ion (Komar et al., 1983).

unprotonated glucosamine is the dominant species, uptake proceeds with the alkalinization of the medium and charge compensating efflux of K^+ (Fig. 9.9A). In contrast, at pH values less than the pK, where the charged, protonated glucosamine is the only species present at appreciable concentration, no proton uptake is observed, but only efflux of H^+ and K^+ (Fig. 9.9B). These observations are most readily interpreted in terms of the protonated form of 2-amino-2-deoxyglucose binding at the site normally occupied by glucose *plus* a proton. As it is more difficult to imagine that a site on a glucose/OH^- antiport for binding the OH^- anion could be replaced by the positively charged protonated 2-amino group, it seems likely that the *C. vulgaris* hexose transport system functions as a proton symport rather than a hydroxide antiport (compare with Chap. 8, section 8.3.4.).

9.6.4 Transport of Negatively Charged Sugars and Carboxylic Acids

The schemes in Fig. 9.10 provide a useful framework for discussing the energetics of H^+ cotransport for neutral (Fig. 9.10A) and anionic substrates (Fig. 9.10B). In Fig. 9.10B, the substrate carrier protein (designated "C") is assumed to be electrically neutral at "low" pH (defined below) and to bind the substrate (of charge $-z$) and an integral number of protons, n (equal to the negative charge of the substrate), on the outside of the membrane. The ternary complex and the transport act are electrically netural because z negative charges (from the A^{-z} anion) and $n = z$ positive charges are transferred from the outside to the inside of the cell, so that $\Delta\psi$ cannot serve as an energy source for anion transport at low pH.

The model of Fig. 9.10B has been tested for a number of anionic substrates including the sugar derivatives gluconate and glucuronate, and also accounts for the energetics of carboxylic acid transport. For example, in *E. coli* mem-

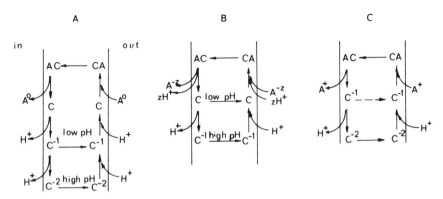

Figure 9.10. Models for H^+/substrate symports in bacteria (Rottenberg, 1976). (A) Neutral substrates; (B) anionic substrates; (C) cationic substrates.

brane vesicles transport of D-lactate at pH 5.5 was inhibited when nigericin was added to eliminate ΔpH but little inhibition was observed when $\Delta\psi$ was decreased by the addition of valinomycin (Ramos and Kaback, 1977a). At an external pH of 7.5, when $\Delta pH = 0$, D-lactate transport was inhibited by valinomycin but nigericin had no effect on transport. Furthermore, the extent of D-lactate accumulation at the steady state was proportional to ΔpH at pH 5.5 and to $\Delta\psi$ at pH 7.5. These results established that D-lactate transport must be transformed from an electroneutral to an electrogenic process as the pH is raised from 5.5 to 7.5. Additional evidence for the model of Fig. 9.10B comes from data on uptake of the anionic substrate L-lactate by *S. faecalis* where ^{31}P-NMR was utilized to measure changes in the internal pH of this bacterium associated with transport (Simpson et al., 1983). By quantitating the L-lactate-dependent proton uptake, the H^+:L-lactate stoichiometry could be measured directly and shown to increase from 1.1:1 at low pH to 2.0:1 at high pH (Fig. 9.11). At intermediate pH values the H^+:L-lactate stoichiometry varied as predicted for a transport protein having a single ionizable group with a $pK = 7.0$.

The model of Fig. 9.10B provides an explanation for the transition from electroneutral to electrogenic transport by postulating an ionizable group on the carrier protein that, in its ionized form, can bind one more proton than the neutral carrier protein. The charge on the carrier protein itself does not contribute to the net charge transferred across the membrane because to complete the catalytic cycle the unloaded carrier, after discharging sugar and proton(s) to the inside aqueous phase, must return to the membrane surface. Thus, at pH values greater than the pK of the ionizable group on the carrier protein, the proton to substrate stoichiometry increases to $(z + 1):z$ from the 1:1 stoichiometry operating at pH values less than the pK of this ionizable group. This increase in H^+:substrate stoichiometry makes substrate transport electrogenic at high pH, as $z + 1$ positive charges but only z negative charges

9.7 Kinetic Studies

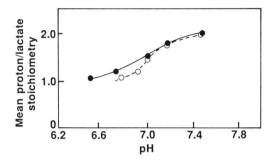

Figure 9.11. The H^+:L-lactate symport stoichiometry as a function of pH during L-lactate uptake by *S. faecalis* cells. [^{14}C]lactate uptake was quantitated after trapping the cells on membrane filters. L-lactate-induced H^+ uptake was measured by the ^{31}P-NMR detected change in internal pH, after correction for the internal buffering capacity. Glucose was supplied as a fermentable source of energy for L-lactate transport. Internal pH (o) external pH (●) (Simpson et al., 1983).

are translocated across the membrane, enabling transport at high pH to be driven by the $\Delta\psi$ maintained by respiration. A similar feature in the model for the transport of neutral substrates (Fig. 9.10A) can explain the increase in H^+:lactose transport stoichiometry observed in *E. coli* vesicles from 1 at low pH to 2 at high pH (Fig. 9.8).

9.7 Kinetic Studies

Some of the kinetic characteristics of lactose uptake in wild-type and transport mutants that establish constraints for possible transport mechanisms will be described. Before discussing these data, it is necessary to define some nomenclature used in such kinetic studies.

(i) *Active transport*. This term is used in kinetic studies in exactly the same fashion as in the definition supplied at the beginning of the chapter. That is, uptake of substrate is followed under conditions where energy is required and supplied. Cases in which the initial concentration of internal substrate is zero are often referred to as "zero-*trans*" conditions.

(ii) *Facilitated diffusion*. Diffusion across the membrane down a concentration gradient, catalyzed by a carrier.

Efflux. Cells or vesicles are loaded with substrate, diluted into a substrate-free medium, and the rate of efflux measured.

Equilibrium exchange (exchange diffusion). The rate of exchange of substrate at equilibrium (i.e., equal substrate concentrations present on both sides of the membrane) is followed using radioactively labeled substrate, initially confined to one side of the membrane. For example, vesicles can be equilibrated with [^{14}C]lactose, concentrated, and then diluted into medium containing an equal concentration of [^{12}C]lactose. It is often found that the rate of exchange exceeds the rate of transport (discussed below).

(v) *Counterflow.* Thermodynamically favorable substrate flow in one direction results in a transient accumulation of substrate flowing in the opposite direction. For example, if *E. coli* vesicles are loaded with a relatively high concentration of [^{12}C]lactose (e.g., 10 mM) and diluted into a medium containing a lower concentration (e.g., 0.5 mM) of [^{14}C]lactose, [^{14}C]lactose will accumulate inside the vesicles for a short period (1–5 min).

Of the above five kinetic processes, only active transport requires energy input or a $\Delta\tilde{\mu}$.

Figure 9.12 illustrates how these different transport modes can be as-

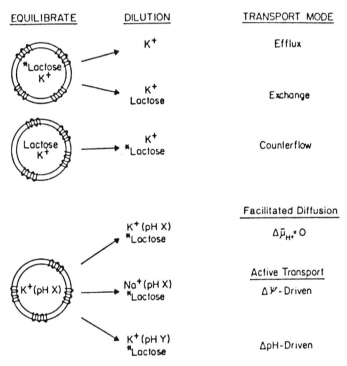

Figure 9.12. Schematic representation of *lac* permease activity assays. Proteoliposomes containing purified *lac* permease and radioactive (*) lactose, or no lactose, incorporated by freezing, thawing, and sonication, were treated with valinomycin. Aliquots of the suspension are then diluted into isoosmotic reaction mixtures containing phosphate buffer (potassium or sodium) with or without labeled or unlabeled lactose (dilution). For efflux and counterflow assays, [lactose]$_{in}$ > [lactose]$_{out}$ at the time of dilution, while the reverse is true for facilitated diffusion, $\Delta\psi$-, and ΔpH-driven uptake. For equilibrium exchange, [lactose]$_{in}$ = [lactose]$_{out}$. Active transport of lactose is driven by $\Delta\psi$ (interior negative) when [K$^+$]$_{in}$ > [K$^+$]$_{out}$ at the time of dilution or by a ΔpH (interior alkaline) when pH$_x$ > pH$_y$ (Viitanen et al., 1986).

9.7 Kinetic Studies

sayed using liposomes into which the purified *lac* carrier protein has been reconstituted.

The usefulness of these different manifestations of the activity of the carrier protein in deducing the mechanism can be better understood from the simple kinetic model for transport shown in Fig. 9.13A, where C represents the carrier protein, S the substrate (lactose in the case of the *lac* carrier), and the subscripts i and o represent the inside and outside of the vesicle, respectively. Binding and release steps for any protons cotransported with lactose are not shown. However, even this simplified model can explain why equilibrium exchange can be faster than transport. For example, in some transport systems the loaded carrier may move across the membrane more rapidly than the unloaded carrier ($k_{-2} > k_4$). In that case, the limiting step during the initial portion of active transport measurements made under zero-*trans* conditions could be the return of the unloaded carrier across the membrane from inside to outside after the dissociation of "S" at the inside membrane face. Under exchange conditions, internal substrate (present at fairly high concentration and ^{14}C-labeled) may bind to the carrier (the reaction characterized by rate constant k_{-3}) immediately after release of ^{12}C-substrate and the exchange be completed by the more rapid k_{-2} and k_{-1} steps, bypassing the slow k_4 state that must be utilized under the conditions of active transport.

The more complicated phenomenon of counterflow can be explained in terms of competitive inhibition. When vesicles have been preloaded with

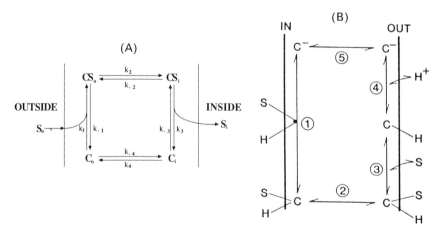

Figure 9.13. (A) Simple kinetic model for active transport (adapted from West, 1983.) (B) Schematic representation of reactions involved in lactose uptake, efflux, exchange and counterflow catalyzed by the lactose carrier protein. Information about the order of lactose and H$^+$ binding on the inner face of the membrane is not implied in this model (Kaback, 1988. Reproduced, with permission, from the Annual Review of Physiology, vol. 50. © 1988 by Annual Reviews Inc.) Compare with models of Lombardi, 1981; Lancaster, 1982; Wright et al., 1986; Page, 1987.

nonradioactive substrate, its high concentration will result in successful competition for the carrier protein at the inner membrane surface with ^{14}C-substrate (present initially at very low concentration) that has been transported from the outside aqueous phase and released to the interior space. This competition inhibits efflux of ^{14}C-substrate (via steps k_{-3}, k_{-2}, and k_{-1}) without inhibiting its influx because the concentration of external ^{12}C-substrate is initially zero, and results in a net initial accumulation of $[^{14}C]$ lactose inside the vesicle. The accumulation of ^{14}C-substrate is a transient phenomenon. Eventually, as efflux of the preloaded ^{12}C-subtrate occurs, the concentration of internal ^{12}C-substrate will no longer exceed that of external ^{12}C-substrate and the net ^{14}C-substrate uptake stops and then reverses. Counterflow experiments not only provide useful kinetic information, but are often used to prove the presence of a carrier-mediated transport or successful reconstitution of a transport system.

Extensive kinetic studies of lactose transport in *E. coli* have not yet produced a single, unambiguous kinetic model for the transport mechanism (Lombardi, 1981; Lancaster, 1982; Wright et al., 1986; Page, 1987). One model (Fig. 9.13B), formulated in the context of the chemiosmotic hypothesis, involves an ordered addition of substrates, with the proton binding to the unloaded carrier protein on the outside (Step 4) prior to lactose binding (Step 3). The fully loaded, neutral carrier ($H^+ \cdot C^- \cdot$ lactose) then crosses the membrane (Step 2) and releases the proton and lactose (Step 1). There is not yet sufficient data to determine whether substrate release is ordered. The catalytic cycle is completed by the return of the negatively charged unloaded carrier to the outside (Step 5).

Data supporting the model of Fig. 9.13B has come from studies of lactose efflux, exchange, and counterflow using *E. coli* vesicles preloaded with [^{14}C]lactose. Efflux, exchange and counterflow do not occur in vesicles prepared from uninduced *E. coli* cells or from *E. coli* y^- mutants and are inhibited by sulfhydryl reagents [e.g., *N*-ethylmaleimide (NEM)] that inactivate the lactose carrier protein. Efflux of lactose from *E. coli* vesicles produces a transient $\Delta\psi$ (inside negative), implying that efflux is an electrogenic process resulting from the net ejection of at least one H^+ for each lactose leaving the vesicles. The rate of exchange showed little dependence on external pH and was much faster than efflux, particularly at pH values < 7.5 (Kaback, 1986). Because, in the specific mechanism shown in Fig. 9.13B, the only steps involved in efflux but not in exchange are the return of the unloaded carrier from the outside to the inside and H^+ dissociation from the fully loaded carrier to the outside aqueous phase, one of these steps must be rate-limiting for efflux. The strong dependence of efflux rate on external pH, decreasing by more than 100-fold as the external pH was lowered from 9.5 to 5.5, implies that deprotonation of the carrier is a likely rate-limiting step. The *Principle of Microscopic Reversibility* (Tinoco et al., 1985) would require that if the mechanism involves an ordered release to the external aqueous phase of lactose followed by H^+ during efflux, then there must be an ordered binding of H^+_{out} followed by lactose$_{out}$ during transport.

9.8 Structure/Function Considerations

The nucleotide sequence of the *lac y* gene has been completely determined and the amino acid sequence of the molecular weight = 46,504 lactose carrier protein deduced from the nucleotide sequence (Wright et al., 1986; Kaback, 1986), allowing a prediction of some aspects of the protein's tertiary structure within the *E. coli* membrane. A hydropathy plot of the *lac* carrier protein suggested that the protein consists of 11 or 12 hydrophobic segments separated by shorter hydrophilic regions (Kaback, 1986). These hydrophobic regions, with an average length of 24 residues, make up 70% of the total polypeptide and are thought to be arranged in α-helical, membrane-spanning domains with an average length of approximately 36 Å, similar to that estimated for the hydrophobic core of the lipid-bilayer membrane. The predicted structure is consistent with a high α-helical content estimated from circular dichroism in the 200–240 nm wavelength region. A proposed structure for the *E. coli lac* carrier protein in the membrane is shown in Fig. 9.14. The sequence data and resulting structural models, together with kinetic studies on lactose uptake in *E. coli*, have led to preliminary structure/function models in this active transport system. A somewhat similar multispan structure has been proposed for the Na^+-melibiose carrier discussed below, and also for the H^+-arabinose and H^+-xylose transport proteins of *E. coli* (Maiden et al., 1988). The sequences of these proteins were found to also have a high degree of similarity to glucose transport proteins from human erythrocytes, a human hepatoma cell line, and rat brain cells (Mueokler et al., 1985; Maiden et al., 1987).

Some topographic experiments have been carried out to test the proposed membrane-spanning nature of the *E. coli* lactose carrier protein: Antibodies against the COOH-terminal region and the hydrophilic sections numbered 5 and 7 in Fig. 9.14 bind preferentially to inside-out vesicles, compared to right-side-out vesicles, suggesting that these segments of the lactose carrier protein are exposed on the side of the protein facing the cytoplasm (Wright et al., 1986; Kaback, 1986). Surface-exposed hydrophilic sections of the lactose carrier protein on both sides of the membrane are sites of protease attack using either right-side-out or inside-out vesicles (Wright et al., 1986; Kaback, 1986).

9.8.1 Modification of Specific Amino Acids of Lac Permease

An understanding of protein mechanisms at the molecular level involves knowledge of the identity of specific amino acid residues involved in the function or catalytic process. The first such studies on the *lac* permease demonstrated inhibition of transport by modification of the sulfhydryl group of Cys-148 (Wright et al., 1986; Kaback, 1986) of the *lac* carrier protein by NEM. Protection by substrates of the *lac* carrier protein against inactivation

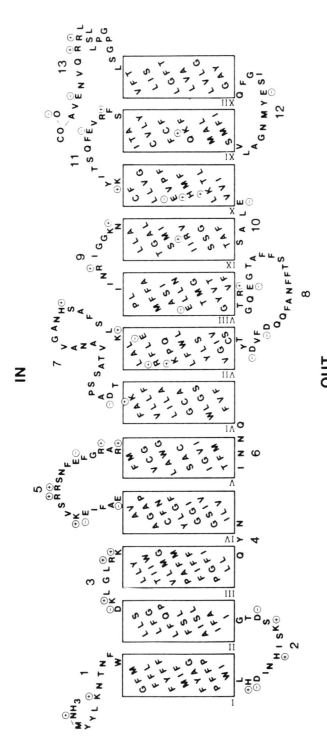

Figure 9.14. Primary and proposed secondary structure of the *lac* permease. Hydrophobic segments are shown in boxes as membrane-spanning helices and are connected by hydrophilic segments (Kaback, 1988. Reproduced, with permission, from the Annual Review of Physiology, vol. 50. © 1988 by Annual Reviews Inc.)

9.8 Structure/Function Considerations

by NEM led to the proposal of an essential cysteine residue at or near the active site of the protein. Using the technique of oligonucleotide-directed, site-specific mutagenesis, Cys-148 has been replaced by glycine or serine with little or no effect on lactose transport (Wright et al., 1986; Kaback, 1986), leading to the conclusion that although Cys-148 is required for substrate protection against NEM-inactivation, this amino acid residue is not essential for operation of the lactose/H^+ symport. Additional mutagenesis experiments indicate that replacement of seven of the eight cysteine residues with glycine had relatively little effect on transport activity (Kaback, 1988). Only alteration of Cys-154, which is located in the middle of proposed membrane spanning helix V in the model of Fig. 9.14, significantly inhibited (70, 90, and 100%, respectively for Val, Ser, and Gly substitutions) lactose transport.

Two amino acid residues have been identified that may be present at the substrate-binding site of the *lac* carrier protein (Brooker and Wilson, 1985). Eighteen independent *lac* carrier protein mutants were isolated that exhibited enhanced ability to transport the disaccharide maltose, a poor substrate for the wild-type carrier protein. All 18 mutations were found to be caused by single base pair changes either at Ala-177 or Tyr-236 and resulted in either the replacement of the former by Val or Thr, of the latter by Phe, His, Ser, or Arg. The mutants at Ala-177, located in the middle of helix VI (Fig. 9.15), gained the ability to transport maltose, but were similar to the wild-type in their ability to transport lactose. In contrast, the mutants of Tyr-236, located on the edge of, or in the outer aqueous phase outside, helix VII (Fig. 9.14), resulted in a marked decrease in glactoside transport. A simple hypothesis would be that Ala-177 and Tyr-236 are both present at the subtrate-binding site. The location of the lactose binding site within the hydrophobic portion of the protein and the membrane is in good agreement with: (i) The lack of accessibility to external proteases of the substrate-binding site of the *lac* carrier protein in right-side-out and inside-out vesicles (Kaback, 1976), and (ii) the absence of collisional quenching, by hydrophilic fluorescence quenchers, of the fluorescence of dansylated lactose analogs (dansylgalactosides), bound to the *lac* carrier protein (Mitaku et al., 1984). If Tyr-236 is to be located in the middle of membrane-spanning helix 7, close to Ala-177, the model of Fig. 9.14 would have to be modified so that helix 7 is altered, placing Tyr-236 near the bilayer center and Ala-177 (Fig. 9.15).

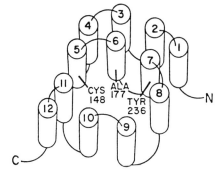

Figure 9.15. Hypothetical tertiary structure of the lactose carrier. The cylinders labeled 1 through 12 represent hydrophobic α-helical transmembrane segments. Cys-148, Ala-177, and Tyr-236 are located on segments, 5, 6, and 7, respectively (Brooker and Wilson, 1985).

9.8.2 Residues Involved in Lactose-H^+ Symport

Chemical modification of histidine residues by the agents diethyl pyrocarbonate or rose bengal had suggested His involvement in the H^+ symport function of the *lac* permease (Kaback, 1986). Three of the four histidine residues of the *lac* permease, those at positions 35, 39, and 205, can be replaced by site-directed mutagenesis without impairing the activity of *lac* permease (Kaback, 1987, 1988), implying that the remaining histidine, His-322, is the critical one. Substitution of Arg for His-322 resulted in the loss of lactose transport, efflux, exchange, and counterflow, activities that require H^+ binding and/or release by the permease. However, *lac* permease containing substitutions at position 322 has been shown to retain the ability to catalyze thermodynamically favorable lactose influx at high substrate concentrations without H^+ cotransport. An Arg-302 → Leu mutation has effects similar to those produced by replacement of His 322, while a Glu 325 → Ala mutation blocks all reactions involving net H^+ translocation. Based on these results, it was proposed that Arg-302, His-322, and Glu-325, located in the center of the membrane bilayer, may participate in a relay system, involved in translocating protons through the *lac* permease (Fig. 9.16). Although there is conflicting evidence about this specific proton relay system, this model has led to the design of informative experiments to test the involvement of specific amino acids in sugar/H^+ cotransport.

In these initial mutagenesis experiments, all steps involving protonation/deprotonation were impaired in mutations that replaced His 322. Thus it was proposed that His 322 is probably involved in Step 1 of the mechanism for lactose transport in *E. coli* outlined in Fig. 9.13B. Replacement of Glu 325 results in a *lac* permease defective in catalyzing all reactions involving net H^+ translocation but capable of catalyzing exchange and counterflow at wild type rates. The most likely explanation is that Glu 325 is involved in Step 4 (The Glu 325 → Ala mutant is unable to lose protons during efflux). The fact that replacement of Arg 302 produces effects on lactose transport similar to those resulting from replacement of His 322, suggests that Arg 302 may also be involved in Step 1, possibly acting prior to His 322 in the proton relay (Kaback, 1988). Additional mutagenesis experiments (Kaback, 1988) have established that His 322 and Glu 325 must have a specific order and configuration within the *lac* carrier protein and that residue 325 must have an acidic side chain (i.e., only aspartate can replace glutamate) for coupled H^+/lactose translocation to occur. However, it has since been demonstrated that His 322 → Phe and His 322 → Tyr mutations, while unable to carry out active accumulation of lactose or alternate substrates, can carry out galactoside-dependent H^+ transport (King and Wilson, 1989a, b). Thus, while His 322 is clearly important for energy transduction by the *lac* permease, it may not be directly involved as a proton carrier. Arg 302, His 322, and Glu 325 appear to be conserved in the *Klebsiella aerogenes* lactose carrier and several other transport proteins (the *E. coli* melibiose, arabinose, xylose, and citrate permeases and the *S. typhimurium* phosphoglycerate permease) have at least one

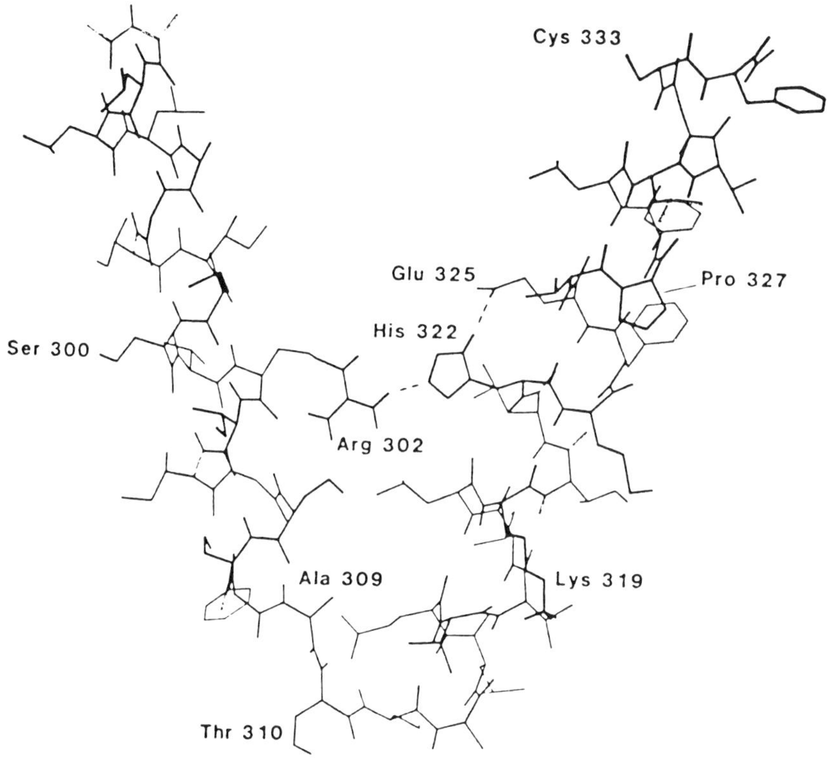

9.9 Amino Acid Transport

Figure 9.16. Molecular model of the proposed H$^+$ relay system in putative helices IX and X from model of Fig. 9.14 (From Kaback, 1988. Reproduced, with permission, from the Annual Review of Physiology, vol. 50. © 1988 by Annual Reviews Inc.)

possible His-Glu(Asp) pair, reinforcing the idea that such ion pairs may play a key role in coupling H$^+$ and substrate translocation.

9.9 Amino Acid Transport

The models of Figs. 9.10A and B that describe the energetics of proton symports for neutral and negatively charged substrates are applicable to amino acids as well as to sugars. There is considerable evidence for proton symport for both negatively charged amino acids (e.g., glutamate transport in *S. aureus*; Mitchell et al., 1979) and neutral amino acids (e.g., glycine and leucine in *E. coli*; Ramos and Kaback, 1977a). Because the experimental evidence for such proton/amino acid symports is quite similar to that described above for sugars and carboxylic acids, this section on amino acid transport will focus instead on a topic not yet covered, active transport of positively charged species.

The model of Fig. 9.10C can be used to treat uptake of any positively charged substrate (not only amino acids), with transport occurring either through a uniport, by direct coupling to $\Delta\psi$ without proton cotransport, or via a proton symport. In the former case, assuming that accumulation of the substrate S (of charge $+z$) is in equilibrium with $\Delta\tilde{\mu}_{H^+}$, the energetic relationship describing substrate accumulation is given (Chap. 1, section 1.11) by:

$$\log_{10} \frac{[S^{+z}]_{in}}{[S^{+z}]_{out}} = \frac{-z\,\Delta\psi}{Z}. \tag{4}$$

Uptake of aminoglycoside antibiotics by *S. aureus* appears to occur via a uniport (Mates et al., 1982).

If transport occurs via a proton symport (the pathway in Fig. 9.10C given by the solid arrows), the relationship between substrate accumulation and driving force would be given by

$$\log_{10} \frac{[S^{+z}]_{in}}{[S^{+z}]_{out}} = n\Delta pH - \frac{(n+z)\Delta\psi}{Z} \tag{5}$$

The $\Delta\tilde{\mu}_S$ of lysine measured at steady-state in *E. coli* vesicles was approximately twice the measured $\Delta\psi$ (Ramos and Kaback, 1977a), inconsistent with uptake of lysine solely via an electrogenic uniport, but in agreement with that predicted by Eq. 5 for the operation of a lysine/H^+ symport.

9.10 Sodium-Dependent Transport

Transport can be coupled to movement of ions other than protons. To date conclusive evidence has been obtained in bacteria for only one cotransported species other than the proton—the Na^+ ion. There have been demonstrations of Li^+/metabolite cotransport in bacteria but in all these cases Li^+ seems to be functioning as a Na^+ analog. Metabolite/Na^+ cotransport can be energetically favorable if the cells utilizing such Na^+ symports maintain a Na^+ gradient ($[Na^+]_{out} > [Na^+]_{in}$) so that the energy released as Na^+ is transported down its concentration gradient compensates for the energy required for substrate accumulation. An additional source of energy can arise from electrogenic coupling of substrate movement to Na^+/cotransport in the presence of $\Delta\psi$ (inside negative), so that the free energy, $\Delta G(Na^+)$, available for Na^+-linked transport, is proportional to the Na^+ electrochemical potential (Chap. 1, section 1.10).

$$\Delta G(Na^+) = n_{Na^+} \cdot \Delta\tilde{\mu}_{Na^+}, \tag{6}$$

where n_{Na^+} is the number of Na^+ ions cotransported per substrate molecule and

$$\Delta\tilde{\mu}_{Na^+} = F\Delta\psi - 2.3RT \cdot \Delta pNa \tag{7}$$

ΔpNa, in analogy with ΔpH, $= -\log_{10} \cdot \frac{[Na^+]_{out}}{[Na^+]_{in}}$.

9.10 Sodium-Dependent Transport

Equations, similar to those previously presented for proton-linked transport systems, can be derived from models analogous to those in Fig. 9.10 relating the extent of accumulation of substrates to the magnitudes of $\Delta\tilde{\mu}_{Na^+}$, $\Delta\psi$ and ΔpNa for substrates of different charge and for different Na^+:substrate stoichiometries. Na^+/substrate symports would function as shown in Fig. 9.4 for H^+/substrate symports, but with Na^+ replacing H^+.

9.10.1 Generation of $\Delta\tilde{\mu}_{Na^+}$

In mammalian cells where Na^+/substrate symports are quite common, both the ΔpNa and $\Delta\psi$ components of $\Delta\tilde{\mu}_{Na^+}$ are maintained largely by a Na^+, K^+-ATPase that couples the efflux of $3Na^+$ and concomitant uptake of $2K^+$ to the hydrolysis of an ATP molecule (Chap. 8, section 8.9). In bacterial cells, the major route for establishing the Na^+ gradient is exchange of internal Na^+ for external H^+ via Na^+/H^+ antiports (Krulwich, 1983). Although the localized vs. delocalized nature of energy-linked H^+ translocation is a matter of discussion (Chap. 3, section 3.11), considerable evidence indicates that Na^+ ions inside cells and organelles are largely free or delocalized (Lanyi, 1979).

9.10.2 Na^+/H^+ Antiports

The existence of Na^+/H^+ antiports in bacteria was suggested by energy-dependent extrusion of Na^+ from cells or vesicles of *S. faecalis*, *E. coli* and other bacteria that was eliminated by protonophoric uncouplers (Krulwich, 1979). *E. coli* vesicles, loaded with $^{22}Na^+$, exhibit Na^+ efflux when energy is supplied by respiration (Fig. 9.17) but not in the absence of a respiratory substrate or in the presence of such uncouplers.

Inverted *E. coli* vesicles have proven useful in the study of the Na^+/H^+ and other cation/H^+ antiports (Beck and Rosen, 1979) because: (i) The K_m for Na^+ can be readily measured since the $[Na^+]$ of the external medium can be conveniently varied. (ii) The reversed sidedness of the vesicles exposes the F_1 portion of the ATPase to the external medium so that the hydrolysis of exogenous ATP can be used as an energy source for Na^+/H^+ exchange. (iii) the metal cation specificity could be determined and it was shown that only Li^+ could substitute for Na^+. (iv) The H^+ movements (in the opposite direction to the Na^+ movements) predicted for the action of an antiport could be followed (Chap. 1, section 1.11). Addition of Na^+ to inverted vesicles partly collapses the ΔpH maintained by respiration. Because Na^+ addition to respiring, inverted *E. coli* vesicles results in H^+ efflux as $^{22}Na^+$ is taken up by the vesicles, the combined data document both of the requirements for a Na^+/H^+ antiport—the movements of the two ions, but in opposite directions. The ability of valinomycin at alkaline pH to collapse $\Delta\psi$ and to eliminate Na^+ efflux by the respiring vesicles (Fig. 9.17B) indicated that the *E. coli* Na^+/H^+ antiport can be electrogenic, with $Na^+:H^+ > 1$.

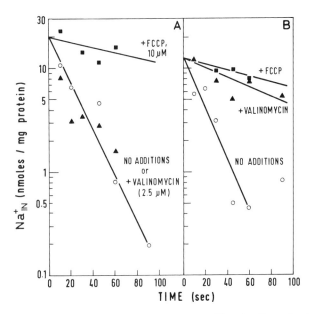

Figure 9.17. Effect of uncoupler and valinomycin on $^{22}Na^+$ efflux, via the Na^+/H^+ antiport in respiring *E. coli* vesicles, with D-lactate as the electron donor. (A) pH 6.6. (B) pH 7.5. (Reprinted with permission from Schuldiner and Fishkes, *Biochemistry*, vol. 17, pp. 706–711. Copyright 1978 American Chemical Society.)

Progress in elucidating the mechanism of the antiport protein in bacterial Na^+ transport should be facilitated by the recent sequencing of a gene that is likely to code for the *E. coli* Na^+/H^+ antiport or one of its subunits (Karpel et al., 1988). The MW = 38,683 protein contains 10 putative transmembrane helical segments.

In addition to its role in creating the Na^+ gradient required to make Na^+/substrate symports energetically favorable, Na^+ efflux via the Na^+/H^+ antiport may also play a major role in maintaining the internal pH relatively constant in many bacteria (Krulwich, 1983; Booth, 1985). Because of the small internal volume of the bacterial cells, continued H^+ efflux by the respiratory pump could lead to an internal pH sufficiently alkaline to interfere with the activity of many cytoplasmic enzymes. Exchange of external protons for internal metal cations would preserve $\Delta\psi$ as an energy source while preventing the internal pH from rising excessively. Using Na^+ as the exchangeable cation, the antiport also effectively converts part of the $\Delta\tilde{\mu}_{H^+}$ produced by respiration or ATP hydrolysis into a $\Delta\tilde{\mu}_{Na^+}$ which is available as an energy source for Na^+-linked active transport. *E. coli* and the marine bacterium *Vibrio alginolyticus* also have K^+/H^+ antiports that may contribute to pH homeostasis (Plack and Rosen, 1980; Nakamura et al., 1984).

Considerable evidence has been obtained supporting the importance of the Na^+/H^+ antiport in regulating the internal pH of the extreme alkalophiles, *Bacillus alcalophilus* (Krulwich, 1983) and *Bacillus firmus* (Krulwich et al.,

9.10 Sodium-Dependent Transport

1982) which grow at external pH values between 9 and 11.5 with growth optima near 10.5. The internal pH varies only from 9.0 to 9.5 over this 2.5 unit range of external pH. Measurements similar to those shown in Fig. 9.17 have documented that *B. alcalophilus* cells and vesicles exhibit an energy-dependent exchange of external H^+ for internal Na^+ characteristic of a Na^+/H^+ antiport (Krulwich, 1983). Nonalkalophilic mutants of *B. alcalophilus* and *B. firmus* have been isolated that grow at external pH values between 5.0 and 9.0, but have lost the ability of the wild-type to grow at pH values >9.0 (Krulwich, 1983; Krulwich et al., 1982). The mutants have also lost energy-dependent Na^+ efflux from cells and Na^+-dependent H^+ efflux from inverted vesicles, and thus have been characterized as Na^+/H^+ antiport mutants. The fact that the loss of Na^+/H^+ antiport activity is accompanied by a loss in the ability to grow at alkaline external pH values is consistent with the key role of the antiport in pH homeostasis. In *E. coli*, an absolute requirement for Na^+ has been demonstrated for growth at pH 8.5 and a lowering of Na^+/H^+ antiport activity by mutation or inhibitor was shown to prevent growth of pH 8.5 McMorrow et al., 1989). A similar requirement for growth and pH homeostasis in *B. firmus* has been found (Krulwich et al., 1982). These results point to the importance of the Na^+/H^+ in pH regulation in neutrophiles such as *E. coli*, as well as in alkalophiles.

9.10.3 Respiration-Linked Na^+ Pump

Respiration appears to be directly coupled to Na^+ efflux without any involvement of $\Delta\tilde{\mu}_{H^+}$ in the marine bacterium *V. alginolyticus*, where respiration-dependent Na^+ efflux and $\Delta\psi$ (inside negative) were inhibited by uncouplers at alkaline pH (Tokuda and Unemoto, 1982), although at neutral or acidic pH Na^+ efflux and $\Delta\psi$ were uncoupler-sensitive. The uncoupler-insensitive Na^+ efflux at alkaline pH was abolished by respiratory inhibitors, while a large decrease in the internal ATP concentration caused by arsenate addition had no effect on Na^+ efflux, indicating that *V. alginolyticus* cells contain both a Na^+/H^+ antiport and a direct, respiration-driven Na^+ pump that operates only at alkaline pH values. The Na^+ translocation coupled to electron transport has been localized in the NADH:ubiquinone oxidoreductase portion of the *V. alginolyticus* electron transport chain (Tokuda and Unemoto, 1984; Chap. 3, section 3.10) and evidence has been obtained for similar Na^+-linked, NADH:ubiquinone oxidoreductases in other halotolerant bacteria (Udagawa et al., 1986; Ken-Dror et al., 1986) and also in the nonhalophilic anaerobe *Klebsiella pneumoniae* (Dimroth and Thomer, 1989).

9.10.4 Na^+-Dependent Metabolite Uptake

The best documented sugar/Na^+ symport in bacteria is that of the disaccharide melibiose. Na^+ causes a 100-fold stimulation in the uptake of TMG, which can serve as a nonmetabolizable analog for melibiose as well as for lactose, by *Salmonella typhimurium* cells (Stock and Roseman, 1971). Although

Li$^+$ could replace Na$^+$ in producing this stimulation of TMG uptake by *S. typhimurium*, all other cations tested were without effect. One effect of Na$^+$ was to lower the K_m of the *S. typhimurium* melibiose transport system for TMG, without any effect on the V_{max} for transport. The initial kinetics of Na$^+$ and TMG uptake could be used to estimate a 1:1 Na$^+$:sugar stoichiometry (Stock and Roseman, 1971). Respiring vesicles prepared from *S. typhimurium*, grown in the presence of melibiose to induce its transport system, accumulate TMG in a reaction stimulated by Na$^+$ or Li$^+$ ions, and also exhibited a TMG-dependent ^{22}Na$^+$ uptake, as would be expected for transport of the melibiose analog via a Na$^+$ symport (Tokuda and Kaback, 1977). TMG-dependent Na$^+$ uptake and a Na$^+$-dependent TMG uptake could be demonstrated with a K$^+$/valinomycin diffusion potential replacing respiration as the energy source for uptake, again as expected for electrogenic uptake of the Na$^+$ or Li$^+$ cation along with the neutral sugar.

Melibiose transport in *E. coli* displays characteristics similar to those observed with *S. typhimurium*. The amino acid sequence of the *E. coli* melibiose carrier protein has been deduced from the nucleotide sequence of the *mel B* gene (Yazyu et al., 1984). The melibiose carrier protein has a molecular weight, calculated from the *mel B* gene sequence, of 52,029 and is very hydrophobic, containing 70% nonpolar amino acids (Yazyu et al., 1984). A hydropathy profile suggests that the protein might be arranged in the membrane with 10 or 11 membrane-spanning, hydrophobic segments (Yazyu et al., 1984; Botfield and Wilson, 1988), an arrangement similar to that proposed for the *lac* carrier protein. Despite such possible structural similarities, there is a low degree of amino acid sequence homology between the two *E. coli* sugar transport proteins, even though the lactose and melibiose transport systems share some common sugar substrates. However, the melibiose carrier protein does contain two putative His-Glu pairs, reminiscent of the His 322-Glu 325 pair found in the *lac* carrier protein. It was observed that when one of these pairs, His 357-Glu 361, was disrupted by converting Glu 361 to either glycine or arginine using site-directed mutagenesis, H$^+$/melibiose counterpart was inactivated (Kaback, 1987).

A unique property of the melibiose carrier protein is its ability to utilize either H$^+$, Na$^+$, or Li$^+$ as the cotransported cation depending on which sugar substrate is transported. Nucleotide sequencing of a series of *mel B* mutants that have lost the ability to utilize H$^+$ as the coupling cation have associated the change in cation specificity to the replacement of Pro-122 by a Ser residue (Yazyu et al., 1985), the first identification of a specific amino acid residue involved in determining the cation specificity of a bacterial transport protein. Replacement of Leu 232 with Phe or Ala 236 with Thr or Val produces a similar phenotype (Kawakami et al., 1988). Amino acid sequences have been deduced for 70 melibiose carrier mutants with impaired TMG recognition. In all but one mutant, the affinity of the carrier for Li$^+$ was also affected (Botfield and Hastings, 1988). The fact that the mutations, which cluster in four regions of the protein, affect both sugar and cation specificity suggests an interaction

9.10 Sodium-Dependent Transport

between the two substrates at their binding site(s). The range of its coupling cations suggests that the melibiose carrier may represent a link between exclusively H^+-coupled and exclusively Na^+-coupled symport systems. Na^+/amino acid symports are dominant in bacteria that live in environments with high external Na^+ ion concentrations (marine bacteria and halophiles such as *Halobacterium halobium*) and in extremely alkaline environments. It is presumably because of the favorable Na^+ gradient ($[Na^+]_{out} > [Na^+]_{in}$) that all twenty common amino acids are transported in a Na^+ ion gradient-dependent manner in *V. alginolyticus* and all amino acids except cysteine are transported via Na^+ symports in *H. halobium* (Lanyi, 1979). As discussed above (section 9.10.2), extreme alkalophiles have growth optima at pH 10.5–11.0, and maintain a ΔpH of the opposite polarity (i.e., inside acidic) to that observed with neutrophiles (Guffanti et al., 1978). Use of Na^+ as the cotransported ion eliminates the unfavorable energy term that would be associated with H^+ cotransport in such cases and it is presumably for this reason that Na^+ symports appear to be so widespread in alkalophilic bacteria.

Alanine uptake, catalyzed by the alanine carrier protein isolated from the thermophilic bacterium PS3 and reconstituted into phospholipid vesicles, can be driven by either $\Delta\tilde{\mu}_{H^+}$ or $\Delta\tilde{\mu}_{Na^+}$ (Hirata et al., 1984). Thus, this permease may be another example of an evolutionary link between H^+- and Na^+-coupled transport systems, as may the isoleucine/valine carrier of the photosynthetic bacterium *Chromatium vinosum*, which acts as an amino acid/H^+ symport at pH values below 7.5 but can function as a Na^+ symport at higher pH values (Cobb and Knaff, 1985). An even more complex symport mechanism appears to apply to the *E. coli* glutamate carrier which appears to form a quaternary carrier·glutamate·H^+·Na^+ complex and catalyze the co-transport of both H^+ and Na^+ with glutamate (Fujimura et al., 1983).

9.10.5 Na^+ Transport Linked to Decarboxylation

A chemically different mechanism for Na^+ transport has been described, with coupling to enzyme-catalyzed decarboxylation reactions. Oxaloacetate decarboxylase from anaerobically-grown cells of the citrate-fermenting bacterium *Klebsiella aerogenes* was shown to be a biotin-containing enzyme (Dimroth, 1982a). All biotin-containing enzymes are known to involve enzyme-bound N^1-carboxybiotin intermediates and isotope exchange studies supported the involvement of a similar intermediate in the *K. aerogenes* oxalacetate decarboxylase (Dimroth, 1982a). The first step of the decarboxylation reaction:

$$\text{Oxaloacetate}^{2-} + \text{E-biotin} \rightleftarrows \text{Pyruvate}^- + \text{E-biotin-CO}_2^-$$

showed no requirement for Na^+ although the overall reaction exhibited an absolute dependence on Na^+. It was demonstrated (Dimroth, 1982a) that Na^+ was required for the second step in the reaction:

$$\text{E-biotin-CO}_2^- + \text{H}^+ \rightarrow \text{E-biotin} + \text{CO}_2$$

The significantly exergonic nature of oxaloacetate decarboxylation ($\Delta G° \simeq -7$ kcal/mol) and the fact that oxaloacetate decarboxylase is a membrane-bound enzyme in *K. aerogenes*, along with the Na^+ dependence of the decarboxylation, suggested that the energy released during the reaction might be utilized for Na^+ transport. As would be expected if such a Na^+ efflux system were present in cells, inverted vesicles prepared from *K. aerogenes* accumulated Na^+ when supplied with oxaloacetate (Dimroth, 1982b). This Na^+ uptake was completely inhibited when oxaloacetate decarboxylation was stopped by blocking the biotin prosthetic group with avidin. Proton-conducting ionophores had no effect on Na^+ uptake linked to oxaloacetate decarboxylation in inverted *K. aerogenes* vesicles, indicating that a proton gradient was not involved. The Na^+ flux catalyzed by this system (Fig. 9.18) is electrogenic and is able to generate a $\Delta\psi = 65$ mV (inside positive in the inverted vesicle, corresponding to inside negative in whole cells) in addition

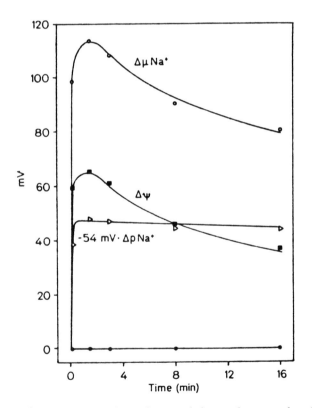

Figure 9.18. Na^+ uptake and $\Delta\psi$ formation coupled to oxaloacetate decarboxylation in inverted vesicles of *K. aerogenes*. The reaction performed at 0°C, was initiated by the addition of oxaloacetate. ΔpNa was measured using $^{22}\text{Na}^+$ and $\Delta\psi$ was calculated from the distribution of [^{14}C]SCN$^-$. The (–●–) represents a control in which oxaloacetate was omitted. (From Dimroth, 1982b.)

to a Na$^+$ gradient with [Na$^+$]$_{in}$ > [Na$^+$]$_{out}$ = 8 (Dimroth, 1982b). The oxaloacetate decarboxylases from *K. aerogenes* and *Klebsiella pneumoniae* have been reconstituted into phospholipid vesicles which could utilize oxaloacetate decarboxylation to generate a Na$^+$ gradient (Dimroth, 1981; Dimroth and Thomer, 1988), and the gene encoding the biotin-containing α-subunit of the *K. pneumoniae* oxaloacetate decarboxylase sequenced (Schwarz et al., 1988). Two additional examples of linkage between Na$^+$ transport and decarboxylation reactions have been identified in the cases of the glutaconyl-CoA decarboxylase in *Acidaminococcus fermentans* and the methylmalonyl-CoA decarboxylase in *Veillonella alcalescens*. Formaldehyde oxidation to CO$_2$ in two species of methanogenic bacteria, although proceeding by different mechanisms than described above for biotin-dependent decarboxylations, has also been shown to be directly coupled to electrogenic Na$^+$ efflux (Kaesler and Schönheit, 1989).

9.10.6 Na$^+$-Motive ATPases

An ability to couple energetically favorable, inward movements of Na$^+$ to ATP synthesis via a Na$^+$-motive ATPase would be of obvious advantage to bacteria that live in environments characterized by relatively high Na$^+$ concentrations and that produce a significant $\Delta\tilde{\mu}_{Na^+}$ by any of the mechanisms described above. The presence of such a reversible Na$^+$-ATPase has been demonstrated in inverted *Propionigenium modestum* vesicles. These vesicles couple Na$^+$ uptake to ATP hydrolysis and can utilize the $\Delta\tilde{\mu}_{Na^+}$ generated by malonyl-CoA decarboxylation to drive ATP formation that is insensitive to protonophoric uncouplers, but is sensitive to the Na$^+$-inophore monensin (Hilpert et al., 1984). High internal [Na$^+$] inhibited respiration-driven ATP synthesis in *V. alginolyticus* while a ΔpNa ([Na$^+$]$_{out}$ > [Na$^+$]$_{in}$) supported synthesis of ATP in the absence of respiration, Δψ and ΔpH, implying that Na$^+$ instead of H$^+$ is the coupling ion for oxidative phosphorylation via and Na$^+$-motive ATPase in *V. alginolyticus* (Dibrov et al., 1986).

9.11 Transport Driven by High-Energy Phosphate Intermediates

Although the transport of a wide variety of substrates is coupled to ion gradients with no direct involvement of high-energy phosphate ($\sim$P) compounds, bacterial transport systems exist that utilize either ATP or phosphoenolpyruvate (PEP) as energy sources. Three types of $\sim$P utilizing transport systems will be described.

9.11.1 ATP-Driven K$^+$ Uptake

Bacteria generally maintain internal K$^+$ concentrations much higher than those in the surrounding medium, and the relatively high internal [K$^+$] levels appear to be required for maintenance of ribosomal structure and the activation of certain enzymes (Walderhaug et al., 1987). *E. coli* contain at least two major K$^+$ uptake systems (Walderhaug et al., 1987) that allow the cells to meet their requirements for high internal [K$^+$], one of which serves the additional function of allowing the cell to adjust its internal osmolarity to changes in the external osmotic strength (Epstein, 1985). Bacteria generally maintain an internal osmolarity higher than that of the surrounding medium, with the higher internal [K$^+$] contributing most of the difference, except at very high external osmolarities (see below). There is considerable evidence indicating that the turgor pressure arising from this difference in osmolarity is essential for normal bacterial growth (Epstein and Laimins, 1980). The cell walls of bacteria are of sufficient strength to enable the cells to maintain large osmotic gradients.

Of the two major *E. coli* K$^+$ uptake systems, perhaps the best studied is the high affinity ($K_m = 2 \mu M$) *Kdp* system. The high affinity of the *Kdp* system for K$^+$ allows the cells to scavenge the K$^+$ necessary for growth under conditions where external [K$^+$] is extremely low. The *E. coli Kdp* system consists of three intrinsic inner membrane proteins, *KdpA*, *KpdB*, and *KpdC*, with molecular weights, calculated from nucleotide sequences, of 59,189, 72,112, and 20,267, respectively (Epstein, 1985). The genes for these three structural proteins are arranged as an operon and two additional genes, *KdpE* and *KdpF*, immediately downstream from the operon, are involved in the regulation of *Kdp* structural gene expression. (Epstein, 1985). Genetic expression of the *Kdp* system is apparently regulated by turgor rather than by external osmolarity per se (Higgins et al., 1987). Loss of the osmolarity gradient (turgor), induced by an increase in external osmolarity, leads to a large uptake in K$^+$ that results from expression of the *Kdp* operon. However, if internal K$^+$ levels were to rise too high, the resulting increase in internal ionic strength and/or [K$^+$] could be deleterious. The cell is protected against this possibility by the fact that high ionic strength shuts down further K$^+$ uptake and induces the expression of the genes coding for the uptake of the "compatible solutes" proline (Csonka, 1982) and betaine (Higgins et al., 1987). Betaine and other compatible solutes, which protect proteins from damage at high ionic strength by poorly understood mechanisms, function as osmoprotectants in plants as well as in bacteria.

Studies with *E. coli Kdp* mutants showing altered K_m values for K$^+$ suggest that the *KdpA* subunit contains the initial binding site for K$^+$ ions that have entered the periplasmic space through access provided by porin in the outer membrane. The energy for the subsequent transport and accumulation of K$^+$ is supplied by the hydrolysis of cytoplasmic ATP (Epstein, 1985). The identification of the *KdpB* subunit as the site of the K$^+$-dependent ATPase activity is based on the observation that a phosphorylated intermediate is formed during ATP hydrolysis and that the γ-phosphoryl residue of ATP is covalently linked to the *KdpB* subunit as an acyl-phosphate intermediate (Epstein, 1985). There is significant local amino acid sequence identity between the *E. coli*

9.11 Transport Driven by High-Energy Phosphate Intermediates

KdpB subunit and the sarcoplasmic reticulum Ca^{2+}-ATPase (Chap. 8, section 8.9). The specific amino acid phosphorylated in the *KdpB* subunit has not yet been identified. However, in the sacroplasmic reticulum Ca^{2+}-ATPase an aspartate residue has been identified as the phosphorylation site and this phosphorylated aspartate lies in a region that displays sequence homology with the *E. coli KdpB* subunit (Fig. 8.25). The sequence homologies raise the possibility of common evolutionary origin and mechanistic similarity between these bacterial and mammalian cation transport systems.

An analysis of the distribution of nonpolar amino acids suggests that both the *KdpA* and *KdpB* proteins contain several membrane-spanning domains and this feature, together with the proposal that the *Kdp* structural proteins may function as a dimer (Epstein, 1985), are incorporated into a schematic representation of the *Kdp* system (Fig. 9.19).

The second major *E. coli* K^+ uptake system, the *TrkA* system, has much lower K^+ affinity ($K_m = 1.5$ mM) than the *Kdp* system, but has a very high turnover number, allowing it to fill the entire K^+ pool of an *E. coli* cell in a few minutes (Epstein and Laimins, 1980). Unlike the repressible *Kdp* system, which is generally not found in cells grown with appreciable amounts of K^+ in the medium, the *TrkA* system is constitutive in *E. coli*. The *TrkA* system is the major vehicle for K^+ uptake by *E. coli* under most conditions. Genetic analysis suggests that as many as six genes are involved in the *E. coli TrkA* system. The energetics of the *E. coli TrkA* system are also complicated, with both $\Delta\tilde{\mu}_{H^+}$ and ATP required for the accumulation of K^+ (Walderhaug et al., 1987).

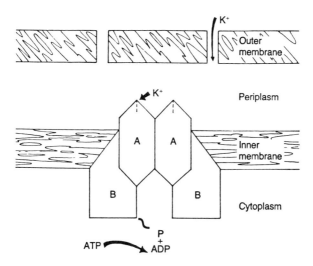

Figure 9.19. Diagram of the *E. coli* Kdp K^+ transport system containing the *KdpA* protein (A) that contain the initial binding site for site for K^+, and the KdpB protein (B) that contains the ATP hydrolysis site, both arranged as a dimer. The *KdpC* subunit is not included because its function is less well studied (Epstein and Laimins, 1980).

9.11.2 The Phosphotransferase System of Sugar Uptake

The overall reaction catalyzed by bacterial phosphotransferase systems (PTS) (Postma and Lengler, 1985) can be summarized by:

$$\text{Sugar}_{out} + \text{Phosphoenolpyruvate}_{in} \rightarrow \text{Sugar-phosphate}_{in} + \text{Pyruvate}_{in}$$

PTS are widespread among different bacterial genera and catalyze the uptake of a number of sugars including glucose, fructose, mannitol, and glucosamine. Phosphorylation of most sugar substrates of PTS occurs at the terminal or primary hydroxyl group (e.g., glucose-6-phosphate arises from glucose as the substrate), but some fructose PTS produce the sugar-1-phosphate (Postma and Lengler, 1985). The reaction occurs in a series of sequential steps, the first two of which occur in the cytoplasm and involve the phosphorylation of a soluble, low-molecular-weight (7.7–9.0 kDa) protein, known as the HPr protein, with PEP serving as the phosphoryl group donor (Postma and Lengler, 1985). The phosphorylation of HPr itself occurs in two sequential steps:

$$\text{PEP}_{in} + \text{E}_{Iin} \rightarrow \text{Pyruvate}_{in} + \text{E}_I \sim \text{P}_{in}$$

$$\text{E}_I \sim \text{P}_{in} + \text{HPr}_{in} \rightarrow \text{E}_{Iin} + \text{HPr} \sim \text{P}_{in}$$

and is catalyzed by "Enzyme I" (E_I). Both E_I and HPr appear to be constitutive in all PTS-containing bacteria.

The site of phosphorylation on HPr is the N-1 nitrogen of the imidazole group of a specific histidine residue of the protein (Postma and Lengler, 1985). The equilibrium constant for the phosphorylation of the HPr protein by PEP has been determined for the *S. aureus* and *S. typhimurium* proteins, allowing the calculation of $\Delta G^{\circ\prime}$ for the hydrolysis of the HPr phosphohistidine residue (Postma and Lengler, 1985). The values determined, e.g., -13.4 kcal/mol for *S. aureus*, classify HPr $\sim$ P as a high-energy compound (Chap. 1, section 1.14). However, because $\Delta G^{\circ\prime}$ for the hydrolysis of PEP is -14.8 kcal/mol, phosphoryl group transfer from PEP to HPr is thermodynamically favorable. The soluble Enzyme I has been highly purified from a number of bacteria, including *E. coli*, *S. typhimurium*, and *S. aureus* (Postma and Lengler, 1985), with M_r values ranging from 57,000 to 70,000. The reaction catalyzed by Enzyme I has been shown to involve a phospho-Enzyme I intermediate, with N-3 of a specific histidine residue in Enzyme I being phosphorylated by PEP and, in turn, phosphorylating HPr.

The sugar specificity of phosphotransferase systems is introduced at the next stage of the process and involves either one or two additional protein components, Enzymes II and III (Fig. 9.20). In PTS for mannitol, only Enzyme II is involved, serving both as the specific sugar recognition component and catalyzing the final transfer of the phosphoryl group from HPr $\sim$ P to sugar as shown below:

$$\text{HPr} \sim \text{P} + \text{Enzyme II} \rightarrow \text{HPr} + \text{phospho-Enzyme II}$$

$$\text{Phospho-Enzyme II} + \text{Sugar}_{out} \rightarrow \text{Enzyme II} + \text{Sugar-phosphate}_{in}$$

9.11 Transport Driven by High-Energy Phosphate Intermediates

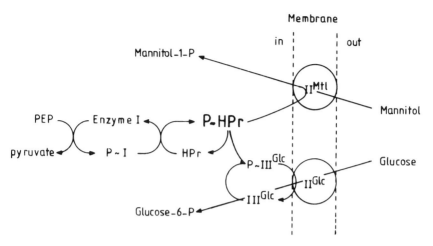

Figure 9.20. The bacterial PTS system. Enzyme I and HPr are the proteins shared by all PTS systems. Only two of the many different enzymes II are shown. II^{Mtl} is specific for mannitol while II^{Glc} and III^{Glc} are specific for glucose. P $\sim$ I, P $\sim$ HPr and P $\sim$ III^{Glc} are the phosphorylated forms of these proteins (Postma and Lengler, 1985).

In PTS that contain only Enzyme II, there appear to be two phosphorylation sites on the enzyme, with N-3 of a histidine at one site being phosphorylated by HPr $\sim$ P. The phosphoryl group is probably then transferred to the N-1 of a histidine at the second site, and subsequently to the sugar substrate (Saier et al., 1988). There exist different Enzyme II's for the different PTS-transported sugars (Postma and Lengler, 1985). Those for sorbitol, mannitol and fructose are inducible in most bacteria, while the E_{II}'s involved in the uptake of glucose, mannose, and, in some bacteria, for fructose are constitutive (Postma and Lengler, 1985). Enzyme II is an integral membrane protein regardless of the bacterial species and sugar involved. The mannitol-specific Enzyme II has been purified to homogeneity in a reconstitutably active form (Jacobsen et al., 1983), and its nucleotide sequence as well as the predicted molecular weight, 67,893, of the gene product have been determined (Lee and Saier, 1983). A number of additional Enzymes II and Enzyme II–III pairs have been sequenced, allowing inferences to be drawn about conserved phosphorylation sites and evolutionary relationships. This analysis suggests that Enzyme II in systems lacking an Enzyme III is the functional equivalent of the Enzyme II plus Enzyme III pairs in systems containing both proteins (Saier et al., 1988).

In the PTS for glucose, the phosphoryl group transfer from HPr to Enzyme II does not occur directly (Fig. 9.20) but involves Enzyme III (Postma and Lengler, 1985):

$$\text{HPr} \sim \text{P} + \text{Enzyme III} \rightarrow \text{HPr} + \text{Phospho-Enzyme III}$$

$$\text{Phospho-Enzyme III} + \text{Enzyme II} \rightarrow \text{Enzyme III} + \text{Phospho-Enzyme II}$$

$$\text{Phospho-Enzyme II} + \text{Sugar}_{out} \rightarrow \text{Enzyme II} + \text{Sugar-phosphate}_{in}$$

Depending on the bacterial species and sugar, Enzyme III can either be a soluble or a peripheral, membrane-associated protein (Postma and Lengler, 1985). The *S. typhimurium* glucose-specific, Enzyme III has been purified to homogeneity (Scholte et al., 1981), the gene for the 18,556 molecular weight protein sequenced (Nelson et al., 1984) and the site of phosphorylation shown to be the N-3 position of His-91 (Postma and Lengler, 1985).

Sequence comparisons of eight PTS permeases (Enzyme II or II–III pairs) reveal homologies in the carboxyl termini that have been interpreted in terms of a common site of interaction on the enzymes for phosphorylated HPr (Saier et al., 1988). A pair of conserved putative phosphorylation sites was detected in the five most similar permeases. One site, proposed to be that in either Enzyme II or Enzyme III phosphorylated by HPr $\sim$ P, was located approximately 80 amino acids from the carboxyl terminus. The second site, proposed to donate the phosphoryl group to the sugar substrate, varied in location from 320 to 440 residues from the carboxyl terminus (Saier et al., 1988).

9.12 Periplasmic Transport Systems

The periplasmic space, located between the cell wall and the plasma membrane in gram-negative bacteria, contains a considerable amount of soluble protein that can comprise as much as 15% of the total cell protein (Ames, 1986). Subjecting the cells of many species of Gram-negative bacteria to relatively mild osmotic shock causes release of these "shockable" periplasmic proteins, including proteins involved in the uptake of a number of sugars (maltose, arabinose, ribose, galactose and xylose), the uptake of sulfate and of some amino acids (e.g., histidine and lysine/arginine/ornithine in *S. typhimurium*, and glutamine and leucine/isoleucine/valine in *E. coli*). These systems all transport their substrates without any covalent modification and can accumulate substrates to concentrations in excess of 10^5-fold greater than that in the external medium (Ames, 1986). In the case of several periplasmic binding-proteins (e.g., galactose, ribose, and maltose), the same protein responsible for initial substrate binding during active transport is also responsible for the initial substrate binding and recognition in the chemotactic response (Koshland, 1981). However, the membrane-bound proteins involved in chemotaxis differ from those involved in transport (Koshland, 1981). For same time, it was thought that both $\Delta\tilde{\mu}_{H^+}$ and a high energy phosphorylated compound, ATP or acetyl phosphate, played a role in supplying energy in such systems (Hong et al., 1979; Ames, 1986). However, data obtained for histidine transport in right-side-out *S. typhimurium* vesicles reconstituted with the histidine binding protein (Prossnitz et al., 1989), for histidine transport in inside-out vesicles (Ames et al., 1989) and for histidine and maltose transport in *E. coli* cells (Jashi et al., 1989) have established that ATP hydrolysis provides the energy for uptake. It has also been shown that at least one of the mem-

9.12 Periplasmic Transport Systems

brane-bound components of the shockable *S. typhimurium* histidine transport system contains an ATP-binding site (Hobson et al., 1984).

The genes for the *S. typhimurium* high-affinity histidine uptake system constitute an operon, and code for four proteins, a single soluble, periplasmic protein responsible for the initial binding of histidine, and three proteins that are membrane-bound (Ames, 1986). This pattern of one soluble protein that accomplishes the initial recognition and binding of the substrate in the periplasmic space and a number of additional membrane-bound proteins appears to be common to many periplasmic transport systems, such as those for maltose, galactose and arabinose, and for branched-chain, nonpolar amino acids. Presumably, the substrate is transferred from its binding site on the periplasmic binding protein to binding sites on the membrane-bound protein components (Fig. 9.21). Such substrate binding sites have been identified on membrane-bound protein components in the *E. coli* maltose transport system (Trepten and Shuman, 1985).

Considerably more is known about the periplasmic, substrate-binding proteins than about the membrane-bound proteins, with both detailed structures and substrate-binding kinetic parameters available for the former. In the case of the *S. typhimurium* periplasmic histidine-binding protein, J, the binding of the substrate L-histidine to the protein is both a rapid (in vitro bimolecular rate constant for binding = 1.0×10^8 M^{-1} s^{-1}) and high-affinity ($K_D = 3 \times 10^{-8}$ M) process (Miller et al., 1983), a common feature of all the periplasmic binding protein systems studied to date (Miller et al., 1983). NMR and fluorescence studies have demonstrated that L-histidine binding results in a conformational change in the J protein (Ames, 1986). Only the conformation present when substrate is bound can interact productively with the membrane-bound components of the histidine transport system (Ames, 1986). Cross-linking experiments (Prossnitz et al., 1988) and genetic analysis suggest the J protein may interact with two of the three membrane-bound compnents of the histidine transport system, the Q and P proteins.

Genetic analysis of a series of *S. typhimurium* mutants indicated that the single polypeptide chain of the J protein has two functional domains, one involved in L-histidine binding and the other in binding to the membrane-bound "P" protein (Ames, 1986) *S. typhimurium* contains another periplasmic transport system for the basic amino acids lysine, arginine and ornithine that contains a single periplasmic binding protein, the LAO protein, that is highly homologous to the J protein and the same three membrane-bound components involved in histidine uptake (Ames, 1986). Regions of the J and LAO proteins implicated in binding to the P protein show a much higher identity (>90%) than the overall amino acid sequences of the proteins (70% identity). The high degree of sequence identity of these two *S. typhimurium* periplasmic binding proteins, and the fact that the genes for the two proteins are closely linked on the bacterial chromosome, indicate that the J and LAO proteins arose via gene duplication and subsequent divergence (Ames, 1986). This also appears to be true for the two periplasmic proteins involved in leucine-binding

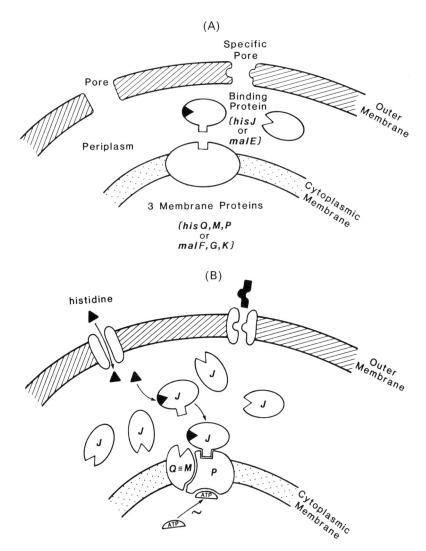

Figure 9.21. Models for "shockable" transport systems. (A) Substrates enter the outer envelop either through a non-specific pore (e.g., histidine) or through a substrate-specific pore (e.g., maltose). (B) Binding of substrate to the periplasmic binding proteins causes a conformational change in the periplasmic binding protein which allows the protein to bind, in turn, to one of the membrane-bound proteins of the transport system. One of the membrane-bound protein constituents has a binding site for ATP and at least one of the other membrane-bound constituents appears to contain a substrate binding site. Measurements of ATP binding suggest that the membrane-bound component, P, is most likely to contain the ATP binding site (From Ames, 1986. Reproduced, with permission, from the Annual Review of Biochemistry, vol. 55. © 1986 by Annual Reviews Inc.)

9.12 Periplasmic Transport Systems

and leucine/isoleucine/valine-binding (Landick et al., 1980). Possible evolutionary relationships between periplasmic transport systems are suggested by the considerable amino acid sequence identities of the corresponding membrane-bound components (Table 9.2).

Considerable detail is avilable on the interactions between substrates and periplasmic binding proteins in the cases of those binding proteins for which high resolution (1.7–3.0 Å) structures are available: (i) The *E. coli* leucine/isoleucine/valine-binding protein (Sack et al., 1989); (ii) the *S. typhimurium* sulfate-binding protein (Pflugarth and Quiocho, 1985); (iii) the *E. coli* L-arabinose-binding protein (Quiocho, 1986); and (iv) the *E. coli* and *S. typhimurium* D-galactose-binding proteins (Quiocho, 1986). The proteins all contain a single polypeptide chain and have similar molecular weights, 31–37 kDa. A striking feature of these proteins is the similarity of their tertiary structures in spite of the small degree of amino acid sequence identity. All of the proteins are elongated (with axial ratios near 2.0) and are composed of two approximately globular domains, containing a central core of β-pleated sheet with at least one pair of helicies on each side of the sheet plane. Another important similarity between all these periplasmic binding proteins is that the single substrate binding site on each of the proteins is located in an essentially identical cleft between the two globular lobes. The cleft is much wider in the structures of proteins crystallized without bound substrate, compared to those crystallized in the presence of substrate. In the case of the *E. coli* arabinose-binding protein, the conformational change associated with substrate binding has been documented by the detection of a 1.0 Å change in the protein's radius of gyration upon substrate binding (Quiocho, 1986). Computer modeling studies suggest that the changes in radius for gyration can best be explained by an arabinose-induced closure of the cleft in which one of the globular lobes rotates as arabinose is bound relative to the other lobe about a hinge deep in the cleft (Fig. 9.22). The substrate in the cleft is almost completely buried and inaccessible to solvent. In the case of the *E. coli* arabinose-binding protein, $<3\%$ of the surface area of the bound substrate is exposed to solvent (Newcomer et al., 1981). Binding of the substrate is accompanied by expulsion of water and the substrates are bound without solvation shells. The absence of water may be of importance in the subsequent transfer of substrate through the membrane via the membrane-bound components of the transport systems. Arabinose binding involves an extensive array of hydrogen bonding interactions between the protein and substrate and the binding site of the protein possesses a novel geometry that can accommodate both the α- and β-anomers of arabinose (Quiocho, 1986). This geometry, which results in virtually identical binding kinetics and affinities for the two anomers, allows utilization of any arabinose present in the growth medium.

In the case of sulfate (Pflugarth and Quiocho, 1985), hydrogen bonds from the peptide backbone predominate in substrate binding so that the charged oxygens of sulfate are stabilized by hydrogen bonds rather than salt bridges. The highly directional nature of hydrogen bonds may, in part, explain their

Table 9.2 Alignment of conserved sequences of membrane-bound components of "shockable" bacterial transport systems

(Position 25 marked above the "-" after rbsA(N) ...L S; Site A and position 50 marked above G K S T M)

rbsA(N)	<1>	M E A L L Q L K G I D K A F P G V K A L S - G A A L N V Y P G R V M A L V G E N G A G K S T M - M K
rbsA(C)	<252>	A P G D I R L K V D N L C G P G V N D V S - - - - F T L R K G E I L G V S G L M G A G R T E L - M K
hisP	<2>	M S E N K L H V I D L H K R Y G G H E V L K G V S L Q A R A G D V I S I I G S S G S G K S T F - L R
malK	<1>	M A S V Q L Q N V T - - K A W G E V V V S K D I N L D I H E G E F V V F V G P S G C G K S T L - L R
pstB	<6>	T A P S K I Q V R N L N F Y Y G K F H A L K N I N L D I A K N Q V T A F I G P S G C G K S T L - L R
oppD	<16>	L L E V N D L R V T F A T P D G D V T A V N D L N F T L R A G E T L G I V G E S G S G K S Q S R L R

Site C

rbsA(N)	V L T G I Y T R D A G T L L W L G K E T T F T G P K S S Q E A - - G I G I I H Q E - - - L N L I P Q - L T
rbsA(C)	V L Y G A L P R T S G Y V T L D G H E V V T R S P Q D G L A N - - G I V Y I S E D R K R D G L V L G - M S
hisP	C I N F L E K P S E G A I I V N G Q N I N L V R D K D G Q L K V A D K N Q L R L L R T R L T M V F Q H F N
malK	M I A G L E T I T S G D L F I - G E K - - - - R M N D T P - - P A E - - - - - - - R G - V G M V F Q S Y A
pstB	T F N K M F E L Y P E Q - R A E G E - I L L - - D G D N I L T N S Q D - - I A L L R A K V G M V F Q K P T
oppD	- L M G L L A T N - G R I G G S A T F N G - - R E I L N L P E R E L N T - - - R R A E Q I S M I F Q D P M

rbsA(N)	I A E N I F - - L G - - R E F V N - R F G K I D W K T M Y A E A D K L L A K L N L R F K S D - - K L V G D
rbsA(C)	V K E N M S - - L T A L R Y F S R - A G G S L K H A D E Q Q A V S D F I L R F N V K T P S M E - Q A I G L
hisP	L W S H M T - - V L E N V M E A P I V L G L S K H - D A R E R A L K Y L A K V G I D E R A Q G - K Y P V H
malK	L Y P H L S - - V A E N - M S F G L P A G A K K E - V I N Q R V N Q - V A E V L Q L A H L L D R K - P K A
pstB	P F P M S I - Y D N I A F G V R L F K L S R A D M D E R V Q W A L T K A A L W N E T K D K L H - Q S G Y S
oppD	T S L N P Y M R V G E Q L M E V L M H K G M S K A E A F E E S V R M L D A V K M P E A R K R M K M Y P H E

9.12 Periplasmic Transport Systems 445

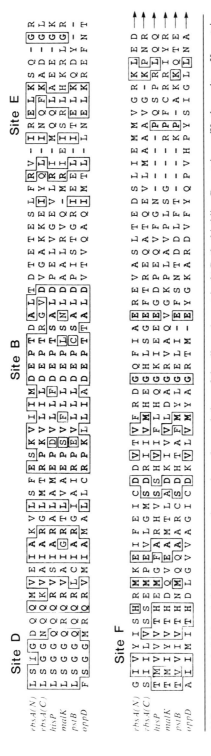

Abbreviations: *rbsA*(N) and *rbsA*(C) represent the two homologous halves of the ribose A protein; *hisP*, the histidine P protein; *malK*, the maltose K protein; *pstB*, the phosphate B protein and *oppD*, the oligopeptide D protein (From Ames, 1986. Reproduced, with permission, from the Annual Review of Biochemistry, vol. 55 © 1986 by Annual Reviews Inc.)

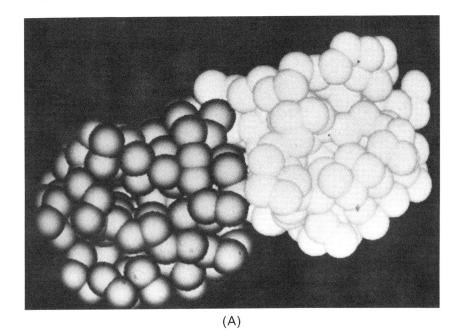

(A)

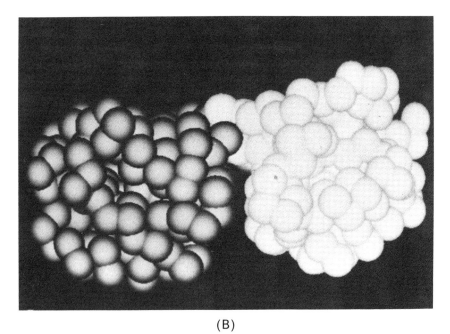

(B)

Figure 9.22. Space-filling models of the *E. coli* arabinose-binding protein. (A) The structure of the binding protein:arabinose complex. (B) The simulated structure for the "open" conformation of the unliganded binding protein (Newcomer et al., 1981).

predominance in that they allow for the necessary geometric specificity of the binding sites. An additional consideration is that hydrogen bond strengths lie in the range where they can provide sufficient stability for tight binding but are not so strong as to preclude rapid binding and reasonably rapid dissociation. Binding of a number of neutral sugars shows second-order rate constants in the range from 1 to 3×10^7 M^{-1} s^{-1} and K_D values in the range from 10^{-6} to 10^{-7} M (Miller et al., 1983). Dissociation rate constants obtained from in vitro measurements indicate that rates of release are also rapid enough to be consistent with the overall transport rates observed in vivo. Such considerations are important not only for rapid, high-affinity substrate accumulation but also for sensitive, rapid-response chemotaxis.

9.13 Motility

Although the detailed biochemistry of motility is in fact quite different in many ways from that of active transport, the two phenomena are at least partially related in the advantages they confer on the cell—if the cell is able to move towards a nutrient for which it possesses an active transport system, the combination of these two functions will clearly benefit the cell. Flagellated bacteria such as *E. coli* move by rotating their flagella as an intact flagellar bundle (Doetsch and Sjoblad, 1980). The flagella behave as semi-rigid propellers and are attached to a basal structure that acts as a motor. The motor consists of two parts, a "rotor" and a "stator" (Fig. 9.23). That stator portion consists of a set of protein subunits, arranged in a circular ring, lying in a plane approximately parallel to that of plasma membrane. The stator appears to be located between the plasma membrane and the peptidoglycan layer and is attached to, or partially embedded in, the membrane. The rotor consists of a second set of protein subunits arranged as a circular "plug" inside of and in close apposition to the stator. The remainder of the flagellum is attached by a central spindle linking the base of the flagellum to the rotor proper. The rotor, as the name implies, is able to rotate within the surrounding "collar" of the stator in the presence of a $\Delta\tilde{\mu}_{H^+}$. Rotation of the rotors, which can be in either the clockwise or counterclockwise direction, occurs about an axis perpendicular to the plane of the plasma membrane. The frequency of flagellar rotation is about 3,000 rpm. A reversal of the direction of motion of the bacterium can be accomplished by a reversal in the direction of rotation of the flagellar bundle.

A number of studies using inhibitors, uncouplers, and F_0F_1 ATPase mutants indicated that $\Delta\tilde{\mu}_{H^+}$, rather than ATP, serves as the energy source for motility in several flagellated bacteria. This was also demonstrated by the ability of a K^+/valinomycin diffusion potential to produce transient motility in approximately 30% of the cells in a population of starved, nonmotile *Bacillus subtilis* (Fig. 9.24). Simultaneous imposition of a ΔpH (outside acidic, created by rapid addition of HCl) increased the motile fraction to greater than 50%.

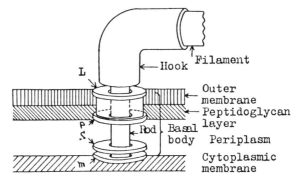

Figure 9.23. The structure of the flagellar basal body (Alexei Glagolev from V.P. Skulachev/P.C. Hinkle, *Chemiosmotic Proton Circuits in Biological Membranes*, © 1981, Addison-Wesley Publishing Co., Inc., Reading, MA. Reprinted with permission.)

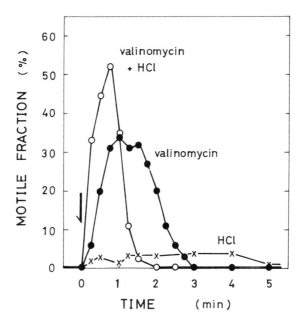

Figure 9.24. Transient motility induced in *B. subtilis* cells by an artificially generated $\Delta\tilde{\mu}_{H^+}$. Starved, nonmotile cells at pH 7.5 were mixed with valinomycin and/or sufficient HCl to lower the pH to 6.0 (Matsuura et al., 1977).

The observation that $\Delta\tilde{\mu}_{H^+}$ and swimming speed remained essentially constant in *B. subtilis* as the external pH is varied from 4.5 to 8.0 implies that both the ΔpH and $\Delta\psi$ components can be equally effective in providing the energy for flagellar rotation, since at pH 4.5 the $\Delta\tilde{\mu}_{H^+}$ of -130 mV consists only of a ΔpH term ($\Delta\psi = 0$), while at pH 7.5, $\Delta pH = 0$ and $\Delta\tilde{\mu}_{H^+}$ consists entirely of $\Delta\psi$ (-150 mV) (Shioi et al., 1980).

While the exact details of the mechanism of flagellar motion are still unknown, it seems clear that the torque developed to turn the flagellar rotor is a direct result of a transmembrane proton flux through one or more of the protein subunit components of the flagellar structure, driven by the $\Delta\tilde{\mu}_{H^+}$ generated by respiration or ATP hydrolysis (Glagotev, 1981; Mitchell, 1984). As in the case for active transport, marine and alkalophilic bacteria may utilize $\Delta\tilde{\mu}_{Na^+}$ instead of $\Delta\tilde{\mu}_{H^+}$ as the energy source for motility (Hirota and Imae, 1983; Chernyak et al., 1983) For example, motility by the marine bacterium *Vibro alginolyticus* has been demonstrated in the absence of $\Delta\tilde{\mu}_{H^+}$ (Chernyak et al., 1983). Under these conditions, motility required the presence of a Na^+ gradient, ($[Na]_{out} > [Na^+]_{in}$). Addition of monensin, which eliminated $\Delta\tilde{\mu}_{Na^+}$, paralyzed the cells. Motility in *Bacillus* was also inhibited by low concentrations of amiloride, with the inhibition competitive with respect to $[Na^+]$. Similar competitive inhibition patterns between Na^+ and amiloride have been observed for Na^+ channels, suggesting that the Na^+ interaction site(s) of the *Bacillus* flagellar motor may have some structural similarity to Na^+ channels (Sugiyama et al., 1988).

9.14 Active Transport in Eukaryotes

9.14.1 Introduction

Active transport in eukaryotic organisms shows many similarities to that in prokaryotes. Substrate/H^+ symports play an important role in eukaryotes and one of them, the mitochondrial phosphate carrier protein, has been discussed earlier (Chap. 3, section 3.9). Na^+/substrate symports are also widespread in eukaryotes. Perhaps the best studied examples come from mammalian amino acid and sugar transport systems operating across the intestinal epithelial membrane and in glucose transport by the kidney. Transport in these systems occurs with $\Delta\tilde{\mu}_{Na^+}$ providing the driving force for transport in precisely the same fashion as in uptake of melibiose by *E. coli* and *S. typhimurium* and in amino acid uptake by *H. halobium*. The only fundamental difference between the bacterial and mammalian systems is that in the former case, the $\Delta\tilde{\mu}_{Na^+}$ usually results from the action of a Na^+/H^+ antiport while in the latter case $\Delta\tilde{\mu}_{Na^+}$ arises from the action of an ATP-driven, electrogenic Na^+/K^+ exchange pump (Chap. 8, section 8.9). The discussion of secondary transport in eukaryotes will be limited to neurotransmitter transport, which has been chosen in order to illustrate some variations on the prokaryote chemiosmotic theme in terms of the coupling of substrate and ion transport.

9.14.2 Neurotransmitter Transport

Active transport of a number of neurotransmitters plays an important role in synaptic transmission in mammalian nervous systems (Kanner, 1983). The process can be viewed as proceeding in three steps beginning with the release of the transmitter into the gap junction and followed by the binding of the transmitter to specific postsynaptic receptors. The third step, the removal of the neurotransmitter from the synapse, is where active transport of the neurotransmitter occurs. These active transport systems are able to achieve and maintain very high concentration gradients, $\sim 10^4$, so that the synaptic junction is kept essentially free of neurotransmitter at times when there is no action potential releasing the neurotransmitter. It is of particular interest that several neurotransmitter active transport systems appear to bind or transport more than one ion along with the neurotransmitter and that at least two ions other than H^+ and Na^+, the Cl^- and K^+ ions, are involved in several of these systems. Another important difference between eukaryotic transport and the prokaryotic systems described above occurs in the chromaffin granules, where substrate transport is coupled to an H^+ antiport rather than H^+ symport. To date, all studies carried out with prokaryotes indicate that protons are exchanged only for inorganic cations and not for organic substrates.

Biogenic Amine Transport in Chromaffin Granules

The presynaptic termini contain specialized organelles for the storage of biogenic amine neurotransmitters such as epinephrine, noradrenaline, and dopamine (Kanner, 1983). Sequestering the biogenic amines in these storage organelles subsequent to their removal from the synaptic junction gap appears to provide several advantages including: (i) facilitation of neurotransmitter removal from the gap by decreasing the concentration gradient of amine across the neuronal membrane; (ii) preserving the transmitter in a storage pool for subsequent release after a later nerve action potential; and (iii) providing an acidic environment in the interior of the storage organelle to chemically stabilize these neurotransmitters. The best-studied of these storage organelles is the catecholamine-storing chromaffin granule of the adrenal medulla. Chromaffin granules have several advantages as experimental material, including the relative simplicity of their metabolism, the ease with which membrane vesicles can be prepared, and the fact that relatively slow rates of transport reactions in chromaffin granules allow the use of measuring techniques with a fairly slow time response.

Chromaffin granules contain a multi-subunit H^+-translocating ATPase (Fig. 8.23) that couples the hydrolysis of external ATP to the transfer of protons from the outside to the inside aqueous phase of the granule (Kanner and Schuldiner, 1987; Moriyama and Nelson, 1987), with a H^+:ATP stoichiometry between 1 and 2. ATP-dependent H^+ translocation is dependent on the presence of external Cl^- or Br^- (Moriyama and Nelson, 1987) and results in the production of a $\Delta\psi$ (inside positive) of some $+50$ mV (Kanner, 1983), compared to the resting potential in the absence of external ATP of approximately -70 mV, and a ΔpH inside more acidic than in the absence of ATP (Fig. 9.25). The chromaffin granule membranes contain an M_r 17,000 hydrophobic, DCCD binding protein (Moriyama and Nelson, 1987), which,

9.14 Active Transport in Eukaryotes

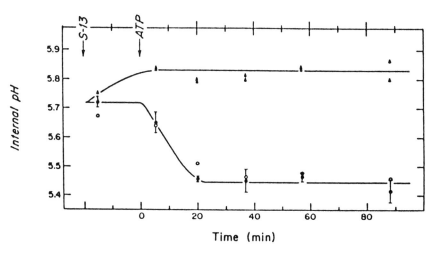

Figure 9.25. ATP-dependent pH changes in the chromaffin granule interior. The external pH was buffered at 6.6 and the ΔpH was measured by [^{14}C]methylamine distribution. In the upper trace, the proton-conducting uncoupler S-13 was added 20 minutes prior to the addition of ATP. In the lower trace, only ATP was added (Reprinted with permission from Casey et al., Biochemistry, vol. 16, pp. 972-977. Copyright 1977 American Chemical Society.)

like the DCCD-binding protein of the F_0F_1 ATPases, may be part of a proton channel.

The physiological function of the chromaffin granule H^+-translocating ATPase is to produce the $\Delta\tilde{\mu}_{H^+}$ that, in turn, provides the energy for uncoupler-sensitive catecholamine accumulation inside the granule to a level of approximate 0.55 M. Transport of catecholamine (Fig. 9.26) in the chromaffin granule appears to occur via a substrate/H^+ antiport, in contrast to the substrate/H^+ symports that prevail in bacterial active transport. As a $\Delta\psi$ (inside positive) across the granule membrane is also capable of energizing catecholamine accumulation, catecholamine transport must be an electrogenic process involving the net efflux of positive charge. Because there is no evidence for the involvement of any co-transported species other than H^+, at least two protons must leave the granule interior space for each catecholamine molecule taken up. It has been demonstrated that both the ΔpH (inside acid) and the $\Delta\psi$ (inside positive) across the chromaffin granule membrane do in fact decrease as catecholamines such as dopamine, epinephrine and serotonin are taken up. The decrease in ΔpH is expected for catecholamine/H^+ exchange, and the decrease in $\Delta\psi$ is consistent with the exchange being electrogenic. This decrease in ΔpH and its reversal by ATP are illustrated in Fig. 9.26 for dopamine uptake by chromaffin granules. Measurements of $\Delta\tilde{\mu}_{amine}$ vs. $\Delta\psi$ and ΔpH show that a maximum of 1 positive charge could be exported per catecholamine molecule taken up. This could occur if two protons were

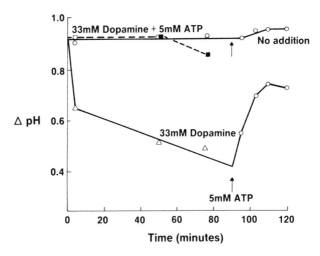

Figure 9.26. The effect of dopamine uptake and ATP on the ΔpH maintained by chromaffin granules. ΔpH was measured by [^{14}C]methylamine distribution (Chap. 3, section 3.4). The reaction mixtures contained chromaffin granules in pH 6.85 buffer (o) and dopamine (△) or dopamine + ATP (■) (Johnson et al., 1978).

exchanged for one molecule of positively charged amine, or if one proton were exchanged for a single molecule of neutral amine. (Fig. 9.27).

The observations that catecholamine transport exhibits saturation kinetics, that transport is stereospecific, with uptake of (−)-catecholamines favored over uptake of the enantiomeric (+) species, and that uptake is blocked by the specific inhibitor reserpine (Kanner and Schuldiner, 1987) implied that a specific catecholamine carrier was present. Reserpine has been shown to act as a competitive inhibitor for the uptake of biogenic amines without affecting the chromaffin granule ATPase. The catecholamine carrier protein has been solubilized by detergent treatment of bovine chromaffin granules (Kanner and Schuldiner, 1987) and used to study uptake of several biogenic amines in liposomes that had the carrier protein incorporated. Uptake in the reconstituted system showed a specificity among the biogenic amines tested and a sensitivity to reserpine that were similar to those found in chromaffin granules. It appears that there is a single carrier protein, perhaps containing several subunits, with rather broad specificity for a number of biogenic amines rather than separate carriers for all the biogenic amines accumulated by the mammalian chromaffin granule.

γ-Aminoburyric Acid (GABA) Transport in Brain

Brain tissue contains a high-affinity (K_m < 10 μM), Na$^+$-dependent GABA transport protein, purified as an M$_r$ 80,000 polypeptide (Radian et al., 1986), that is inhibited by proton-conducting uncouplers capable of collapsing Δψ

9.14 Active Transport in Eukaryotes

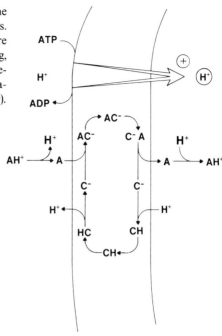

Figure 9.27. A model for catecholamine uptake by isolated chromaffin granules. The broad arrow at the top of the figure represents the proton translocating, chromaffin granule ATPase. "C" represents the reserpine-sensitive catecholamine carrier (Johnson and Scarpa, 1979).

and by the Na^+, K^+-ATPase inhibitor ouabain (Kanner, 1983; Kanner and Schuldiner, 1987). GABA transport in right-side-out membrane vesicles prepared from rat brain can be driven by a Na^+ gradient ($[Na^+]_{out} > [Na^+]_{in}$), and was inhibited by monensin. It was not affected by ATP nor inhibited by ouabain (Kanner, 1983; Kanner and Schuldiner, 1987), because in the presence of an imposed Na^+ gradient, the ouabain-sensitive, ATP-dependent Na^+, K^+-ATPase is not required to produce a Na^+ gradient. Inhibition by an externally positive K^+/valinomycin diffusion potential, along with stimulation by an inside negative $\Delta\psi$ and the above data, implied that GABA transport occurred via an electrogenic Na^+/GABA symport (Kanner, 1983; Kanner and Schuldiner, 1987). GABA transport in the brain also exhibited an requirment for Cl^-, consistent with a model for GABA transport based on GABA/Na^+/Cl^- cotransport (Fig. 9.28). Assuming that GABA is taken up in its neutral, zwitterionic form, the stoichiometry of the other two species being transported must be $Na^+:Cl^- > 1:1$ in order for the total transport process to be electrogenic, in agreement with estimates that $2Na^+$ ions and $1Cl^-$ ion are cotransported with each neutral GABA zwitterion (Kanner, 1983; Kanner and Schuldiner, 1987).

Serotonin(5-Hydroxytryptamine) Transport

Transport of the biogenic amine serotonin (Kanner, 1983), one of the major mammalian neurotransmitters, has been demonstrated in brain slices and in

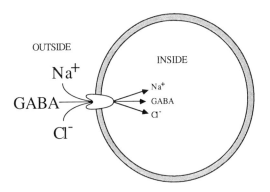

Figure 9.28. GABA cotransport with Na^+ and Cl^- in rat brain synaptosomes (Kanner, 1983).

synaptosomes (pinched off nerve cell terminals purified from homogenized brain tissue), and studied in greatest detail in platelet plasma membrane preparations (Keyes and Rudnick, 1982). Uptake of this biogenic amine by both platelet and brain preparations is Na^+-dependent, and the demonstration that serotonin uptake in vesicles prepared from platelet plasma membranes could be driven by a Na^+ gradient ($[Na^+]_{out} > [Na^+]_{in}$), and inhibited by monensin, suggests a Na^+-serotonin symport mechanism (Kanner and Schuldiner, 1987). Serotonin uptake in platelets is also stimulated by internal K^+ and, as is the case for the brain GABA system described above, exhibits an absolute requirement for external Cl^- ions (Kanner, 1983; Keyes and Rudnick, 1982). Internal Cl^- in the platelet membrane-derived vesicles inhibited serotonin uptake in a manner consistent with the expected if the Cl^- gradient ($[Cl^-]_{out} > [Cl^-]_{in}$) contributes to the net driving force for serotonin accumulation. External K^+ inhibited serotonin uptake and stimulated the efflux of serotonin from vesicles preloaded with the neurotransmitter. These observations suggest that serotonin uptake is coupled to the movement of three inorganic ions, with K^+ efflux in addition to Na^+ and Cl^- uptake, accompanying the movement of the neurotransmitter. Kinetic data are consistent with a model in which internal K^+ binds to the unloaded serotonin carrier inside the vesicles and facilitates the return of unloaded serotonin carrier across the membrane to the outside face where K^+ release to the outside aqueous space favors binding of serotonin, Na^+, and Cl^-. As the mammalian Na^+, K^+-ATPase maintains $[K^+]_{in} > [K^+]_{out}$, K^+ efflux coupled to serotinin uptake would contribute to thermodynamically favorable uptake of the neurotransmitter.

Other Neurotransmitters

In addition to the three Na^+-dependent neurotransmitter transport systems discussed above, a large number of other Na^+-coupled systems for neurotransmitters have been at least partially characterized. In several cases, evidence has been obtained for additional ion requirements (i.e., K^+ and/or Cl^-). A partial list of these systems is shown in Table 9.3.

Table 9.3. Sodium-coupled neurotransmitter transport systems

System	Additional ions	
	Cl^-	K^+
γ-Aminobutyric acid	+	−
Glutamate/aspartate, brain	−	+
Serotonin, platelet	+	+
Serotonin, rat basophilic leukemia cells	+	+
Serotonin, brain	+	Not determined
Glutamate, kidney	−	+
Glutamate, liver	−	+
Glutamate, crab nerve	−	Not determined
γ-Aminobutyric acid, insect nervous syst	+	−
Glycine, brain	+	−
Glycine, erythrocytes	+	−
Choline, brain	+	Not determined
Norepinephrine, brain rat heart nerve	+	Not determined
Dopamine, brain	+	Not determined
Amino acids, fish intestine	+	−

From Kanner (1983).

9.15 Transport or Translocation of Macromolecules

9.15.1 Viral Infection of Bacteria

The preceding discussion of transport has focussed on processes that result in the uptake of some substance useful to the cell. Although uptake of an infecting virus (phage) can hardly be considered a beneficial process for the individual bacterium since infection often leads to cell death, this process can contribute to gene transfer in the population and is of interest in the context of $\Delta\tilde{\mu}_{H^+}$-dependent uptake of macromolecules. Diversion of the bacterial energy source to accomplish the viral infection also provides an example of the virus' ability to exploit the bacterial metabolism for its own ends.

Proton-conducting uncouplers and K^+-valinomycin were effective inhibitors of infection of *E. coli* by T4 phase, while ATP depletion by arsenate treatment and nigericin had little effect (Kalasauskaite et al., 1983). These results show that the $\Delta\psi$ component of $\Delta\tilde{\mu}_{H^+}$ has the dominant role in supplying energy for phage infection. At relatively high phage:bacterial cell ratios, the percentage of *E. coli* cells infected by T4 is approximately proportional to $\Delta\psi$ over the range from -90 to -190 mV (Grinius, 1981).

It may seem counterproductive for bacteria to have evolved specific receptor and/or transport proteins that allow infection by foreign DNA that ultimately leads to the death of the bacteria. In all likelihood, these bacterial proteins evolved for the purpose of transporting small metabolites and have been exploited by phages and colicins. The injection of DNA in transformation or during sexual conjugation is part of the "normal" life of bacterial cells, and transformation shares with phage infection a requirement for energy derived from the $\Delta\tilde{\mu}_{H^+}$ (van Niewhoven et al., 1982; Grinius, 1987).

9.15.2 Energy Sources for Protein Import and Secretion

Translocation of nuclear-encoded polypeptides made in the cytoplasm across the organelle membrane is a fundamental process in the biogenesis of chloroplasts and mitochondria. Many of the protein complexes (cytochrome $b-c_1$, b_6-f, oxidase, or F_0F_1 ATPase) are assembled from both nuclear- and organelle-encoded polypeptides. Similarly, biogenesis of the bacterial cell envelope requires secretion of proteins made in the cytoplasm into and across the inner or cytoplasmic membrane, and into the outer membrane of Gram-negative organisms. In addition, it is likely that the processes of import and secretion have much in common with the mechanism of translocation of colicins and toxins (e.g., diphtheria toxin).

Protein import into chloroplasts and mitochondria requires synthesis of a precursor protein (with an NH_2-terminal extension) in the cytoplasm, targeting of the precursor protein to the outer surface of the outer organelle membrane, energy-dependent translocation, and enzymatic removal of the NH_2-extension. Posttranslational import (subsequent to protein synthesis) of proteins into the chloroplasts and mitochondria, and export from the bacterial cytoplasm, requires energy, along with the leader peptide and other protein factors, to accomplish the polypeptide conformational changes that must accompany the transition from *cis*-polar aqueous phase (e.g., cytoplasm) to the nonpolar membrane bilayer phase, or to the *trans*-polar phase, and for the transfer of individual ionic amino acids through the nonpolar membrane bilayer. The energy sources for protein translocation (i) from the cytoplasm across the chloroplast envelope to the stroma, (ii) from the cytoplasm to the mitochondrial matrix, and (iii) for export from the bacterial cytoplasm, have been found, respectively, to be (i) ATP, (ii) ATP and $\Delta\psi$, and (iii) $\Delta\psi$ or $\Delta\tilde{\mu}_{H^+}$, and ATP (Table 9.4).

The utilization of ATP exclusively as the energy source for import across the chloroplast envelope is not surprising considering that the magnitude of the $\Delta\tilde{\mu}_{H^+}$ across the chloroplast envelope double membrane is known to be much smaller than that across the mitochondrial and *E. coli* inner membranes. However, an ATPase in the chloroplast envelope has not been characterized. If one is present, it would be unusual in its inability to generate a substantial $\Delta\psi$. The ATP requirement for import into chloroplasts and mitochondria may instead act through a kinase to phosphorylate import receptors in the chloroplast envelope or mitochondrial outer membrane as part of the translocation mechanism, somewhat in analogy to the phosphotransferase pathway for sugar transport in bacteria (section 9.11.2). Although it appears that a $\Delta\tilde{\mu}_{H^+}$

9.15 Transport or Translocation of Macromolecules

Table 9.4 Energy sources for protein import and export

Membrane	Protein	Energy source	Reference
Import			
1. Chloroplast envelope and thylakoid (e)[a], membrane (e + t).	SSU[b] (e)	ATP	Flügge and Hinz (1986), Pain and Blöbel (1987)
	Plastocyanin (e + t); ferredoxin(e);	ATP	Theg et al. (1989).
	LHCII(e)[c]	ATP	Chitnis et al. (1987); Cline (1986).
2. Mitochondria	Many ADP/ATP carrier COIV-DHFR[d] $MF_1 \beta^e$	$\Delta\psi$ $\Delta\psi$ ATP, GTP, $\Delta\psi$ ATP, $\Delta\psi$	Schatz and Butow (1983) Pfanner and Neupert (1985) Eilers et al. (1987) Pfanner and Neupert (1986)
Export			
1. E. coli (Gram-negative)	M13 phage procoat β-lactamase pro-$ompA^f$	$\Delta\psi$ $\Delta\tilde{\mu}_{H^+}$ ($\Delta\psi$ or ΔpH) $\Delta\psi$, ATP	Wickner (1983) Bakker and Randall (1984) Chen and Tai (1985), Geller et al. (1986), Yamane et al. (1987).
2. B. amyloliquefaciens (Gram-positive)	α-Amylase	$\Delta\tilde{\mu}_{H^+}$ ($\Delta\psi$ or ΔpH)	Muren and Randall (1985)

e and (e + t) refer, respectively, to translocation across the chloroplast outer envelope membrane, and thylakoid membrane as well.
small subunit of RUBP carboxylase.
apoprotein of light-harvesting complex of photosystem II, (PS II).
leader peptide of cytochrome oxidase subunit IV attached to the cytosolic enzyme dihydrofolate-reductase.
subunit of mitochondrial F_1 ATPase.
omp, outer membrane protein.

is not required for import of plastocyanin across the thylakoid membrane, one must say that this is surprising since the ΔpH is utilized for import in bacteria (Table 9.4), and thylakoids maintain a substantial ΔpH.

A second role proposed for ATP in protein import is (i) to energize unfolding, a mechanism emphasized for mitochondria (Eilers et al., 1987) or (ii), as emphasized for secretion in *E. coli* (Randall et al., 1987), to maintain a partly folded state of the protein that can be coupled to components of the secretion apparatus. It is plausible that proteins must undergo a large structural change to pass from the polar cytoplasm to the apolar membrane (Wickner, 1983, 1988). The leader peptidase (Wickner, 1988) and pre-maltose binding protein (Randall, 1987) that are exported from the *E. coli* cytoplasm undergo a change before import from a protease-sensitive relaxed conformation to a protease-

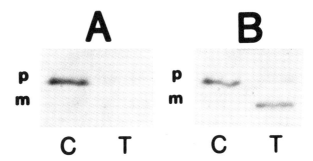

Figure 9.29. Two states of precursor maltose binding protein (MBP) as defined by protease accessibility and sensitivity. (A) Conformationally relaxed state defined by protease accessibility to many sites, obtained from cells labeled for 15 s with [^{35}S]methionine. (B) Tightly folded "mature" conformation in which few sites, near the NH$_2$-leader region, were accessible to protease; sample taken after an additional incubation at 37°C for 5 min to allow folding to a protease-insensitive state. P, precursor MBP, M$_r$ 41,000; m, mature protein after removal of leader peptide, M$_r$ 38,000. C, control; T, treated with protease. (From Randall and Hardy, 1986. Copyright by Cell Press.)

resistant form (Fig. 9.29A and B). Similarly, uptake by mitochondria of the cytochrome oxidase subunit IV—dihydrofolate reductase (DHFR) fusion protein requires a partly unfolded conformation that is blocked by the DHFR inhibitor methotraxate, and which is dependent upon the presence of ATP (Eilers et al., 1987). A sequential effect of energy from the membrane potential and ATP has been proposed for insertion of an NH$_2$-terminal segment by $\Delta\psi$ at junction sites of the mitochondrial inner and outer membranes, followed by conformational change or maintenance of an unfolded or mobile conformation, and translocation of the rest of the protein (Fig. 9.30). The model shown in Fig. 9.30 would place the site of the ATP requirement for import external to the membrane.

9.15.3 Mechanisms for Utilization of $\Delta\psi$ and ΔpH (Table 9.5)

Utilization of the $\Delta\psi$ for protein translocation has been ascribed (i) to an electrophoresis mechanism by which electrostatic energy is provided for movement of charged dipole segments across the membrane (Daniels et al., 1981). Some conceptual difficulties with such an electrophoretic mechanism are: (a) protein import to mitochondria and secretion by *E. coli* proceed in opposite directions although the membrane potential has the same sign, inside-negative, for both membranes; (b) the $\Delta\psi$ and ΔpH components function interchangeably in the export process of *E. coli* (Table 9.4B); (c) the translocation of an uncompensated charge through the low dielectric medium

9.15 Transport or Translocation of Macromolecules

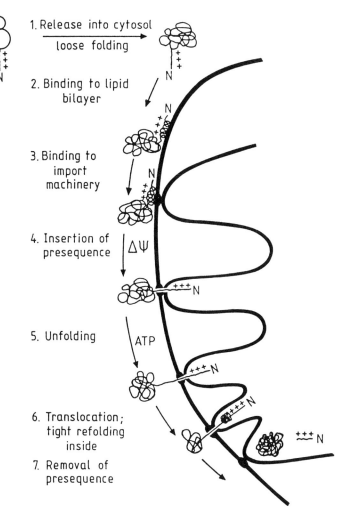

Figure 9.30. Model for sequential action of $\Delta\psi$ and ATP for protein translocation in mitochondria (Verner and Schatz, 1987, 1988; compare with Schleyer and Neupert, 1985; Hurt, 1987).

of the membrane bilayer center requires a much larger amount of energy, ca. 40 kcal/mol, than can be supplied by electrophoretic movement of charge across a potential of ~0.1 V. Problems (a) and (b) imply that the membrane potential may cause structural changes in the membrane that facilitate the translocation, such as (ii) destabilization of the bilayer structure or transient formation of inverted micelles (Schatz and Butow, 1983), or (iii) enhanced probability of junction formation between inner and outer membranes. The latter mechanism cannot apply, however, to secretion in Gram-positive bac-

Table 9.5. Suggested functions and mechanisms of energy sources used for protein translocation in secretion and import

Energy source	Proposed function and mechanisms
ATP	May be required for all post-translational secretion and import. Mechanisms: phosphorylation of receptor; "unfoldase" to generate relaxed conformation of translocated protein, or anti- "foldase" to prevent folding into a more compact or immobile form, leader peptide may also retard folding.
$\Delta\psi$	Required for translocation across inner membranes of both bacteria and mitochondria. For mitochondria, required to move NH_2-part of leader across the membranes: Mechanisms: electrophoresis of charged or dipolar protein segments, destabilization of membrane bilayer or transient formation of inverted micelles, fusion of inner and outer membranes, protonation of protein Asp and Glu residues.
ΔpH	(2.3 $RT \cdot \Delta pH$) energetically equivalent to $F\Delta\psi$; protonation of protein Asp and Glu residues.
Protein folding:	For mitochondria: Folding of protein on *trans*-side may provide free energy source for translocation of entire precursor and mature protein
ATP and $\Delta\psi$ acting sequentially (mitochondria; Fig. 9.30).	Unfolding/translocation (ATP); insertion of NH_2-terminal domains ($\Delta\psi$).

teria that do not have an outer membrane (Murën and Randall, 1986). Utilization of polar or aqueous channels for translocation has been postulated in order to solve thermodynamic problem (c) discussed above (Singer et al., 1987). (iv) The presence of a negative membrane potential may also increase the pK of aspartic and glutamic acid residues by approximately one pK unit for every 60 mV of potential, facilitating the protonation and neutralization of these acidic residues. It seems likely that more than one of mechanisms (i)–(iv) (Table 9.5) may operate concurrently to facilitate protein translocation.

Utilization of ΔpH, as well as a $\Delta\psi$, could also facilitate insertion into the membrane by protonation of charged acidic residues of the translocated protein, and neutralization of the net electrostatic charge of these residues. Utilization of a ΔpH, along with $\Delta\psi$, for protein translocation has been demonstrated in the Gram-negative *E. coli* for secretion of β-lactamase (Bakker and Randall, 1984), and in the Gram-positive organism, *B. anyoliquefaciens*. The large net negative charge of the central segment of the M13 phage

9.15 Transport or Translocation of Macromolecules

Model of Procoat Membrane Assembly

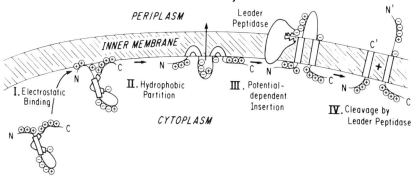

Figure 9.31. Stages in the translocation of M13 phage procoat protein by an electrophoretic mechanism: I. electrostatic binding using (+) charge domains; this mechanism could involve neutralization of the negatively charged central domain of the protein; II. hydrophobic partition into the bilayer; III. insertion into the bilayer using an electrophoretic mechanism; IV. cleavage by leader peptidase on the *trans*-side (Reprinted with permission from Wickner, *Biochemistry*, vol. 27, pp. 1081–1086. Copyright 1988 American Chemical Society.)

procoat suggests the possibility that both $\Delta\psi$ and ΔpH could increase protonation of some of the acidic residues in this segment, thus neutralizing its net negative charge and facilitating its approach to the negatively charged membrane surface (Fig. 9.31).

9.15.4 Leader Peptide

The properties of the signal or leader peptide that is present on the NH_2-terminus of translocatable "precursor" proteins are summarized in Table 9.6. As part of the translocation mechanism, the leader peptide extension is cleaved by a leader peptidase to form the "mature" translocated protein. Leader peptides for these proteins typically range in size from about 15–70 amino acids. Formation of the mature protein from the precursor can thus be detected on an SDS-PAGE gel by a decrease of two thousand or more in molecular mobility (Fig. 9.29B).

The information needed for plastocyanin, which is targeted to the chloroplast lumen, to cross three chloroplast membranes, is contained in the extra length of its 66 residue leader peptide (Smeekens et al., 1986). The NH_2-terminal half of the transit peptide is used in translocation across the two outer envelope membranes. It is then cleaved in the stroma to expose the COOH-terminal half of the leader that targets the protein to the thylakoid lumen. A similar conclusion was reached from an analogous situation with the 65

Table 9.6. Properties of precursor proteins of (I) secreted proteins from bacteria and eukaryotes, and (II) for import into mitochondria and chloroplasts

Membrane	Signal or leader peptide
Bacterial and eukaryotic secreted proteins	$\sim 13-30$ residues; three domains: ($+$) charged NH_2-terminal region, central hydrophobic core ≥ 10 residues, and polar COOH-terminal region ending with cleavage site. Residues (A, C, S, T, G) at positions 1 and 3 before cleavage site must have small side chains; minimum length of 13 residues: 10 for hydrophobic span, 1 or 2 charges at the NH_2-terminus and an initiator fMet that is uncharged; acidic residues seldom found in prokaryotes, with the most common charge, $+2$, as in eukaryotes where the NH_2-terminus contributes $+1$. One function of ($+$) charges is to bind to ($-$) membrane surface.
Mitochondria and chloroplasts	Up to 90 residues. No extended hydrophobic segment. Many Lys, Arg, Thr, Ser residues; few acidic residues; predicted to be amphiphilic in either α or β conformation, with β more pronounced for chloroplasts.

residue leader peptide of the chloroplast photosystem II 33-kDa extrinsic protein (Chap. 6, section 6.8.4). In contrast, the analogous protein in the cyanobacterium, *Synechococcus* sp. 7942, which does not have the envelope membranes of the higher plant chloroplast has only a 28 residue leader peptide (Kuwabara et al., 1987).

The function(s) of the leader peptide in the translocation process may include preservation of a conformationally altered precursor state (Park et al., 1988), and targeting of the protein to a particular site in the mitochondrion or chloroplast. The information in the leader peptide (e.g., mitochondria, Fig. 9.32) that determines whether an imported protein can (i) cross the outer and inner membranes and be delivered to the matrix, (ii) can cross only the outer membrane, or (iii) is targeted to reside in the inner or outer membrane, is contained in the sequence of basic, hydroxylated, and aliphatic amino acids, the distribution of hydrophobic and hydrophilic residues, and the peptide conformation (von Heijne, 1986; Briggs and Gierasch, 1986; Hurt and van Loon, 1986; Allison and Schatz, 1986; Roise et al., 1988).

Polypeptide translocation may occur through an aqueous channel provided by an intrinsic translocator or receptor protein. This channel would provide the necessary polar environment for translocation of charged amino acids, with the amphipathic signal peptide providing part of this transmembrane channel (Fig. 9.33).

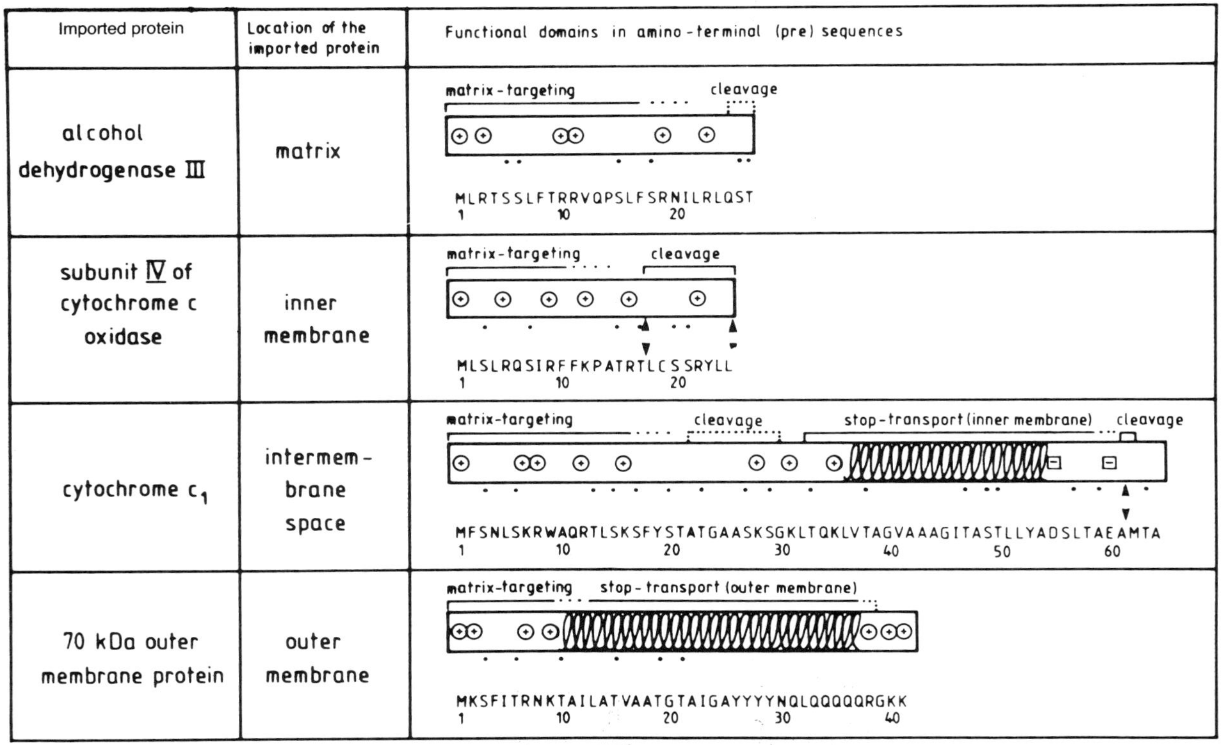

9.15 Transport or Translocation of Macromolecules

Imported protein	Location of the imported protein	Functional domains in amino-terminal (pre) sequences
alcohol dehydrogenase III	matrix	matrix-targeting ... cleavage MLRTSSLFTRRVQPSLFSRNILRLQST 1 10 20
subunit IV of cytochrome c oxidase	inner membrane	matrix-targeting ... cleavage MLSLRQSIRFFKPATRTLCSSRYLL 1 10 20
cytochrome c_1	intermembrane space	matrix-targeting ... cleavage ... stop-transport (inner membrane) ... cleavage MFSNLSKRWAQRTLSKSFYSTATGAASKSGKLTQKLVTAGVAAAGITASTLLYADSLTAEAMTA 1 10 20 30 40 50 60
70 kDa outer membrane protein	outer membrane	matrix-targeting ... stop-transport (outer membrane) MKSFITRNKTAILATVAATGTAIGAYYYNQLQQQQRGKK 1 10 20 30 40

Figure 9.32. Amino terminal leader sequences of imported yeast mitochondrial proteins: alcohol dehydrogenase, subunit IV of cytochrome oxidase, cytochrome c_1, and 70-kDa outer membrane protein. (+) and (−), basic and acidic amino acids; coil, putative α-helical membrane spanning segment; dots, serines or threonine; arrows, proteolytic sites (Hurt and van Loon, 1986).

464　　9: Active Transport

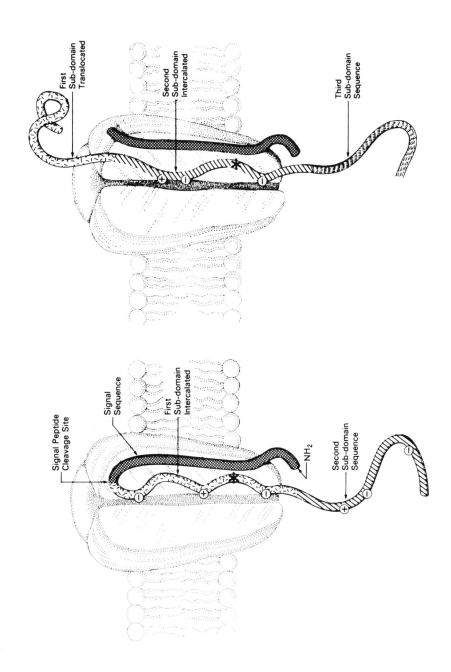

9.16 Summary

Bacteria utilize the $\Delta\tilde{\mu}_{H^+}$ (outside positive and acidic) that results from H^+ translocation during respiration or ATP hydrolysis as an energy source for the uptake and accumulation of a large number of useful compounds (e.g., sugars and amino acids). These secondary active transport systems utilize a single, membrane-bound protein to catalyze the movement of a specific transportable compound, or group of structurally related compounds, across the plasma membrane and are capable of coupling the unfavorable movement of the transportable substrate against a concentration gradient to the favorable movement of protons down the $\Delta\tilde{\mu}_{H^+}$ gradient. Both the ΔpH and $\Delta\psi$ components of $\Delta\tilde{\mu}_{H^+}$ can provide energy for substrate accumulation via these proton/substrate symports.

Bacteria exchange protons excreted during respiration and/or ATP hydrolysis for Na^+ and other cations such as K^+. These H^+/cation antiports appear to play a major role in the regulation of internal pH by bacteria and, in the case of the Na^+/H^+ antiport, contribute to the establishment of a ΔpNa^+ ($[Na^+]_{out} > [Na^+]_{in}$). In some bacteria, $\Delta\tilde{\mu}_{Na^+}$ can also result from the direct coupling of Na^+ efflux to respiration or to biotin-dependent decarboxylation reactions. $\Delta\tilde{\mu}_{Na^+}$ can be used in turn for the active transport of sugars (e.g., melibiose) and amino acids via Na^+/substrate symports. Substrate cotransport with Na^+ is also common in eukaryotes. Bacterial cells maintain $[K^+]_{in} > [K^+]_{out}$, with ATP supplying the energy for K^+ uptake. K^+ uptake appears to play a central role in osmoregulation by bacteria.

Another mode of active transport by bacteria involves initial substrate recognition and binding by a soluble, periplasmically located protein. Subsequent transport across the membrane involves a series of additional membrane-bound proteins. The energy source for such transport systems is ATP hydrolysis. An important additional transport process for the uptake of sugars by bacteria involves concomitant transport and phosphorylation via the phosphotransferase system, with phosphoenolpyruvate (PEP) serving as the phosphoryl group donor and energy source.

Ion gradients and trans-membrane electrical potentials also serve as energy sources for other diverse processes including motility in bacteria, the transport of neurotransmitters in mammals, the uptake of viral and bacterial DNA by bacteria and, along with ATP, for the secretion and import of proteins across bacterial, chloroplast, and mitochondrial membranes.

Figure 9.33. Two-stage scheme for polypeptide translocation through a channel defined by a multisubunit "translocator protein" and the leader peptide. (From Singer et al., 1987). (Left) The hydrophobic segment of the leader peptide is inserted into the bilayer prior to cleavage by the leader peptidase on the *trans* side of the membrane. An inserted sequence of 20–30 residues following the leader is shown facing the aqueous channel provided by the translocator protein. If the signal peptide and the subsequent inserted segment form an α-helical hairpin, the formation of this hairpin from a random coil in the polar phase can provide a large $-\Delta G$ to drive the insertion process (Engelman and Steitz, 1981; see Chap. 1, Problem 19). (∗) denotes particular site in channel that interacts with a residue in successive domains of the translocated polypeptide.

Appendix I
Answers to Problems

Chapter 1: 2. (a) $\Delta W = -20$ kJ, (b) $\Delta W = -10$ kJ, with more work done in (a); 4. $2.3R = 4.6$ cal/°K $= 19.1$ J/°K; 5. (a) $\Delta S = 11$ J/°K, (b) volume expansion ratio $= 14$; 7. (a) $K_{eq} = 3.2 \times 10^5$, (b) $\Delta S° = -30$ e.u.; 8. 373°K; 9. $\Delta G° = -4.2$ kcal/mol; 10. $T = 44°C$. 11. $\Delta G = -23.3$ kJ $= -5.5$ kcal; 12. (a) $\Delta S° = -15.9$ e.u., due to the order of H_2O molecules in contact with nonpolar molecules in aqueous solvent, (b) $\Delta S° = -72.5$ e.u.; 13. $K_{eq} = 0.5, 1.1, 1.4, 1.5, 1.1,$ and 1.01 for G, A, L, V, E, and K; 14. $\Delta G°$ (kcal/mol) $= 1.8, 1.3, 0.65, 0, -0.55, -1.15, -1.5,$ and -2.1 at T (°C) $= 10, 20, 30, 40, 50, 60, 70,$ and 80; 15. Using $R = 0.082$ atm-L/mol-°K, p $= 2.4$ atm; 16. (a) $\Delta(\Delta G°) = -1.37$ kcal/mol, (b) K_{eq} increases by 10-fold; 17. $\Delta G°$ (kcal/mol) $=$ (a) -1.25, (b) -2.5, (c) 0, (d) $+1.25$; 18. $\Delta \mu° = -3,476$ and $-5,789$ cal/mol; 19. (a) $\Delta G = +4.5$ kcal/mol, (b) $\Delta G° = -105$ kcal/mol; 20. $\Delta S° = +13.4$ e.u.; 21. $\Delta G° = -7.36$ kcal/mol. 22. 10^{-12} M; 23. (a) 100, (b) 10^4; 24. (b) from 10 to 1,000; 26. $\Delta G° = -8.2$ kcal/mol; 27. 2.79; 28. 23.06 kcal/mol.

Chapter 2: 29. 54.2, 59.1, and 61.5 mV ($n = 1$) and 27.1, 29.6, and 30.8 mV ($n = 2$); 31. (a) -7.3 kcal, (b) -14.6 kcal; 32. 7.5×10^{-4} M; 36. (b) $+140$ mV, (c) decrease of 118 mV; 38. $E_{m1} = +48.5$ mV, $E_{m2} = +131.5$ mV; 39. -1.07 V. 40. 3.1×10^7 M^{-1} s^{-1}; 41. Fastest, c_{551}, 4.4×10^7 M^{-1} s^{-1}, and slowest, horse cyt c, 4.8×10^4 M^{-1} s^{-1}; 43. (a) $\Delta G^{\ddagger} = 0$, (c) 20.1 Å; 44. $\Delta pK = -6.0$. 45. 5×10^{-8} s, if $m = 10^{-4}$ cm$^2 \cdot$ V$^{-1} \cdot$ s^{-1}.

Chapter 3: 46. $\Delta pH = 3$; 47. $\Delta[K^+] = 3 \times 10^{-4}$; 48. 1.9×10^4; 49. 730 Å; 50. 2.7 nm; 51. H$^+$/ATP $= 3.1$; 52. (a) basal rate $= 90$ μmol/mg Chl-h, (b) ATP/O $=$ ATP/2e$^- = 1.0$ or 0.67 depending on whether the electron transport rate is corrected for the basal rate.

Chapter 4: 53. $\Delta A = 5.1 \times 10^{-4}$; 54. (a) $g = 6.09$, (b) $g = 3.35$, (c) $H = 3,344$ gauss $= 0.3344$ T; 55. (a) 4.9×10^{-31} J/gauss, (b) 2.1×10^8 Hz; 56. (a) 3.1×10^{-20} J, (b) 7.5 times kT at 300 K; 57. (a) $E_m = -80$ mV, (b) $K_s = 2.05 \times 10^{-2}$.

Chapter 5: 59. $\sqrt{2}$; 60. 10-fold greater; 61. (a) $D_T = 8.5 \times 10^{-9}$ cm^2-s^{-1}, $D_R = 1.02 \times 10^5$ radians^2-s^{-1}, (c) 10^{-2}; 62. (a) $D_T = 1.2 \times 10^{-8}$ cm^2-s^{-1} (larger than any measured D_T for an integral protein), (b) $D_R = 2.3 \times 10^5$ radian$^2 \cdot$s^{-1}.

Chapter 6: 63. zero; 64. (a) 64/65, (b) 1/65; 65. (e) 1/4, (f) 3×10^9 s^{-1} and 4.7×10^{-7} s^{-1}, (g) $R = 2R_o = 60$ Å; 67. (a) $R_o = 36.2$ Å, (b) 29.9 Å; 68. $\phi_T = 0.97$; 69. 0.34 ATP/CO$_2$.

Chapter 7: 70. (a) 10 H$^+$/bR-s for bR, and about half that rate for the fusion protein, (b) reconstituted rate about 1/20 that in the natural membrane, (c) 380; 71. A decrease of 120 e.u.; 72. 4.2; 73. $K_{eq} = 4.7$.

Chapter 8: 74. $\Delta H°$ is positive; 75. $\Delta \psi = +3.5$ mV, -144 mV, and -173 mV in chromatophores, mitochondria, and *E. coli* membranes; 76. ΔpH $= 3.0$; 77. $n_{H^+} = 3$; 78. H$^+$/ATP $= 2.9$; 79. (a) ATP/ADP $= 1.4 \times 10^3$, (b) ATP/ADP $= 5.9 \times 10^{-5}$, showing that the ATPase with $n_{H^+} = 3$ can work as a synthase, but the enzyme with $n_{H^+} = 1$ cannot; 80. (a) 1,200 H$^+$/CF$_0$F$_1$-s, (b) 6.2×10^5; 81. 72 H$^+$/F$_0$F$_1$-s; 82. (a) 1.2×10^4/s, (b) 9.9×10^4 kcal/cell-s.

Appendix II
Physical, Chemical, and Biochemical Constants

A. Fundamental Constants

Constant	Symbol	SI Units	Value	cgs Units
Avogadro's number	N		6.02×10^{23} mol^{-1}	
Speed of light in vacuum	c	$3 \cdot 10^8$ m s^{-1}		$3 \cdot 10^{10}$ cm s^{-1}
Bottzmann's constant	k	$1.38 \cdot 10^{-23}$ J K^{-1}		$1.38 \cdot 10^{-16}$ erg K^{-1}
Gas constant (Nk)	R	8.31 J K^{-1} mol^{-1}		1.99 cal K^{-1} mol^{-1}
Elementary charge	e	$1.6 \cdot 10^{-19}$ coulomb		4.8×10^{-10} statcoulo ($g^{1/2}$ cm$^{3/2}$ s^{-1})
Faraday constant (Ne)	F	$9.65 \cdot 10^4$ C mol^{-1}		
Planck's constant	h	$6.63 \cdot 10^{-34}$ J s		$6.63 \cdot 10^{-27}$ J s
	$\hbar = h/2\pi$			
Electron mass	m_e	$9.11 \cdot 10^{-31}$ kg		$9.11 \cdot 10^{-28}$ g
Proton mass	m_p	$1.67 \cdot 10^{-27}$ kg		$1.67 \cdot 10^{-24}$ g
Bohr magneton, electron	β	$e\hbar/2m =$ $9.27 \cdot 10^{-24}$ J Tesla^{-1}		$e\hbar/2mc =$ $9.27 \cdot 10^{-21}$ erg gauss^{-1}
g value, free electron	g		2.00232	

B. Conversion of Energy Units
 1 Joule (SI) = 10^7 ergs (cgs)
 1 electron-volt (eV) = 96.48 J mol^{-1} (SI) = $1.6 \cdot 10^{-12}$ erg (cgs) = 23.06 kcal $\cdot$ mol^{-1}.
 1 calorie = 4.184 Joule

C. Electrical and Magnetic Units
 1 coulomb = 3×10^9 statcoulombs
 1 volt = (1/300) $\cdot$ statvolt
 1 ampere = 1 coulomb $\cdot$ s^{-1} = $6.2 \cdot 10^{18}$ charges $\cdot$ s^{-1}
 1 siemen = 1 amp/volt

1 Debye = $3.3 \cdot 10^{-30}$ coulomb-meter
1 Tesla = 10^4 Gauss
D. Useful Physical-Chemical and Biochemical Data
$2.3RT$ = 1.36 kcal·mol^{-1} = 5.69 J mol^{-1} (25°C)
$2.3RT/F$ = 59.1 mV (25°C)
Viscosity of H_2O = 0.01 poise (g cm^{-1} s^{-1}) [20°C]
Solubility of O_2 in H_2O = 275 μM (20°C).
Saturating intensity of heat-filtered white light, plant photosynthesis:
  5×10^2 J·m^{-2}·s^{-1}
Solar constant, light intensity incident on top of atmosphere:
  2 cal/cm^2-min = 1.4×10^3 J·m^{-2}·s^{-1}
Energy content of light: 42 kcal/Einstein at 680 nm, 64 kcal/Einstein at 450 nm
E. Amino Acid Code and Molecular Weights

Amino Acid	Code	MW, pH 7.0, in Proteins
Alanine	A	71
Asparagine	N	114
Aspartic Acid (−)	D	114
Arginine (+)	R	157
Cysteine	C	103
Glutamic Acid (−)	E	128
Glutamine	Q	128
Glycine	G	57
Histidine	H	137
Isoleucine	I	113
Leucine	L	113
Lysine (+)	K	129
Methionine	M	131
Phenylalanine	F	147
Proline	P	97
Serine	S	87
Threonine	T	101
Tryptophan	W	186
Tyrosine	Y	163
Valine	V	99

Appendix III
Prediction of Protein Folding in Membranes

Estimation of the nonpolar/polar or hydrophobic/hydrophilic character of the amino acids is of interest in connection with calculation of the folding pattern of polypeptide chains in the low dielectric hydrophobic core of biological membranes. Chemical potential changes have been calculated for the transfer of individual amino acids from water to solvents of lower polarity such as ethanol or dioxane (Cohn and Edsall, 1943; Nozaki and Tanford, 1971; Tanford, 1980), using known solubilities of the amino acids in water and ethanol (Cohn and Edsall, 1943). These transfer free energy values are summarized in Table A.1, along with data on the partitioning of the amino acids between water and the vapor phase as another estimate of the transfer potentials from the polar water phase to the nonpolar vapor phase (Hine and Mookerjee, 1975; Wolfenden et al., 1981). Although there is disagreement about the $\Delta\mu^o$ values for several of the residues such as tryptophan, the water to vapor phase transfer potentials correlate fairly well with another measure of hydrophobicity, the extent to which a given residue has been found statistically to be buried in the nonpolar interior of a globular protein (Chothia, 1976) (Table A.2). Data such as these have been used to derive indices and scales of relative hydrophobicity, two of which, the "hydropathy" (strong feeling about water) scale (Kyte and Doolittle, 1982) and the normalized consensus scale of Eisenberg (1984), are shown in Table A.2. These scales are frequently used to derive predictions of the folding of membrane proteins in the hydrophobic bilayer of the membrane. In order to account for the presence of charged amino acids in the bilayer in the case of proteins such as bacteriorhodopsin, it has also been noted that the membrane insertion energy of charged amino acids (Arg, Lys, Asp, Glu) arranged as neutral or ion pairs (e.g., Asp–COOH–NH_2–Lys or Asp–$COOH^-$–NH_3^+–Lys) may be much smaller than the energies for the individual charged amino acids (Honig and Hubbell, 1984).

It is useful to test the similarity of hydropathy plots for a family of proteins

Appendix III: Prediction of Protein Folding in Membranes

Table A.1. Cross-section of data on which amino acid hydrophobicity scales are based: (i) chemical potential changes for transfer for the side chains of the amino acids from H_2O to ethanol or vapor phase; (ii) accessibility of amino acids in soluble proteins.[a]

Side-chain (one letter code)	$\Delta\mu°$ for transfer (kcal/mol)		Side chains, 95% buried[c]
	Water into condensed vapor	Water into ethanol[b]	
Leucine (L)	−3.2	−2.4	2.8
Isoleucine (I)	−3.1	—	4.2
Valine (V)	−2.8	−1.7	4.2
Alanine (A)	−2.3	−0.7	1.6
Phenylalanine (F)	−0.2	−2.6	3.5
Methionine (M)	+0.6	−1.3	1.9
Cysteine (C)	+0.6	—	3.2
Threonine (T)	+4.2	−0.4	−1.0
Serine (S)	+4.6	−0.05	−1.0
Tryptophan (W)	+4.8	−3.2	−0.3
Tyrosine (Y)	+5.1	−2.4	−2.2
Lysine (K)	+8.1	—	−4.2
Glutamine (Q)	+8.6	+0.1	−3.6
Asparagine (N)	+9.0	+0.0	−2.7
Glutamic acid (E)	+9.1	+2.9	−1.7
Histidine (H)	+9.6	−0.45	−1.9
Aspartic acid (D)	+10.0	+3.4	−2.3

[a] Data compiled by Kyte and Doolittle (1982).
[b] Nozaki and Tanford (1971).
[c] Scale of +4.5 − −4.5 of Kyte and Doolittle (1982).

with similar functions and extensive sequence homology. The *b* cytochromes of complex III of energy-transducing membranes (Chap. 7, section 7.4) constitute such a family.

Fig. A.1A–E show the existence for the mitochondrial cytochrome *b* from yeast and maize, bacteriorhodopsin and the *L* and *M* subunits of the *Rps. viridis* photosynthetic reaction center of a number of hydrophobic domains separated by hydrophilic regions. To determine whether these plots are quantitatively similar and whether the degree of similarity of hydropathy plots is greater between two proteins from the same family relative to dissimilar proteins, the hydropathy functions can be mathematically compared (Shiver et al., 1989) by calculation of a cross-correlation coefficient between (i) two functions derived from the same family and (ii) between a function derived from one family (e.g., *b* cytochromes, Chap. 7.4) and one from an unrelated membrane protein of approximately the same size. The cross-correlation coefficient C is defined as:

$$C = \frac{\sum_{j=1}^{n}(X_j - \bar{X})(Y_j - \bar{Y})}{\left[\sum_{j=1}^{n}(X_j - \bar{X})^2 \sum_{j=1}^{n}(Y_j - \bar{Y})^2\right]^{1/2}},$$

Table A.2. Two amino acid data bases used to determine hydrophobicity indices of membrane proteins

Side chain	Hydropathy index[a]	Normalized hydrophobicity index[b]
Isoleucine	4.5	1.4
Valine	4.2	1.1
Leucine	3.8	1.1
Phenylalanine	2.8	1.2
Cysteine/cystine	2.5	0.29
Methionine	1.9	0.64
Alanine	1.8	0.62
Glycine	−0.4	0.48
Threonine	−0.7	−0.05
Tryptophan	−0.9	0.81
Serine	−0.8	−0.18
Tyrosine	−1.3	0.26
Proline	−1.6	0.12
Histidine	−3.2	−0.40
Glutamic acid	−3.5	−0.74
Glutamine	−3.5	−0.85
Aspartic acid	−3.5	−0.90
Asparagine	−3.5	−0.78
Lysine	−3.9	−1.5
Arginine	−4.5	−2.5

[a] From Kyte and Doolittle (1982).
[b] From Eisenberg (1984).

for comparison of the functions over n residues of polypeptides X and Y, with $\bar{X}$ and $\bar{Y}$ the average value of the hydrophobicity function over the length, n, of the polypeptide, corrected for the averaging interval, m. X_j and Y_j are the average values over the sampling interval of the hydrophobicity index for the two functions to be compared.

$$\bar{X} = \frac{1}{n} \cdot \sum_{j=1}^{n} X_j; \quad \bar{Y} = \frac{1}{n} \cdot \sum_{j=1}^{n} Y_j. \quad X_j = \frac{1}{m} \cdot \sum_{i=1}^{m} h_i; \quad Y_j = \frac{1}{m} \cdot \sum_{i=1}^{m} h_i.$$

in which h_i is the hydrophobicity of residue i. The above cross-correlation function in normalized to a value of 1.0 for autocorrelation (the hydropathy function compared to itself). Values of the correlation coefficient close to 1.0 and 0.0 are obtained for functions that are, respectively, similar in amplitude and phase, or not related to each other in phase and/or amplitude. The correlation coefficients between cytochrome b_6 (Fig. 7.17A) and the aligned b cytochromes from mouse and maize are 0.76 and 0.80, whereas that with cytochrome f is 0.13 (Shiver et al., 1989).

Appendix III: Prediction of Protein Folding in Membranes 473

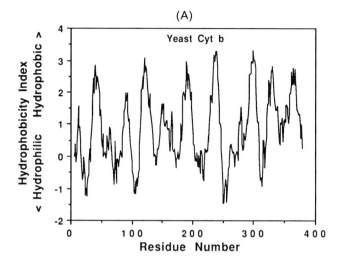

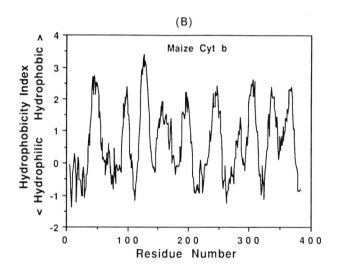

Figure A.1. Hydropathy plots for cytochrome *b* from the yeast (A) and maize (B) complex III, bacteriorhodopsin (C), and the *Rps. viridis* L (D) and M (E) polypeptides, using the amino acid hydrophobicity data base of Kyte and Doolittle (1982). The hydrophobicity index was averaged over an interval of 11 amino acids and the plots were generated by sliding this 11 residue interval along the length of the polypeptide one residue at a time. The use of different amino acid data bases is discussed in Shiver et al. (1989).

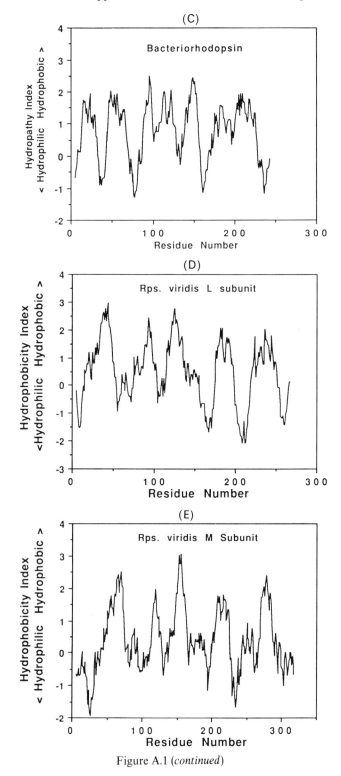

Figure A.1 (*continued*)

References

Abbott, M. S., Czernecki, J. J., and Selman, B. R. (1984) Localization of the high-affinity site for ATP on the membrane-bound chloroplast ATP synthase. *J. Biol. Chem.*, 259, 12271–12278.
Ackrell, B. A. C., Kearney, E. B., Mims, W. B., Peisach, J., and Beinert, H. (1984) Iron-sulfur cluster 3 of beef heart succinate-ubiquinone oxidoreductase is a 3-iron cluster. *J. Biol. Chem.*, 259, 4015–4018.
Addison, R. (1986) Primary structure of the *Neurospora* plasma membrane H^+-ATPase deduced from the gene sequence. *J. Biol. Chem.*, 261, 14896–14901.
Admon, A., and Hammes, G. G. (1987) Amino acid sequence of the nucleotide binding region of chloroplast coupling factor I. *Biochemistry*, 26, 3193–3197.
Adolfsen, R., and Moudrianakis, E. N. (1973) Roles for metal ions in the hydrolysis of adenosine triphosphate by the 13S coupling factors of bacterial and mitochondrial oxidative phosphorylation. *Biochemistry*, 12, 2926–2933.
Adolfsen, R., and Moudrianakis, E. N. (1976) Binding of adenine nucleotides to the purified 13S coupling factor of bacterial phosphorylation. *Arch. Biochem. Biophys.*, 172, 425–433.
Åkerlund, H. E., and Andersson, B. (1983) Quantitative separation of spinach thylakoids into photosystem II-enriched inside-out vesicles and photosystem I-enriched right-side-out vesicles. *Biochim. Biophys. Acta*, 725, 34–40.
Alam, J., and Krogmann, D. W. (1984) The primary structure of cytochrome *f* from spinach and cyanobacteria. *Advances in Photosynthesis Research I* (C. Sybesma, ed.), pp. 521–524, Martinus Nijhoff/W. Junk.
Alam, J., Whitaker, R. A., Krogmann, D. W., and Curtis, S. E. (1986) Isolation and sequence of the gene for ferredoxin I from the cyanobacterium, *Anabaena sp.* strain PCC 7120. *J. Bacteriol.*, 168, 1265–1271.
Alberty, R. A. (1969) Standard Gibbs free energy, enthalpy, and entropy changes as a function of pH and pMg for several reactions involving adenosine phosphates. *J. Biol. Chem.*, 244, 3290–3302.
Albracht, S. P. J. (1980) The prosthetic groups in succinate dehydrogenase. Number and stoichiometry. *Biochim. Biophys. Acta*, 612, 11–28.

Allen, J. P., Feher, G., Yeates, T. O., Komiya, H., and Rees, D. C. (1987a) Structure of the reaction center from Rb. sphaeroides R-26: the protein subunits. Proc. Natl. Acad. Sci. USA, 84, 6162–6166.

Allen, J. P., Feher, G., Yeates, T. O., Komiya, H., and Rees, D. C. (1987b) Structure of the reaction center from Rb. sphaeroides R-26: the cofactors. Proc. Natl. Acad. Sci. USA, 84, 5730–5734.

Allen, J. P., Feher, G., Yeates, T. O., Komiya, H., and Rees, D. C. (1988) Structure of the reaction center from Rb. sphaeroides R-26: protein-cofactors (quinones and Fe^{2+}) interactions. Proc. Natl. Acad. Sci. USA, 85, 8487–8491.

Allison, D. S., and Schatz, G. (1986) Artificial mitochondrial presequences. Proc. Natl. Acad. Sci. USA, 83, 9011–9015.

Alt, J., and Herrmann, R. G. (1984) Nucleotide sequence of the gene for preapocytochrome f in the spinach plastid chromosome. Curr. Genet., 8, 551–557.

Alt, J., Morris, Westhoff, P., and Herrmann, R. G. (1984) Nucleotide sequence of the clustered genes for the 44 kDa chlorophyll a apoprotein and the "32 kd"-like protein of the photosystem II reaction center in the spinach plastid chromosome. Curr. Genet., 8, 597–606.

Altendorf, K., Harold, F. M., and Simoni, R. D. (1974) Impairment and restoration of the energized state in membrane vesicles of E. coli lacking adenosine triphosphatase. J. Biol. Chem., 249, 4587–4593.

Ames, G. F.-L. (1986) Bacterial periplasmic transport systems: Structure, mechanism and evolution. Ann. Rev. Biochem., 55, 397–425.

Ames, G. F.-L., Nikaido, K., Groarke, J. and Petithory, J. (1989) Reconstitution of periplasmic transport in inside-out membrane vesicles. Energization by ATP. J. Biol. Chem., 264, 3998–4002.

Amzel, L. M., McKinney, M., Narayanan, P., and Pedersen, P. L. (1982) Structure of the mitochondrial F_1 ATPase at 9-Å resolution. Proc. Natl. Acad. Sci. USA, 79, 5852–5856.

Amzel, L. M., Narayanan, P., and Pedersen, P. (1984) Quarternary structure of F_1-ATPase in H^+-ATPase. Structure, Function, and Biogenesis of the F_0F_1 Complex of Coupling Membranes (S. Papa, N. Altendorf, L. Ernster, L. Packer, eds.), pp. 125–134. Adriatica Editrice, Bari, Italy.

Andersen, O. S., Durkin, J. T., and Koeppe, R. E. (1988) Do amino acid substitutions alter the structure of gramicidin channels? Chemistry at the single molecule level. In Transport through Membranes: Carriers, Channels, and Pumps (A. Pullman, ed.), pp. 115–132, D. Reidel, Dordrecht.

Anderson, G. P., Draheim, J. E., and Gross, E. L. (1985) Plastocyanin conformation: the effect of the oxidation state on the pK_a of nitrotyrosine-83. Biochim. Biophys. Acta, 810, 123–131.

Anderson, J. M. (1981) Consequences of spatial separation of photosystems 1 and 2 in thylakoid membranes of higher plant chloroplasts. FEBS Lett., 142, 1–10.

Anderson, J. (1986) Photoregulation of the composition, function, and structure of thylakoid membranes. Annu. Rev. Plant Physiol., 37, 93–136.

Anderson, J. M., and Melis, A. (1983) Localization of different photosystems in separate regions of the thylakoid membrane. Proc. Natl. Acad. Sci. USA, 80, 745–749.

Andersson, B., and Haehnel, W. (1982) Location of photosystem I and photosystem II reaction centers in different thylakoid regions of stacked chloroplasts. FEBS Lett., 146, 13–17.

Andersson, B., Sundby, C., and Albertsson, P. (1980) A mechanism for the formation

of inside-out vesicles. *Biochim. Biophys. Acta, 599*, 301–402.
Andrews, K. M., Crofts, A. R., and Gennis, R. B. (1988) Spectroscopic characterization of purified ubiquinol cytochrome *c* reductase from *Rb. sphaeroides. Biophys. J., 53*, 32a.
Anraku, Y. and Gennis, R. B. (1987) The aerobic respiratory chain of *Escherichia coli. Trends. Biochem. Sci., 12*, 262–266.
Aprille, J. (1988) Regulation of the mitochondrial adenine nucleotide pool size in liver: mechanism and metabolic role. *FASEB J., 2*, 2547–2556.
Aquila, H., Link, T. A., and Klingenberg, M. (1985) The uncoupling protein from brown fat mitochondria is related to the mitochondrial ADP/ATP carrier. Analysis of sequence homologies and of folding of the protein in the membrane. *EMBO J., 4*, 2369–2376.
Aris, J. P., Klionsky, D. J., and Simoni, R. D. (1985) The subunits of the *Escherichia coli* F_1F_0-ATPase are sufficient to form a functional proton pore. *J. Biol. Chem., 260*, 11207–11215.
Arnon, D. I. (1984) The discovery of photosynthetic phosphorylation. *Trends. Biochem. Sci., 9*, 258–262.
Atta-Asafo-Adjei, E., and Dilley, R. A. (1985) Plastocyanin stimulation of whole chain and photosystem I electron transport in inside-out thylakoid vesicles. *Arch. Biochem. Biophys., 243*, 660–667.
Attardi, G., Chomyn, A., Doolittle, R. F., Mariottini, P., and Ragan, C. I. (1986) Seven unidentified reading frames of human mitochondrial DNA encode subunits of the respiratory chain NADH dehydrogenese. *Cold Spring Symp. Quant. Biol., 51*, 103–114.
Au, D. C. T., Lorence, R. M., and Gennis, R. G. (1985) Isolation and characterization of an *E. coli* mutant lacking the cytochrome *o* terminal oxidase. *J. Bacteriol., 161*, 123–127.
Axerod, D., Koppel, D. K., Schlessinger, J., Elson, E., and Webb, W. W. (1976) Mobility measurement by analysis of fluorescence photobleaching recovery kinetics. *Biophys. J., 16*, 1055–1069.
Babcock, G. T. (1986) The photosynthetic oxygen evolving process. *New Comprehensive Biochemistry—Photosynthesis* (J. Amesz, ed.), pp. 125–158, Elsevier, Amsterdam.
Babcock, G. T., and Callahan, P. M. (1983) Redox-linked hydrogen bond changes in cytochrome *a*: implications for a cytochrome oxidase proton pump. *Biochemistry, 22*, 2314–2318.
Babcock, G. T., Widger, W. R., Cramer, W. A., Oertling, W. A., and Metz, J. C. (1985) Axial ligands of chloroplast cytochrome *b*-559: identification and requirement for a heme-cross-linked polypeptide structure. *Biochemistry, 24*, 3638–3645.
Babior, B. M. (1987) The respiratory burst oxidase. *Trends Biochem. Sci., 12*, 241–243.
Baccarini-Melandri, A., Casadio, R., and Melandri, B. A. (1977) Thermodynamics and kinetics of photophosphorylation in bacterial chromatophores and their relation with the transmembrane electrochemical difference of protons. *Eur. J. Biochem., 78*, 389–402.
Baccarini-Melandri, A., Gabellini, N., Melandri, B. A., Hurt, E., and Hauska, G. (1980) Structural requirements of quinone coenzymes for endogenous and dye-mediated coupled electron transport in bacterial photosynthesis. *J. Bioenerg. Biomem., 12*, 95–110.
Baird, B., Pick, U., and Hamnes, G. G. (1979) Structural investigation of reconstituted

chloroplast ATPase with fluorencence measurements. *J. Biol. Chem.*, *254*, 3818–3825.
Bakker, E. P., and Randall, L. L. (1984) The requirement for energy during export of β-lactamase in *Escherichia coli* is fulfilled by the total protonmotive force. *EMBO J.*, *3*, 895–900.
Baldwin, J. M., Henderson, R., Beckman, E., and Zemlin, F. (1988) Images of purple membrane at 2.8 Å resolution obtained by cryo-electron microscopy. *J. Mol. Biol.*, *202*, 585–591.
Barber, J. (1982) Influence of surface charges on thylakoid structure and function. *Annu. Rev. Plant Physiol.*, *33*, 261–295.
Barber, J. (1983) Photosynthetic electron transport in relation to thylakoid membrane composition and organization. *Plant Cell Envir.*, *6*, 311–322.
Barber, J. (1987) Photosynthetic reaction centers: a common link. *Trends Biochem. Sci.*, *12*, 321–326.
Barr, R., and Crane, F. L. (1971) Quinones in algae and higher plants. In *Methods Enzymol., Vol. 23* (A. San Pietro, ed.), pp. 372–408, Academic Press, New York.
Barry, B. A., and Babcock, G. T. (1987) Tyrosine radicals are involved in the photosynthetic O_2-evolving system. *Proc. Natl. Acad. Sci. USA*, *84*, 7099–7103.
Batie, C. J., and Kamin, H. (1984) Ferredoxin:$NADP^+$ oxidoreductase, Equilibria in binary and ternary complexes with $NADP^+$ and ferredoxin. *J. Biol. Chem.*, *259*, 8832–8839.
Bechtold, R., and Bosshard, H. R. (1985) Structure of an electron transfer Complex II. Chemical modification of carboxyl groups of cytochrome *c* peroxidase in the presence and absence of cytochrome *c*. *J. Biol. Chem.*, *260*, 5191–5200.
Beck, J. C., and Rosen, B. P. (1979) Cation/proton antiport systems in *Escherichia coli*: properties of the sodium/proton antiporter. *Arch. Biochem. Biophys.*, *194*, 208–214.
Beinert, H., and Albracht, S. P. J. (1982) New insights, ideas, and unanswered questions concerning iron-sulfur clusters in mitochondria. *Biochim. Biophys. Acta*, *683*, 245–277.
Belknap, W. R., and Haselkorn, R. (1987) Cloning and light regulation of expression of the phycocyanin operon of the cyanobacterium *Anabaena*. *EMBO J.*, *6*, 871–884.
Bennett, J., Steinback, K. E., and Arntzen, C. J. (1980) Chloroplast phosphoproteins: regulation by excitation energy transfer of phosphorylation of thylakoid membrane polypeptides. *Proc. Natl. Acad. Sci. USA*, *77*, 5253–5257.
Bennett, J., Shaw, E. K., and Bakr, S. (1987) Phosphorylation of thylakoid proteins and synthetic peptide analogs. *FEBS Lett.*, *210*, 22–26.
Bensasson, R., and Land, E. J. (1973) Optical and kinetic properties of semireduced plastoquinone and semiquinone electron acceptor in photosynthesis. *Biochim. Biophys. Acta*, *325*, 175–181.
Benz, R., and McLaughlin, S. (1983) The molecular mechanism of action of the proton ionophore FCCP (carbonylcyanide-*p*-trifluoromethoxyphenylhydrazone). *Biophys. J.*, *41*, 381–398.
Benzinger, T., Kitzinger, C., Hems, R., and Burton, K. (1959) Free-energy changes of the glutaminase reaction and the hydrolysis of the terminal pyrophosphate bond of adenosine triphosphate. *Biochem. J.*, *71*, 400–407.
Berden, J. A., and Slater, E. C. (1972) The allosteric binding of antimycin A to cytochrome *b* in the mitochondrial membrane. *Biochim. Biophys. Acta*, *256*, 199–215.
Bergström, J. (1985) The EPR spectrum and orientation of cytochrome *b*-563 in the

chloroplast thylakoid membranes. *FEBS Lett.*, *183*, 87–90.

Bernardi, P., and Azzone, G. F., (1981) Cytochrome c as an electron shuttle between the outer and inner mitochondrial membranes. *J. Biol. Chem.*, *256*, 7187–7192.

Berry, E. A., and Hinkle, P. C. (1983) Measurement of the electrochemical gradient in submitochondrial particles. *J. Biol. Chem.*, *258*, 1471–1486.

Berry, E. A., and Trumpower, B. L. (1985) Isolation of ubiquinol oxidase from *Paracoccus denitrificans* and resolution into cytochrome b–c_1 and c–aa_3 complexes. *J. Biol. Chem.*, *260*, 2458–2467.

Berthold, D. A., Babcock, G. T., and Yocum, C. F. (1981) A highly resolved, oxygen-evolving photosystem II preparation from spinach thylakoid membranes. *FEBS Lett.*, *134*, 231–234.

Binder, A. (1982) Respiration and photosynthesis in energy-transducing membranes of cyanobacteria. *Bioenerg. Biomembr.*, *14*, 271–286.

Bisson, R., and Capaldi, R. A. (1981) Binding of arylazido cytochrome c to yeast cytochrome c peroxidase. *J. Biol. Chem.*, *256*, 4362–4367.

Black, M. T., Widger, W. R., and Cramer, W. A. (1987) Large-scale purification of active cytochrome b_6/f complex from spinach chloroplasts. *Arch. Biochem. Biophys.*, *252*, 655–661.

Blackwell, M. F., Gounaris, K., Zara, S. J., and Barber, J. (1987) A method for estimating lateral diffusion coefficients in membranes from steady-state fluorescence quenching studies. *Biophys. J.*, *51*, 735–744.

Blair, D. F., Ellis, W. R., Wang, H., Gray, H. B., and Chan, S. I. (1986) Spectroelectrochemical study of cytochrome c oxidase: pH and temperature dependence of the cytochrome potentials. *J. Biol. Chem.*, *261*, 11524–11537.

Blanck, A. and Oesterhelt, D. (1987) The halo-opsin gene. II. Sequence, primary structure of halorhodopsin and comparison with bacteriorhodopsin. *EMBO J.*, *6*, 265–273.

Blankenship, R. E. (1984) Primary photochemistry in green photosynthetic bacteria. *Photochem. Photobiol.*, *40*, 801–806.

Blankenship, R. E., and Prince, R. C. (1985) Excited-state redox potentials and the Z scheme of photosynthesis. *Trends Biochem. Sci.*, *10*, 382–384.

Blankenship, R. E., Trost, J. T., and Mancino, L. J. (1988) Properties of reaction centers from the green photosynthetic bacterium, *Chloroflexus aurantiacus*. In *Structure of Bacterial Reaction Centers* (J. Breton and A. Vermeglio, eds.), pp. 119–127, Plenum, New York.

Blasie, J. K., Harbette, L. G., Pascolini, G., Skita, V., Pierce, D. H. and Scarpa, A. (1985) Time-resolved x-ray diffraction studies of the sarcoplasmic reticulum membrane during active transport. *Biophys. J.*, *48*, 9–18.

Boekema, E. J., Berden, J. A., and van Heel, M. G. (1986) Structure of mitochondrial F_1-ATPase studied by electron microscopy and image processing. *Biochim. Biophys. Acta*, *851*, 353–360.

Boekema, E. J., Schmidt, G., Gräber, P., and Berden, J. A. (1987) Structure of the ATP synthase from chloroplasts and mitochondria studied by electron microscopy. *Z. Naturforsch.*, *43c*, 219–225.

Bogner, W., Aquila, H., and Klingenberg, M. (1986) The transmembrane arrangement of the ADP/ATP carrier as elucidated by the lysine reagent pyridoxal-5-phosphate. *Eur. J. Biochem.*, *161*, 611–620.

Bohm, D. (1951) *Quantum Theory*, Chap. 19, Prentice-Hall, Englewood Cliffs, NJ.

Bolli, R., Nalecz, K. A., and Azzi, A. (1985) The aggregation state of bovine cytochrome *c* oxidase and its kinetics in monomeric and dimeric forms. *Arch. Biochem. Biophys.*, 240, 102–116.

Booth, I. R. (1985) Regulation of cytoplasmic pH in bacteria. *Microbiol. Rev.*, 49, 359–378.

Booth, I. R., Mitchell, W. J., and Hamilton, W. A. (1979) Quantitative analysis of proton-linked transport systems. The lactose permease of *Escherichia coli*. *Biochem. J.*, 182, 687–696.

Bosshard, H. R., Bänziger, T. Hasler, and T. L. Poulos (1984) The cytochrome *c* peroxidase–cytochrome *c* electron transfer complex. The role of histidine residues. *J. Biol. Chem.*, 259, 5683–5690.

Bosshard, H. R., Davidson, M. W., Knaff, D. B., and Millett, F. (1986) Complex formation and electron transfer between mitochondrial cytochrome *c* and flavocytochrome *c* from *Chromatium vinosum*. *J. Biol. Chem.*, 261, 190–193.

Bosshard, H. R., Wynn, R. M., and Knaff, D. B. (1987) Binding site on *R. rubrum* cytochrome c_2 for *R. rubrum* cytochrome bc_1 complex. *Biochemistry*, 26, 7688–7693.

Botfield, M. C. and Wilson, T. H. (1988) Mutations that simultaneously alter both sugar and cation specificity in the melibiose carrier of *Escherchia coli*. *J. Biol. Sci.* 263, 12909–12515.

Boulay, F., Dalban, P., and Vignais, P. V. (1985) Photoaffinity labeling of mitochondrial ATPase by 2-azidoadenosine 5'-[α-^{2}P]diphosphate. *Biochemistry*, 24, 7372–7379.

Bowyer, J. R. and Ohnishi, T. (1985) EPR spectroscopy in the study of ubiquinones in redox chains, in *Coenzyme Q* (G. Lenaz, ed.), pp. 409–432, John Wiley, Chichester.

Bowyer, J. R., and Trumpower, B. L. (1981) Rapid reduction of cytochrome c_1 in the presence of antimycin and its implication for the mechanism of electron transfer in the cytochrome b–c_1 segment of the mitochondrial respiratory chain. *J. Biol. Chem.*, 256, 2245–2251.

Bowyer, J. R., Edwards, C. A., Ohnishi, T., and Trumpower, B. L. (1982) An analogue of ubiquinone which inhibits respiration by binding to the iron-sulfur protein of the cytochrome b–c_1 segment of the mitochondrial respiratory chain. *J. Biol. Chem.*, 257, 8321–8330.

Boyer, P. D. (1983) How cells make ATP. In *Biochemistry of Metabolic Processes* (B. L. F. Lennen, F. W. Stratman, and R. M. Zahiter, eds.), pp. 465–477, Elsevier, Amsterdam.

Boyer, P. D., Cross, R. L., and Momsen, W. (1973) A new concept for energy coupling in oxidative phosphorylation based on a molecular explanation of the oxygen exchange reactions. *Proc. Natl. Acad. Sci. USA*, 70, 2837–2839.

Braiman, M., Mogi, T., Stern, L. J., Hackett, N. R., Chao, B. H., Khorana, H. G., and Rothschild, K. J. (1988) Vibrational spectroscopy of bacteriorhodopsin mutants: I. Tyrosine-185 protonates and deprotonates during the photocycle. *Proteins*, 3, 219–229.

Brand, M. O., Reynafarje, B., and Lehninger, A. L. (1976) Stoichiometric relationship between energy-dependent proton ejection and electron transport in mitochondria. *Proc. Natl. Acad. Sci. USA*, 73, 437–441.

Brandl, C. J., Green, N. M., Korczak, B., and MacLennan, D. H. (1986) Two Ca^{2+} ATPase genes: homologies and mechanistic implications of deduced amino acid sequences. *Cell*, 44, 597–607.

Brasseur, R. (1988) Calculation of the three-dimensional structure of *Saccharomyces cerevisiae* cytochrome *b* inserted into a lipid matrix. *J. Biol. Chem.*, 263, 12571–12575.

Brautigan, D. L., Ferguson-Miller, S., and Margoliash, E. (1978). Mitochondrial cytochrome c. Preparation and modification of native and chemically modified cytochromes c. *Methods Enzymol.*, 53, (S. Fleischer and L. Packer, eds.), pp. 128–164. Academic Press, New York.

Bridger, W. A., and Henderson, J. F. (1983) *Cell ATP*, 170 pp., John Wiley, New York.

Briggs, M. C., and Gierasch, L. M. (1986) Molecular mechanisms of protein secretion. *Adv. Prot. Chem.* (C. B. Anfinsen, J. T. Edsall, and F. M. Richards, eds.), 38, 109–180, Academic Press, New York.

Bröger, C., Salardi, S., and Azzi, A. (1983) Interaction between isolated cytochrome c_1 and cytochrome c. *Eur. J. Biochem.*, 131, 349–352.

Brooker, R. T., and Wilson, T. H. (1985) Isolation and nucleotide sequencing of lactose carrier mutants that transport maltose. *Proc. Natl. Acad. Sci. USA*, 82, 3959–3963.

Brudvig, G. W., and Crabtree, R. H. (1986) Mechanism for photosynthetic O_2 evolution. *Proc. Natl. Acad. Sci. USA*, 83, 4586–4588.

Brune, D. C., Nozawa, T., and Blankenship, R. E. (1987) Antenna organization in green photosynthetic bacteria. 1. Oligomeric bacteriochlorophyll c in *Chloroflexus aurianticus* chlorosomes. *Biochemistry*, 26, 8644–8652.

Brunisholz, R. A., Zuber, H., Valentine, J., Lindsay, J. G., Wooley, K. G., and Cogdell, R. A. (1986) The membrane location of the B890-complex from *R. rubrum* and the effect of carotenoid on the conformation of its two apoproteins at the cytoplasmic surface. *Biochim. Biophys. Acta*, 849, 295–303.

Brunschwig, B. A., Deluive, P. J., English, A. M., Goldberg, M., Gray, H. B., Mayo, S. L., and Sutin, N. (1985) Kinetics of mechanisms of electron transfer between blue copper proteins and electronically excited chromium and ruthenium polypyridine complexes. *Inorg. Chem.*, 24, 3743–3749.

Bullough, D. A., and Allison, W. S. (1986a) Three copies of the β subunit must be modified to achieve complete inactivation of the bovine mitochondrial F_1 ATPase by 5'-p-fluorosulfonylbenzoyladenosine. *J. Biol. Chem.*, 261, 5722–5730.

Bullough, D. A., and Allison, W. S. (1986b) Inactivation of the bovine heart mitochondrial F_1-ATPase by 5'-p-fluorosulfonylbenzoyl [^{3}H]inosine is accompanied by modification of tyrosine-345 in a single β subunit. *J. Biol. Chem.*, 261, 14171–14177.

Bürgi, R., Suter, F., and Zuber, H. (1987) Arrangement of the light-harvesting chlorophyll a/b protein complex in the thylakoid membrane. *Biochim. Biophys. Acta*, 890, 346–351.

Butler, W. F., Calvo, R., Fredkin, D. R., Isaacson, R. A., Okamura, M. Y., and Feher, G. (1984) The electronic structure of Fe^{2+} in reaction centers from *Rb. sphaeroides*. *Biophys. J.*, 45, 947–973.

Butlin, J. D., Cox, G. D., and Gibson, F. (1971) Oxidative phosphorylation in *Escherichia coli* K12: mutations affecting magnesium ion- or calcium ion-stimulated adenosine triphosphatase. *Biochem. J.*, 124, 75–81.

Bylina, E., and Youvan, D. (1988) Directed mutations affecting spectroscopic and electron transfer properties of the primary donor in the photosynthetic reaction center. *Proc. Natl. Acad. Sci. USA*, 85, 7226–7230.

Cain, B. D., and Simoni, R. D. (1986) Impaired proton conductivity resulting from mutations in the a subunit of the F_0F_1 ATPase in *Escherichia coli*. *J. Biol. Chem.*, 261, 10043–10050.

Cain, B. D., and Simoni, R. D. (1988) Interaction between Glu-219 and His-245 within the a subunit of F_1F_0-ATPase in *E. coli*. *J. Biol. Chem.*, 263, 6606–7712.

Cammack, R. (1973) Super-reduction of *Chromatium* high-potential iron-sulfur protein in the presence of dimethyl sulfoxide. *Biochem. Biophys. Res. Commun.*, 54, 548–554.

Canaani, O., Barber, J., and Malkin, S. (1984) Evidence that phosphorylation and dephosphorylation regulate the distribution of excitation energy between the two photosystems *in vivo*: photoacoustic and fluorimetric study of an intact leaf. *Proc. Natl. Acad. Sci. USA*, 81, 1614–1618.

Cantley, L. C. (1981) Structure and mechanism of the (Na, K)-ATPase. *Curr. Top. Bioenerg.*, 11, 201–237.

Cantrell, A., and Bryant, D. A. (1987) Molecular cloning and nucleotide sequence of the *psaA* and *psaB* gene of the cyanobacterium *Synechococcus sp.* PCC 7002. *Plant Mol. Biol.*, 9, 453–468.

Capaldi, R. (1982) Arrangement of proteins in the mitochondrial inner membrane. *Biochim. Biophys. Acta*, 654, 291–706.

Capaldi, R. A. (1988) Mitochondria myopathies and respiratory chain proteins. *Trends Biochem. Sci.*, 13, 144–148.

Capaldi, R. A., Malatesta, F., and Darley-Usmar, V. M. (1983) Structure of cytochrome *c* oxidase. *Biochim. Biophys. Acta*, 726, 135–148.

Caplan, S. R. (1971) Non-equilibrium thermodynamics and its application to bioenergetics. *Curr. Top. Bioenerg.*, 4, (D. R. Sanadi, ed.), pp. 1–79, Academic Press, New York.

Carafoli, E., and Roman, I. (1980) Mitochondria and Disease. *Molecular aspects of medicine*. pp. 295–429, Pergamon Press, Oxford.

Carrillo, N., and Vallejos, R. H. (1987) Ferredoxin-NADP$^+$ reductase. In *The Light Reactions* (J. Barber, ed.), pp. 527–560, Elsevier, Amsterdam.

Carter, C. W., Kraut, J., Freer, S. T., Alden, R. A., Sieker, L. C., Adman, E., and Jensen, L. H. (1972) A comparison of Fe_4S_4 clusters in high potential iron protein and in ferredoxin. *Proc. Natl. Acad. Sci. USA*, 69, 3526–3529.

Carter, K. R., Tsai, A., and Palmer, G. (1981). The coordination environment of mitochondrial cytochromes. *FEBS Lett.*, 132, 243–246.

Casey, R. P., Njus, D., Radda, G. K., and Sehr, P. A. (1977) Active proton uptake by chromaffin granules: Observed by amine distribution and phosphorus-31 nuclear magnetic resonance techniques. *Biochemistry*, 16, 972–977.

Cattell, K. J., Lindop, C. R., Knight, I. G., and Beechey, R. B. (1971) The identification of the site of action of N, N'-dicyclohexylcarbodiimide as a proteolipid in mitochondrial membranes. *Biochem. J.*, 125, 169–177.

Chan, R. L., Carillo, N., and Vallejos, R. H. (1985) Isolation and sequencing of an active-site peptide from spinach ferredoxin-NADP$^+$ oxidoreductase after affinity labeling with periodate-oxidized NADP$^+$. *Arch. Biochem. Biophys.*, 240, 172–177.

Chan, T.-M., Ulrich, E. L. and Markley, J. L. (1983) Nuclear magnetic resonance studies of two-iron-two-sulfur ferredoxins. 4. Interactions with redox partners. *Biochemistry*, 22, 6002–6007.

Chance, B., and Powers, L. (1985) Structure of cytochrome oxidase redox centers in native and modified forms: a redox study. In *Current Topics in Bioenergetics* (C. P. Lee, ed.), 14, 1–19, Academic Press, New York.

Chance, B., and Williams, G. B. (1956) The respiratory chain and oxidative phosphorylation. *Adv. Enzymol.*, 17, 65–135.

Chandrasekhar, S. (1957) *An Introduction to the Study of Stellar Structure*. Chap. 1, Dover, New York.

Chang, C.-H., Tiede, D. M., Tang, J., Smith, U., Norris, J. R., and Schiffer, M. (1986)

Structure of *Rb. sphaeroides* R-26 reaction center. *FEBS Lett.*, 205, 82–86.
Chen, L., and Tai, P. C. (1985) ATP is essential for protein translocation into *Escherichia coli* membrane vesicles. *Proc. Natl. Acad. Sci. USA*, 82, 4384–4388.
Chernyak, B. V., and Kozlov, I. A. (1986) Regulation of H^+-ATPases in oxidative- and photophosphorylation. *Trends Biochem. Sci.*, 11, 32–35.
Chernyak, B. V., Dibrov, P. A., Glagolev, A. N., Sherman, M. Yu, and Skulachev, V. R. (1983) A novel type of energetics in a marine alkalitolerant bacterium. $\Delta\mu_{Na^+}$- driven motility and sodium cycle. *FEBS Lett.*, 164, 38–42.
Chiang, G., and Dilley, R. A. (1987) Evidence for Ca^{++}-gated proton fluxes in chloroplast thylakoid membranes: Ca^{++} controls a localized to delocalized proton gradient switch. *Biochemistry*, 26, 4911–4916.
Chitnis, P. R., and Thornber, J. P. (1988) The major light-harvesting complex of photosystem II: aspects of its molecular and cell biology. *Photosyn. Res.*, 16, 41–63.
Chitnis, P. R., Nechushtai, R., and Thornber, J. P. (1987) Insertion of the precursor of the light-harvesting chlorophyll a/b protein into the thylakoid requires the presence of a developmentally regulated stromal factor. *Plant Mol. Biol.*, 10, 3–11.
Chomyn, A., Mariottini, P., Cleater, M. W. J., Rogen, C. I., Satsuno-Yagi, A., Hatefi, Y., Doolittle, R. F., and Attardi, G. (1985) Six unidentified reading frames of human mitochondrial DNA encode components of the respiratory-chain NADH dehydrogenase. *Nature*, 314, 592–597.
Chothia, C. (1976) The nature of the accessible and buried surfaces in proteins. *J. Mol. Biol.*, 105, 1–14.
Clark, A. J., Cotton, N. P. J., and Jackson, J. B. (1983) The relation between membrane ionic current and ATP synthesis in chromatophores from *Rps. capsulata*. *Biochim. Biophys. Acta*, 723, 440–453.
Clark, R. D., and Hind, G. (1983a) Isolation of a five polypeptide cytochrome $b-f$ complex from spinach chloroplasts. *J. Biol. Chem.*, 250, 10348–10354.
Clark, R. D., and Hind, G. (1983b) Spectrally distinct cytochrome b-563 components in a chloroplast b_6-f complex: interaction with a hydroxyquinoline-N-oxidase. *Proc. Natl. Acad. Sci. USA*, 80, 6249–6253.
Clark, W. M. (1960) *Oxidation–Reduction Potentials of Organic Systems*. 584 pp. Williams Wilkins, Baltimore.
Clayton, R. K. (1980) *Photosynthesis: Physical Mechanism and Chemical Patterns*, 281 pp., Cambridge Univ. Press, Cambridge.
Cline, C. (1986) Import of proteins into chloroplasts. *J. Biol. Chem.*, 261, 14804–14810.
Cline, J. F., Hoffman, B. M., Mims, W. B., LaHaie, E., Ballou, D. P. and Fee, J. A. (1985), Evidence for N coordination of Fe in the [2Fe-2S] clusters of *Thermus* Rieske protein and phthalate dioxygenase from *Pseudomonas*. *J. Biol. Chem.*, 260, 3251–3254.
Closs, G. R., and Miller, J. R. (1988) Intramolecular long-distance electron transfer in organic molecules. *Science*, 240, 440–447.
Cobb, A. D., and Knaff, D. B. (1985) Active transport of nonpolar amino acids in *Chromatium vinosum*. *Arch. Biochem. Biophys.*, 238, 97–110.
Cohn, E. J. and Edsall, J. T. (1943) *Proteins, Amino Acids and Peptides as Ions and Dipolar Ions*, Chaps. 8 and 9, Reinhold, New York.
Colbeau, A., Nachbaur, J. and Vignais, P. M. (1971) Enzymic characterization and lipid composition of rat liver subcellular membranes. *Biochim. Biophys. Acta*, 249, 462–492.

Cole, J., Yachandra, V. K., Guiles, R. D., McDermott, A. E., Britt, R. D., Dexheiner, S. L., Sauer, K., and Klein, M. P. (1987) Assignment of the g = 4.1 EPR signal to manganese in the S_2 state of the photosynthetic OEC: an EXAFS study. *Biochim. Biophys. Acta*, 890, 395–398.

Colman, P. M., Freeman, H. C., Guss, J. M., Murata, N., Norris, V. A., Ramshaw, J. A. M., and Venkatappa, M. P. (1978) x-ray crystal structure analysis of plastocyanin at 2.7 Å resolution. *Nature*, 272, 319–324.

Cornell, B. A., Keniry, M. A., Post, A., Robertson, R. N., Weir, L. E., and Westerman, P. W. (1987) Location and activity of ubiquinone 10 and ubiquinone analogues in model and biological membranes. *Biochim. Biophys. Acta*, 26, 7702–7707.

Coughlan, S. J., and Hind, G. (1987) A protein kinase that phosphorylates lightharvesting complex is autophosphorylated and is associated with photosystem II. *Biochemistry*, 26, 6515–6521.

Cox, G. B., Fimmel, A. L., Gibson, F., and Hatch, L. (1986) The mechanism of ATP synthesis: a reassessment of the function of the *b* and *a* subunits. *Biochim. Biophys. Acta*, 849, 62–68.

Cozens, A. L., and Walker, J. E. (1987) The organization and sequence of the genes for ATP synthase subunits in the cyanobacterium *Synechococcus* 6301. Support for an endosymbiotic origin of chloroplasts *J. Mol. Biol.*, 194, 359–383.

Cozens, A. L., Walker, J. E., Phillips, A. L., Huttly, A. K., and Gray, J. C. (1986) A sixth subunit of ATP synthase, an F_0 component, is encoded in the pea chloroplast genome. *EMBO J.*, 5, 217–222.

Cramer, W. A., and Butler, W. L. (1969) Potentiometric titration of the fluorescence yield of spinach chloroplasts. *Biochim. Biophys. Acta*, 172, 503–510.

Cramer, W. A., and Crofts, A. R. (1982) Electron and proton transport. in *Photosynthesis: Energy Conversion by Plants and Bacteria*, vol. I, pp. 387–467. Academic Press, New York.

Cramer, W. A., Widger, W. R., Herrmann, R. G., and Trebst, A. (1985) Topography and function of thylakoid membrane proteins. *Trends Biochem. Sci.*, 10, 325–329.

Cramer, W. A., Theg, S. M., and Widger, W. R. (1986) On the structure and function of cytochrome *b*-559. *Photosyn. Res.*, 10, 393–403.

Cramer, W. A., Widger, W. R., Black, M. T., and Girvin, M. (1987) Structure and function of photosynthetic bc_1 and $b_6 f$ complexes, in *The Light Reactions, Vol. 8* (J. Barber, ed.), pp. 447–493, Elsevier, Amsterdam.

Crane, F. L. (1977) Hydroquinone dehydrogenases. *Annu. Rev. Biochem.*, 46, 439–469.

Crane, F. L., and Barr, R. (1985) Chemical structure and properties of coenzyme Q and related compounds, in *Coenzyme Q* (G. Lenaz, ed.), pp. 1–37, John Wiley, Chichester.

Crane, F. L., Hatefi. Y., Lester, R. L., and Widmer, C. (1957) Isolation of a quinone from beef heart mitochondria. *Biochim. Biophys. Acta*, 25, 220–221.

Creighton, S., Hwang, J.-K., Warshel, A., Parson, W. W., and Norris, J. (1988) Simulating the dynamics of the primary charge separation process in bacterial photosynthesis. *Biochemistry*, 27, 774–781.

Critchley, C. (1985) The role of chloride in photosystem II. *Biochim. Biophys. Acta*, 811, 33–46.

Crofts, A. R. (1985) The mechanism of the ubiquinol:cytochrome *c* oxido-reductases of mitochondria and *Rb. sphaeroides*, *The Enzymes of Biological Membranes* (A. Mortonisi, ed.), 2nd ed., vol. 4, pp. 347—382.

Crofts, A. R., Meinhardt, S. W., Jones, K. R., and Snozzi, M. (1983) The role of the quinone pool in the cyclic electron transfer chain of *Rb. sphaeroides. Biochim. Biophys. Acta*, 723, 202–218.

Crofts, A. R., Robinson, H. H., and Snozzi, M. (1984) Reactions of quinones at catalytic sites: a diffusional role in H-transfer. In *Advances in Photosynthesis Research* (C. Sybesma, ed.), I, pp. 461–468, M. Nijhoff/W. Junk, The Hague.

Cross, R. L., and Boyer, P. D. (1975) The rapid labeling of adenosine triphosphate and the exchange of phosphate oxygens as related to conformational coupling in oxidative phosphorylation. *Biochemistry*, 14, 392–398.

Cross, R. L., and Nalin, C. M. (1982) Adenine nucleotide binding sites on beef heart F_1-ATPase. *J. Biol. Chem.*, 257, 2874–2881.

Cross, R. L., Grubmeyer, C., and Penefsky, H. S. (1982) Mechanism of ATP hydrolysis by beef heart mitochondrial ATPase: rate enhancements resulting from cooperative interactions between multiple catalytic sites. *J. Biol. Chem.*, 257, 12101–12105.

Cross, R. L., Cunningham, D., Miller, C. G., Xue, Z., Zhou, J.-M., and Boyer, P. D. (1987) Adenine nucleotide binding sites on beef heart F_1 ATPase: Photoaffinity labeling of β-subunit Tyr-368 at a noncatalytic site. *Proc. Natl. Acad. Sci. USA*, 84, 5715–5719.

Csonka, L. N. (1982) A third L-proline permease in *S. typhimurium* which functions in media of elevated osmotic strength. *J. Bacteriol.*, 151, 1433–1443.

Daldal, F. (1987) Molecular genetic approaches to studying the structure and function of the cytochrome c_2 and the cytochrome bc_1 complex of *Rb. capsulatus*. In *Cytochrom System: Molecular Biology and Bioenergetics* (S. Papa, B. Chance, and L. Ernster, eds.), pp. 23–34, Plenum Press, New York.

Daldal, F., Cheng, J., Applebaum, J., Davidson, E., and Prince, R. C. (1986) Cytochrome c_2 is not essential for photosynthetic growth of *Rb. capsulatus. Proc. Natl. Acad. Sci. USA*, 83, 2012–2016.

Daldal, F., Davidson, E., and Cheng, S. (1987) Isolation of the structural genes for the Rieske Fe-S protein, cytochrome b and cytochrome c_1 of *Rb. capsulatus. J. Mol. Biol.*, 195, 1–12.

Damoder, R., Klimov, V. V., and Dismukes, G. C. (1986) The effect of Cl^- depletion and X^- reconstitution on the O_2 evolution rate, the yield of the multiline manganese EPR signal and EPR signal II in the isolated photosystem II complex. *Biochim. Biophys. Acta*, 848, 378–391.

Daniels, C. J., Bole, D. G., Quay, S. C., and Oxender, D. L. (1981) Role for membrane potential in the secretion of protein into the periplasm of *E. coli. Proc. Natl. Acad. Sci. USA*, 78, 5396–5400.

Davidson, E., and Daldal, F. (1987a) Primary structure of the bc_1 complex of *Rhodopseudomonas capsulata*. Nucleotide sequence of the *pet* operon encoding the Rieske, cytochrome b, and cytochrome c_1 apoproteins. *J. Mol. Biol.*, 195, 13–24.

Davidson, E., and Daldal, F. (1987b) *fbc* operon encoding the Rieske Fe-S protein, cytochrome b and c_1 apoproteins previously described from *Rb. sphaeroides*, is from *Rb. capsulata. J. Mol. Biol.*, 195, 25–29.

Davis, D. J., Frame, M. K., and Johnson, D. A. (1988) Resonance Raman spectroscopy indicates a lysine as the sixth iron ligand in cytochrome *f*. *Biochim. Biophys. Acta*, 936, 61–66.

Davis, M. S., Forman, A., and Fajer, J. (1979) Ligated chlorophyll cation radicals: their function in photosystem II of plant photosynthesis. *Proc. Natl. Acad. Sci. USA, 76*, 4170–4174.

de Kouchkovsky, Y., Haraux, F., and Sigalet, C. (1982) Effect of (^{1}H-^{2}H) exchange on energy-coupled processes in thylakoids. *FEBS Lett., 139*, 245–249.

de Kouchkovsky, Y., Haraux, F., and Sigalat, C. (1984) A microchemiosmotic interpretation of energy-dependent processes in biomembranes based on the photosynthetic behavior of thylakoids. *Bioelectrochem. Bioenerg., 13*, 143–162.

de Lorimier, R., Bryant, D. A., Porter, R. D., Lin, W.-Y., Jay, E., and Stevens, S. E. (1984) Genes for the α and β subunits of phycocyanin. *Proc. Natl. Acad. Sci. USA, 81*, 7946–7950.

de Meis, L., Behrens, M. I., and Petretski, J. H. (1985) Contribution of water to the free energy of hydrolysis of pyrophosphate. *Biochemistry 24*, 7783–7789.

de Paula, J. C., Beck, W. F., and Brudvig, G. W. (1986) Magnetic properties of manganese in the photosynthetic O_2-evolving complex. 2. Evidence for a manganese tetramer. *J. Am. Chem. Soc., 108*, 4002–4009.

de Vitry, C., and Diner, B. A. (1984) Photoaffinity labeling of the atrazine receptor site in reaction centers of *Rb. sphaeroides*. *FEBS Lett., 167*, 327–331.

de Vries, S., Berden, J. A., and Slater, E. C. (1980) Properties of a semiquinone anion located in the QH_2:cytochrome *c* oxidoreductase segment of the mitochondrial respiratory chain. *FEBS Lett., 122*, 143–149.

de Vries, S., Albracht, S. P. J., Berden, S. A., and Slater, E. C. (1981) A new species of bound ubisemiquinone anion in QH_2:cytochrome *c* oxidoreductase. *J. Biol. Chem., 256*, 11996–11998.

Dean, G. E., Nelson, P. J., and Rudnick, G. (1986) Characterization of native and reconstituted hydrogen ion pumping ATPase from chromaffin granules. *Biochemistry, 25*, 4918–4925.

Deatherage, J. F., Henderson, R., and Capaldi, R. A. (1982) Three dimensional structure of cytochrome *c* oxidase vesicle crystals in negative stain. *J. Mol. Biol., 158*, 487–499.

Debus, R. J., Barry, B. A., Babcock, G. T., and McIntosh, L. (1988) Site-directed mutagenesis identifies a tyrosine radical involved in the photosynthetic oxygen-evolving system. *Proc. Natl. Acad. Sci. USA, 85*, 427–430.

Deckers-Hebestreit, G., Schmid, R., Klitz, H., and Altendorf, K. (1987) F_0 portion of *Escherichia coli* ATP synthase: orientation of subunit *c* in the membrane. *Biochemistry, 26*, 5486–5492.

Deisenhofer, J., Michel, H., and Huber, R. (1985a) The structural basis of photosynthetic light reactions in bacteria. *Trends Biochem. Sci., 10*, 243–248.

Deisenhofer, J., Epp, O., Miki, K., Huber, R., and Michel, H. (1985b) Structure of the protein subunits in the photosynthetic center of *Rhodopseudomonas viridis* at 3 Å resolution. *Nature, 318*, 618–624.

Dekker, J. P., Plijter, J. J., Ouwehand, L., and Van Gorkum, H. J. (1984a) Kinetics of manganese redox transitions in the oxygen-evolving apparatus of photosynthesis. *Biochim. Biophys. Acta, 767*, 176–179.

Dekker, J. P., Van Gorkum, H. J., Wensink, J., and Ouwehand, L. (1984b) Absorbance difference spectra of the successive redox states of the oxygen-evolving apparatus of photosynthesis. *Biochim. Biophys. Acta, 767*, 1–9.

DeVault, D. (1984) *Quantum-Mechanical Tunneling in Biological Systems*. 207 pp., Cambridge University Press, Cambridge.

DeVault, D. (1986) Vibronic coupling to electron transfer and the structure of the *R. viridis* reaction center. *Photosyn. Res.*, *10*, 125–137.

DeVault, D., and Chance, B. (1966) Studies of photosynthesis using a pulsed laser. I. Temperature dependence of cytochrome oxidation rate in *Chromatium*. Evidence for tunneling. *Biophys. J.*, *6*, 825–847.

di Rago, J.-P., and Colson, A.-M. (1988) Molecular basis for resistance to antimycin and diuron, Q cycle inhibitors acting at the Q_i site in the mitochondrial ubiquinol-cytochrome *c* reductase in *S. cerevisiae. J. Biol. Chem. 263*, 12564–12570.

Dibrov, P. A., Kostyrko, V. A., Lazarova, R. L., Skulachev, V. P., and Smirnova, I. A. (1986) The sodium cycle. I. Na^+-dependent mobility and modes of membrane energization in the marine alkali-tolerant *Vibrio olginolyticus*. *Biochim. Biophys. Acta*, *850*, 449–457.

Dibrov, P. A., Lazarova, R. L., Skulachev, V. P., and Verkhovskaya, M. L. (1986) The sodium cycle II. Na^+-coupled oxidative phosphorylation in *vibrio alginolyticus* cells. *Biochim. Biophys. Acta*, *850*, 458–465.

Dickerson, R. E. (1980a) Cytochrome *c* and the evolution of energy metabolism. *Sci. Am.* *242*, 136–153.

Dickerson, R. E. (1980b) Evolution and gene transfer in photosynthetic bacteria. *Nature*, *283*, 210–212.

Dickerson, R. E., and Timkovich, R. (1975) Cytochromes *c*, in *The Enzymes*, XI (P. D. Boyer, ed.), pp. 397–547, Academic Press, New York.

Dilley, R. A. (1971) Coupling of ion and electron transport in chloroplasts. *Curr. Top. Bioenerg.*, *4*, 237–271.

Dilley, R. A., Theg, S. M., and Beard, W. A. (1987) Membrane-proton interactions in chloroplast bioenergetics. Localized proton domains. *Annu. Rev. Plant Physiol.* 347–389.

Dimroth, P. (1981) Reconstitution of sodium transport from purified oxaloccetate decarboxylase and phospholipid vesicles. *J. Biol. Chem.*, *256*, 11974–11976.

Dimroth, P. (1982a) The role of biotin and sodium in the decarboxylation of oxaloacetate by the membrane-bound oxaloacetate decarboxylase from *Klebsiella aerogenes. Eur. J. Biochem.*, *121*, 435–441.

Dimroth, P. (1982b) The generation of an electrochemical gradient of sodium ions upon decarboxylation of oxaloacetate by the membrane-bound and Na^+-activated oxaloacetate decarboxylase from *Klebsiella aerogenes. Eur. J. Biochemistry*, *121*, 443–449.

Dimroth, P., and Thomer, A. (1988) Dissociation of the sodium-ion-translocating oxalocetate decarboxylase of *Klebsiella pneumoniae* and reconstitution of the active complex from the isolated subunits. *Eur. J. Biochem.*, *175*, 175–180.

Diner, B., and Petrouleas, V. (1987) Light-induced oxidation of the acceptor side Fe(II) of photosystem II by exogenous quinones acting through the Q_B binding site. II. Blockage by inhibitors and their effects on the Fe(III) EPR spectra. *Biochim. Biophys. Acta*, *893*, 138–148.

Diner, B. A., Ries, D. F., Cohen, B. N., and Metz, J. G. (1988) COOH-terminal processing of polypeptide DI in the photosystem II reaction center of *S. obliquus* is necessary for the assembly of the O_2-evolving complex. *J. Biol. Chem, 263*, 8972–8980.

Dismukes, G. C., and Siderer, Y. (1981) Intermediates of a polynuclear manganese center involved in photosynthetic oxidation of water. *Proc. Natl. Acad. Sci. USA*, *78*, 274–278.

Dixit, B. P. S. N., Waring, A. J., Wells, K. O., Wong, P. S., Woodrow, G. V., and Vanderkooi, J. M. (1982) Rotational motion of cytochrome c derivatives bound to membranes measured by fluorescence and phosphorescence anisotropy. *Eur. J. Biochem.*, 126, 1–9.

Dockter, M. E., Steinemann, A., and Schatz, G. (1978) Mapping of yeast cytochrome c oxidase by fluorescence resonance energy transfer. *J. Biol. Chem.*, 253, 311–317.

Doetsch, R. N., and Sjoblad, R. D. (1980) Flagellar structure and function in eubacteria. *Annu. Rev. Microbiol.*, 34, 69–108.

Dolla, A., and Bruschi, M. (1988) The cytochrome c_3-ferredoxin electron transfer complex: cross-linking studies. *Biochim. Biophys. Acta*, 932, 26–32.

Douce, R. (1985) *Mitochondria in Higher Plants: Structure, Function, Biogenesis.* Academic Press, Orlando.

Douce, R., and Neuberger, M. (1987) Specific properties of plant mitochondria. In *Plant Membranes: Structure, Function, Biogenesis* (eds., C. Leaver and H. Sze), pp. 3–26, Alan R. Liss, New York.

Doyle, M. P., and Yu, C.-A. (1985) Preparation and reconstitution of a phospholipid deficient cytochrome b_6–f complex from spinach chloroplasts. *Biochem. Biophys. Res. Commun.*, 131, 700–706.

Doyle, M. P., Li, L.-B., Yu, L., and Yu, C.-H. (1989) Identification of a $M_r = 17,000$ protein as the plastoquinone-binding protein in the cytochrome b_6–f complex from spinach chloroplasts. *J. Biol. Chem.*, 264, 1387–1392.

Driessen, M., Postma, P. W., and van Dam, K. (1987) Energetics of glucose uptake in *Salmonella typhimurium*. *Arch. Microbiol.*, 146, 358–361.

Dunahay, T. G., Staehelin, L. A., Siebert, M., Ogilvie, P. D., and Berg, S. P. (1984) Structural, biochemical, and biophysical characterization of four oxygen-evolving photosystem II preparations from spinach. *Biochim. Biophys. Acta*, 764, 179–193.

Dunn, R. J., McCoy, J., Simsek, M., Mahnudar, A., Chang, S. H., Rajbhandary, U. L., and Khorana, H. G. (1981) The bacteriorhodopsin gene. *Proc. Natl. Acad. Sci. USA*, 78, 6744–6748.

Dunn, R. J., Hackett, N. R., McCoy, J. M., Chao, B. H., Kimura, K., and Khorana, H. G. (1987) Structure-function studies in bacteriorhodopsin. I. Expression of the bacterio-opsin gene in *E. coli J. Biol. Chem.*, 262, 9246–9254.

Dupuis, A., Zaccai, G., and Satre, M. (1983) Optical properties and small-angle neutron scattering of bovine heart mitochondrial oligomycin sensitivity conferring protein. *Biochemistry*, 22, 5951–5956.

Dus, K., Tedro, S. and Bartsch, R. G. (1973) The complete amino acid sequence of *Chromatium* high potential iron-sulfur proteins. *J. Biol. Chem.*, 248, 7318–7331.

Dutton, P. L. (1978) Redox potentiometry: determination of midpoint potentials of oxidation-reduction components of biological electron transport systems. *Methods Enzymol.*, LIV, 411–435.

Dutton, P. L., Kihara, T., McCray, J. A., and Thornber, J. P. (1971) Cytochrome c-553 and bacteriochlorophyll interaction at 77°K in chromatophores and a sub-chromatophore preparation from *Chromatium* D. *Biochim. Biophys. Acta*, 226, 81–87.

Dux, L., Taylor, K. A., Beall, H. P. T., and Martonosi, A. (1985) Crystallization of sarcoplasmic reticulum by calcium and lanthanide ions. *J. Biol. Chem.*, 260, 11730–11743.

References

Egnius, H., Heber, U., Matthiesen, V., and Kirk, M. (1975) Reduction of oxygen by the electron transport chain of chloroplasts during assimulation of carbon dioxide. *Biochim. Biophys. Acta*, 408, 252–268.

Eilers, M., Oppliger, W., and Schatz, G. (1987) Both ATP and an energized membrane are required to import a purified precursor protein into mitochondria. *EMBO J.*, 6, 1073–1077.

Eisenberg, D. (1984) Three dimensional structure of membrane and surface proteins. *Annu. Rev. Biochem.*, 53, 595–623.

Eisenberg, D., and Crothers, D. (1979) *Physical Chemistry with Applications to the Life Sciences*, Chaps. 2 and 9, Benjamin/Cummings.

Eisenberg, D., Weiss, R. M., and Terwilliger, T. C. (1984) The hydrophobic moment detects periodicity in protein hydrophobicity. *Proc. Natl. Acad. Sci. USA*, 81, 140–144.

Elferink, M. G. L., Hellingwerf, K. J., Nano, F. E., Kaplan, S., and Konings, W. N. (1983) The lactose carrier of *Escherichia coli* functionally incorporated in *R. sphaeroides* obeys the regulatory conditions of the phototrophic bacterium. *FEBS Lett.* 164, 185–190.

Engelhard, M., Gerwert, K., Hess, B., Kreutz, W., and Siebert, F. (1985) Light-driven protonation changes of internal aspartic acids of bacteriorhodopsin. An investigation by static and time-resolved infrared difference spectroscopy using [4-^{13}C] aspartic acid labeled purple membrane. *Biochemistry*, 24, 400–407.

Engelman, D. M., and Steitz, T. A. (1981) The spontaneous insertion of proteins into and across membranes: the helical hairpin hypothesis. *Cell*, 23, 411–422.

Engelman, D. M., Steitz, T. A., and Goldman, A. (1986) Identifying nonpolar transbilayer helices in amino acid sequences of membrane protein. *Annu. Rev. Biophys. Biophys. Chem.*, 15, 321–353.

Epstein, W. (1985) The Kdp system: A bacterial K^+ transport ATPase. In *Current Topics in Membranes and Transport, Vol. 23* (E. A. Adelberg and C. W. Slayman, eds.) pp. 153–175, Academic Press, New York.

Epstein, W., and Laimins, L. (1980) Potassium transport in *Escherichia coli*: diverse systems with common control by osmotic forces. *Trends Biochem. Sci.*, 5, 21–23.

Erecínska, M., D. F. Wilson, and J. K. Blasie (1978) Studies on the orientation of the mitochondrial redox carriers II. Orientation of the mitochondrial chromophores with respect to the plane of the membrane in hydrated, oriented mitochondrial multilayers. *Biochim. Biophys. Acta*, 501, 63–71.

Fajer, J., Brune, D. C., Davis, M. S., and Spaulding, L. D. (1975) Primary charge separation in bacterial photosynthesis: oxidized chlorophyll and reduced pheophytin. *Proc. Natl. Acad. Sci. USA*, 72, 4956–4960.

Falk, G., and Walker, J. E. (1988) DNA sequence of a gene cluster coding for subunits of the F_0 membrane sector of ATP synthase in *Rhodospirillum rubrum*. *Biochem. J.*, 254, 109–122.

Falk, J. E. (1964) *Porphyrins and Metalloporphyrins*, Chaps. 1 and 5, Elsevier, Amsterdam.

Farchaus, J. W., Widger, W. R., Cramer, W. A., and Dilley, R. A. (1982) Kinase-induced changes in electron transport rates of spinach chloroplasts. *Arch. Biochem. Biophys.*, 217, 362–367.

Fato, R., Battino, M., degli Esposti, M., Parenti-Castelli, G., and Lenaz, G. (1986) Determination of partition and lateral diffusion coefficients of ubiquinones by fluorescence quenching of n-(9-anthroyloxy) stearic acids in phospholipid vesicles and mitochondrial membranes. *Biochemistry*, 25, 3378–3390.

Fee, J. A., Kuila, D., Mather, M. W., and Yoshida, T. (1986) Respiratory proteins from extremely thermophilic aerobic bacteria. *Biochim. Biophys. Acta*, 853, 153–185.

Feher, G. (1969) Electron paramagnetic resonance with applications to selected problems in biology. In *Physical Problems in Biological Systems*, Les Houches School, pp. 253–365. Gordon and Breach, New York.

Feher, G., Arno, T. R., and Okamura, M. Y. (1988) The effect of an electric field on the charge recombination rate of $D^+ Q_A^- \to DQ_A$ in reaction center from *Rb. sphaeroides* R-26 in *The Photosynthetic Bacterial Reaction Center*, (J. Breton and A. Vermeglio, eds.), pp. 271–287, Plenum, New York.

Feick, R. G. and Fuller, R. C. (1984) Topography of the photosynthetic apparatus of *Chloroflexus aurantiacus*. *Biochemistry*, 23, 3693–3700.

Felle, H., Porter, J. S., Slayman, C. L., and Kaback, H. R. (1980) Quantitative measurements of membrane potential in *Escherichia coli*. *Biochemistry*, 19, 3585–3590.

Fenton, J. M., Pellin, M. J., Govindjee, and Kaufman, K. J. (1979) Primary photochemistry of the reaction center of photosystem I. *FEBS Lett.*, 100, 1–4.

Ferguson, S. J. (1985) Fully delocalized chemiosmotic or localized proton flow pathways in energy coupling? A scrutiny of experimental evidence. *Biochim. Biophys. Acta*, 811, 47–95.

Ferguson, S. J. (1986) The ups and downs of P/O ratios (and the question of nonintegral coupling stoichiometries for oxidative phosphorylation and related processes). *Trends Biochem. Sci.*, 11, 351–353.

Ferguson-Miller, S., Brautigan, D. L., and Margoliash, E. (1978) Definition of cytochrome *c* binding domains by chemical modification. III. Kinetics of reaction of carboxynitrophenyl cytochrome *c* with cytochrome *c* oxidase. *J. Biol. Chem.*, 257, 4426–4437.

Ferro-Luzzi Ames, G. (1986) Bacterial periplasmic transport systems: structure, mechanism and evolution. *Annu. Rev. Biochem.*, 55, 397–425.

Fillingame, R. H. (1984) F_0 sector of *E. coli* ATP synthase. H^+ *ATPase: Structure, Function, Biogenesis of the F_0F_1 Complex of Coupling Membranes* (S. Papa, K. Altendorf, L. Ernster, and L. Packer, eds.) pp. 109–118, Adriatica Editrice, Bari.

Fillingame, R. H., Peters, L. K., White, L. K., Mosher, M. E., and Paule, C. R. (1984) Mutations altering Asp-61 of the Ω subunit of *E. coli* H^+-ATPase differ in effect on coupled ATP hydrolysis. *J. Bacteriol.*, 158, 1078–1083.

Fimmel, A. L., Jans, D. A., Langman, L., James, L. B., Downie, J. A., Ash, G. R., Senior, A. E., Gibson, F., and Cox, G. B. (1983) The F_1F_0-ATPase of *Escherichia coli*. Substitution of proline by leucine at position 64 in the *c*-subunit causes loss of oxidative phosphorylation. *Biochem. J.*, 213, 451–458.

Fimmel, A. L., Jans, D. A., Hatch, L., James, L. B., Gibson, F., and Cox, G. B. (1985) The F_1F_0-ATPase of *E. coli*. The substitution of alanine by threonine at position 25 in the *c* subunit affects function but not assembly. *Biochim. Biophys. Acta*, 808, 252–258.

Finel, M., and Wikström, M. K. F. (1986) Studies on the role of the oligomeric state and subunit III of cytochrome oxidase in proton translocation. *Biochim. Biophys. Acta*, 851, 99–108.

Finel, M., Haltin, T., Holm, L., Jalli, T., Metso, T., Puuatinen, A., Raitio, M. Saraste, M., and Wikström, M. K. F. (1987) Universal features in cytochrome oxidase. In

Cytochrome Systems, Molecular Biology and Bioenergetics (S. Papa, B. Chance, and L. Ernster, eds.), pp. 247–252, Plenum Press, New York.

Finzel, B. C., Poulos, T. L., and Kraut, J. (1984) Crystal structure of yeast cytochrome *c* peroxidase refined at 1.7 Å resolution. *J. Biol. Chem.*, *259*, 13027–13036.

Fish, L. E., Kuck, V., and Bogorad, L. (1985) Two partially homologous adjacent light-inducible maize chloroplast genes encoding polypeptides of the P700 chlorophyll *a*-protein complex of photosystem I. *J. Biol. Chem.*, *260*, 1413–1421.

Fleming, G. R., Martin, J. L., and Breton, J. (1988) Rates of primary electron transfer in photosynthetic reaction centres and their mechanistic implications. *Nature, 333*, 190–192.

Flores, S., and Ort, D. R. (1984) Investigation of the apparent inefficiency of the coupling between photosystem II electron transport and ATP formation. *Biochim. Biophys. Acta, 766*, 289–302.

Flügge, V. I., and Hinz, G. (1986) Energy dependence of protein translocation into chloroplasts. *Eur. J. Biochem.*, *160*, 563–570.

Förster, T. (1959) Transfer mechanisms of electronic excitation. *Disc. Faraday Soc.*, *27*, 7–17.

Förster, V., and Junge, W. (1985) Stoichiometry and kinetics of proton release upon photosynthetic water oxidation. *Photochem. Photobiol.*, *41*, 183–190.

Foster, D. L., and Fillingame, R. H. (1982) Stoichiometry of subunits in the H^+-ATPase complex of *Escherichia coli*. *J. Biol. Chem.*, *257*, 2009–2015.

Fragata, M., Ohnishi, S., Asada, K., Ito, T., and Takahashi, M. (1984) Lateral diffusion of plastocyanin in multilamellar mixed-lipid bilayer studied by fluorescence recovery after photoleaching. *Biochemistry, 23*, 4044–4051.

Fry, D. C., Kuby, S. A., and Mildvan, A. S. (1986) ATP-binding site of adenylate kinase: mechanistic implications of its homology with *ras*-encoded p21, F_1-ATPase and other nucleotide-binding proteins. *Proc. Natl. Acad. Sci. USA, 83*, 907–911.

Fujimura, T., Yamato, I., and Anraku, Y. (1983) Mechanism of glutamate transport in *Escherichia coli* B. 2. Kinetics of glutamate transport driven by artificially imposed proton and sodium ion gradients across the cytoplasmic membrane. *Biochemistry, 22*, 1959–1965.

Fukuyama, K., Nogahera, Y., Tsukihara, T., Katsube, Y., Hase, T., and Matsubara, H. (1988) Tertiary structure of *Bacillus thermoproteolyticus* [4Fe-4S] ferredoxin. Evolutionary implications for bacterial ferredoxins. *J. Mol. Biol.*, *199*, 183–193.

Fuller, S. D., Capaldi, R. A., and Henderson, R. (1982) Preparation of two-dimensional arrays from purified beef heart cytochrome *c* oxidase. *Biochemistry, 21*, 2525–2529.

Furbacher, P. N., Girvin, M. E., and Cramer, W. A. (1989) On the question of interheme electron transfer in the chloroplast cytochrome b_6 *in situ*. Biochemistry, *28*, in press.

Futai, M. (1977) Reconstitution of ATPase activity from isolated alpha, beta, and gamma subunits of the coupling factor, F_1, of *Escherichia coli*. *Biochem. Biophys. Res. Commun.*, *79*, 1231–1237.

Futami, A., Hurt, E., and Hauska, G. (1979) Vectorial redox reactions of physiological quinones. I. Requirements of a minimal length of the isoprenoid chain. *Biochim. Biophys. Acta, 547*, 583–596.

Futami, A., and Hauska, G. (1979) Vectorial redox reactions of physiological quinones. II. A study of transient semiquinone formation. *Biochim. Biophys. Acta, 547*, 597–608.

Gabellini, N., and Sebald, W. (1986) Nucleotide sequence and transcription of the *fbc* operon from *Rhodopseudomonas sphaeroides*. Evaluation of the deduced amino acid sequences of the FeS protein, cytochrome b, and cytochrome c_1. *Eur. J. Biochem.*, *194*, 569–579.

Gabellini, N., Bowyer, J. R., Hurt, E., Melandri, B. A., and Hauska, G. (1982) A cytochrome b/c_1 complex with ubiquinol-cytochrome c_2 oxidoreductase activity from *R. sphaeroides*. *Eur. J. Biochem.*, *126*, 105–111.

Gal, A., Shahak, Y., Schuster, G., and Ohad, I. (1987) Specific loss of LHC II phosphorylation in the *Lemna* mutant 1703 lacking the cytochrome b_6–f complex. *FEBS Lett.*, *221*, 205–210.

Gantt, E. (1980) Structure and function of phycobilisomes: light harvesting pigment complexes in red and blue-green algae. *Int. Rev. Cytol.*, *66*, 45–80.

Garboczi, D. N., Fox, A. H., Gerring, S. L., and Pedersen, P. L. (1988) β subunit of rat liver mitochondrial ATP synthase–cDNA cloning, amino acid sequence, expression in *E. coli*, and structural relationship to adenylate kinase. *Biochemistry*, *27*, 553–560.

Garin, J., Boulay, F., Issartel, J. P., Lunardi, J., and Vignais, P. V. (1986) Identification of amino acid residues photolabeled with 2-azido [α-^{32}P] adenosine diphosphate in the β subunit of beef heart mitochondrial F_1-ATPase. *Biochemistry*, *25*, 4431–4437.

Garland, P. D., Clegg, R. A., Boxer, D., Downie, J. A., and Haddock, B. A. (1975) Proton-translocating nitrate reductase of *Escherichia coli*. In *Electron Transfer Chains and Oxidative Phosphorylation* (Quagliariello, E., Papa, S., Palmieri, F., Slater, E. C., and Siliprandi, N., eds.), pp. 351–358, Elsevier/North Holland, Amsterdam.

Geller, B. L., Movva, N. R., and Wickner, W. (1986) Both ATP and the electrochemical potential are required for optimal assembly of pro-Omp A into *Escherichia coli* inner membrane vesicles. *Proc. Natl. Acad. Sci. USA*, *83*, 4219–4222.

Gelles, J., Blair, D. F., and Chan, S. I. (1986) The proton-pumping site of cytochrome c oxidase: a model of its structure and mechanism. *Biochim. Biophys. Acta*, *853*, 205–236.

George, G. N., Prince, R. C., and Cramer, S. P. (1989) The manganese site of the photosynthetic water-splitting enzyme. *Science* *243*, 789–791.

George, P., Witonsky, R. J., Trachtman, M., Wu, C., Douwart, W., Richman, L. and Lentz, B. (1970) "Squiggle-H_2O." An inquiry into the importance of solvation effects in phosphate ester and anhydride reactions. *Biochim. Biophys. Acta*, *233*, 1–15.

Ghanotakis, D. F., Babcock, G. T., and Yocum, C. F. (1984) Calcium reconstitutes high rates of O_2 evolution in polypeptide-depleted photosystem II preparations. *FEBS Lett.* *167*, 127–130.

Ghanotakis, D. F., Waggoner, C. M., Bowlby, N. R., Demetriou, D. M., Babcock, G. T., and Yocum, C. F. (1987) Comparative structural and catalytic properties of O_2-evolving PS II preparation. *Photosyn. Res.*, *14*, 191–200.

Giersch, G. (1983) Nigericin-induced stimulation of photophosphorylation in chloroplasts. *Biochim. Biophys. Acta*, *725*, 309–319.

Girvin, M. (1985) Electron and proton transfer in the quinone–b_6/f region of chloroplasts. Ph.D. thesis, 140 pp., Purdue University.

Girvin, M., and Cramer, W.A. (1984) A redox study of the electron transfer pathway responsible for the generation of the slow electrochromic phase in chloroplasts. *Biochim. Biophys. Acta*, *767*, 29–38.

Glagotev, A. N. (1981) Proton circuits of bacterial flagella. In *Chemiosmotic Proton Circuits in Biological Membranes* (V. P. Skulachev and P. C. Hinkle, eds.), pp. 577–600, Academic Press, New York.

Glaser, E. G., and Crofts, A. R. (1984) A new electrogenic step in the ubiquinol: cytochrome c oxidoreductase complex of $Rb.$ $sphaeroides.$ $Biochim.$ $Biophys.$ $Acta,$ 766, 322–333.

Glazer, A. N. (1985) Light harvesting by phycobilisomes. $Annu.$ $Rev.$ $Biophys.$ $Biophys.$ $Chem.,$ 14, 47–77.

Golbeck, J. K., Parrett, K. G., and McDermott, A. E. (1987) Photosystem I charge separation in the absence of centers A and B. III. Biochemical characterization of a reaction center peptide containing P-700 and F_x. $Biochim.$ $Biophys.$ $Acta,$ 893, 149–160.

Golden, S. S., Brusslan, J., and Haselkorn, R. (1986) Expression of a family of $psbA$ genes encoding a photosystem II polypeptide in the cyanobacterium $Anacystis$ $nidulans$ R2. $EMBO$ $J.,$ 5, 2789–2698.

Gonzalez-Halphen, D., Lindonfer, H. A., and Capaldi, R. A. (1988) Subunit arrangement in beef heart complex III. $Biochemistry,$ 27, 7021–7031.

Goodin, D. B., Yachandra, V. K., Britt, R. D., Sauer, K, and Klein, M. P. (1984) The state of manganese in the photosynthetic apparatus. 3. Light-induced changes in x-ray absorption (K-edge) energies of manganese in photosynthetic membranes. $Biochim.$ $Biophys.$ $Acta,$ 767, 209–216.

Gouffeau, A., and Slayman, C. W. (1981) The proton-translocating ATPase of the fungal plasma membrane. $Biochim.$ $Biophys.$ $Acta,$ 639, 197–223.

Gould, J. M., and Cramer, W. A. (1977) Relationship between oxygen-induced proton efflux and membrane energization in cells of $Escherichia$ $coli.$ $J.$ $Biol.$ $Chem.,$ 252, 5875–5882.

Gounaris, K., Sundby, C., Andersson, B., and Barber, J. (1983) Lateral heterogeneity of polar lipids in the thylakoid membranes of spinach chloroplasts. $FEBS$ $Lett.,$ 156, 170–174.

Govindjee, R., Ebrey, T. G., and Crofts, A. R. (1980) The quantum efficiency of proton pumping by the purple membrane of $H.$ $halobium.$ $Biophys.$ $J.,$ 30, 231–246.

Govindjee, Kambara, T., and Coleman, W. (1985) The electron donor side of photosystem II: the O_2 evolving complex. $Photochem.$ $Photobiol.,$ 42, 187–210.

Graan, T., and Ort, D. R. (1983) Initial events in the regulation of electron transfer in chloroplasts: the role of the membrane potential. $J.$ $Biol.$ $Chem.,$ 258, 2831–3836.

Graan, T., and Ort, D. R. (1986) Quantitation of DBMIB binding sites in chloroplast membranes. Evidence for functional dimer of the cytochrome b_6–f complex. $Arch.$ $Biochem.$ $Biophys.,$ 248, 445–451.

Gray, H. (1986) Long-range electron transfer in blue copper proteins. $Chem.$ $Soc.$ $Rev.,$ 15, 17–30.

Green, B. (1988) The chlorophyll-protein complexes of higher plant photosynthetic membranes. $Photosyn.$ $Res.,$ 15, 3–32.

Gresser, M. J., Myers, J. A., and Boyer, P. D. (1982) Catalytic site cooperativity of beef heart MF_1 ATPase. $J.$ $Biol.$ $Chem.,$ 257, 12030–12038.

Grinius, L. (1981) Proton current and DNA transport. In $Chemiosmotic$ $Proton$ $Circuits$ in $Biological$ $Membranes$ (V. P.Skulachev and P. C. Hinkle, eds.), pp. 551–565, Academic Press, New York.

Grinius, L., Slusnyté, R., and Griniuvienè, B. (1975) ATP synthesis driven by proton motive force imposed across $Escherichia$ $coli$ cell membranes. $FEBS$ $Lett.,$ 57, 290–293.

Gromet-Elhanon, Z., Khananshvilli, D., Weiss, S., Kanazawa, H., and Futai, M. (1985) ATP synthesis and hydrolysis by a hybrid system reconstituted from the β subunit

of *Escherichia coli* and the β-less chromatophores of *Rhodospirilum rubrum*. *J. Biol. Chem.*, 260, 12635–12640.
Grubmeyer, C., Cross, R. L., and Penefsky, H. S. (1982) Mechanism of ATP hydrolysis by beef heart mitochondrial ATPase. Rate constants for elementary steps in catalysis at a single site. *J. Biol. Chem.*, 257, 12092–12100.
Guffanti, A., and Krulwich, T. (1984) A transmembrane electrical potential generated by respiration is not equivalent to a diffusion potential of the same magnitude for ATP synthesis by *B. firmus* RAB. *J. Biol. Chem.*, 259, 2971–2975.
Guffanti, A. A., Susman, P., Blanco, R., and Krulwich, T. A. (1978) The proton-motive force and γ-aminoisobutyric acid transport in an obligately alkalophilc bacterium. *J. Biol. Chem.*, 253, 708–715.
Gunner, M. R., and Dutton, P. L. (1988) Temperature and $-\Delta G°$ dependence of the electron transfer from BPhe$^-$ to Q_A in reaction center protein from *Rb. sphaeroides* with different quinones as Q_A. *J. Am. Chem. Soc. 111*, 3400–3412, 1989.
Gupte, S., Wu, E. S., Hoechli, L., Hoechli, M., Jacobsen, K., Sowers, A. E., and Hackenbrock, C. R. (1984) Relationship between lateral diffusion, collision frequency, and electron transfer of mitochondrial inner membrane oxidation-reduction components. *Proc. Natl. Acad. Sci. USA*, 81, 2606–2610.
Guss, M., and Freeman, H. C. (1983) Structure of oxidized plastocyanin at 1.6 Å resolution. *J. Mol. Biol.*, 169, 521–563.
Gutman, M., Bonomi, F., Pagani, S., Cerletti, P. and Kroneck, P. (1980) Modulation of the flavin redox potential as a mode of regulation of succinate dehydrogenase activity. *Biochim. Biophys. Acta*, 591, 400–406.
Guynn, R. W., and Veech, R. L. (1973) The equilibrium constants of the adenosine triphosphate hydrolysis and the adenosine triphosphate-citrate lyase reactions. *J. Biol. Chem.*, 248, 6966–6972.
Hackenbrock, C. R. (1981) Lateral diffusion and electron transfer in the mitochondrial inner membrane. *Trends Biochem. Sci.*, 6, 151–154.
Hackett, N. R., Stern, L. J., Chao, B. H., Kronis, K. A., and Khorana, H. G. (1987) Structure–function studies on bacteriorhodopsin. V. Effects of amino acid substitutions in the putative helix F. *J. Biol. Chem.*, 262, 9277–9284.
Hackney, D. D., and Boyer, P. D. (1978) Subunit interaction during catalysis. Implications of O_2 concentration exchanges accompanying oxidative phosphorylation for alternating site cooperativity. *J. Biol. Chem.*, 253, 3164–3170.
Hager, K. M., Mandala, S. M., Davenport, J. W., Speicher, D. W., Benz, E. J., and Slayman, C. W. (1986) Amino acid sequence of the plasma membrane ATPase of *N. crassa*: deduction from genomic and cDNA sequences. *Proc. Natl. Acad. Sci. USA*, 83, 7693–7697.
Hales, B. (1976) Temperature dependence of the rate of electron transport as a function of protein motion. *Biophys. J.*, 16, 471–480.
Hall, D. O., Cammack, R., Rao, K. K., Evans, M. C. W., and Mullinger, R. (1975) Ferredoxins, blue-green algae, and evolution. *Biochem. Soc. Trans.*, 3, 361–368.
Hall, J., Zha, X., Durham, B., O'Brien, P., Viera, B., Davis, D., Okamura, M., and Millet, F. (1987a) Reaction of cytochromes c and c_2 with the *Rhodobacter sphaeroides* reaction of cytochromes c and c_2 with the *Rhodobacter sphaeroides* reaction center involves the heme crevice domain. *Biochemistry*, 26, 4494–4500.
Hall, J., Za, X., Yu, L., Yu, C.-A., and Mitchell, F. (1987b) The binding domain on horse cytochrome c and *Rh. sphaeroides* cytochrome c_2 for the *Rh. sphaeroides* cytochrome bc_1 complex. *Biochemistry*, 26, 4501–4504.
Hall, J., Ayres, M, Zha, X., O'Brien, P., Durham, B., Knaff, D., and Millett, F. (1987c)

The reaction of cytochromes c and c_2 with the *Rhodospirillum rubrum* reaction center involves the heme crevice domains. *J. Biol. Chem.*, 262, 11046–11051.
Hall, J., Kriaucionas, A., Knaff, D., and Millett, F. (1987d) The reaction domain on *R. rubrum* cytochrome c_2 and horse cytochrome c for the *R. rubrum* cytochrome bc_1 complex. *J. Biol. Chem.*, 262, 14005-14009.
Halliwell, B. (1987) Oxidants and human disease: some new concepts. *FASEB J.*, 1, 358–363.
Hamamoto, T., Ohno, K., and Kagawa, Y. (1982) Net ATP synthesis driven by an external electric field in rat liver mitochondria. *J. Biochem.*, 91, 1759–1766.
Hangarter, R. P., and Good, N. E. (1982) Energy thresholds for ATP synthesis in chloroplasts. *Biochim. Biophys. Acta*, 681, 396–404.
Hangarter, R. P., Jones, R. W., Ort, D. R., and Whitmarsh, J. (1987a) Stoichiometries and energetics of proton translocation coupled to electron transport in chloroplasts. *Biochim. Biophys. Acta*, 890, 106–115.
Hangarter, R., Grandoni, P., and Ort, D. R. (1987b) The effects of chloroplast coupling factor reduction on the energetics of activation and on the energetics and efficiency of ATP formation. *J. Biol. Chem.*, 262, 13513–13519.
Hanks, S. K., Quinn, A. M., and Hunter, T. (1988) The protein kinase family: conserved features and deduced phylogeny of the catalytic domains. *Science*, 241, 42–52.
Hansson, O., Andréasson, L.-E., and Vanngard, T. (1986) Oxygen from water is coordinated to water in the S_2 state of photosystem II. *FEBS Lett.*, 195, 151–154.
Hansson, O., Aasa, R., and Vänngard, T. (1987) The origin of the multiline and g = 4.1 EPR signal from the oxygen-evolving complex of photosystem II. *Biophys. J.*, 51, 825–832.
Haraux, F. (1985) Localized or delocalized protons and ATP synthesis in biomembranes. *Physiol. Veg.*, 23, 397–410.
Haraux, F., and de Kouchkovsky, Y. (1980) Measurement of chloroplast internal protons with 9-aminoacridine. *Biochim. Biophys. Acta,* 592, 153–168.
Harnisch, U., Weiss, H., and Sebald, W. (1985) The primary structure of the iron-sulfur subunit of ubiquinol cytochrome c reductase from *Neurospora*, determined by cDNA and gene sequencing. *Eur. J. Biochem.*, 149, 95–99.
Harold, F. M., Baarda, J. R., Baron, C., and Abrams, A. (1969) Inhibition of membrane-bound adenosine triphosphatase and of cation transport in *Streptococcus faecalis* by N, N'-dicyclohexylcarbodiimide. *J. Biol. Chem.*, 244, 2261–2268.
Hartung, A., and Trebst, A. (1985) New inhibitors of the cytochrome b_6-f complex in ferredoxin-catalyzed cyclic photophosphorylation. *Physiol. Veg.*, 23, 605–628.
Hatefi, Y. (1985) The mitochondrial electron transport chain and oxidative phosphorylation. *Annu. Rev. Biochem.*, 54, 1015–1069.
Hatefi, Y., and Galante (1980) Isolation of cytochrome b_{566} from complex II and its reconstitution with succinate dehydrogenase. *J. Biol. Chem.*, 255, 5530–5537.
Hatefi, Y., Ragan, I. I., and Galante, Y. M. (1985) The enzymes and enzyme complexes of the mitochondrial oxidative phosphorylation system. In *The Enzymes of Biological Membranes* (A. Martonosi, ed.), *Vol. 4*, pp. 1–70. Planum Press, New York.
Hauska, G. (1986) Preparation of electrogenic, proton-transporting cytochrome complexes of the b_6-f type (chloroplasts and cyanobacteria) and bc_1 type (*R. sphaeroides*). *Method Enzymol.*, 126, 273–285.
Hauska, G., and Hurt, E. (1982) Pool function behavior and mobility of isoprenoid quinones. In *Function of Quinones in Energy Conserving Systems* (B. Trumpower, ed.), pp. 87–110, Academic Press, New York.

Hauska, G., Hurt, E., Gabellini, N., and Lockau, W. (1983) Comparative aspects of quinol-cytochrome c/plastocyanin oxidoreductases. *Biochim. Biophys. Acta*, 726, 97–133.

Hauska, G., Nitschke, W., and Herrmann, R. G. (1988) Amino acid identities in the three redox center-carrying polypeptides of cytochrome bc_1/b_6f complexes. *J. Bioenerg. Biomembr.*, 20, 211–228.

Heinemeyer, W., Alt, J., and Herrmann, R. G. (1984) Nucleotide sequence of the clustered genes for apocytochrome b_6 and subunit 4 of the cytochrome b_6/f complex in the spinach plastid chromosome. *Curr. Genet.*, 8, 543–549.

Helgerson, S., Requadt, C. and Stoeckenius, W. (1983) *Halobacterium halobium* photophosphorylation: illumination dependent increase in the adenylate energy charge and phosphorylation potential. *Biochemistry*, 22, 5746–5753.

Helgerson, S. L., Methew, M. K., Dirin, D. B., Wolber, P. K., Heinz, E., and Stoeckenius, W. (1985) Coupling between the bacteriorhodopsin photocycle and the proton-motive force in *H. halobium* cell envelope vesicles. *Biophys. J.*, 48, 709–719.

Henderson, R., and Unwin, P. N. T. (1975) Three dimensional model of purple membrane obtained by electron microscopy. *Nature*, 257, 28–33.

Herrmann, R. G., Alt, J., Schiller, B., Widger, W. R., and Cramer, W. A. (1984) Nucleotide sequence of the gene for apocytochrome b-559 on the spinach plastid chromosome; implication for the structure of the membrane protein. *FEBS Lett.*, 176, 239–244.

Herweijer, M. A., Berden, J. A., Kemp, A., and Slater, E. C. (1986) Inhibition of energy-transducing reactions by 8-nitreno-ATP covalently bound to bovine heart submitochondrial particles: direct interaction between ATPase and redox enzymes. *Biochim. Biophys. Acta*, 809, 81–89.

Herzberg, G. (1950) *Spectra of Diatomic Molecules*, Chap. 4 (1957 printing) D. Van Nostrand, Princeton.

Hesse, J. E., Wieczorek, L., Altendorf, K., Reicin, A. S., Dorus, E., and Epstein, W. (1984) Sequence homology between two membrane transport ATPases, the Kdp-ATPase of *Escherichia coli* and the Ca^{++}-ATPase of sarcoplasmic reticulum. *Proc. Natl. Acad. Sci. USA*, 81, 4746–4750.

Heyn, M. P., Westerhausen, J., Wallat, I., and Seiff, F. (1988) High sensitivity neutron diffraction of membranes. Location of the Schiff base end of the chromophore of bacteriorhodopsin. *Proc. Natl. Acad. Sci. USA*, 85, 2146–2150.

Hicks, D. B., and Krulwich, T. A. (1986) The membrane ATPase of alkalophilic *Bacillus firmus* RAB is an F_1-ATPase. *J. Biol. Chem.*, 261, 12896–12902.

Higgins, C. F., Cairney, J., Stirling, D. A., Sutherland, L., and Booth, I. R. (1987) Osmotic regulation of gene expression: ionic strength as an intracellular signal *Trends Biochem. Sci.*, 12, 339–344.

Higuchi, Y., Kurunoki, M., Matsuura, Y., Yasuoka, N., and Kakudo, M. (1984) Refined structure of cytochrome c_3 at 1.8 Å resolution. *J. Mol. Biol.*, 172, 109–139.

Hill, T. L. (1966) *Lectures on Matter and Equilibrium*, Chap. 8, W. A. Benjamin, New York.

Hill, T. L. and Morales, M. F. (1951) On "high energy phosphate bonds" of biochemical interest. *J. Am. Chem. Soc.*, 73, 1656–1660.

Hiller, R., and Carmeli, C. (1985) Cooperativity among manganese-binding sites in the H^+-ATPase of chloroplasts. *J. Biol. Chem.*, 260, 1614–1617.

Hilpert, W., Schink, B., and Dimroth, P. (1984) Life by a new decarboxylation-dependent energy conservation mechanism with Na^+ as coupling ion. *EMBO J.*, 3, 1665–1670.

Hine, J. and Mookerjee, P. K. (1975) The intrinsic hydrophobic character of organic compounds. Correlation in terms of structural contributions. *J. Org. Chem.*, 40, 292–298.

Hinkle, P. C., and Yu, M. L. (1979) The phosphorus/oxygen ratio of mitochondrial oxidative phosphorylation. *J. Biol. Chem.*, 254, 2450–2455.

Hinkle, P., and Mitchell, P. (1970) Effect of membrane potential on equilibrium poise between cytochromes *a* and *c* in rat liver mitochondria, *J. Bioenerg.*, 1, 45–60.

Hirasawa, M., Boyer, J. M., Gray, K. A., Davis, D. J., and Knaff, D. B. (1986) The interaction of ferredoxin with chloroplast ferredoxin-linked enzymes. *Biochim. Biophys. Acta*, 851, 23–28.

Hirata, H., Altendorf, K., and Harold, F. M. (1973) Role of an electrical potential in the coupling of metabolic energy to active transport by membrane vesicles of *Escherichia coli*. *Proc. Natl. Acad. Sci. USA*, 70, 1804–1808.

Hirata, H., Kambe, T., and Kagawa, Y. (1984) A purified alanine carrier composed of a single polypeptide from the thermophilic bacterium PS3 driven by either proton or sodium ion gradient. *J. Biol. Chem.*, 259, 10653–10656.

Hirata, H., Ohno, K., Sone, N., Kagawa, Y., and Hamamoto, T. (1986) Direct measurement of the electrogenicity of the H^+-ATPase from thermophilic bacterium PS3 reconstituted in planar phospholipid bilayers. *J. Biol. Chem.*, 261, 9839–9843.

Hirota, N., and Imae, Y. (1983) Na^+-driven flagellar motors of an alkalophilic *Bacillus* strain YN-1. *J. Biol. Chem.*, 258, 10577–10581.

Hirschberg, J., and McIntosh, L. (1983) Molecular basis of herbicide resistance in *Amaranthus hybridus*. *Science*, 222, 1346–1329.

Ho, C., and Russu, I. M. (1987) How much do we know about the Bohr effect of hemoglobin? *Biochemistry*, 26, 6299–6305.

Hobson, A. C., Gho, D., and Müller-Hill, B. (1977) Isolation, genetic analysis and characterization of *Escherichia coli* mutants with defects in the *lac y* gene. *J. Bacteriol.*, 131, 830–838.

Hobson, A. C., Weatherwax, R., and Ames, G. (1984) ATP-binding sites in the membrane components of histidine permease, a periplasmic transport system. *Proc. Natl. Acad. Sci. USA*, 81, 7333–7337.

Hochman, J., Ferguson-Miller, S., and Schindler, M. (1985) Mobility in the mitochondrial electron transport chain. *Biochemistry*, 24, 2509–2516.

Høj, P. B., and Lindberg-Møller, B. (1986) The 100 kDa reaction center protein of photosystem I, P700-chlorophyll *a*-protein I, is an iron-sulfur protein. *J. Biol. Chem.*, 261, 14292–14300.

Høj, P. B., Svendsen, I., Scheller, H. V., and Lindberg-Møller, B. (1987) Identification of a chloroplast-encoded 9-kDa polypeptide as a 2[4Fe-4S] protein carrying centers A and B of photosystem I. *J. Biol. Chem.*, 262, 12676–12684.

Holm, L., Saraste, M., and Wikström, M. K. F. (1987) Structural models of the redox centres in cytochrome oxidase. *EMBO J.*, 6, 2819–2823.

Holten, D., Windsor, M. W., Parson, W. W., and Thornber, J. P. (1978) Primary photochemical processes in isolated reaction centers of *Rps. viridis*. *Biochim. Biophys. Acta*, 501, 112–126.

Holwerda, R. A., Read, R. A., Scott, R. A., Wherland, S., Gray, H. B., and Millett, F. (1978) Electron transfer reactions of trifluoroacetylated horse heart cytochrome *c*. *J. Am. Chem. Soc.*, 100, 5028–5033.

Homann, P. H. (1986) The relations between the chloride calcium and polypeptide requirements of photosynthetic water oxidation. *J. Bioenerg. Biomem.*, 19, 105–113.

Homann, P. H., Gleiter, H., Ono, T., and Inoue, Y. (1986) Storage of abnormal oxidants

Σ_1, Σ_2, and Σ_3 in photosynthetic water oxidases inhibited by Cl$^-$ removal. *Biochim. Biophys. Acta, 850*, 10–20.

Hong, J. H., Hunt, A. G., Masters, P. S., and Lieberman, M. A. (1979) Requirement of acetylphosphate for the binding protein-dependent transport systems in *Escherichia coli*. *Proc. Natl. Acad. Sci. USA, 76*, 1213–1217.

Honig, B., and Hubbell, W. L. (1984) Stability of salt bridges in membrane proteins. *Proc. Natl. Acad. Sci. USA, 81*, 5412–5416.

Hopfer, V., Lehninger, A. L., and Thompson, T. E. (1968) Protonic conductance across phospholipid biolayer membranes induced by uncoupling agents for oxidative phosphorylation. *Proc. Natl. Acad. Sci. USA, 59*, 484–490.

Hopfield, J. J. (1974) Electron transfer between biological molecules by thermally activated tunneling. *Proc. Natl. Acad. Sci. USA, 71*, 3640–3644.

Hoppe, J., and Sebald, W. (1984) The proton conducting F_0 part of bacterial ATP synthesis. *Biochim. Biophys. Acta, 768*, 1–27.

Horner, R. D., and Moudrianakis, E. N. (1983) The effect of permeant buffers on initial ATP synthesis by chloroplasts using rapid mix-quench techniques. *J. Biol. Chem., 258*, 11643–11647.

Horton, P., and Black, M. T. (1980) Activation of adenosine-5'-triphosphate-induced quenching of chlorophyll fluorescence by reduced plastoquinone. *FEBS Lett., 119*, 141–144.

Hosler, J. P., and Yocum, C. (1985) Evidence for two cyclic photophosphorylation reactions concurrent with ferredoxin-catalyzed non-cyclic electron transport. *Biochim. Biophys. Acta, 808*, 21–31.

Howell, N., and Gilbert, K. (1987) Sequence analysis of mammalian cytochrome *b* mutants. In *Cytochrome Systems: Molecular Biology and Bioenergetics*, (S. Papa, B. Chance, and L. Ernster, eds.), pp. 79–86, Plenum Press, New York.

Howitt, S. M., Gibson, F., and Cox, G. B. (1988) The proton pore of the F_0 of the ATP synthase of *Escherichia coli*: Ser-206 is not required for proton translocation. *Biochim. Biophys. Acta, 936*, 74–80.

Hsu, S.-Y., Senda, M., Kanazawa, H., Tsuchiya, T., and Futai, M. (1984) Comparison of F_1's of oxidative phospharylation from *E. coli* and *S. typhimurium* and demonstration of the interchangeability of their subunits. *Biochemistry, 23*, 988–993.

Huang, K. S., Ramachandran, R., Bayley, H., and Khorana, H. G. (1982) Orientation of retinal in bacteriorhodopsin as studied by cross-linking using a photosensitive analog of retinal. *J. Biol. Chem., 257*, 13616–13623.

Hunter, C. N., van Grondelle, R., and Olsen, J. D. (1989) Photosynthetic antenna proteins: 100 ps before photochemistry starts. *Trends Biochem. Sci., 14*, 72–75.

Huppert, D., Gutman, M., and Kaufman, K. J. (1981) Laser studies of proton transfer. *Adv. Chem. Phys., 47*, 643–679.

Hurt, E. C. (1987) Unravelling the role of ATP in post-translational protein translocation. *Trends Biochem. Sci., 12*, 369–370.

Hurt, E., and Hauska, G. (1981) A cytochrome f/b_6 complex of five polypeptides with plastoquinol-plastocyanin-oxidoreductase activity from spinach chloroplasts. *Eur. J. Biochem., 117*, 591-599.

Hurt, E. C., and van Loon, A. P. G. M. (1986) How proteins find mitochondria and intramitochondrial compartments. *Trends Biochem. Sci., 11*, 204–207.

Hutton, R. L., and Boyer, P. D. (1979) Subunit interaction during catalysis. Alternating site cooperativity of mitochondrial adenosine triphosphatase. *J. Biol. Chem., 254*, 9990–9993.

Ikeuchi, M., and Inoue, Y. (1988) A new photosystem II reaction center component (4.8 kDa protein) encoded by chloroplast genome. *FEBS Lett.*, 241, 98–104.

Ikeuchi, M., Koike, H., and Inoue, Y. (1988) Iodination of D1 herbicide-binding protein is coupled with photooxidation of ^{125}I-associated with Cl$^-$-binding site in photosystem II water oxidation system. *Biochim. Biophys. Acta*, 932, 160–169.

Ingledew, W. J. and Ohnishi, T. (1977) The probable site of action of thenoyltrifluoroacetone in the respiratory chain. *Biochem. J.*, 164, 617–620.

Ingledew, W. J. and Poole, R. K. (1984) The respiratory chain of *E. coli*. *Microbiol. Rev.*, 48, 222–271.

Ito, A. (1980) Cytochrome b_5-like hemoprotein of outer mitochondrial membrane of outer membrane cytochrome *b*. II. Contribution of outer membrane cytochrome *b* to rotenone-insensitive NADH-cytochrome *c* reductase activity. *J. Biochem.*, 87, 73–80.

Itoh, S., Yerkes, C. T., Koike, H., Robinson, H. H., and Crofts, A. R. (1984) Effects of Cl$^-$ depletion on electron donation from the H_2O oxidizing complex to the PS II reaction center as measured by the μs rise of chlorophyll fluorescence. *Biochim. Biophys. Acta*, 766, 612–622.

Jackson, J. B., and Dutton, P. L. (1973) The kinetic and redox potentiometric resolution of the carotenoid shifts in *Rb. sphaeroides* chromatophores: their relationship to electric field alterations in electron transport and energy coupling. *Biochim. Biophys. Acta*, 325, 102–113.

Jacobson G. R., Lee, C. A., Leonard, J. E., and Saier, Jr., M. H. (1983) Mannitol-specific Enzyme II of the bacterial phosphotransferase system I. Properties of the purified enzyme. *J. Biol. Chem.*, 258, 10748–10756.

Jagendorf, A., and Uribe E. (1966) ATP formation caused by acid–base transition of spinach chloroplasts. *Proc. Natl. Acad. Sci. USA*, 55, 170–177.

Jans, D. A., Fimmel, A. L., Langman, L., James, L. B., Downie, J. A., Senior, A. E., Ash, G. R., Gibson, F., and Cox, G. B. (1983) Mutations in the *unc*E gene affecting assembly of the *c* subunit of the ATPase of *E. coli*. *Biochem. J.*, 211, 717–726.

Jans, D. A., Fimmel, A. L., Hatch, L., Gibson, F., and Cox, G. B. (1984a) An additional acidic residue in the membrane portion of the *b*-subunit of the energy transducing ATPase of *E. coli* affects both assembly and function. *Biochem. J.*, 223, 43–51.

Jans, D. A., Hatch, L., Fimmel, A. L., Gibson, F., and Cox, G. B. (1984b) An acidic or basic amino acid at position 26 of the *b* subunit of *E. coli* F_1F_0-ATPase impairs membrane proton permeability: suppression of the *unc* F469 nonsense mutation. *J. Bacteriol.*, 160, 764–770.

Jans, D. A., Hatch, L., Fimmel, A. L., Gibson, F., and Cox, G. B. (1985) Complementation between *unc* F alleles affecting assembly of the F_1F_0-ATPase complex of *E. coli*. *J. Bacteriol.*, 162, 420–426.

Jaraush, J., and Kadenbach, B. (1985a) Structure of the cytochrome *c* oxidase complex of rat liver. Studies on nearest neighbor relationship of polypeptides with cross-linking reagents. *Eur. J. Biochem.*, 146, 211–217.

Jaraush, J., and Kadenbach, B. (1985b) Structure of the cytochrome *c* oxidase complex of rat liver. Topological orientation of polypeptides in the membrane as studied by proteolytic digestion and immunoblotting. *Eur. J. Biochem.*, 146, 219–225.

Jencks, W. P. (1976) In *Handbook of Biochemistry and Molecular Biology* (G. D. Fasman, ed.), Vol. *I*, p. 302, CRC Press, Cleveland.

Jencks, W. P. (1980) The utilization of binding energy in coupled vectorial processes. *Adv. Enzymol.*, 51, 75–106.

Johnson R. G., Carlson, N. J., and Scarpa, A. (1978) ΔpH and catecholamine distribution in isolated chromaffin granules. *J. Biol. Chem.*, *253*, 1512–1521.

Johnson, R. G. and Scarpa, A. (1979) Protonmotive force and catecholamine transport in isolated chromaffin granules. *J. Biol. Chem.*, *254*, 3750–3760.

Joliot., P., and Delosme, R. (1974) Flash-induced 519 nm absorption change in green algae. *Biochim. Biophys. Acta*, *357*, 267–284.

Joliot, P., and Joliot, A. (1986) Proton pumping and electron transfer in the cytochrome b/f complex of algae. *Biochim. Biophys. Acta*, *849*, 211–222.

Joliot, P., and Kok, B. (1975) Oxygen evolution in photosynthesis. In *Bioenergetics of Photosynthesis* (Govindjee, ed.), pp. 387–412, Academic Press, New York.

Joliot, P., Barbieri, G., and Chabaud, R. (1969) Un nouveau modele des centres photochimiques du systeme II. *Photochem. Photobiol.*, *10*, 309–329.

Jones, R. W., and Whitmarsh, J. (1985) Origin of the electrogenic reaction in the chloroplast b/f complex. *Photobiochem. Photobiophys.*, *9*, 119–127.

Jortner, J. (1976) Temperature dependent activation energy for electron transfer between biological molecules. *J. Chem. Phys.*, *64*, 4860–4867.

Joshi, A. K., Ahmed, S. and Ames, G. F.-L. (1989) Energy coupling in bacterial perplasmic transport systems. *J. Biol. Chem.*, *264*, 2126–2133.

Junesch, V., and Gräber, P. (1985) The rate of ATP synthesis as a function of ΔpH in normal and dithiothreitol-modified chloroplasts. *Biochim. Biophys. Acta*, *809*, 429–434.

Junesch, V., and Gräber, P. (1987) Influence of the redox state and the activation of the chloroplast ATP synthase on proton transport-coupled ATP synthesis/hydrolysis. *Biochim. Biophys. Acta*, *893*, 275–288.

Junge, W. (1987) Complete tracking of transient proton flow through active chloroplast ATP synthase. *Proc. Natl. Acad. Sci. USA*, *84*, 7084–7088.

Junge, W., and Jackson, J. B. (1982) The development of $\Delta\tilde{\mu}_{H^+}$ across photosynthetic membranes. In *Photosynthesis: Energy Conversion by Plants and Bacteria* (Govindjee, ed.), Vol. 1, pp. 589–647, Academic Press, New York.

Junge, W., Schönknecht, G., and Förster, V. (1986) Neutral red as an indicator of pH transients in the lumen of thylakoids—some answers to criticism. *Biochim. Biophys. Acta*, *852*, 852, 93–99.

Kaback, H. R. (1976) Molecular biology and energetics of membrane transport. *J. Cell. Physiol.*, *89*, 575–594.

Kaback, H. R. (1986) Active transport in *Escherichia coli*: passage to permease. *Annu. Rev. Biophys. Biophys. Chem.*, *15*, 279–319.

Kaback, H. R. (1987) Use of site-directed mutagenesis to study the mechanism of a membrane transport protein. *Biochemistry*, *26*, 2071–2076.

Kaback, H. R. (1988) Site-directed mutagenesis and ion-gradient driven active transport: on the path of the protein. *Annu. Rev. Physiol.*, *50*, 243–256.

Kadenbach, B., and Merle, P. (1981) On the function of multiple subunits of cytochrome oxidase from higher eukaryotes. *FEBS Lett.*, *135*, 1–11.

Kaesler, B., and Schonheit, P. (1989) The role of sodium ions in methanogenesis. *Eur. J. Biochem.*, *184*, 223–232.

Kagawa, Y. (1978) Reconstitution of the energy transformer, gate and channel: subunit reassembly, crystalline ATPase and ATP synthesis. *Biochim. Biophys. Acta*, *505*, 45–93.

Kagawa, Y. (1984) Proton-motive ATP synthesis. In *Bioenergetics* (L. Ernster, ed.), pp. 149–186, Elsevier, Amsterdam.

Kalasauskaite, E. V., Kadisaite, D. L., Daugelavicius, R. J., Grinius, L. L., and Jasaitis,

A. A. (1983) Studies on energy supply for genetic processes. Requirement for membrane potential in Escherichia coli infection by phage T4. Eur. J. Biochem., 130, 123–130.

Kanner, B. I. (1983) Bioenergetics of neurotransmitter transport. Biochim. Biophys. Acta, 726, 293–316.

Kanner, B. I., and Schuldiner, S. (1987) Mechanism of transport and storage of neurotransmitters. CRC Crit. Rev. in Biochem., 22, 1–38.

Kaplan, J. H., Forbush, B., and Hoffman, J. F. (1978) Rapid photolytic release of adenosine 5'-triphosphate from a protected analogue: utilization by the Na:K pump of human red blood cell ghosts. Biochemistry, 17, 1929–1935.

Kaplan, R. S., Pratt, R. D., and Pedersen, P. L. (1986) Purification and characterization of reconstitutively active phosphate transporter from rat liver mitochondria. J. Biol. Chem., 261, 12767–12773.

Kaplan, W. (1953) Advanced Calculus, pp. 439–445, Addison-Wesley, Cambridge.

Karpel, R., Olami, Y., Taglicht, D., Schuldiner, S., and Padan, E. (1988) Sequencing the gene ant which affects the Na^+/H^+ antiporter activity in Escherichia coli. J. Biol. Chem., 263, 10408–10414.

Karplus, P. A., Walsh, K. A., and Herriott, J. R. (1984) Amino acid sequence of ferredoxin: $NADP^+$ oxidoreductase. Biochemistry, 23, 6576–6583.

Kashket, E. R. (1982) Stoichiometry of the H^+-ATPase of growing and resting, aerobic Escherichia coli. Biochemistry, 21, 5534–5538.

Kashket, E. R. (1983) Stoichiometry of the H^+-ATPase of E. coli cells during anaerobic growth. FEBS Lett., 154, 343–346.

Kashket, E. R., and Wilson, T. H. (1973) Proton-coupled accumulation of galactoside in Streptococcus lactis 7962. Proc. Natl. Acad. Sci. USA, 70, 2866–2869.

Katre, N. V., Kimura, Y., and Stroud, R. M. (1986) Cation binding sites on the projected structure of bacteriorhodopsin. Biophys. J., 50, 277–284.

Katsikas, H., and Quinn, P. J. (1982) The polyisoprenoid chain length influences the interaction of ubiquinones with phospholipid bilayers. Biochim. Biophys. Acta, 689, 363–369.

Kawakami, T., Akizawa, Y., Ishikawa, T., Shimamoto, T., Tsuda, M. and Tsuchiya, T. (1988) Amino acid substitutions and alteration in cation specificity in the melibiose carrier of Escherichia coli. J. Biol. Chem., 263, 14276–14280.

Kayalar, C., Rosing, J., and Boyer, P. D. (1977) An alternating site sequence for oxidative phosphorylation suggested by measurement of substrate binding patterns and exchange reaction inhibition. J. Biol. Chem., 252, 2486–2491.

Keegstra, K., Werner-Washbourne, M., Cline, K., and Andrews, J. (1984) The chloroplast envelope: is it homologous with the double membrane of mitochondria and gram-negative bacteria? J. Cell. Biochem., 24, 55–68.

Keilin, D., and Keilin, J. (1966) The History of Cell Respiration and Cytochromes. Cambridge University Press, London.

Ken-Dror, S., Lanyi, J. K., Schobert, B., Silver, B., and Avi-Dor, Y. (1986) An NADH:quinone oxidoreductase of the halotolerant bacterium Ba_1 is specifically dependent on sodium ions. Arch. Biochem. Biophys., 244, 766–771.

Kennaway, N. G., Buist, N. G. M., Darley-Usmar, V. M., Papadimitriou, A., Dimauro, S., Kelley, R. I., Capaldi, R. A., Blank, N. K., and D'Agostino, A. (1984) Lactic acidosis and mitochondrial myopathy associated with deficiency of several components of complex III of the respiratory chain. Pediatr. Res., 18, 991–999.

Keyes, S. R., and Rudnick, G. (1982) Coupling of transmembrane proton gradients to platelet serotonin transport. J. Biol. Chem., 257, 1172–1176.

Khananshvilli, D., and Gromet-Elhanon, Z. (1985) Evidence that the Mg-dependent low affinity site for ATP and Pi demonstrated on the isolated β subunit of the F_0F_1 ATP synthase is a catalytic site. *Proc. Natl. Acad. Sci. USA*, 82, 1886–1890.

Khorana, H. G. (1988) Bacteriorhodopsin, a membrane protein that uses light to translocate proteins. *J. Biol. Chem.*, 263, 7439–7442.

Kihara, T., and McCray, J. A. (1973) Water and cytochrome oxidation–reduction reactions. *Biochim. Biophys. Acta*, 292, 297–309.

King, S. C. and Wilson T. H. (1989a) Galactoside-dependent proton transport by mutants of the *Escherichia coli* lactose carrier. *J. Biol. Chem.*, 264, 7390–7394.

King, S. C., and Wilson, T. H. (1989b) Galactoside-dependent protein uptake by mutants of the *Escherichia coli* lactose carrier: Substitution of tyrosine for histidine-322 and of leucine for serine-306. *Biochim. Biophys. Acta*, 982, 253–264.

Kirmaier, C., and Holten, D. (1987) Primary photochemistry of reaction centers from the photosynthetic purple bacteria. *Photosyn. Res.*, 13, 225–260.

Kirmaier, C., Holten, D., and Parson, W. W. (1985) Picosecond-photodichroism studies of the transient states in *Rb. sphaeroides* reaction centers at 5°K. Effects of electron transfer on the six bacteriochlorin pigments. *Biochim. Biophys. Acta*, 810, 49–61.

Kirmaier, C., Holten, D., Debus, R. J., Feher, G., and Okamura, M. Y. (1986) Primary photochemistry of iron-depleted and zinc-reconstituted reaction centers from *Rb. sphaeroides*. *Proc. Natl. Acad. Sci. USA*, 83, 6407–6411.

Kirmaier, C., Holten, D., Bylina, E. J., and Youvan, D. C. (1988) Electron transfer in a genetically modified bacterial reaction center containing a heterodimer. *Proc. Natl. Acad. Sci. USA*, 85, 7562–7566.

Kironde, F. A. S., and Cross, R. L. (1987) Adenine nucleotide binding sites on F_1-ATPase. *J. Biol. Chem.*, 262, 3488–3495.

Klimov, V. V., and Krasnovskii, A. A. (1982) Pheophytin as the primary electron acceptor in photosystem 2 reaction centers. *Photosynthetics*, 15, 592–609.

Klingenberg, M. (1985) The ADP/ATP carrier in mitochondrial membranes. In *The Enzymes of Biological Membranes* (A. Martonosi, ed.), Vol. IV, pp. 511–533, Plenum Press, New York.

Klingenberg, M., and Rottenberg, H. (1977) Relation between the gradient of the ADP/ATP ratio and the membrane potential across the mitochondrial membrane. *Eur. J. Biochem.*, 73, 125–130.

Klotz, I. M. (1967) *Energy Changes in Biochemical Reactions*, Chap. 6, Academic Press, New York.

Klotz, I. M. (1986) *Introduction to Biomolecular Energetics*, Academic Press, Orlando.

Knapp, E. W., Fischer, S. F., Zinth, W., Sanler, M., Kaiser, W., Deisenhofer, J., and Michel, H. (1985) Analysis of optical spectra from single crystals of *Rps. viridis* reaction centers. *Proc. Natl. Acad. Sci. USA*, 82, 8463–8467.

Kohorn, B. D., and Tobin, E. (1987) Amino acid charge distribution influences the assembly of apoproteins into light harvesting complex II. *J. Biol. Chem.*, 262, 12897–12899.

Kok, B., Forbush, B., and McGloin, M. (1970) Cooperation of charges in photosynthetic O_2 evolution.—A linear four step mechanism. *Photochem. Photobiol.*, 11, 457–475.

Komar, E., Schobert, C., and Cho, B.-H. (1983) The hexose uptake system of *Chlorella*: is it a proton symport or a hydroxyl ion antiport system? *FEBS Lett.*, 156, 6–10.

Kometani, T., Kinosita, K., Furuno, T., Kouyama, T., and Ikegami, A. (1987) Transmembrane location of retinal in purple membrane. Fluorescence energy transfer in maximally packed donor-acceptor systems. *Biophys. J.*, *52*, 509–517.

Konings, W. N. and Booth, I. R. (1981) Do the stoichiometries of ion-linked transport systems vary? *Trends Biochem. Sci.*, *6*, 257–262.

Konings, W. N., and Robillard, G. N. (1982) Physical mechanism for regulation of proton solute symport in *Escherichia coli*. *Proc. Natl. Acad. Sci. USA*, *79*, 5480–5484.

Koppenol, W. H., and Margoliash, E. (1982) The asymmetric distribution of charges on the surface of horse cytochrome *c*. Functional implications. *J. Biol. Chem.*, *257*, 4426–4437.

Koshland, Jr., D. E. (1981) Biochemistry of sensing and adaptation in a simple bacterial system. *Annu. Rev. Biochem.*, *50*, 765–782.

Kostić, N. M., Morgolit, R., Che, C.-M., and Gray, H. B. (1983) Kinetics of long distance ruthenium-to-copper electron transfer in [pentaamine ruthenium histidine 83-azurin]. *J. Am. Chem. Soc.*, *105*, 7765–7767.

Krab, K., and Wikström, M. K. F. (1987) Principles of coupling between electron transfer and proton translocation with special reference to proton translocation mechanisms in cytochrome oxidase. *Biochim. Biophys. Acta*, *895*, 25–39.

Kranz, R. G., and Gennis, R. B. (1985) Immunological investigation of the distribution of cytochromes related to the two terminal oxidases of *E. coli* in other gram-negative bacteria. *J. Bacteriol.*, *161*, 709–713.

Kretschmann, H., Dekker, J. P., Säygin, Ö., and Witt, H. T. (1988) An agreement on the quaternary oscillation of ultraviolet absorption changes accompanying the water splitting in isolated photosystem II complexes from the cyanobacterium *Synecococcus sp*. *Biochim. Biophys. Acta*, *932*, 358–361.

Krinner, M., Hauska, G., Hurt, E., and Lockau, W. (1982) A cytochrome $f b_6$ complex with plastoquinol-cytochrome *c* reductase activity from *Anabaena variabilis*. *Biochim. Biophys. Acta*, *681*, 110–117.

Krishnamoorthy, G., and Hinkle, P. (1984) Non-ohmic proton conductance of mitochondria and liposomes. *J. Biol. Chem.*, *23*, 1640–1645.

Kröger, A., and Klingenberg, M. (1973) The kinetics of the redox reactions of ubiquinone related to the electron transport activity in the respiratory chain. *Eur. J. Biochem.*, *34*, 358–368.

Krulwich, T. A. (1983) Na^+/H^+ antiporters. *Biochim. Biophys. Acta*, *726*, 245–264.

Krulwich, T. A., Guffanti, A. A., Bornstein, R. F., and Hoffstein, J. (1982) A sodium requirement for growth, solute transport and pH homeostasis in *Bacillus firmus* RAB. *J. Biol. Chem.*, *257*, 1885–1889.

Kühlbrandt, W. (1984) Three dimensional structure of the light-harvesting chlorophyll *a/b* complex. *Nature*, *307*, 478–480.

Kühlbrandt, W. (1988a). Studies of light-harvesting chlorophyll *a/b* protein complex from plant photosynthetic membrane at 7 Å resolution in projection. *J. Mol. Biol.*, *202*, 849–864.

Kühlbrandt, W. (1988b). Three dimensional crystallization of membrane proteins. *Quart. Rev. Biophys.*, *21*, 429–477.

Kuhn, A., Wickner, W., and Kreil, G. (1986) The cytoplasmic carboxy terminus of M13 procoat is required for the membrane insertion of its central domain. *Nature*, *322*, 335–339.

Kuhn-Nentwig, L., and Kadenbach, B. (1985) Isolation and properties of cytochrome *c* oxidase from rat liver and quantification of immunological differences between isozymes from various rat tissues with subunit-specific antisera. *Eur. J. Biochem.*, *149*, 147–158.

Kuila, D., and Fee, J. A. (1986) Evidence for a redox-linked ionizable group associated with the [2Fe-2S] cluster of *Thermus* Rieske protein. *J. Biol. Chem.*, *261*, 2768–2771.

Kumamoto, C., and Simoni, R. D. (1987) A mutation of the *c* subunit of the *Escherichia coli* proton-translocating ATPase that suppresses the effects of a mutant *b* subunit. *J. Biol. Chem.*, *262*, 3060–3064.

Kunst, M. and Warman, J. M. (1980) Proton mobility in ice. *Nature*, *288*, 465–467.

Kurowski, B., and Ludwig, B. (1987) The genes of the *Paracoccus denitrificans* bc_1 complex. *J. Biol. Chem.*, *262*, 13805–13811.

Kuwabara, T., and Murata, N. (1982) Inactivation of photosynthetic O_2 evolution and concomitant release of three polypeptides in the photosystem II particles of spinach chloroplasts. *Plant Cell Physiol.*, *23*, 533–539.

Kuwabara, T., Miyao, M., Murata, T., and Murata, N. (1985) The function of 33 kDa protein in the photosynthetic oxygen-evolution system studied by reconstitution experiments. *Biochim. Biophys. Acta*, *806*, 283–289.

Kuwabara, T., Reddy, K. J., and Sherman, L. A. (1987) Nucleotide sequence of the gene from the cyanobacterium *Anacystis nidulans* encoding the Mn-stabilizing protein involved in photosystem II water oxidation. *Proc. Natl. Acad. Sci. USA*, *84*, 8230–8234.

Kyte, J. and Doolittle, R. F. (1982) A simple method for displaying the hydrophobic character of a protein. *J. Mol. Biol.*, *157*, 105–132.

Lancaster, Jr., J. R. (1982) Mechanism of lactose-proton cotransport in *Escherichia coli*. Kinetic results in terms of the site exposure model. *FEBS Lett.*, *150*, 9–18.

Landick, R., Anderson, J. J., Mayo, M. M. Gunsalus, Mavromara, P., Daniels, C. J., and Oxender, D. L. (1980) Regulation of high affinity leucine transport in *Escherichia coli*. *J. Supramol. Struct.*, *14*, 527–537.

Lanyi, J. K. (1979) The role of Na^+ in transport processes of bacterial membranes. *Biochim. Biophys. Acta*, *559*, 377–397.

Lanyi, J. (1988) Mechanism of chloride transport by halorhodopsin. In *Transport Through Membranes: Carriers, Channels, and Pumps* (A. Pullman, ed.), pp. 429–440, D. Reidel, Dordrecht.

Larsson, C. (1983) Partition in aqueous polymer two phase systems. In *Isolation of Membranes*, pp. 277–309, Academic Press, London.

Larsson, U. K., Sundby, C., and Andersson, B. (1987) Characterization of two different subpopulations of spinach light-harvesting chlorophyll *a/b*-protein complex (LHC II): polypeptide composition, phosphorylation pattern, and association with photosystem II. *Biochim. Biophys. Acta*, *894*, 59–68.

Läuger, P. (1972) Carrier-mediated ion transport. *Science*, *178*, 24–30.

Lavergne, J. (1987) Optical-difference spectra of the S-state transitions in the photosynthetic oxygen-evolving complex. *Biochim. Biophys. Acta*, *894*, 91–107.

Lawrence, G. D., and Sawyer, D. T. (1978) The chemistry of biological manganese. *Coord. Chem. Rev.*, *27*, 173–193.

Lee, C. A., and Saier, Jr., M. H. (1983) Mannitol-specific Enzyme II of the bacterial phosphotransferase system III. The nucleotide sequence of the permease gene. *J. Biol. Chem.*, *258*, 10761–10767.

Lehninger, A. L. (1971) *Bioenergetics*, Chap. 7, W. A. Benjamin, Menlo Park.

Leifer, D., and Henderson, R. (1983) Three-dimensional structure of orthorhombic purple membrane at 6.5 Å resolution. *J. Mol. Biol.*, *163*, 451–466.

Lemaire, C., Girault, G., and Galmiche, J. M. (1985) Flash-induced ATP synthesis in pea chloroplasts in relation to proton flux. *Biochim. Biophys. Acta, 803*, 285–292.

Lemaire, C., Girard-Basceau, J., Wollman, F.-A., and Bennoun, P. (1986) Studies on the cytochrome b_6–f complex. I. Characterization of the complex subunits in *C. reinhardtii*. *Biochim. Biophys. Acta, 851*, 229–238.

Lemasters, J. J. (1984) The ATP to oxygen stoichiometries of oxidative phosphorylation by rat liver mitochondria. *J. Biol. Chem., 259*, 13123–13130.

Lemasters, J. J., and Billica, W. H. (1981) Non-equilibrium thermodynamics of oxidative phosphorylation by inverted inner membrane vesicles of rat liver mitochondria. *J. Biol. Chem., 256*, 12949–12957.

Lenaz, G. and de Santis, A. (1985) Function and specificity of ubiquinone in the respiratory chain. In *Coenzyme Q* (G. Lenaz, ed.), pp. 165–199, Wiley, New York.

Leonard, K., Wingfield, P., Arad, T., and Weiss, H. (1981) Three-dimensional structure of ubiquinol: cytochrome c reductase from Neurospora mitochondria determined by electron microscopy of membrane crystals. *J. Mol. Biol., 149*, 259–274.

Lewin, R. (1976) *Prochlorophyta* as a proposed new division of algae. *Nature, 261*, 697–698.

Lewin, R. (1987) The unmasking of mitochondrial Eve. *Science, 238*, 24–26.

Lewis, A., Spoonhower, J., Bogolmolni, R. A., Lozier, R. H., and Stoeckenius, W. (1974) Tunable laser resonance Raman spectroscopy of bacteriorhodopsin. *Proc. Natl. Acad. Sci. USA, 71*, 4462–4466.

Liang, N., Pielak, G. J., Mauk, A. G., Smith, M., and Hoffman, B. M. (1987) Yeast cytochrome c with phenylalanine or tyrosine at position 87 transfers electrons to (zinc cytochrome c peroxidase) at a rate ten thousand times that of the serine-87 or glycine-87 variants. *Proc. Natl. Acad. Sci. USA, 84*, 1249–1252.

Liberman, E. A., and Topaly, V. P. (1968) Selective transport of ions through bimolecular phospholipid membranes. *Biochim. Biophys. Acta, 163*, 125–136.

Lightowlers, R. N., Howitt, S. M., Hatch, L., Gibson, F., and Cox, G. B. (1987) The proton pore in the *E. coli* F_0F_1-ATPase: a requirement for arginine at position 210 of the *a* subunit. *Biochim. Biophys. Acta, 894*, 399–406; *ibid* (1988) Substitution of glutamate by glutamine at position 219 of the *a* subunit prevents F_0-mediated proton permeability. *Biochim. Biophys. Acta, 933*, 241–248.

Lill, H., Althoff, G., and Junge, W. (1987) Analysis of ionic channels by a flash spectrophotometric technique applicable to thylakoid membranes: CF_0, the proton channel of the chloroplast ATP synthase and, for comparison, gramicidin. *J. Membr. Biol., 98*, 69–78.

Lin, S.-L., Ormos, P., Eisenstein, L., Govindjee, R., Konno, K., and Nakanishi, K. (1987) Deprotonation of tyrosines in bacteriorhodopsin as studied by FTIR with deuterium and nitrate labelling. *Biochemistry, 26*, 8327–8331.

Link, T. A., Schägger, H., and Von Jagow, G. (1986) Analysis of the structures of the subunits of the bc_1 complex from beef heart mitochondria. *FEBS Lett., 204*, 9–15.

Lockhart, D. J., and Boxer, S. G. (1988) Stark effect spectroscopy of *Rb. sphaeroides* and *Rps. viridis* reaction centers. *Proc. Natl. Acad. Sci. USA, 85*, 107–111.

Lombardi, F. J. (1981) Lactose-$H^+(OH^-)$ transport system of *Escherichia coli*. Multistate gated pore model based on half-sites stoichiometry for high-affinity substrate binding in a symmetrical dimer. *Biochim. Biophys. Acta, 649*, 661–679.

Lorusso, M., Galti, C., Marzo, M., and Papa, S. (1985) Effect of papain digestion on redox-linked proton translocation in bc_1 complex from beef heart reconstituted into liposomes. *FEBS Lett., 182*, 370–374.

Louis, G. V., Pielak, G. J., Smith, M., and Brayer, G. D. (1988) Role of phenylalanine-82 in yeast iso-1-cytochrome c and remote conformational changes induced by a serine residue at this position. *Biochemistry, 27*, 7870–7878.

Lozier, R. H., Niederberger, W., Bogolmolni, R. A., Hwang, S. B., and Stoeckenius, W. (1976) Kinetics and stoichiometry of light-induced proton release and uptake from purple membrane fragments, cell envelopes, and phospholipid vesicles containing oriented purple membranes. *Biochim. Biophys. Acta, 440*, 545–556.

Luft, R., Ikkos, D., Palmieri, G., Ernster, L., and Alzelius, B. (1962) A study of severe hypermetabolism of non-thyroid origin with a defect in the maintenance of mitochondrial respiratory control: a correlated clinical, biochemical, and morphological study. *J. Clin. Invest., 41*, 1776–1804.

Lugtenburg, J., Mathies, R. A., Griffen, R. G., and Herzfeld, J. (1988) Structure and function of rhodopsins from solid state and resonance Raman spectroscopy of isotopic retinal derivatives. *Trends Biochem. Sci., 13*, 388–393.

Lundegårdh, H. (1945) Absorption and exudation of inorganic ions by roots. *Ark. Bot., 32A, 12*, I.

MacLennan, D. H., Brandl, C. J., Korczak, B., and Green, N. M. (1985) Amino acid sequence of a $Ca^{++} + Mg^{++}$-dependent ATPase from rabbit muscle sarcoplasmic reticulum, deduced from its complementary DNA sequence. *Nature, 316*, 696–700.

Magnusson, K., Hederstedt, L., and Rutberg, L. (1985) Cloning and expression in *Escherichia coli* of *sdh*A, the structural gene for cytochrome b_{558} of the *Bacillus subtilis* succinate dehydrogenase complex. *J. Bacteriol., 162*, 1180–1185.

Maguire, J. J., Johnson, M. K., Morningstar, J. E., Ackrell, B. A. C., and Kearney, E. B. (1985) EPR studies of mammalian succinate dehydrogenase. *J. Biol. Chem., 260*, 10909–10912.

Mahan, B. H. (1964) *Elementary Chemical Thermodynamics*, Chap. 2, W. A. Benjamin, New York.

Maiden, M. C. J., Davis, E. O., Baldwin, S. A., Moore, D. C. M., and Henderson, P. J. F. (1987) Mammalian and bacterial sugar transport proteins are homologous. *Nature, 325*, 641–643.

Maiden, M. C. J., Jones-Mortimer, M. C., and Henderson, P. J. F. (1988) The cloning, DNA sequence, and overexpression of the gene *ara*E coding for the arabinose-proton symport in *Escherichia coli* K12. *J. Biol. Chem., 263*, 8003–8010.

Malkin, R. (1982) Interaction of photosynthetic electron transport inhibitors and the Rieske iron-sulfur center in chloroplast and b_6–f complex. *Biochemistry, 21*, 2945–2950.

Malkin, R. (1987) Photosystem I. In *The Light Reactions* (J. Barber, ed.), pp. 495–525, Elsevier, Amsterdam.

Malkin, R. and Rabinowitz, J. C. (1966) The reconstitution of *Clostridial* ferredoxin. *Biochem. Biophys. Res. Commun., 23*, 822–827.

Malmström, B. G. (1985) Cytochrome c oxidase as a proton pump. A transition-state mechanism. *Biochim. Biophys. Acta, 811*, 1–12.

Mandel, M., Moriyama, Y., Hulmes, J. D., Pon, Y. E., Nelson, H., and Nelson, N. (1988) cDNA sequence encoding the 16-kDa proteolipids of chromaffin granules implies gene duplication in the evolution of H^+-ATPases. *Proc. Natl. Acad. Sci. USA, 85*, 5521–5524.

Mao, D., Wachter, E., and Wallace, B. A. (1982) Folding of the mitochondrial proton adenosine triphosphatase proteolipid channel in phospholipid vesicles. *Biochemistry, 21*, 4960–4968.

Marcus, R. A. (1965) On the theory of electron transfer reactions. VI. Unified treatment for homogeneous and electrode reactions. *J. Chem. Phys.*, *43*, 679–701.

Marcus, R. A. and Sutin, N. (1985) Electron transfers in chemistry and biology. *Biochim. Biophys. Acta*, *811*, 265–322.

Margoliash, E., and H. R. Bosshard (1983) Guided by electrostatics, a textbook protein comes of age. *Trends Biochem. Sci.*, *8*, 316–320.

Margoliash, E., Barlow, C. H., and Byers, V. (1970). Differential binding properties of cytochrome *c*: possible relevance for mitochondrial ion transport. *Nature*, *288*, 723–726.

Marinetti, T. D., Okamura, M. Y., and Feher, G. (1979) Localization of the primary quinone binding site in reaction centers from *Rb. sphaeroides* R-26 by photoaffinity labeling. *Biochemistry*, *10*, 3126–3133.

Maróti, P., and Wraight, C. A. (1988) Flash-induced H^+ binding by bacterial photosynthetic reaction centers: influences of the redox state of the acceptor quinones and primary quinone. *Biochim. Biophys. Acta*, *934*, 329–347.

Maróti, P., Kirmaier, C., Wraight, C., Holten, D., and Pearlstein, R. M. (1985) Photochemistry and electron transfer in borohydride-treated photosynthetic reaction centers. *Biochim. Biophys. Acta*, *810*, 132–139.

Mates, S. M., Eisenberg, E. S., Mandel, L. J., Patel, L., Kaback, H. R., and Miller, M. H. (1982) Membrane potential and gentamycin uptake in *Staphylococcus aureus*. *Proc. Natl. Acad. Sci. USA*, *79*, 6693–6697.

Mathies, R. A., Brito Cruz, C. H., Pollard, W. T., and Shank, C. V. (1988) Direct observation of the femtosecond excited-state *cis–trans* isomerization in bacteriorhodopsin. *Science*, *240*, 777–779.

Matthews, B. W., Fenna, R. E., Bolognesi, M. C., Schmidt, M. F., and Olson, J. M. (1979) Structure of a bacteriochlorophyll *a*–protein from the green photosynthetic bacterium *Prosthecochloris aestuarii*. *J. Mol. Biol.*, *131*, 259–285.

Matsumo-Yagi, A., and Hatefi, Y. (1988) Estimation of the turnover number of bovine heart F_0F_1 complexes for ATP synthesis. *Biochemistry*, *27*, 335–340.

Matsushita, N., Patel, L., and Kaback, H. R. (1984) Cytochrome *o* type oxidase from *Escherichia coli*. Characterization of the enzyme and mechanism of electrochemical proton gradient generation. *Biochemistry*, *23*, 4703–4714.

Matsuura, S., Shioi, J., and Imae, Y. (1977) Motility in *Bacillus subtilis* driven by an artificial proton motive force. *FEBS Lett.*, *82*, 187–190.

Matsuura, K., O'Keefe, D. P., and Dutton, P. L. (1983) A reevaluation of the events leading to the electrogenic reaction and proton translocation in the ubiquinol-cytochrome *c* oxidoreductase of *Rb. sphaeroides*. *Biochim. Biophys. Acta*, *722*, 12–22.

Mauk, M. R., and Mauk, A. G. (1986) Electrostatic analysis of the interaction of cytochrome *c* with native and dimethyl ester heme substituted cytochrome b_5. *Biochemistry*, *25*, 7085–7091.

Mayo, S. L., Ellis, W. R., Crutchley, R. J., and Gray, H. B. (1986) Long range electron transfer in heme proteins. *Science*, *233*, 948–952.

McCarty, R. E. (1978) The stoichiometry of the proton-translocating ATPase of chloroplast thylakoids. In *The Proton and Calcium Pumps* (G. F. Azzone, M. Avron, J. C. Metcalfe, E. Quagliloriello, and N. Siliprandi, eds.), pp. 65–70. Elsevier/North-Holland.

McCarty, R. E., and Hammes, G. C. (1987) Molecular architecture of chloroplast coupling factor I. *Trends Biochem. Sci.*, *12*, 234–237.

McCray, J. A., Herbette, L., Kihara, T., and Trentham, D. R. (1980) A new approach to time-resolved studies of ATP-requiring biological systems: Laser flash photolysis of caged ATP. *Proc. Natl. Acad. Sci. USA*, 77, 7237–7241.

McElroy, J. D., Feher, G., and Mauzerall, D. (1969) On the nature of the free radical formed during the primary process of bacterial photosynthesis. *Biochim. Biophys. Acta*, 172, 180–183.

McMorrow, Jr, Shuman, H. A., Size, D., Wilson, D. M., and Wilson, T. H. (1989) Sodium/proton antiport is required for growth of *Escherichia coli* at alkaline pH. *Biochim. Biophys. Acta*, 981, 21–26.

McPherson, P. H., Okamura, M. Y., and Feher, G. (1988) Light-induced proton uptake by photosynthetic reaction centers from *Rb. sphaeroides* R-26. I Protonation of the one-electron states $D^+Q_A^-$, DQ_A^-, $D^+Q_A Q_B^-$, and $DQ_A Q_B^-$. *Biochim. Biophys. Acta*, 934, 348–368.

Meinhardt, S. W., and Crofts, A. R. (1982) Kinetic and thermodynamic resolution of cytochromes c_1 and c_2 from *Rb. sphaeroides*. *FEBS Lett.*, 149 217–222.

Meinhardt, S. W., and Crofts, A. R. (1983) The role of cytochrome b-566 in the electron transfer chain of *Rb. sphaeroides*. *Biochim. Biophys. Acta*, 723, 219–230.

Meinhardt, S. W., Yang, X., Trumpower, B. L., and Ohnishi, T. (1987) Identification of a stable ubisemiquinone and characterization of the effects of ubiquinone oxidation-reduction status on the Rieske iron-sulfur protein in the three subunit UQH_2-Cyt c oxidoreductase complex of *Paracoccus denitrificans*. *J. Biol. Chem.*, 262, 8702–8706.

Merchant, S., and Bogorad, L. (1986a) Rapid degradation of apoplastocyanin in Cu(II)-deficient cells of *Chlamydomonas reinhardtii*. *J. Biol. Chem.*, 261, 15850–15853.

Merchant, S., and Bogorad, L. (1986b) Regulation by copper of the expression of plastocyanin and cytochrome c_{552} in *Chlamydomonas reinhardtii*. *Mol. Cell Biol.*, 6, 462–469.

Metzner, H., Fischer, K. and Balzar, O. (1979) Isotope ratios in photosynthetic oxygen. *Biochim. Biophys. Acta*, 548, 287–295.

Meyer, T. E., and Kamen, M. D. (1982) New perspectives on c-type cytochromes. *Adv. Prot. Chem.* (C. B. Anfinsen, J. T. Edsall, and F. M. Richards, eds.), 35, 105–212, Academic Press, New York.

Michel, H. (1983) Crystallization of membrane proteins. *Trends Biochem. Sci.*, 8, 56–59.

Michel, H., and Deisenhofer, J. (1988a) Relevance of the photosynthetic reaction center from purple bacteria to the structure of photosystem II. *Biochemistry*, 27, 1–7.

Michel, H., and Deisenhofer, J. (1988b) The structure of the photosynthetic reaction center from *Rps. viridis*. *Bull. Inst. Pasteur*, 86, 37–45.

Michel, H., and Oesterhelt, D. (1980) Electrochemical proton gradient across the cell membrane of *Halobacterium halobium*: comparison of the light-induced increase with the increase of intracellular adenosine triphosphate under stready-state illumination. *Biochemistry*, 19, 4607–4614.

Michel, H., Epp, O., and Deisenhofer, J. (1986b) Pigment–protein interactions in the photosynthetic reaction centre from *Rps. viridis*. *EMBO J.*, 5, 2445–2451.

Michel, H., Weyer, K. A., Gruenberg, H., Dunger, I., Oesterhelt, D., and Lottspeich, F. (1986a) The "light" and "medium" subunits of the photosynthetic reaction centre from *Rps. viridis*: isolation of the genes, nucleotide, and amino acid sequence. *EMBO J.*, 5, 1149–1158.

Michel-Beyerle, M. E., Plato, M., Deisenhofer, J., Michel, H., Bixon, M., and Jortner,

J. (1988) Unidirectionality of charge separation in reaction centers of photosynthetic bacteria. *Biochim. Biophys. Acta*, 932, 52–70.
Miki, J., Takeyama, M., Noumi, T., Kanazawa, H., Maeda, M., and Futai, M. (1986) *Escherichia coli* H^+-ATPase: loss of the carboxyl terminal region of the γ subunit causes defective assembly of the F_1 portion. *Arch. Biochem. Biophys.*, 251, 458–464.
Miles, C. D., and Jagendorf, A. T. (1970) Evaluation of electron transport as the basis of adenosine triphosphate synthesis after acid-base transition by spinach chloroplasts. *Biochemistry*, 9, 429–434.
Miller, D. M., Olson, J. S., Pflugarth, J. S., and Quiocho, F. A. (1983) Rates of ligand binding to periplasmic proteins involved in bacterial transport and chemotaxis. *J. Biol. Chem.*, 258, 13665–13672.
Miller, J. R., Beitz, J. V., and Huddleston, R. K. (1984a) Effect of free energy on rates of electron transfer between molecules. *J. Am. Chem. Soc.*, 106, 5057–5068.
Miller, J. R., Calcaterra, L. T., and Closs, G. L. (1984b) Intramolecular long-distance electron transfer in radical anions. The effects of free energy and solvent on the reaction rates. *J. Am. Chem. Soc.*, 106, 3047–3049.
Millett, F., De Jong, K., Paulson, L., and Capaldi, R. (1983) Identification of specific carboxylate groups on cytochrome c oxidase that are involved in binding cytochrome c. *Biochemistry* 22, 546–552.
Millner, P. A., and Barber, J. (1984) Plastoquinone as a mobile redox carrier in the thylakoid membrane. *FEBS Lett.*, 169, 1–6.
Millner, P. A., Widger, W. R., Abbott, M. S., Cramer, W. A., and Dilley, R. A. (1982) The effect of adenine nucleotides on inhibition of the thylakoid protein kinase by sulfhydryl-directed reagents. *J. Biol. Chem.*, 257, 1736–1742.
Mills, J. D., and Mitchell, P. (1982) Modulation of coupling factor ATPase in intact chloroplasts. *Biochim. Biophys. Acta*, 679, 75–83.
Mitaku, S., Wright, J. K., Best, L., and Jähnig, F. (1984) Localization of the galactoside binding site in the lactose carrier of *Escherichia coli*. *Biochim. Biophys. Acta*, 776, 247–258.
Mitchell, P. (1961) Coupling of phosphorylation to electron and proton transfer by a chemi-osmotic type of mechanism. *Nature*, 191, 144–148.
Mitchell, P. (1966) Chemiosmotic coupling in oxidative and photosynthetic phorphorylation. *Biol. Rev.*, 41, 445–502.
Mitchell, P. (1976) Possible molecular mechanisms of the protonmotive function of cytochrome systems. *J. Theor. Biol.*, 62, 327–367.
Mitchell, P. (1984) Bacterial flagellar motors and osmoelectric molecular rotation by an axially transmembrane well and turnstile mechanism. *FEBS Lett.*, 176, 287–294.
Mitchell, P. (1985) Molecular mechanisms of protonmotive $F_0 F_1$ ATPases. *FEBS Lett.*, 182, 1–7.
Mitchell, P., and Moyle, J. (1965) Stoichiometry of proton translocation through the respiratory chain and adenosine triphosphatase systems of rat liver mitochondria. *Nature*, 208, 147–151.
Mitchell, P., Mitchell, R., Moody, A. J., West, I. C., Baum, H., and Wrigglesworth, J. M. (1985) Chemiosmotic coupling in cytochrome oxidase, possible protonmotive O loop and O cycle mechanisms. *FEBS Lett.*, 188, 1–7.
Mitchell, W. J., Booth, I. R., and Hamilton, W. A. (1979) Quantitative analysis of proton-linked transport systems. Glutamate transport in *Staphylococcus aureus*. *Biochem. J.*, 184, 441–449.
Miyoshi, H., and Fujita, T. (1987) Quantitative analyses of uncoupling activity of

SF6847 [2,6-di-*t*-4-(2,2-dicyanovinyl) phenol] and its analogs with spinach chloroplasts. *Biochim. Biophys. Acta*, *894*, 339–345.
Mogi, T., Stern, L. J., Hackett, N. R., and Khorana, H. G. (1987) Bacteriorhodopsin mutants containing single tyrosine to phenylalanine substitutions are all active in proton translocation. *Proc. Natl. Acad. Sci. USA*, *84*, 5595–5599.
Mogi, T., Stern, L. J., Marti, T., Chao, B. H., and Khorana, H. G. (1988) Aspartic acid substitutions affect proton translocation by bacteriorhodopsin. *Proc. Natl. Acad. Sci. USA*, *85*, 4148–4152.
Mohraz, M., and Smith, P. R. (1988) Three-dimensional structure of Na^+,H^+-ATPase and a model for the oligomeric form and the mechanism of the Na,K pump. In *The Na^+, K^+ Pump. Part A: Molecular Aspects*, pp. 99–106, Alan, R. Liss, New York.
Moody, A. J., Mitchell, R., West, I. C., and Mitchell, P. (1987) Protonmotive stoichiometry of rat liver cytochrome *c* oxidases: determination by a new rate/pulse method. *Biochim. Biophys. Acta*, *894*, 209–227.
Moore, G. R., and Williams, R. J. P. (1977) Structural basis for the variation in midpoint potential of cytochromes. *FEBS Lett.*, *79*, 229–232.
Moore, J. M., Case, D. A., Chazin, W. J., Gippert, G. P., Havel, T. F., Powls, R., and Wright, P. F. (1988) Three-dimensional solution structure of plastocyanin from the green alga, *Scenedesmus obliquus*. *Science*, *240*, 314–317.
Moriyama, Y., and Nelson, N. (1987a) Nucleotide binding sites and chemical modification of the chromaffin granule ATPase. *J. Biol. Chem.*, *262*, 14723–14729.
Moriyama, Y., and Nelson, N. (1987b) The purified ATPase from chromaffin granule membranes is an anion-dependent proton pump. *J. Biol. Chem.*, *262*, 9175–9180.
Morris, J., and Herrmann, R. C. (1984) Nucleotide sequence of the gene for the P680 chlorophyll *a* apoprotein of the photosystem II reaction center from spinach. *Nucleic Acids. Res.*, *12*, 2837–2850.
Mortenson, L. E., Valentine, R. C., and Carnahan, J. E. (1962) An electron transport factor from *Clostridium pasteurianum* Biochem. *Biophys. Res. Comm.*, *7*, 448–452.
Mosher, M. E., White, L. K., Hermolin, J., and Fillingame, R. H. (1985) H^+-ATPase of *Escherichia coli*. An *unc* E mutation impairing coupling between F_1 and F_0, but not F_0-mediated H^+ translocation. *J. Biol. Chem.*, *260*, 4807–4814.
Moss, D. A., and Bendall, D. S. (1984) Cyclic electron transport in chloroplasts. The Q-cycle and the site of action of antimycin. *Biochim. Biophys. Acta*, *767*, 389–395.
Mueckler, M., Caruso, C., Baldwin, S. A., Panico, M., Blench, I., Morris, H. R., Allard, W. J., Lienhard, G. J., and Lodish, H. (1985) Sequence and structure of a human glucose transporter. *Science*, *229*, 941–945.
Mullet, J. E. (1983) The amino acid sequence of the polypeptide segment which regulates membrane adhesion (grana stacking) in chloroplasts. *J. Biol. Chem.*, *258*, 9941–9948.
Murata, N., and Miyao, M. (1987) Oxygen-evolving complex of photosystem II in higher plants. *Prog. Photosyn. Res.* (J. Biggins, ed.), *I*, pp. 453–462, Martinus Nijhoff, Dordrecht.
Murén, E., and Randall, L. L. (1985) Export of α-amylase by *B. amyloliquefaciens* requires protonmotive force. *J. Bacteriol.*, *164*, 712–716.
Nachliel, E., and Gutman, M. (1984) Kinetic analysis of proton transfer between reactants adsorbed to the same micelle. The effect of proximity on the rate constants. *Eur. J. Biochem.*, *143*, 83–88.
Nadeau, J. G. (1985) The chemical synthesis and physical characteristics of several

oligodeoxyribonucleotides d(CGAGTTTGACGP) forms an unusually stable hairpin loop structure. Ph.D. Thesis, Purdue University, 94 pp.

Nageswara Rao, B. D., Cohn, M., and Scopes, R. K. (1978) ^{31}P NMR study of bound reactants and products of yeast 3-phospho-glycerate kinase at equilibrium and the effect of sulfate ion. *J. Biol. Chem.*, *253*, 8056–8060.

Nageswara Rao, B. D., Kayne, F. J., and Cohn, M. (1979) ^{31}P NMR studies of enzyme-bound substrates of rabbit muscle pyruvate kinase. *J. Biol. Chem.*, *254*, 2689–2696.

Nagle, J. F., and Dilley, R. A. (1986) Models of localized energy coupling. *J. Bioenerg. Biomembr.*, *18*, 55–64.

Nagle, J. F., and Morowitz, H. (1978) Molecular mechanisms for proton transport in membranes. *Proc. Natl. Acad. Sci. USA*, *75*, 298–302.

Nagle, J. F., and Tristram-Nagle, S. (1983) Hydrogen bonded chain mechanisms for proton conduction and proton pumping. *J. Membr. Biol.*, *74*, 1–14.

Nakamura, T., Tokuda, H., and Unemoto, T. (1984) K^+/H^+ antiporter functions as a regulator of cytoplasmic pH in a marine bacterium, *Vibrio alginolyticus*. *Biochim. Biophys. Acta*, *776*, 330–336.

Nakashima, R. A., Mangar, P. S., Colombini, M., and Pedersen, P. L. (1986) Hexokinase receptor protein in hepatoma mitochondria: evidence from N,N'-dicyclohexylcarbodiimide-labeling studies for the involvement of pore-forming protein VDAC. *Biochemistry*, *25*, 1015–1021.

Nalecz, K. A., Bolli, R., Ludwig, B., and Azzi, A. (1985) The role of subunit III in bovine cytochrome *c* oxidase. Comparison between native, subunit III-depleted and *Paracoccus denitrificans* enzymes. *Biochim. Biophys. Acta*, *808*, 259–272.

Nanba, O., and Satoh, K. (1987) Isolation of a photosystem II reaction center consisting of D-1 and D-2 polypeptides and cytochrome b-559. *Proc. Natl. Acad. Sci. USA*, *84*, 109–112.

Nelson, N. (1988) Structure, function, and evolution of proton ATPases. *Plant Physiol.*, *86*, 0001–0003.

Nelson, S. O., Schuitema, A. R. J., Benne, R., van der Ploeg, L. H. T., Plijter, J. J., Aan, F., and Postma, P. W. (1984) Molecular cloning, sequencing and expression of the *ccr* gene: the structural gene for III^{Glc} of the bacterial PEP:glucose phosphotransferase system. *EMBO J.*, *3*, 1587–1593.

Neumann, J., and Jagendorf, A. T. (1964) Light-induced pH changes related to phosphorylation by chloroplasts. *Arch. Biochem. Biophys.*, *107*, 109–119.

Newcomer, M. E., Gilliland, G. L. and Quiocho, F. A. (1981) L-Arabinose-binding protein-sugar complex at 2.4 Å resolution. Stereochemistry and evidence for a structural change. *J. Biol. Chem.*, *256*, 13213–13217.

Nicholls, D. G. (1982) *Bioenergetics: An Introduction to Chemiosmotic Theory*, 190 pp., Academic Press, London.

Nicholls, D. G., and Rial, E. (1984) Brown fat mitochondria. *Trends Biochem. Sci.*, *9*, 489–491.

Nicolay, K., Lolkema, J., Hellingwerf, K., Kaptein, R., and Konings, W. N. (1981) Quantitative agreement between the values for the light-induced ΔpH in *Rhodopseudomonas sphaeroides* measured with automated flow-analysis and ^{31}P NMR. *FEBS Lett.*, *123*, 319–323.

Nobrega, F. G., and Tzagoloff, A. (1980) Assembly of the mitochondrial membrane system. DNA sequence and organization of the cytochrome *b* gene in *Saccharomyces cerevisiae*. *J. Biol. Chem.*, *255*, 9828–9837.

Nocera, D. G., Winkler, J. R., Yocum, K. M., Bordignon, E., and Gray, H. B. (1984) Kinetics of intramolecular electron transfer from Ru^{II} to Fe^{III} in ruthenium-modified cytochrome c. J. Am. Chem. Soc., 106, 5145–5150.

Norris, J. R., Uphaus, R. A., Crespi, H. G., and Katz, J. J. (1971) Electron spin resonance of chlorophyll and the origin of signal I in photosynthesis. Proc. Natl. Acad. Sci. USA, 68, 625–629.

Norris, K. H., and Butler, W. L. (1961) Techniques for obtaining absorption spectra on intact biological samples. IRE Trans. Bio-Med Elect., 8, 153–157.

Northrup, S. H., Boles, J. O., and Reynolds, J. C. L. (1988) Brownian dynamics of cytochrome c and cytochrome c peroxidase association. Science, 241, 67–70.

Nostic, N. K., Margalit, R., Che, C.-M., and Gray, H. B. (1983) Kinetics of long distance ruthenium-to-copper electron transfer in [pentaamineruthenium histidine-83] azurin. J. Am. Chem. Soc., 105, 7765–7767.

Noumi, T., Futai, M., and Kanazawa, H. (1984a) Replacement of serine 373 by phenylalanine in the α subunit of Escherichia coli F_1-ATPase results in loss of steady-state catalysis by the enzyme. J. Biol. Chem., 259, 10076–10079.

Noumi, T., Mosher, M. E., Natori, S., Futai, M., and Kanazawa, H. (1984b) A phenylalanine for serine substitution in the β subunit of Escherichia coli F_1-ATPase affects dependence of its activity on divalent cations. J. Biol. Chem., 259, 10071–10075.

Noumi, T., Oka, N., Kanazawa, H., and Futai, M. (1986a) Mutational replacements of conserved amino acid residues in the β subunit resulted in defective assembly of H^+-translocating ATPase $(F_0 F_1)$ in Escherichia coli. J. Biol. Chem., 261, 7070–7075.

Noumi, T., Taniai, M., Kanazawa, H., and Futai, M. (1986b) Replacement of arginine 246 by histidine in the β subunit of Escherichia coli H^+-ATPase resulted in loss of multisite ATPase activity. J. Biol. Chem., 261, 9196–9201.

Nozaki, Y. and Tanford, C. (1971) The solubility of amino acids and two glycine peptides in aqueous ethanol and dioxane solutions. J. Biol. Chem., 246, 2211–2217.

Nyrén, P., Nore, B. F., and Baltscheffsky, M. (1986) Studies on photosynthetic inorganic pyrophosphate formation in Rhodospirillum rubrum chromatophores. Biochim. Biophys. Acta, 851, 276–282.

O'Keefe, D. P. (1983) Sites of cytochrome b-563 reduction, and the mode of action of DNP-INT and DBMIB in the chloroplast cytochrome b_6-f complex. FEBS Lett., 162, 349–354.

O'Neal, C. C., and Boyer, P. D. (1984) Assessment of the rate of bound substrate interconversion and of ATP acceleration of production release during catalysis by mitochondrial adenosine triphosphate. J. Biol. Chem., 259, 5761–5767.

Oesper, P. (1950) Source of the high energy content in energy-rich phosphates. Arch. Biochem. Biophys., 27, 255–270.

Oesterhelt, D., and Hess, B. (1973) Reversible photolysis of the purple complex in the purple membrane of Halobacterium halobium. Eur. J. Biochem., 37, 316–326.

Oettmeier, W., Masson, K., Soll, H.-J., Hurt, E., and Hauska, C. (1982) Photoaffinity labelling of plastoquinone binding sites in the chloroplast b_6-f complex. FEBS Lett., 144, 313–317.

Oettmeier, W., Masson, K., and Olchewski, E. (1983) Photoaffinity labeling of chloroplast cytochrome b_6-f complex by an inhibitor azido derivative. FEBS Lett., 155, 241–244.

Oettmeier, W., Godde, D., Kunze, B., and Höfle, G. (1985) Stigmatellin: a dual type inhibitor of photosynthetic electron transport. Biochim. Biophys. Acta, 807, 216–219.

Ogawa, T., Miyano, A., and Inoue, Y. (1985) Photosystem I-driven inorganic carbon transport in the cyanobacterium, *Anacystis nidulans*. *Biochim. Biophys. Acta*, 808, 77–84.

Oh-oka, H., Tanaka, S., Wada, K., Kuwabara, T., and Murata, N. (1986) Complete amino acid sequence of 33 kDa protein isolated from spinach photosystem II particles. *FEBS Lett.*, 197, 63–66.

Ohnishi, T. (1979) Mitochordrial iron-sulfur flavodehydrogenases. In *Membrane Proteins in Energy Transduction* (Capaldi, R. A., ed.), pp. 1–87, Marcel Dekker, New York.

Ohnishi, T. (1987) Structure of the Succinate-ubiquinone oxidoreductase (complex II). *Curr. Top. Bioenerg.* (C. P. Lee, ed.), 15, 37–65.

Ohnishi, T., and Trumpower, B. L. (1980) Differential effects of antimycin on ubisemiquinone bound in different environments in isolated succinate cytochrome c reductase complex. *J. Biol. Chem.*, 255, 3278–3284.

Ohnishi, T., Salerno, J. C., Winter, D. B., Lim, J., Yu, C., Yu, L., and King, T. E. (1976) Thermodynamic and EPR characteristics of two ferredoxin type iron-sulfur centers in the succinate-ubiquinone reductase segment of the respiratory chain. *J. Biol. Chem.*, 251, 2094–2104.

Ohnishi, T., King, T. E., Salerno, J. C., Blum, H., Bowyer, J. R. and Maida, T. (1981) Thermodynamic and electron paramagnetic resonance characterization of flavin in succinate dehydrogenase. *J. Biol. Chem.*, 256, 5577–5582.

Ohnishi, T., LoBrutto, R., Salerno, J. C., Bruckner, C., and Frey, T. G. (1982) Spatial relationship between cytochrome a and a_3. *J. Biol. Chem.*, 257, 14821–14825.

Ohyama, K., Fukuzawa, H., Kohchi, T., Shirai, H., Sano, T., Sano, S., Umesono, K., Shiki, Y., Takeuchi, M., Chang, Z., Aota, S., Inokuchi, H., and Ozeki, H. (1986) Chloroplast gene sequence of liverwort *Marchantia polymorpha* chloroplast DNA. *Nature*, 322, 572–574.

Ohyama, K., Kohchi, T., Sano, T., and Yamada, Y. (1988) Newly identified groups of genes in chloroplasts. *Trends Biochem. Sci.*, 13, 19–22.

Okamura, M. Y., Feher, G., and Nelson, N. (1982). Reaction centers In *Photosynthesis: Energy Conversion by Plants and Bacteria* (Govindjee, ed.), pp. 195–264, Academic Press, New York.

Ono, T-A., and Inoue, Y. (1983) Mn-preserving extraction of 33- 24-, and 15 kDa proteins from O_2-evolving PS II particles by divalent salt washing. *FEBS Lett.*, 164, 255–260.

Ort, D. R. (1986) Energy transduction in oxygenic photosynthesis: an overview of structure and mechanism. *Encyclop. Plant Physiol.*, New Series, Vol. 19, (L. A. Staehelin and C. J. Arntzen, eds.), pp. 143–196, Springer-Verlag, Berlin.

Ort, D. R., and Good, N. E. (1988) Textbooks ignore photosystem II-dependent ATP formation: is the Z scheme to blame. *Trends Biochem. Sci.*, 13, 467–469.

Ort, D. R, and Parson, W. W. (1979) The quantum yield of flash-induced proton release by bacteriorhodopsin-containing membrane fragments. *Biophys. J.*, 25, 341–352.

Ort, D. R., and Parson, W. W. (1979) Enthalpy changes during the photochemical cycle of bacteriorhodopsin. *Biophys. J.*, 25, 355–364.

Ort, D. R., Dilley, R. A., and Good, N. E. (1976) Photophosphorylation as a function of illumination time. II. Effects of permeant buffers. *Biochim. Biophys. Acta*, 449, 108–124.

Ovchinnikov, Y. A. (1987) Probing the folding of membrane proteins. *Trends Biochem. Sci.*, 12, 434–438.

Ovchinnikov, Y. A., Abdulaev, N. G., Feigina, M. Y., Kiselev, A. V., and Lobanov, N. A. (1979) The structural basis of the functioning of bacteriorhodopsin: an overview. *FEBS Lett.*, 100, 219–224.

Ovchinnikov, Y. A., Abdulaev, N. G., Vasilev, R. G., Hurina, I. Y., Kurgatove, A. B., and Kiselev, A. V. (1985) The antigenic structure and topography of bacteriorhodopsin in purple membranes as determined by interaction with monoclonal antibodies. *FEBS Lett.*, 179, 343–350.

Ovchinnikov, Y. A., Arzamazova, N. M., Arystarkhova, E. A., Gevondyan, N. M., Aldanova, N. A., and Modyanov, N. N. (1987) Detailed structural analysis of exposed domains of membrane-bound Na^+, K^+-ATPase. A model of transmembrane arrangement. *FEBS Lett.*, 217, 269–274.

Ovchinnikov, Y. A., Abdulaev, N. G., Zolotarev, A. S., Shmukler, B. E., Zargarov, A. A., Kutuzov, M. A., Telezhinskaya, I. N., and Levina, N. B. (1988a) Photosynthetic reaction centre of *Chloroflexus aurantiacus*. Primary structure of L-subunit. *FEBS Lett.*, 231, 237–242.

Ovchinnikov, Y. A., Abdulaev, N. G., Shmukler, B. E., Zargarov, A. A., Kutuzov, M. A., Telezhinskaya, I. N., Levina, N. B., and Zolotarev, A. S. (1988b) Photosynthetic reaction centre of *Chloroflexus aurantiacus*. Primary structure of M-subunit. *FEBS Lett.*, 232, 364–368.

Overfield, R. E., and Wraight, C. A. (1980) Oxidation of cytochromes c and c_2 by bacterial photosynthetic reaction centers in phospholipid vesicles. 2. Studies with negative membranes. *Biochemistry*, 19, 3328–3334.

Owen, P., and Kaback, H. R. (1979) Antigenic architecture of membrane vesicles from *E. coli Biochemistry*, 18, 1422–1426.

Padan, E., Zilberstein, D., and Rottenberg, H. (1976) The proton electrochemical gradient in *Escherichia coli* cells. *Eur. J. Biochem.*, 63, 533–544.

Page, M. G. P. (1987) The role of proteins in the mechanism of galactoside transport via the lactose permease of *Escherichia coli*. *Biochim. Biophys. Acta*, 897, 112–126.

Pain, D., and Blöbel, G. (1987) Protein import into chloroplast requires a chloroplast ATPase. *Proc. Natl. Acad. Sci. USA*, 84, 3288–3292.

Palmer, G. (1977) Current insights into the active center of spinach ferredoxin and other iron-sulfur proteins. In *Iron-Sulfur Proteins* (W. Lovenberg, ed.), Vol. 3, pp. 285–325, Academic Press, New York.

Palmer, G. (1985) The electron paramagnetic resonance of metalloproteins. *Biochem. Soc. Trans.*, 13, 548–560.

Park, S., Liu, G., Topping, T. B., Cover, W. H., and Randall, L. L. (1988) Modulation of folding pathways of exported protein by the leader sequence. *Science*, 239, 1033–1035.

Parson, W. W. (1978) Thermodynamics of the primary reactions of photosynthesis. *Photochem. Photobiol.*, 28, 389–393.

Pauling, L. (1960) *Nature of the Chemical Bond*, p. 321, Cornell University Press, Ithaca.

Pearlstein, R. (1982) Chlorophyll singlet excitons. In *Photosynthesis: Energy Conversion by Plants and Bacteria*, (Govindjee, ed.). Vol. I, pp. 293–330, Academic Press, New Uork.

Pedersen, P. L., and Amzel, L. M. (1985) Structure of ATPases of the F_0F_1 type: chemical asymmetry and implications for mechanism. In *Achievements and Perspectives of Mitochondrial Research* (E. Quagliariello, E. C. Slater, F. Palmieri, C. Saccore, and A. M. Kroon, eds.), pp. 169–188, Elsevier, Amsterdam.

Pedersen, P. L., and Carafoli, E. (1987a) Ion motive ATPases. I. Ubiquity, properties and significance to cell function. *Trends Biochem. Sci.*, 12, 146–150; II. (1987b) Energy coupling and work output. *Trends Biochem. Sci.*, 12, 186–189.

Penefsky, H. S. (1985a) Mechanism of inhibition of mitochondrial adenosine triphosphatase by dicyclohexylcarbodiimide and oligomycin: Relationship to ATP synthesis. *Proc. Natl. Acad. Sci. USA*, 82, 1589–1593.

Penefsky, H. S. (1985b) Reaction mechanism of the membrane-bound ATPase of submitochondrial particles from beef heart. *J. Biol. Chem.*, 260, 13728–13734.

Penefsky, H. S. (1985c) Energy-dependent dissociation of ATP from high affinity catalytic sites of beef heart mitochondrial adenosine triphosphatase. *J. Biol. Chem.*, 260, 13735–13741.

Penefsky, H. S., Pullman, M. E., Datta, A., and Racker, E. (1960) Partial resolution of the enzymes catalyzing oxidative phosphorylation: I. Purification and properties of soluble, dinitrophenol-stimulated adenosine triphosphatase. *J. Biol. Chem.*, 235, 3322–3329.

Perlin, D. S., San Francioco, M. J. D., Slayman, C. W., and Rosen, B. P. (1986) H^+/ATP stoichiometry of proton pumps from *Neurospora crassa* and *Escherichia coli*. *Arch. Biochem. Biophys.*, 248, 53–61.

Perutz, M. (1978) Electrostatic effects in proteins. *Science*, 201, 1187–1191.

Perutz, M. F., Kilmartin, J. V., Nishikura, K., Fogg, J. R., Butler, P. J. G., and Rolleana, H. S. (1980) Identification of residues contributing to the Bohr effect of human hemoglobin. *J. Mol. Biol.*, 138, 649–670.

Peter, G., and Thornber, J. P. (1988) The antenna components of PS II with emphasis on the major pigment protein, LHC IIb. In *Photosynthetic Light-Harvesting Systems* (H. Scheer and S. Schneider, eds.), pp. 175–186, W. de Gruyter, Berlin.

Peterson-Kennedy, S. E., McGourty, J. L., and Hoffman, B. M. (1984) Temperature dependence of long-range electron transfer in [Zn, Fe^{III}] hybrid hemoglobin. *J. Am. Chem. Soc.*, 106, 5010–5012.

Petrouleas, V., and Diner, B. (1987) Light-induced oxidation of the acceptor side Fe(II) of photosystem II by exogeneous quinones through the Q_B binding site. I. Quinones, kinetics, and pH dependence. *Biochim. Biophys. Acta*, 893, 126–137.

Pfanner, N., and Neupert, W. (1985) Transport of proteins into mitochondria: a potassium diffusion potential is able to drive the import of ADP/ATP carrier. *EMBO J.*, 4, 2819–2825.

Pfanner, N., and Neupert, W. (1986) Transport of F_1-ATPase subunit β into mitochondria depends on both a membrane potential and nucleoside triphosphates. *FEBS Lett.*, 209, 152–156.

Pfister, K., Steinback, K. E., Gardner, G., and Arntzen, C. J. (1981) Photoaffinity labeling of a herbicide receptor protein in chloroplast membranes. *Proc. Natl. Acad. Sci. USA*, 78, 981–985.

Pflugarth, J. W., and Quiocho, F. A. (1985) Sulphate sequestered in the sulphate-binding protein of *Salmonella typhimurium* is bound solely by hydrogen bonds. *Nature*, 314, 257–260.

Pichersky, E., Bernatsky, R., Tanusley, S. D., Breidenbach, R. B., Kousch, A. S., and Cashmore, A. R. (1985) Molecular characterization and genetic mapping of two clusters of gene encoding chlorophyll *a/b* binding proteins in *L. esculentum* (tomato). *Gene*, 40, 247–258.

Pick, U., and Racker, E. (1979) Purification and reconstitution of the N,N'-dicyclohexylcarbodiimide-sensitive ATPase complex from spinach chloroplasts. *J. Biol. Chem.*, 254, 2793–2799.

Pick, V., Weiss, M., and Rottenberg, H. (1987) Anomalous uncoupling of photophosphorylation by palmitic acid and gramicidin D. *Biochemistry*, 26, 8295–8302.

Pierce, D. H., Scarpa, A., Topp, M. R., and Blasie, J. K. (1983) Kinetics of calcium uptake by isolated sarcoplasmic reticulum vesicles using flash photolysis of caged adenosine-5′-triphosphate. *Biochemistry*, 22, 5254–5261.

Pierrot, M., Haser, R., Frey, M., Payan, F., and Astier, J.-P. (1982) Crystal structure and electron transfer properties of cytochrome c_3. *J. Biol. Chem.*, 257, 14341–14348.

Plack, Jr., R. H., and Rosen, B. P. (1980) Cation/proton antiport systems in *E. coli*. Absence of potassium/proton antiporter activity in a pH-sensitive mutant. *J. Biol. Chem.*, 255, 3824–3825.

Poole, R. K. (1983) Bacterial cytochrome oxidases. *Biochim. Biophys. Acta*, 726, 205–243.

Porter, A. C. G., Kumamoto, C., Aldape, K., and Simoni, R. D. (1985) Role of the *b* subunit in the *Escherichia coli* proton-transducing ATPase. A mutagenic analysis. *J. Biol. Chem.*, 260, 8182–8187.

Porter, G., Tredwell, C. J., Searle, G. F. W., and Barber, J. (1978) Picosecond time-resolved energy transfer in *Porphyridium cruentum*. *Biochim. Biophys. Acta*, 501, 232–245.

Postma, P. W., and Lengler, J. W. (1985) Phosphoenolpyruvate: carbohydrate phosphotransferase system of bacteria. *Microbiol. Rev.*, 49, 232–269.

Poulos, T. L., and Finzel, B. C. (1984) Heme enzyme structure and function. In *Peptide and Protein Reviews* (M. T. W. Hearn, ed.), Marcel Dekker, New York.

Pressman, B. C. (1976) Biological applications of ionophores. *Annu. Rev. Biochem.*, 45, 501–530.

Price, E. M., and Lingrel, J. B. (1988) Structure-function relationships in the Na,K-ATPase α subunit: site-directed mutagenesis of glutamine-111 to arginine and asparagine-122 to aspartic acid generates a ouabain-resistant enzyme. *Biochemistry*, 27, 8400–8408.

Prince, R. C. (1988) The proton pump of cytochrome oxidase. *Trends Biochem. Sci.*, 13, 159–160.

Prince, R. C., Matsuura, K., Hurt, E., Hauska, G., and Dutton, P. L. (1982) Reduction of cytochromes b_6 and f in isolated plastoquinol-plastocyanin oxidoreductase driven by photochemical reaction centers from *Rb. sphaeroides*. *J. Biol. Chem.*, 257, 3379–3381.

Prochaska, L. J., and Reynolds, K. A. (1986) Characterization of electron transfer and proton-translocation activities in bovine heart mitochondrial cytochrome *c* oxidase deficient in subunit III. *Biochemistry*, 25, 781–787.

Prossnitz, E., Gee, A. and Ames, G. F.-L. (1989) Reconstitution of the histidine periplasmic transport system in membrane vesicles. *J. Biol. Chem.*, 264, 5006–5014.

Prossnitz, E., Nikaido, K. Ulbrich, S. J. and Ames, G. F.-L. (1988) Formaldehyde and photoactivated cross-linking of the periplasmic binding protein to a membrane component of the histidine transport system of *Salmanella typhimurium*. *J. Biol. Chem.*, 263, 17917–17920.

Quiocho, F. A. (1986) Carbohydrate-binding proteins: tertiary structures and protein-sugar interactions. *Annu. Rev. Biochem.*, 55, 287–315.

Rabinowitch, E. (1945) *Photosynthesis and Related Processes, Vol. I*, p. 14, Interscience, New York.

Rabinowitch, E., and Govindjee (1969) *Photosynthesis*, Chaps. 11–18, John Wiley, New York.

Rabon, E., Gunther, R. D., Soumarmon, A., Bassilian, S., Levin, M., and Sachs, G. (1985) Solubilization and reconstitution of the gastric H-K-ATPase. *J. Biol. Chem.*, *260*, 10200–10207.

Racker, E. (1965) *Mechanisms in Bioenergetics*, Chapt. 2 and 3. Academic Press, New York.

Racker, E., and Stoeckenius, W. (1974) Reconstitution of purple membrane vesicles catalyzing proton uptake and adenosine triphosphate formation. *J. Biol. Chem.*, *249*, 662–663.

Radian, R., Bendahan, A., and Kanner, B. I. (1986) Purification and identification of the functional sodium and chloride-coupled γ-aminobutyric acid transport glycoprotein from rat brain. *J. Biol. Chem.*, *261*, 15437–15441.

Radmer, R., and Cheniae, G. (1977) Mechanisms of oxygen evolution. In *Primary Processes of Photosynthesis* (J. Barber, ed.), pp. 303–348, Elsevier/North Holland, Amsterdam.

Radmer, R., and Ollinger, O. (1980) Isotopic composition of photosynthetic O_2 flash yields in the presence of $H_2{}^{18}O$ and $HC^{18}O_3{}^-$. *FEBS Lett.*, *110*, 57–61.

Ragan, I. (1985) Structure and function of respiratory complex I. *Coenzyme Q* (G. Lenaz, ed.), pp. 315–336, Wiley, New York.

Ragan, I. (1987) Structure of NADH-ubiquinone reductase (complex I). *Curr. Topics in Bioenergetics* (C. P. Lee, ed.), *15*, 1–36, Academic Press, San Diego.

Raitio, M., Jalli, T., and Saraste, M. (1987) Isolation and analysis of the genes for cytochrome *c* oxidase in *Paracoccus denitrificans*. *EMBO J.*, *6*, 2825–2833.

Ramos, S., and Kaback, H. R. (1977a) pH-Dependent changes in proton: substrate stoichiometries during active transport in *Escherichia coli* membrane vesicles. *Biochemistry*, *16*, 4271–4275.

Ramos, S., and Kaback, H. R. (1977b) The relationship between the electrochemical proton gradient and active transport in *Escherichia coli* membrane vesicles. *Biochemistry*, *16*, 854–859.

Randall, L. L., and Hardy, S. J. S. (1986) Correlation of competence for export with lack of tertiary structure of the mature species: A study *in vivo* of maltose-binding protein in *E. coli*. *Cell*, *46*, 421–428.

Randall, L. L., Hardy, S. J. S., and Thom, J. R. (1987) Export of proteins: a biochemical view. *Annu. Rev. Microbiol.*, *41*, 507–541.

Randall, S. K., and Sze, H. (1986) Properties of the purified tonoplast H^+-pumping ATPase from oat roots. *J. Biol. Chem.*, *261*, 1364–1371.

Rao, R., Perlin, D. O., and Senior, A. E. (1987) The defective proton-ATPase of *uncA* mutants of *Escherichia coli*: ATP-binding and ATP-induced conformational changes in mutant α-subunits. *Arch. Biochem. Biophys.*, *255*, 309–315.

Renger, G., and Weiss, W. (1986) Studies on the nature of the water oxidizing enzyme. III. Spectral characterization of the intermediary redox states in the water-oxidizing enzyme system Y. *Biochim. Biophys. Acta*, *850*, 184–196.

Renger, G., Wacker, U., and Völker, M. (1987) Studies on the protolytic reactions coupled with water cleavage in photosystem II membrane fragments from spinach. *Photosyn. Res.*, *13*, 167–189.

Ricard, J., Nari, J. and Diamantidis, G. (1980) Complex-forming properties of spinach $NADP^+$ reductase with ferredoxin, ferrocyanide and NADP. *Eur. J. Biochem.*, *108*, 55–66.

Rich, P. R. (1984) Electron and proton transfers through quinones and cytochrome *bc* complexes. *Biochim. Biophys. Acta*, *768*, 53–79.

Rich, P. R. (1985) Mechanism of quinol oxidation in photosynthesis. *Photosyn. Res.*, 6, 335–348.

Rich, P. R. (1988) A critical examination of the supposed variable proton stoichiometry of the chloroplast *bf* complex. *Biochim. Biophys. Acta*, 932, 33–42.

Rich, P. R., and Bendall, D. S. (1980) The kinetics and thermodynamics of the reduction of cytochrome *c* by substituted *p*-benzoquinols in solution. *Biochim. Biophys. Acta*, 592, 506–518.

Rich, P. R., and Wikström, M. K. F. (1986) Evidence for a mobile semiquinone in the redox cycle of the mammalian cytochrome bc_1 complex. *FEBS Lett.*, 194, 176–182.

Rieder, R., and H. R. Bosshard (1980) Comparison of the binding sites on cytochrome *c* for cytochrome *c* oxidase, cytochrome bc_1, and cytochrome c_1. Differential acetylation of lysyl residues in free and complexed cytochrome *c*. *J. Biol. Chem.*, 255, 4732–4739.

Rieske, J. S. (1976) Composition and function of complex III of the respiratory chain. *Biochim. Biophys. Acta*, 456, 195–247.

Rieske, J. S., and Ho, S. H. K. (1985) Respiratory complex III. Structure-function relationships. In *Coenzyme Q* (G. Lenaz, ed.), pp. 337–368, John Wiley and Sons, Chichester.

Robertson, D. E., Prince, R. C., Bowyer, J. R., Matsuura, K., Dutton, P. L. and Ohnishi, T. (1984) Thermodynamic properties of the semiquinone and its binding site in the ubiquinol-cytochrome *c* (c_2) oxidoreductase of respiratory and photosynthetic systems. *J. Biol. Chem.*, 259, 1758–1763.

Rochaix, J.-D., and Erickson, J. (1988) Function and assembly of photosystem II: genetic and molecular analysis. *Trends Biochem. Sci.*, 13, 56–59.

Rochaix, J. D., Dron, M., Rahire, M., and Malnoe, P. (1984) Sequence homology between the 32 kD protein and the D2 chloroplast membrane polypeptides of *C. reinhardtii*. *Plant Mol. Biol.*, 3, 363–370.

Roepe, P., Ahl, P. L., Das Gupta, S. K., Herzfeld, J., and Rothschild, K. J. (1987a) Tyrosine and carboxyl protonation changes in the bacteriorhodopsin photocycle. 1. M_{413} and L_{550} intermediates. *Biochemistry*, 26, 6696–6707.

Roepe, P., Scherrer, P., Ahl, P. L., Das Gupta, S. K., Bogolmolni, R. A., Herzfeld, J., and Rothschild, K. J. (1987b) Tyrosine and carboxyl protonation changes in the bacteriorhodopsin photocycle. Tyrosines-26 and -64. *Biochemistry*, 26, 6708–6717.

Roise, D., Theiler, F., Horvath. S. J., Tomich, J. M., Richards, J. H., Allison, D. H., and Schatz, G. (1988) Amphipathicity is essential for mitochondrial presequence function. *EMBO J.*, 7, 649–653.

Rosen, B. P. and E. Kashket (1978) *Energetics of Active Transport*. In *Bacterial Transport* (B. P. Rosen, ed.), pp. 559–620. Marcel Dekker, New York.

Rosing, J., and Slater, E. C. (1972) The value of the $\Delta G°$ for the hydrolysis of ATP. *Biochim. Biophys. Acta*, 267, 275–290.

Rossmann, M. G., Moras, D., and Olsen, K. W. (1974) Chemical and biological evolution of a nucleotide-binding protein. *Nature*, 250, 194–199.

Rottenberg, H. (1976) The driving force for proton(s)-metabolite cotransport in bacterial cells. *FEBS Lett.*, 66, 159–163.

Rottenberg, H. (1979) Non-equilibrium thermodynamics of energy conversion in bioenergetics. *Biochim. Biophys. Acta*, 549, 225–253.

Rottenberg, H. (1983) Uncoupling of oxidative phorphorylation in rat liver mitochondria by general anaesthetics. *Proc. Natl. Acad. Sci. USA*, 80, 3313–3317.

Rottenberg, H. (1986) Energetics of proton transport and secondary transport. *Methods Enzymol.*, Vol. 125 (S. Fleisher, ed.), pp. 3–15, Academic Press, New York.

Rottenberg, H., Grunwald, T., and Avron, M. (1972) Determination of ΔpH in chloroplasts. *Eur. J. Biochem.*, 25, 54–63.

Rottenberg, H., Robertson, D. E., and Rubin, E. (1985) The effect of temperature and chronic ethanol feeding on the proton electrochemical potential and phosphate potential in rat liver mitochondria. *Biochim. Biophys. Acta*, 809, 1–10.

Ruben, S., Randall, M., Kamen, M., and Hyde, J. L. (1941) Heavy oxygen (^{18}O) as a tracer in the study of photosynthesis. *J. Am. Chem. Soc.*, 63, 877–879.

Rutherford, A. W., and Acker, S. (1986) Orientation of the primary donor in isolated photosystem II reaction centers studied by EPR. *Biophys. J.*, 49, 101–102.

Rutherford, A. W., and Evans, M. C. W. (1980) Direct measurement of the redox potential of the primary and secondary quinone electron acceptors in *Rb. sphaeroides* (wild-type) by EPR spectrophotometry. *FEBS Lett.*, 110, 257–261.

Rutherford, A. W., and Zimmerman, J. L. (1984) A new EPR signal attributed to the primary plastosemiquinone signal in photosystem II. *Biochim. Biophys. Acta*, 767, 168–175.

Sack, J. S., Soper M. A. and Quiocho, F. A. (1989) Periplasmic binding protein structure and function. Refined x-ray structure of the leucine/isoleucine/valine-binding protein and its complex with leucine. *J. Mol. Biol.*, 206, 171–191.

Sadler, I., Suda, K., Schatz, G., Kandewitz, F., and Haid, A. (1984) Sequencing of the nuclear gene for the yeast cytochrome c_1 precursor reveals an unusually complex amino-terminal presequence. *EMBO J.*, 3, 2137–2143.

Saffman, P. G., and Delbrück, M. (1975) Brownian motion in biological membranes. *Proc. Natl. Acad. Sci. USA*, 72, 3111–3113.

Saier, Jr., M. H., Yamada, M., Erni, B., Suda, K., Lengler, J., Ebner, R., Argos, P., Rak, B., Schnetz, K., Lee, C. A., Stewart, G. C., Breidt, Jr., F., Waygood, G. P., Peri, K. G., and Doolittle, R. F. (1988) Sugar permeases of the bacterial phosphoenolpyruvate-dependent phosphotransferase system: sequence comparisons. *FASEB J.*, 2, 199–208.

Salach, J., Walker, W. H., Singer, T. P., Emenberg, A., Hemmerich, P., Ghisla, S., and Hartman, V. (1972) Studies on succinate dehydrogenase. Site of attachment of the covalently bound flavin to the peptide chain. *Eur. J. Biochem.*, 26, 267.

Salemme, F. R. (1977) Structure and function of cytochrome *c*. *Annu. Rev. Biochem.*, 46, 299–330.

Salerno, J. and Ohnishi, T. (1980) Studies on the stabilized ubisemiquinone species in the succinate-cytochrome *c* reductase segment of the intact mitochondrial membrane system. *Biochem. J.*, 192, 769–781.

Santos, H., Moura, J. J. G., LaGall, J., and Xavier, A. V. (1984) NMR studies of electron transfer mechanisms in a protein with interacting redox centres: *Desulfovibrio gigas* cytochrome c_3. *Eur. J. Biochem.*, 141, 283–296.

Saphon, S., and Crofts, A. R. (1977) Protolytic reactions in photosystem II: a new model for the release of protons accompanying the photooxidation of water. *Z. Naturforsch.*, 326, 617–626.

Saraste, M. (1984) Location of haem-binding sites in the mitochondrial cytochrome *b*. *FEBS Lett.*, 166, 367–372.

Saygin, Ö., and Witt, H. T. (1987) Optical characterization of intermediates in the water splitting enzyme system of photosynthesis—possible states and configurations of manganese and water. *Biochim. Biophys. Acta*, 893, 452–469.

Sayre, R. T., Andersson, B., and Bogorad, L. (1986) The topology of a membrane protein: the orientation of the 32 kD Q_b-binding chloroplast thylakoid binding protein. *Cell*, 47, 601–608.

Schägger, H., Borchart, U., Macheldt, W., Link, T. A., and von Jagow, G. (1987) Isolation and amino acid sequence of the "Rieske" iron sulfur protein of beef heart ubiquinol:cytochrome c reductase. *FEBS Lett.*, 219, 161–168.

Schatz, G., and Butow, R. A. (1983) How are proteins imported into mitochondria? *Cell*, 32, 316–318.

Schatz, G. H., Brock, H., and Holzworth, A. R. (1988) A kinetic and energetic model for the primary processes in photosystem II. *Biophys. J.*, 54, 397–406.

Schejter, A., and Margolit, R. (1970) The redox potentials of cytochrome c: Ion binding and oxidation state as linked functions. *FEBS Lett.*, 10, 179–181.

Schirmer, T., Bode, W., and Huber, R. (1987) Refined three-dimensional structures of cyanobacterial C-phycocyanine at 2.1 and 2.5 Å resolution. *J. Mol. Biol.*, 196, 677–695.

Schleyer, M., and Neupert, W. (1985) Transport of proteins into mitochondria: translocational intermediates spanning contact sites between outer and inner membranes. *Cell*, 43, 339–350.

Schneider, E., and Altendorf, K. (1987) Bacterial adenosine 5′-triphosphate synthase (F_1F_0): purification and reconstitution of F_0 complexes and biochemical and functional characterization of their subunits. *Microbiol. Rev.*, 51, 477–497.

Schneider, H., Lemasters, J. J., and Hackenbrock, C. R. (1985) Membrane fluidity and mobility of ubiquinone. In *Coenzyme Q* (G. Lenaz, ed.), pp. 201–214, John Wiley, New York.

Scholes, T. A., and Hinkle, P. A. (1984) Energetics of ATP-driven reverse electron transfer from cytochrome c to fumarate and from succinate to NAD in submitochondrial particles. *Biochemistry*, 23, 3341–3345.

Scholte, B. J., Schuitema, A. R., and Postma, P. W. (1981) Isolation of III^{Glc} of the phosphoenolpyruvate-dependent glucose phosphotransferase system of *Salmonella typhimurium*. *J. Bacteriol.*, 148, 257–264.

Schuldiner, S., and Fishkes, H. (1978) Sodium-proton antiport in isolated membrane vesicles of *Escherichia coli*. *Biochem.*, 17, 706–711.

Schuldiner, S., and Kaback, H. R. (1975) Membrane potential and active transport in membrane vesicles from *Escherichia coli*. *Biochemistry*, 14, 5451–5461.

Schuldiner, S., Rottenberg, H., and Avron, M. (1972) Determination of ΔpH in chloroplasts. 2. Fluorescent amines as a probe for the determination of ΔpH in chloroplasts. *Eur. J. Biochem.*, 25, 64–70.

Schwarz, E., Oesterhelt, D., Reinke, H., Bayreuther, K., and Dimroth, P. (1988) The sodium-ion-translocating oxalacetate decarboxylase of *Klebsiella pneumoniae*. *J. Biol. Chem.*, 263, 9640–9645.

Schwerzmann, K., and Pedersen, P. L. (1986) Regulation of the mitochondrial ATP Synthase/ATPase complex. *Arch. Biochem. Biophys.*, 250, 1–18.

Scott, J. (1985) Distances between the functional sites of the ($Ca^{2+} + Mg^{2+}$)-ATPase of sarcoplasmic reticulum. *J. Biol. Chem.*, 260, 14421–14423.

Scott, R. A., Mauk, A. G., and Gray, H. B. (1985) Experimental approaches to studying biological electron transfer. *J. Chem. Ed*, 62, 932–938.

Seelig, J., MacDonald, P. M., and Scherer, P. G. (1987) Phospholipid head groups as sensors of electric charge in membranes. *Biochemistry*, 26, 7535–7541.

Seely, G. R. (1978) The energetics of electron-transfer reactions of chlorophyll and other compounds. *Photochem. Photobiol.* 27, 639–654.

Seiff, F., Wallat, I., Ermann, P., and Heyn, M. P. (1985) A neutron diffraction study on the location of the polyene chain in bacteriorhodopsin. *Proc. Natl. Acad. Sci. USA*, 82, 3227–3231.

Selak, M. A., and Whimarsh, J. (1982) Kinetics of the electrogenic step and cytochrome b_6 and f redox changes. Evidence for a Q cycle. *FEBS Lett.*, 150, 286–292.

Senior, A. E. (1983) Secondary and tertiary structure of membrane proteins involved in energy transduction. *Biochim. Biophys. Acta*, 726, 81–95.

Senior, A. E. (1985) The proton-ATPase of *Escherichia coli*. In *Current Topics in Membranes and Transport* (E. A. Adelberg and C. W. Slayman, eds.), vol. 23, 135–151, Academic Press, New York.

Senior, A. E., and Wise, J. G. (1983) The proton ATPase of bacteria and mitochondria. *J. Membr. Biol.*, 73, 105–124.

Serrano, R., Kanner, B. I., and Racker, E. (1976) Purification and properties of the proton-translocating adenosine triphosphatase complex of bovine heart mitochondria. *J. Biol. Chem.*, 251, 2453–2461.

Serrano, R., Keilland-Brandt, M. C., and Fink, G. R. (1986) Yeast plasma membrane ATPase is essential for growth and has homology with $(Na^+ + K^+)$, K^+-, and Ca^{2+}-ATPases. *Nature*, 319, 689–693.

Shahak, Y. (1982) Activation and deactivation of H^+-ATPase in intact chloroplasts. *Plant Physiol.*, 70, 87–91.

Shavit, N. (1980) Energy transduction in chloroplasts: structure and function of the ATPase complex. *Annu. Rev. Biochem.*, 49, 111–138.

Sheriff, S. and Herriott, J. R. (1981) Structure of ferredoxin-$NADP^+$ oxidoreductase and the location of the $NADP^+$ binding site. *J. Mol. Biol.*, 145, 441–451.

Shioi, J., Matsuura, S., and Imae, Y. (1980) Quantitative measurements of proton motive force and motility in *Bacillus subtilis*. *J. Bacteriol.*, 144, 891–897.

Shiver, J. W., Peterson, A. A., Widger, W. R., Furbacher, P. N., and Cramer, W. A. (1989) *Methods Enzymol.* (B. Fleischer and S. Fleischer, eds.), 172, Chap. 25, pp. 439–461. Academic Press, Orlando, FL.

Shull, G. E., Schwartz, A., and Lingrel, J. B. (1985) Amino acid sequence of the catalytic subunit of the $(Na^+ + K^+)$ ATPase deduced from a complementary DNA. *Nature*, 316, 691–695.

Shulman, R. G., Brown, T. R., Ugurbil, K., Ogawa, S., Cohen, S. M., and den Hollander, J. A. (1979) Cellular applications of ^{31}P and ^{13}C nuclear magnetic resonance. *Science*, 205, 160–166.

Shultheiss, H. P., and Bolte, H. D. (1985) Immunological analysis of autoantibodies against the adenine translocator in dilated cardiomyopathy. *J. Mol. Cell. Cardiol.*, 17, 603–617.

Shultheiss, H. P., Berg, P. A., and Klingenberg, M. (1983) The mitochondrial adenine nucleotide translocator is an antigen in primary biliary cirrhosis. *Clin. Exp. Immunol.*, 54, 648–654.

Shultheiss, H. P., Berg, P. A., and Klingenberg, M. (1984) Inhibition of the adenine translocator by organ specific autoantibodies in primary biliary cirrhosis. *Clin. Exp. Immunol.*, 58, 596–602.

Shuvalov, V. A., Nuijs, A. M., van Gorkom, H. J., Smit, H. W. J., and Duysens, L. N. M. (1986) Picosecond absorbance charges upon selective excitation of the primary donor P-700 in photosystem I. *Biochim. Biophys. Acta*, 850, 319–323.

Siedow, J. N., Power, S., de la Rosa, F. F., and Palmer, G. (1978) The preparation and characterization of highly purified enzymatically active complex III from Baker's yeast. *J. Biol. Chem.*, 253, 2392–2399.

Siedow, J. N., Vickery, L. E., and Palmer, G. (1980) The nature of the axial ligands of spinach cytochrome *f*. *Arch. Biochem. Biophys.*, 203, 101–107.

Sigalat, C., Haraux, F., de Kouchkovsky, F., Hung, S. P. N., and de Kouchkovsky, Y. (1985) Adjustable microchemiosmotic character of the proton gradient generated by system I and II for photosynthetic phosphorylation in thylakoids. *Biochim. Biophys. Acta*, 809, 403–413.

Siggel, U., Renger, G., Stiehl, H. H., and Rumberg, B. (1972) Evidence for electronic and ionic interaction between electron transport chains in chloroplasts. *Biochim. Biophys. Acta*, 256, 328–335.

Sigrist-Nelson, K., and Azzi, A. (1980) The proteolipid subunit of the chloroplast adenosine triphosphatase complex: rrconstitution and demonstration of proton-conductive properties. *J. Biol. Chem.*, 255, 10638–10643.

Simonis, W., and Urbach, W. (1973) Photophosphorylation *in vivo*. *Annu. Rev. Plant Physiol.*, 24, 89–114.

Simpson, D. (1979) Freeze-fracture studies on barley plastid membranes. III. Location of the light-harvesting chlorophyll protein. *Carlberg Res. Commun.*, 44, 305–336.

Simpson, S. J., Bendall, M. R., Egan, A. F., Vink, R., and Rogers, P. J. (1983) High-field phosphorous NMR studies of the stoichiometry of the lactate/proton carrier in *Streptococcus faecalis*. *Eur. J. Biochem.*, 136, 63–69.

Sims, P. J., Waggoner, A. S., Wang, C.-H., and Hoffman, J. F. (1974) Studies on the mechanism by which cyanine dyes measure membrane potential in red blood cells and phosphatidylcholine vesicles. *Biochemistry*, 13, 3315–3330.

Singer, S. J., Maher, P. A., and Yaffe, M. P. (1987) On the translocation of proteins across membranes. *Proc. Natl. Acad. Sci. USA*, 84, 1015–1019.

Skulachev, V. P. (1988) *Membrane Bioenergetics*. 442 pp. Springer-Verlag, Heidelberg/New York.

Smeekens, S., Bauerle, C., Hageman, J., Keegstra, K., and Weisbeek, P. (1986) The role of the transit peptide in the routing of precursors toward different chloroplast compartments. *Cell*, 46, 365–375.

Smith, J. M., Smith, W. H. and Knaff, D. B. (1981) Electrochemical titrations of a ferredoxin-ferredoxin:$NADP^+$ oxidoreductase complex. *Biochim. Biophys. Acta*, 635, 405–411.

Smith, M. B., Stonehuerner, J., Ahmed, A. J., Staudenmeyer, N., and Millett, F. (1980) Use of specific trifluoroacetylation of lysine residues in cytochrome *c* to study the reaction with cytochrome b_5, cytochrome c_1, and cytochrome oxidase. *Biochim. Biophys. Acta*, 592, 303–313.

Smith, S. O., Myers, A. B., Pardoen, J. A., Winkel, C., Mulder, P. P. J., Lugtenberg, J., and Mathies, R. A. (1984) Determination of retinal Schiff base configuration in bacteriorhodopsin. *Proc. Natl. Acad. Sci. USA*, 81, 2055–2059.

Snyder, B., and Hammes, G. G. (1985) Structural organization of chloroplast coupling factor. *Biochemistry*, 24, 2324–2331.

Solioz, M., Mathews, S., and Fürst, D. (1987) Cloning of the K^+-ATPase of *Streptococcus faecalis*. *J. Biol. Chem.*, 262, 7358–7352.

Sone, N., Hamamoto, T., and Kagawa, Y. (1981) pH dependence of H^+ conduction through the membrane moiety of the H^+-ATPase and effects of tyrosyl residue modification. *J. Biol. Chem.*, 256, 2873–2877.

Sorgato, M. C., Caliazzo, F., Panato, L., and Ferguson, S. J. (1982) Estimation of H^+ translocation stoichiometry of mitochondrial ATPase by comparison of proton-motive forces with clamped phosphorylation potentials in submitochondrial particles. *Biochim. Biophys. Acta*, 682, 184–188.

Spudich, J. L., and Bogomolni, R. A. (1988) Sensory rhodopsins of halobacteria. *Annu. Rev. Biophys. Biophys. Chem.*, 17, 193–215.

Srere, P. A. (1987) Complexes of sequential metabolic enzymes. *Annu. Rev. Biochem.*, 56, 89–124.

Staehelin, L. A. (1986) Chloroplast structure and supramolecular organization of photosynthetic membranes. In *Encylopedia of Plant Physiology.*, New Series, Vol. 19 (L. A. Staehelin and C. J. Arntzen, eds.), pp. 1–84, Springer-Verlag.

Steinback, K. E., McIntosh, L., Bogorad, L., and Arntzen, C. J. (1981) Identification of the triazine receptor protein as a chloroplast gene product. *Proc. Natl. Acad. Sci. USA*, 78, 7463–7467.

Steppuhn, J., Rother, C., Hermans, J., Jansen, T., Salnikow, J., Hauska, G., and Herrmann, R. G. (1987) The complete amino acid sequence of the Rieske FeS-precursor protein from spinach chloroplasts deduced from cDNA analysis. *Mol. Genet.*, 210, 171–177.

Stewart, A. C., and Bendall, D. S. (1979) Preparation of an active, oxygen-evolving photosystem 2 particle from a blue-green alga. *FEBS Lett.*, 107, 308–312.

Stidham, M. A., McIntosh, T. J., and Siedow, J. N. (1984) On the localization of ubiquinone in phosphatidylcholine bilayers. *Biochim. Biophys. Acta*, 767, 423–431.

Stock, J., and Roseman, S. (1971) A sodium-dependent sugar to-transport system in bacteria. *Biochim. Biophys. Res. Commun.*, 44, 132–138.

Stoeckenius, W. (1985) The rhodopsin-like pigments of halobacteria. *Trends Biochem. Sci.*, 10, 483–485.

Stonehuerner, J., O'Brien, P., Geren, L., Millett, F., Steidl, J., Yu, L., and Yu, C.-A. (1985) Identification of the binding site on cytochrome c_1 for cytochrome c. *J. Biol. Chem.*, 260, 5392–5398.

Storey, B. T., Scott, D. M., and Lee, C. P. (1980) Energy-linked quinacrine fluorescence changes in submitochondrial particles from skeletal muscle mitochondria. *J. Biol. Chem.*, 255, 5224–5229.

Stout, G. H., Turley, S., Sieker, L. C., and Jensen, L. H. (1988) Structure of ferredoxin I from *Azotobacter vinelandii*. *Proc. Natl. Acad. Sci. USA*, 85, 1020–1022.

Stroobant, P., and Kaback, H. R. (1979) Reconstitution of ubiquinone-linked functions in membrane vesicles from a double quinone mutant of *Escherichia coli*. *Biochemistry*, 18, 226–231.

Stryer, L., and Haugland, R. P. (1967) Energy transfer: a spectroscopic ruler. *Proc. Natl. Acad. Sci. USA*, 58, 719–726.

Sugiura, M., Shinozaki, K., Zaita, N., Kusuda, M., and Kumano, M. (1986) Clone bank of the tobacco (*N. tabacum*) chloroplast genome as a set of overlapping restriction endonuclease fragments: mapping of eleven ribosomal protein genes. *Plant Sci.* 44, 211–216.

Sugiyama, S., Cragoe, E. J., and Imae, Y. (1988) Amiloride, a specific inhibitor of the Na^+-driven flagellar motors of alkalophilic *Bacillus*. *J. Biol. Chem.*, 263, 8215–8219.

Sussman, M. R., Strickler, J. E., Hager, K. M., and Slayman, C. W. (1987) Location of a dicyclohexylcarbodiimide-reactive residue in the *N. crassa* plasma membrane ATPase. *J. Biol. Chem.*, 262, 4569–4573.

Szczepaniak, A., and Cramer, W. A. (1989) On the topography and function of the cytochrome b_6-f complex. *Zeit für Naturforsch*, 44c, 123–131.

Tae, G.-S., Black, M. T., Cramer, W. A., Vallon, O., and Bogorad, L. (1988) Thylakoid membrane protein topography: trans-membrane orientation of the chloroplast cytochrome b-559 *psb*E gene product. *Biochemistry*, 27, 9075–9080.

Tagawa, K., Tsujimoto, H. Y., and Arnon, D. I. (1963) Role of chloroplast ferredoxin in the energy conversion process of photosynthesis. *Proc. Natl. Acad. Sci. USA*, 49, 567–572.

Takahashi, Y., Hase, T., Wada, K. and Matsubare, H. (1983) Ferredoxins in developing spinach cotyledons: the presence of two molecular species. *Plant Cell Physiol.*, 24, 185–198.

Takamiya, K., and Dutton, P. L. (1977) The influence of trans-membrane potentials on the redox equilibrium between cytochrome c_2 and the reaction center in *Rb. sphaeroides* chromatophores. *FEBS Lett.*, 80, 279–284.

Takano, T., and Dickerson, R. E. (1980) Redox conformation changes in refined tuna cytochrome c. *Proc. Natl. Acad. Sci. USA*, 77, 6371–6375.

Takeda, K., Hirano, M., Kanazawa, H., Nukiwa, N., Kagawa, Y., and Futai, M. (1982) Hybrid ATPases formed from subunits of coupling factor F_1 of *Escherichia coli* and thermophilic bacterium PS3. *J. Biochem.*, 91, 695–701.

Tamura, N., and Cheniae, G. (1985) Effects of photosystem II extrinsic proteins on microstructure of the OEC and its reactivity to water analogs. *Biochim. Biophys. Acta*, 809, 245–259.

Tanford, C. (1980) *The Hydrophobic Effect*, Chaps. 2 and 13. Wiley Interscience, New York.

Tavan, P., Schulten, K., Gärtner, W., and Oesterhelt, D. (1985) Substituents at the C_{13} position of retinal and their influence on the function of bacteriorhodopsin. *Biophys. J.*, 47, 349–355.

Tedeschi, H. (1980) The mitochondrial membrane potential. *Biol. Rev.*, 55, 171–206.

Telser, J., Hoffman, B. M., LoBrutto, R., Ohnishi, T., Tsai, A.-L., Simpkin, D., and Palmer, G. (1987) Evidence for N coordination to Fe in the [2Fe-2S] center in yeast mitochondrial complex III. *FEBS Lett.*, 214, 117–121.

Terada, H. (1981) The interaction of highly active uncouplers with mitochondria. *Biochim. Biophys. Acta*, 639, 225–242.

Terada, H., Shirakawa, K., Kametari, F., and Yoshikawa, K. (1983). Acceleration of H^+ exchange between octanol and water by 2,4-dinitrophenol, determined by ^1H NMR spectrometry. *Biochim. Biophys. Acta*, 725, 254–260.

Terada, H., Ikuno, M., Shirohara, Y., and Yoshikawa, K. (1984a) Mechanism of the Mg^{2+}-facilitated specific cleavage of the terminal phosphoryl group of adenosine-5′-triphosphate. *Biochim. Biophys. Acta*, 767, 648–650.

Terada, H., Kumazawa, N., Ju-Ichi, M., and Yoshikawa, K. (1984b) Molecular basis of the protonophoric and uncoupling activities of the potent uncoupler SF-6847 and derivatives: regulation of their electronic structures by restricted intramolecular motion. *Biochim. Biophys. Acta*, 767, 192–199.

Thayer, W. S., and Hinkle, P. C. (1975) Synthesis of adenosine triphosphate by an artificially imposed electrochemical proton gradient in bovine heart submitochondrial particles. *J. Biol. Chem.*, 250, 5330–5335.

Theg, S. M., Jursinic, P. A., and Homann, P. A. (1984) Studies on the mechanism of chloride action on photosynthetic water oxidation. *Biochim. Biophys. Acta,* 766, 636–646.

Theg, S. M., Bauerle, C., Olsen, L. J., Selman, B. R., and Keegstra, K. (1989) ATP is the only energy requirement for the translocation of precursor protein across chloroplast membranes. *J. Biol. Chem.,* 264, 6730–6736.

Thelen, M. O'Shea, P., and Azzi, A. (1985) New insights into the cytochrome c oxidase proton pump. *Biochem. J.,* 227, 163–167.

Thompson, L., and Brudvig, G. (1988) Cytochrome b-559 may function to protect photosystem II from photoinhibition. *Biochemistry,* 27, 6653–6658.

Thornber, J. P. (1986) Biochemical characterization and structure of pigment-proteins of photosynthetic organisms. *Encyclopedia of Plant Physiology, New Series, Vol. 19* (L. A. Staehelin and C. J. Arntzen, eds.), pp. 98–142, Springer-Verlag, Heidelberg.

Tiede, D. M. (1987) c-cytochrome orientation in electron transfer complexes with photosynthetic reaction centers of *Rh. sphaeroides*, and when bound to the surface of negatively charged membranes: characterization by optical linear dichroism. *Biochemistry,* 26, 397–410.

Tiede, D. M., Bodil, D. E., Tang, J., El-Kalbani, O., Norris, J. R., Chang, C.-H., and Scheffer, M. (1988) Symmetry-breaking structures involved in the docking of cytochrome c and primary electron transfer in reaction centers of *Rb. sphaeroides*. In *Structure of Bacterial Reaction Centers*, (J. Breton and A. Vermeglio, eds), pp. 13–20, Plenum Press, New York.

Tiedge, H., Lünsdorf, H., Schäfer, G., Schairer, H. U. (1985) Subunit stoichiometry and juxtaposition of the photosynthetic coupling factor I: Immunoelectron microscopy using monoclonal antibodies. *Proc. Natl. Acad. Sci. USA,* 82, 7874–7878.

Timkovich, R., Cork, M. S., Gennis, R. B., and Johnson, P. V. (1985) Proposed structure of heme d, a prosthetic group of bacterial cytochrome oxidases. *J. Am. Chem. Soc.,* 107, 6069–6075.

Tinoco, Jr., T., Sauer, K., and Wang, J. C. (1985) *Physical Chemistry: Principles and Applications in Biological Sciences*, p. 309, Prentice-Hall, Englewood Cliffs, NJ.

Tokuda, H., and Kaback, H. P. (1977) Sodium-dependent methyl 1-thio-α-D-galactopyranoside transport in membrane vesicles isolated from *Salmonella typhimurium*. *Biochemistry,* 16, 2130–2136.

Tokuda, H., and Unemoto, T. (1982) Characterization of the respiration-dependent Na^+ pump in the marine bacterium *Vibrio alginolyticus*. *J. Biol. Chem.,* 257, 10007–10014.

Tokuda, H., and Unemoto, T. (1984) Na^+ is translocated at NADH:quinone oxidoreductase segment in the respiratory chain of *Vibrio alginolyticus*. *J. Biol. Chem.,* 259, 7785–7790.

Tollin, G., Cheddar, G., Watkins, J. A., Meyer, T. E., and Cusanovich, M. A. (1984) Electron transfer between flavodoxin semiquinone and c-type cytochromes: correlation between electrostatically corrected rate constants, redox potentials, and surface topologies. *Biochemistry,* 23, 6345–6349.

Trebst, A. (1986) The topology of plastoquinone and herbicide binding peptides in photosystem II in the thylakoid membrane. *Z. Naturforsch.,* 41C, 240–245.

Trebst, A. (1987) The three-dimensional structure of the herbicide binding niche on the reaction center polypeptides of photosystem II. *Z. Naturforsch.,* 42c, 742–750.

Trebst, A., Wietoska, H., Draber, W., and Krops, H. J. (1978) The inhibition of photosynthetic electron flow in chloroplasts by the dinitrophenylether of bromo- or iodo-nitrothynol. *Z. Naturforsch.*, *33C*, 919–927.

Trepten, N. A., and Shuman, H. A. (1985) Genetic evidence for substrate and periplasmic-binding protein recognition by the MalF and MalG proteins, cytoplasmic membrane components of the *Escherichia coli* maltose transport system. *J. Bacteriol.*, *163*, 654–660.

Trewhella, S., Anderson, J., Fox, R., Gogol, E., Khan, S., Engelman, D., and Zaccai, G. (1983) Assignment of segments of the bacteriorhodopsin sequence to positions in the structural map. *Biophys. J.*, *42*, 233–241.

Trissl, H. (1985) Primary electrogenic processes in bacteriorhodopsin probed by photoelectric measurements with capacitative metal electrodes. *Biochim. Biophys. Acta*, *806*, 124–135.

Trissl, H-W, Leibl, W., Depret, J., Dobek, J., and Breton, J. (1987) Trapping and annihilation in the antenna system of photosystem I. *Biochim. Biophys. Acta*, *893*, 320–332.

Tronrud, D. E., Schmid, M. F., and Mathews, B. W. (1986) Structure and X-ray amino acid sequence of a bacteriochlorophyll *a* protein derived from *P. aestuarii* refined at 1.9 Å resolution. *J. Mol. Biol.*, *188*, 443–454.

Trumpower, B. (1981a) New concepts on the role of ubiquinone in the mitochondrial respiratory chain. *J. Bioenerg. Biomembr.*, *13*, 1–24.

Trumpower, B. L. (1981b) Function of the iron-sulfur protein of the cytochrome b-c_1 segment in electron transfer and energy-conserving reactions of the mitochondrial respiratory chain. *Biochim. Biophys. Acta*, *639*, 129–155.

Tsukihara, T., Fukuyama, K., Nakamura, M., Katsube, Y., Tanaka, N., Kakudo, M., Wada, K., Hase, T., and Matsubara, H. (1981) X-ray analysis of a [2Fe-2S] ferredoxin from *Spriulina platensis*. Main chain fold and location of side chains at 2.5 Å resolution. *J. Biochem.*, *90*, 1763–1773.

Tsygannik, I. N., and Baldwin, J. M. (1987) Three-dimensional structure of deoxycholate-treated purple membrane at 6 Å resolution and molecular averaging of three crystal forms of bacteriorhodopsin. *Eur. J. Biophys.*, *14*, 263–272.

Tujimura, T., Yamato, I., and Anraku, Y. (1983) Mechanism of glutamate transport in *Escherichia coli* B. 2. Kinetics of glutamate transport driven by artificially imposed proton and sodium ion gradients across the cytoplasmic membrane. *Biochemistry*, *22*, 1959–1965.

Turro, N. (1975) *Molecular Photochemistry*, Chapt. III. W. A. Benjamin, Reading, Mass.

Tzagoloff, A. (1982) *Mitochondria*, 342 pp. Plenum Press, New York.

Udagawa, T, Unemoto, T., and Tokuda, H. (1986) Generation of Na^+ electrochemical potential by the NADH oxidase and Na^+/H^+ antiport system of a moderately halophilic *Vibrio costicola*. *J. Biol. Chem.*, *261*, 2616–2622.

Ulrich, E. L., Girvin, M. E., Cramer, W. A., and Markley, J. L. (1985) Location and mobility of ubiquinones of different chain lengths in artificial membrane vesicles. *Biochemistry*, *24*, 2501–2508.

Ulstrup, J., and Jortner, J. (1975) The effect of intramolecular quantum modes on free energy relationships for electron transport reactions. *J. Chem. Phys.*, *63*, 4358–4368.

Urban, P. F., and Klingenberg, M. (1969) On the redox potentials of ubiquinone and cytochrome *b* in the respiratory chain. *Eur. J. Biochem.*, *9*, 519–525.

Vallejos, R. H., Ceccarelli, E., and Chan, R. (1984) Evidence for the existence of a thylakoid intrinsic protein that binds ferredoxin-NADP$^+$ oxidoreductase. *J. Biol. Chem.*, *259*, 8048–8051.

van den Berg, W. H., Prince, R. C., Bashford, C. L., Takamiya, K., Bonner, W. D., and Dutton, P. L. (1979) Electron and proton transport in the ubiquinone cytochrome $b-c_1$ oxidoreductase of *Rb. sphaeroides*. *J. Biol. Chem.*, *254*, 8595–8604.

van der Waal, H. N., van Grondelle, R., Millett, F., and Knaff, D. B. (1987) Oxidation of cytochrome c_2 and cytochrome c by reaction centers of *R. rubrum* and *Rb. sphaeroides*. The effect of ionic strength and of lysine modification on reaction rates. *Biochim. Biophys. Acta*, *893*, 490–498.

Van Nieuwhoven, M. H., Hellingwerf, K. J., Venema, G., and Konings, W. N. (1982) Role of protonmotive force in genetic transformation of *Bacillus subtilis*. *J. Bacteriol.*, *151*, 771–776.

Van Rensen, J. J. S., Tonk, W. J. M., and de Bruijn, S. M. (1988) Involvement of bicarbonate in the protonation of the secondary quinone electron acceptor of photosystem II via the non-heme iron of the quinone-iron acceptor complex. *FEBS Lett.*, *226*, 347–351.

Vanderkooi, J. M., Landesberg, R., Hayden, G. W., and Owen, C. S. (1978) Metal-free and metal-substituted cytochromes c. Use in characterization of the cytochrome c binding site. *Eur. J. Biochem.*, *81*, 339–347.

Váró, G., and Keszthelyi, L. (1985) Arrhenius parameters of the bacteriorhodopsin photocycle in dried oriented samples. *Biophys. J.*, *47*, 243–246.

Velthuys, B. R. (1979) Electron flow through plastoquinone and cytochromes b_6 and f in chloroplasts. *Proc. Natl. Acad. Sci. USA*, *76*, 2765–2769.

Velthuys, B. (1981) Electron transport dependent competition between plastoquinone and inhibitors for binding to photosystem II. *FEBS Lett.*, *126*, 277–281.

Vermaas, W. F. J., Hansson, O., and Rutherford, A. W. (1988) Site-directed mutagenesis in photosystem II of the cyanobacterium *Synechocystis sp.* PCC 6803: Donor D is a tyrosine residue in the D2 protein. *Proc. Natl. Acad. Sci. USA*, *85*, 8477–8481.

Verner, K., and Schatz, G. (1987) Import of an incompletely folded precursor protein into isolated mitochondria requires an energized inner membrane, but no added ATP. *EMBO J.*, *6*, 2449–3456.

Verner, K., and Schatz, G. (1988) Protein translocation across membranes. *Science*, *241*, 1307–1313.

Vieira, B. J., and Davis, D. J. (1986) Interaction of ferredoxin with FNR: effects of chemical modification of ferreodxin. *Arch. Biochem. Biophys.*, *247*, 140–146.

Vignais, P. V., Block, M. R., Boulay, F., Brandolis, G., and Lauquin, G. J. M. (1985) Molecular aspects of structure-function relationships in mitochondrial adenine nucleotide carrier. *Structure and Properties of Cell Membranes* (G. Bangha, ed.), *Vol. II*, pp. 139–179, CRC Press.

Vignais, P. V., and Lunardi, J. (1985) Chemical probes of the mitochondrial ATP synthesis and translocation. *Annu. Rev. Biochem.*, *54g*, 977–1014.

Viitanen, P., Newman, M. J., Foster, D. L., Wilson, T. H., and Kaback, H. R. (1986) Purification, reconstitution and characterization of the *lac* permease of *Escherichia coli*. In *Methods in Enzymology*, Vol. *125* (S. Fleischer, ed.), pp. 429–452, Academic Press, New York.

Villars, F. M. H., and Benedek, G. B. (1974) *Physics with Illustrative Examples from Medicine and Biology, Vol. II*, Chap. II, Addison-Wesley, Reading, MA.

Vincent, J. B., and Christou, G. (1987) A molecular "double-pivot" mechanism for water oxidation. *Inorg. Chim. Acta*, *136*, 241–243.

Vincent, J. B., Christmas, C., Huffman, J. C., Christou, G., Chang, H.-R., and Henrickson, D. N. (1987) Modelling the photosynthetic water oxidation center. Synthesis, structure, and magnetic properties of $[Mn_4O_2(OAc)_7(bipy)_2](ClO_4 \cdot 3H_2O)$. *J. Chem. Soc. Chem. Commun.*, 236–238.

Vink, R., Bendall, M. R., Simpson, S. J., and Rogers, P. J. (1984) Estimation of H^+ to adenosine 5'-triphosphate stoichiometry of *Escherichia coli* ATP synthase using ^{31}P NMR. *Biochemistry*, 24, 3667–3675.

von Heijne, G. (1986) Mitochondrial targeting sequences may form amphipathic helices. *EMBO J.*, 5, 1335–1341.

Von Jagow, G., Ljungdahl, P. O., Grab, P., Ohnishi, T., and Trumpower, B. L. (1984) An inhibitor of mitochondrial respiration which binds to cytochrome *b* and displaces quinone from the iron-sulfur center of the cytochrome bc_1 complex. *J. Biol. Chem.*, 259, 6318–6326.

Wagner, R., Engelbrecht, S., and Andreo, C. S. (1985) Quarternary structure of chloroplast F_1-ATPase in solution. *Eur. J. Biochem.*, 147, 163–170.

Wainio, W. W. (1970) *The Mammalian Mitochondrial Respiratory Chain*, 499 pp. Academic Press, New York.

Wakabayashi, S., Matsubara, H., Kim, C. H., and King, T. E. (1982) Structural studies of bovine heart cytochrome c_1. *J. Biol. Chem.*, 257, 9335–9344.

Wald, G. (1969) Life in the second and third periods: Or why phosphorus and sulfur for high-energy bonds? In *Biological Phosphorylations* (H. Kalckar, ed.), pp. 156–167. Prentice-Hall, Englewood Cliffs.

Waldemeyer, B., and Bosshard, H. R. (1985) Structure of an electron transfer complex. I. Covalent cross-linking of cytochrome *c* peroxidase and cytochrome *c*. *J. Biol. Chem.*, 5184–5190.

Waldemeyer, B., R. Bechtold, H. R. Bosshard and T. L. Poulos (1982) The cytochrome *c* peroxidase cytochrome *c* electron transfer complex. Experimental support of a theroretical model. *J. Biol. Chem.*, 257, 6073–6076.

Walderhaug, M. O., Post, R. L., Saccomani, G., Leonard, R. T., and Briskin, D. P. (1985) Structural relatedness of three adenosine triphosphatases around their active sites of phosphorylation. *J. Biol. Chem.*, 260, 3852–3859.

Walderhaug, M. O., Dosch, D. C., and Epstein, W. (1987) Potassium transport in bacteria. In *Ion Transport in Bacteria* (B. P. Rosen and S. Silver, eds.), Academic Press, New York pp. 85–130.

Walker, F. A., Huynh, B. H., Scheidt, W. R., and Osvath, S. R. (1986) Models of the cytochrome *b*. The effect of axial ligand plane orientation on the EPR and Mössbauer spectra of low-spin ferrihemes. *J. Am. Chem. Soc.*, 108, 5288–5297.

Walker, J. E., Saraste, M., Runswick, M. J., and Gay, N. J. (1982) Distantly related sequences in the alpha- and beta-subunits of ATP synthase, myosin, kinases and other ATP-requiring enzymes and a common nucleotide binding fold. *EMBO J.*, 1, 945–951.

Walker, J. E., Saraste, M., and Gay, N. J. (1984) The *unc* operon: nucleotide sequence, regulation, and structure of ATP synthase. *Biochim. Biophys. Acta*, 768, 164–200.

Walker, J. E., Fearnley, I. M., Gay, N. J., Gibson, B. W., Northrop, F. D., Powell, S. J., Runswick, M. J., Saraste, M., and Tybulewicz, V. L. J. (1985) Primary structure and subunit stoichiometry of F_1-ATPase from bovine mitochondria. *J. Mol. Biol.*, 184, 677–701.

Wallace, B., and Ravikumar, K. (1988) The gramicidin gene: crystal structures of a cesium complex. *Science*, 241, 182–187.

Wallace, D. C. (1982) Structure and evolution of organelle genomes. *Microbiol. Rev.*, 46, 208–240.
Walz, D. (1979) Thermodynamics of oxidation–reduction reactions and its application to bioenergetics. *Biochim. Biophys. Acta*, 505, 279–353.
Wanders, R. J. A., and Westerhoff, H. V. (1988) Signoidal relation between mitochondrial respiration and log ($[ATP/ADP])_{out}$ under conditions of extramitochondrial ATP utilization. Implication for the control and thermodynamics of oxidative phosphorylation. *Biochemistry*, 27, 7832–7840.
Wanders, R. J. A., Groen, A. K., van Roermund, C. W. T., and Tager, J. M. (1984) Factors determining the relative contribution of the adenine nucleotide translocator and the ADP-regenerating system to the control of oxidative phosphorylation in isolated rat-liver mitochondria. *Eur. J. Biochem.*, 142, 417–424.
Wang, J. (1986) Functionally distinct β subunits in F_1-adenosinetriphosphatase. *J. Biol. Chem.*, 260, 1374–1377.
Waring, R. B., Davies, R. W., Lee, S., Grisi, E., McPhail-Berks, M., and Scazzachio, C. (1981) The mosaic organization of the apocytochrome b gene of *A. nidulans* revealed by DNA sequencing. *Cell*, 27, 4–11.
Wasielewski, M. R., and Tiede, D. M. (1986) Primary photoreactions in *Rps. viridis* following excitation with sub-picosecond flashes at 950 nm. 204, 368–372.
Wasielewski, M. R., Fenton, J. M., and Govindjee (1987) The rate of formation of $P700^+$-A_0^- in PS I particles from spinach as measured by picosecond transient absorption spectroscopy. *Photosyn. Res.*, 12, 181–190.
Wechsler, T., Suter, F., Fuller, R. C., and Zuber, H. (1985) The complete amino acid sequence of the bacteriochlorophyll c binding polypeptide from chlorosomes of the green photosynthetic bacterium *C. aurianticus*. *FEBS Lett.*, 181, 173–178.
West, I. C. (1983) *The Biochemistry of Membrane Transport*, Chapman and Hall, London.
West, I., and Mitchell, P. (1972) Proton-coupled β-galactoside translocation in nonmetabolizing *Escherichia coli*. *J. Bioenerg*, 3, 445–462.
West, I. C., Mitchell, P., and Rich, P. R. (1988) Electron conduction between b cytochromes of the mitochondrial respiratory chain in the presence of antimycin plus myxathiazol. *Biochim. Biophys. Acta*, 933, 35–41.
Westerhoff, H. V., Melandri, B. A., Venturoli, G., Azzone, G. F., and Kell, D. B. (1984) Mosaic protonic coupling hypothesis for free energy transduction. *FEBS Lett.*, 165, 1–5.
Wherland, S., and Pecht, I. (1978) Protein-protein electron transfer. A Marcus theory analysis of reactions between c-type cytochromes and blue copper proteins. *Biochemistry*, 17, 2585–2591.
Whitmarsh, J. (1986) Mobile electron carriers in thylakoids. In *Encyclopedia of Plant Physiology*, New Series, Vol. 19 (L. A. Staehelin and C. J. Arntzen, eds.), pp. 508–527, Springer-Verlag.
Whitmarsh, J., Bowyer, J. R., and Crofts, A. R. (1982) Modification of the apparent redox reaction between cytochrome f and the Rieske iron-sulfur protein. *Biochim. Biophys. Acta*, 682, 404–412.
Wickner, W. (1988) Mechanism of membrane assembly: general lessons from the study of M13 coat protein and *E. coli* leader peptidase. *Biochemistry*, 27, 1081–1086.
Widger, W. R., W. A. Cramer, R. G. Herrmann, and A. Trebst (1984) Sequence homology and structural similarity between cytochrome b of mitochondrial com-

plex III and the chloroplast b_6-f complex: position of the cytochrome b hemes in the membrane. *Proc. Natl. Acad. Sci. USA, 81,* 674–678.

Widger, W. R., Cramer, W. A., Hermodson, M., and Herrmann, R. G. (1985) Evidence for a hetero-oligomeric structure of the chloroplast cytochrome b-559. *FEBS Lett., 191,* 186–190.

Wikström, M. K. F. (1987) Insight into the mechanism of cellular respiration from its partial reversal in mitochondria. *Chem. Scripta. 27B,* 53–58.

Wikström, M. K. F., and Berden, J. A. (1972) Oxidoreduction of cytochrome b in the presence of antimycin. *Biochim. Biophys. Acta, 283,* 403–420.

Wikström, M. K. F., and Saraste, M. (1984) The mitochondrial respiratory chain. In *Bioenergetics* (L. Ernster, ed.), pp. 49–94, Elsevier, Amsterdam.

Wikström, M., Krab, K., and Saraste, M. (1981) *Cytochrome Oxidase: A Synthesis.* 198 pp. Academic Press, London.

Wikström, M. K. F., Saraste, M., and Penttilä, T. (1985) Relationships between structure and function in cytochrome oxidase. In *The Enzymes of Biological Membranes* (A. Martonosi, ed.), pp. 111–148, Plenum Press, New York.

Willey, D. L., and Gray, J. C. (1988) Synthesis and assembly of the cytochrome $b-f$ complex in higher plants. *Photosyn. Res., 17,* 125–144.

Willey, D. L., Auffet, A. D., and Gray, J. C. (1984) Structure and topology of cytochrome f in pea chloroplast membranes. *Cell, 36,* 556–562.

Williams, J. C., Steiner, L. A., Ogden, R. C., Simon, M. I., and Feher, G. (1983) Primary structure of the reaction center from *Rb. sphaeroides. Proc. Natl. Acad. Sci. USA, 80,* 6505–6509.

Williams, J. C., Steiner, L. A., Feher, G., and Simon, M. I. (1984) Primary structure of the L subunit of the reaction center from *Rb. sphaeroides. Proc. Natl. Acad. Sci. USA, 81,* 7303–7307.

Williams, J. C., Steiner, L. A., and Feher, G. (1986) Primary structure of the reaction center from *Rb. sphaeroides. Proteins, 1,* 312–325.

Williams, R. J. P. (1961) Possible functions of chains of catalysts. *J. Theor. Biol., 1,* 1–17.

Williams, R. J. P. (1962) Possible functions of chains of catalysis II. *J. Theor. Biol., 3,* 209–229.

Williams, R. J. P. (1978) The multifarious couplings of energy transduction. *Biochim. Biophys. Acta, 505,* 1–44.

Williams, R. J. P. (1985) The nature of proteins in membranes. In *Recent Advances in Biological Membrane Studies* (L. Packer, ed.), pp. 17–28, Plenum, New York.

Wilson, D. F., and Forman, N. (1982) Mitochondrial transmembrane pH and electrical gradients: evaluation of their energy relationships with respiratory rate and ATP synthesis. *Biochemistry, 21,* 1438–1444.

Wilson, D. F., Erecínska, M., and Owens, C. S. (1976a) Some properties of the redox components of cytochrome c oxidase and their interactions. *Arch. Biochem. Biophys., 175,* 160–172.

Wilson, D. M., Alderete, J. F., Maloney, P. C., and Wilson, T. H. (1976b) Proton motive force as the source of energy for adenosine-5'-triphosphate synthesis in *Escherichia coli. J. Bacteriol., 126,* 327–337.

Wilson, E., Farley, T. M., and Takemoto, T. Y. (1985) Photoaffinity labeling of an antimycin binding site in *Rb. sphaeroides. J. Biol. Chem., 260,* 10288–10292.

Wise, J. G., Latchney, L. R., Ferguson, A. M., and Senior, A. E. (1984) Defective proton ATPase of UncA mutants of *Escherichia coli.* 5' Adenylyl imidophosphate binding and ATP hydrolysis. *Biochemistry, 23,* 1426–1432.

Witt, H. T. (1979) Energy conversion in the functional membrane of photosynthesis.

Analysis by light pulse and electric pulse methods. *Biochim. Biophys. Acta*, 505, 355–427.

Witt, H. T. (1988) Some recent functional and structural contributions to the molecular mechanism of photosynthesis. In *The Roots of Modern Chemistry* (Kleinkaag, van Döhren, and Jaenicke, eds.), pp. 713–720, Walter de Gruyter, Berlin.

Wolfenden, R., Andersson, L., Cullis, P. M. and Southgate, C. C. B. (1981) Affinities of amino acid side chains for solvent water. *Biochemistry*, 20, 849–855.

Wollman, F.-A., and Lemaire, C. (1988) Studies on kinase-controlled state transitions in photosystem II and b_6–f mutants from *C. reinhardtii* which lack quinone-binding proteins. *Biochim. Biophys. Acta*, 933, 95–94.

Wood, P. M. (1978) Interchangeable copper and iron proteins in algal photosynthesis. *Eur. J. Biochem.*, 87, 9–19.

Wood, P. M. (1984) Bacterial proteins with CO-binding b or c-type haem functions. *Biochim. Biophys. Acta*, 768, 293–317.

Wood, P. M. (1987) The two redox potentials for oxygen reduction to superoxide. *Trends Biochem. Sci.*, 12, 250–251.

Woodbury, N. W., Becker, M., Middendorf, D., and Parson, W. W. (1985) Picosecond kinetics of the initial photochemical electron transfer reaction in bacterial photosynthetic reaction centers. *Biochemistry*, 24, 7516–7521.

Woods, T. A., Decker, G. L., and Pedersen, P. L. (1977) Antihyperlipidemic drugs—*in vitro* effect on the function and structure of rat liver mitochondria. *J. Mol. Cell. Cardiol.* 9, 807–822.

Wraight, C. A. (1982) The involvement of stable semiquinones in the two electron gates of plant and bacterial photosynthesis. In *Function of Quinones in Energy-Conserving Systems* (B. Trumpower, ed.). pp. 181–197.

Wraight, C. A., Cogdell, R. J., and Chance, B. (1978) Ion transport and electrochemical gradients in photosynthetic bacteria. In *The Photosynthetic Bacteria* (R. K. Clayton and W. R. Sistrom, eds.), pp. 471–511, Plenum, New York.

Wright, J. K., Seckler, R., and Overath, P. (1986) Molecular aspects of sugar:ion co-transport. *Annu. Rev. Biochem.*, 55, 225–248.

Wynn, R. N., and Malkin, R. (1988) Interaction of plastocyanin with photosystem I: A chemical cross-linking study of a polypeptide that binds plastocyanin. *Biochemistry*, 27, 5863–5869.

Xue, Z., Zhon, J.-M, Malese, T., Cross, R. L., and Boyer, P. D. (1987) Chloroplast F_1 ATPase has more than three nucleotide binding sites, and 2-azido-ADP or 2-azido-ATP at both catalytic and non-catalytic sites labels the β subunit. *Biochemistry*, 26, 3749–3753.

Yachandra, V. K., Guiles, R. D., McDermott, Britt, D. R., Deyheimer, S. L., Sauer, K., and Klein, M. P. (1986) The state of manganese in the photosynthetic apparatus. 4. Structure of the manganese complex in photosystem II studied using EXAFS spectroscopy. The S_1 state of the O_2-evolving photosystem II complex from spinach. *Biochim. Biophys. Acta*, 850, 324–332.

Yamane, K., Ichihara, S., and Mizushima, S. (1987) *In vitro* translocation of protein across *E. coli* membrane vesicles requires both the protonmotive force and ATP. *J. Biol. Chem.*, 262, 2358–2362.

Yang, F.-O, Yu, L., Yu, C.-A., Lorence, R., and Gennis, R. B. (1986) Use of an azido-ubiquinone derivative to identify subunit V as the ubiquinol binding site of the cytochrome d terminal oxidase complex of *Escherichia coli*. *J. Biol. Chem.*, 261, 14987–14990.

Yang, X., and Trumpower, B. (1986) Purification of a three subunit ubiquinol-cytochrome c oxidoreductase complex from *Paracoccus denitrificans*. *J. Biol. Chem.*, *261*, 12282-12289.

Yazyu, H., Shiota-Niiya, S., Shinanitim, T., Kanazawa, H., Futai, M., and Tsuchiya, T. (1984) Nucleotide sequence of the *mel B* gene and characteristics of deduced amino acid sequence of the melibiose carrier in *Escherichia coli*. *J. Biol. Chem.*, *259*, 4320-4326.

Yazyu, H., Shiota, S., Futai, M., and Tsuchiya, T. (1985) Alteration in cation specificity of the melibiose transport carrier of *Escherichia coli* due to replacement of proline 122 with serine. *J. Bacteriol.*, *162*, 933-937.

Yeates, T. O., Komiya, H., Rees, D. C., Allen, J. P., and Feher, G. (1987) Structure of the reaction center from *Rb. sphaeroides* R-26: membrane-protein interactions. *Proc. Natl. Acad. Sci. USA*, *84*, 6438-6442.

Yeates, T. O., Komiya, H., Chirino, A., Rees, D. C., Allen, J. P., and Feher, G. (1988) Structure of the reaction center from *Rb. sphaeroides* R-26 and 2.4.1: protein-cofactor (bacteriochlorophyll, bacteriopheophytin, and carotenoid) interactions. *Proc. Natl. Acad. Sci. USA*, *85*, 7993-7997.

Yoch, D. C. and Carrithers, R. P. (1979) Bacterial iron-sulfur proteins. *Microbiol. Rev.*, *43*, 384-421.

Yocum, C., Yerkes, C. T., Blankenship, R. E., Sharp, R. F., and Babcock, G. T. (1981) Stoichiometry, inhibitor sensitivity, and organization of manganese associated with photosynthetic oxygen evolution. *Proc. Natl. Acad. Sci. USA*, *78*, 7507-7511.

Yoshida, M., and Allison, W. S. (1986) Characterization of the catalytic and noncatalytic ADP binding sites of the F_1-ATPase from the thermophilic bacterium, PS3. *J. Biol. Chem.*, *261*, 5714-5721.

Yoshida, M., Sone, N., Hirata, H., and Kagawa, Y. (1977) Reconstitution of adenosine triphosphatase of thermophilic bacterium from purified individual subunits. *J. Biol. Chem.*, *252*, 3480-3485.

Youvan, D. C., Bylina, E. J., Alberti, M., Begusch, H., and Hearst, J. E. (1984) Nucleotide and deduced polypeptide sequences of the photosynthetic reaction center, B870 antennae, and flanking polypeptide from *Rb. capsulata*. *Cell*, *37*, 949-957.

Ysern, X., Amzel, L. M., and Pedersen, P. L. (1988) ATP synthases—structure of the F_1 moiety and its relationship to function and mechanism. *J. Bioenerg. Biomembr.*, *20*, 423-450.

Yu, L., and Yu, C.-A. (1980) Resolution and reconstitution of succinate-cytochrome c reductase. *Biochim. Biophys. Acta*, *591*, 409-420.

Yu, L., and Yu, C.-A. (1987a) Identification of cytochrome *b* and *a* molecular weight 12 kDa protein as the ubiquinone-binding proteins in the cytochrome $b-c_1$ complex of a photosynthetic bacterium, *Rb. sphaeroides* R-26. *Biochemistry*, *26*, 3658-3664.

Yu, C.-A., and Yu, L. (1987b) The nature of ubiquinone binding sites in the mitochondrial electron transfer complexes. In *Bioenergetics: Structure and Function of Energy-Transducing Systems* (T. Uzawa and S. Papa, eds.), pp. 81-99, Japan Sci. Soc. Press, Springer-Verlag, Berlin.

Yu, L., Mei, Q.-C., and Yu, C.-A. (1984) Characterization of purified cytochrome bc_1 complex from *Rb. sphaeroides* R-26. *J. Biol. Chem.*, *259*, 5752-5760.

Yu, L., Xu, J.-X., Haley, P. E., and Yu, C.-A. (1987) Properties of bovine heart mitochondrial cytochrome b_{560}. *J. Biol. Chem.*, *262*, 1137-1143.

Zanetti, G. (1976) A lysyl residue at the NADP binding site of ferredoxin-NADP reductase. *Biochim. Biophys. Acta*, *445*, 14-24.

Zanetti, G., Morelli, D., Ronchi, S., Negri, A., Aliverti, A., and Curti, B. (1988) Struc-

tural studies on the interaction between ferredoxin and ferredoxin-NADP$^+$ reductase. *Biochemistry*, 27, 3753–3759.
Zhang, Y.-Z., Lindorfer, M. A., and Capaldi, R. A. (1988) Orientation of cytoplasmically made subunits of beef heart cytochrome c oxidase determined by antibody digestion and antibody binding experiments. *Biochemistry*, 27, 1389–1394.
Zilberstein, D., Schuldiner, S., and Padan, E. (1979) Proton electrochemical gradient in *Escherichia coli* cells and its relation to active transport of lactose. *Biochemistry*, 18, 669–673.
Zimmerman, B. H., Nitschke, C. I., Fee, J. A., Rusnak, F., and Munck, E. (1988) Properties of a copper containing cytochrome b–a_3: a second terminal oxidase from the extreme thermophile, *T. thermophilus*. *Proc. Natl. Acad. Sci. USA*, 85, 5779–5783.
Zimmerman, J.-L., and Rutherford, A. W. (1986) EPR properties of the S_2 state of the oxygen-evolving complex of photosystem II. *Biochemistry*, 25, 4609–4615.
Zoratti, M., Favaron, M., Pietrobon, D., and Azzone, G. F. (1986) Intrinsic uncoupling of mitochondrial proton pumps. 1. Non-ohmic conductance cannot account for non-linear dependence of static head respiration on $\Delta\tilde{\mu}_{H^+}$. *Biochemistry*, 25, 760–767.
Zubay, G. (1983) *Biochemistry*, p. 623, Addison-Wesley, Reading, Mass.
Zuber, H. (1986) Structure of light-harvesting antenna complexes of photosynthetic bacteria, cyanobacteria, and red algae. *Trends Biochem. Sci.*, 11, 414–419.
Zurawski, G., Bohnert, H. J., Whitfeld, P. R. and Bottomley, W. (1982) Nucleotide sequence of the gene for the M_r 32,000 thylakoid membrane protein from *Spinacia oleracea* and *Nicotiana debneyi* predicts a totally conserved primary translation product of M_r 38,950. *Proc. Natl. Acad. Sci. USA*, 79, 7699–7703.

Glossary of Abbreviations

A	absorbance
ADP	adenosine-5'-diphosphate
ATP	adenosine-5'-triphosphate
BChl	bacteriochlorophyll
bR	bacteriorhodopsin
CCP	cytochrome c peroxidase
Chl	chlorophyll
CL	cardiolipin
Cyt	cytochrome
D	diffusion constant
DBMIB	2,5-dibromo-3-methyl-6-isopropylbenzoquinone
DCCD	dicyclohexylcarbodiimide
DCMU	3-(3,4-dichlorophenyl)-1,1-dimethylurea
DGDG	digalactosyldiacylglycerol
DNA	deoxyribonucleic acid
DNP	2,4-dinitrophenol
ϵ_M	molar extinction coefficient
EPR	electron paramagnetic resonance
F	Faraday constant
FAD	flavin adenine dinucleotide
FCCP	carbonyl cyanide p-trifluoromethoxyphenylhydrazone
FMN	flavin mononucleotide
FNR	ferredoxin: $NADP^+$ oxido reductase
FRAP	fluorescence recovery after photobleaching
Gibbs	free energy
H(N)QNO	2-n-heptyl(nonyl)-4-hydroxyquinoline N-oxide
hR	halorhodopsin

Glossary of Abbreviations

ISp	iron-sulfur protein
λ (lambda)	wavelength
LHCP	light harvesting chlorophyll protein
kb	kilobase
kDa	kilodalton
MGDG	monogalactosyldiacylglycerol
MQ	menaquinone
M_r	relative mobility or molecular weight
mV	millivolt
MW	molecular weight
NAD^+	nicotinamide-adenine dinucleotide
$NADP^+$	nicotinamide-adenine dinucleotide phosphate
NEM	N-ethylmaleimide
NMR	nuclear magnetic resonance
OEC	oxygen evolving complex
PEG	polyethyleneglycol
PC	phosphatidylcholine or plastocyanin, depending on context
PE	phosphatidylethanolamine
PG	phosphatidylglycerol
PI	phosphatidylinositol
PQ	plastoquinone
PTS	phosphotransferase system for sugar transport
PSI, PSII	photosystems I, II
SDH	succinate dehydrogenase
SHAM	salicylhydroxamic acid
SMP	sub-mitochondrial particles
TCA	trichloroacetic acid
TMPD	tetramethylphenylenediamine
UHDBT	5-n-undecyl-6-hydroxy-4,7-dioxobenzothiazole
UQ	ubiquinone
V	volt

Index*

A

Adenine nucleotide
 measurement, 27
 translocator, 120-121, 123
ADP/ATP carrier, molecular mechanism, 122
Amine transport, chromaffin granules and, 450-452, 453
Amino acid transport, 427, *428*
Antiport reactions, 20, 21
ATP
 caged, 28-29
 evolution and, 29
 potassium uptake and, 435-437
 proton electrochemical potential and, 388-390
ATP hydrolysis
 ATPase and, 386-388
 Gibbs Free Energy and, *26-27*
 kinetics, 383-388
 thermodynamics, 383-388
ATP synthase
 evolution and, 369
 F type, 394
 function, 362-363, 366
 structure, 360-361
 function, 355-360
 nucleotide release, 379, 381-383
 proton requirement, 118-123
ATP synthesis, 119-120
 bound nucleotides and, 378-379
 chemiosmotic hypothesis and, 105-112
 energy requirement, 379, 381-383
 Gibbs Free Energy and, 26-27
 membrane potential and, 107-109, 111
 proton electrochemical potential and, *21*, 356-357
 rate, 358
 sodium transport and, 435
 stoichiometry, 348-349
 thylakoids, 109-110
 transmembrane pH gradient and, 107-109, 111
ADP-ATP translocator, 357
ATPases
 calcium transporting, 400, 401, 402
 F type, 355-357, 359-360, 364, 365, 402, 403
 ATP hydrolysis and, 386-388
 crystallographic studies, 368
 electric currents and, 368
 electron micrograph, 369
 folding patterns, 377, 378
 mapping, 246-248

---

*Page numbers for equations are in italics.

ATPases (*cont.*)
F type (*cont.*)
reconstitution, 366-367
subunit *a*, 373-374
subunit *b*, 374, 375
subunit *c*, 374, 376
folding arrangement, 393
orientation, 356
P type, 355, 390-391, 402, 403-404
amino acid identities, 395
characterization, 394, 396-400
hydropathy plot, 396
properties, 392
regulation, 358, 359-360
V type, 355, 390-392, 394, 402-403

B
Bacteria
Gram-negative, cell envelope, 409-410
membrane structure, 78, 79, 81-82, 86, 135-136
photosynthetic antenna complexes, 252-257
photosynthetic reaction centers, 269-278
proton electrochemical potential, 410-412
transport systems, 406-407, 465
viral infections, 455-456
Bacteriochlorophyll, antenna complexes and, 252, 254
Bacteriorhodopsin
folding pattern, 304, 306
halorhodopsin and, 309-311
hydropathy plot, 474
photocycle, 301, 310
proton pump, 299-311
proton translocation, 72-73
spectral intermediates, 307-309
structure, 300, 302, 303, 304
Biliproteins, spectra, 250
Bound nucleotides, ATP synthesis and, 378-379
Brown fat mitochondria
protein uncouplers, 115
thermogenesis and, 115, 137

C
Calcium, photosystem II and, 290-291
ATPases and, 400-402
Carboxylic acids, transport, 417-419
Cardiomyopathy, dilated, 121, 123
Carotenoid band shift, 100, 102
Chemical work
chemical potential and, *17-18*
Gibbs Free Energy and, *17-18*
Chemiosmotic hypothesis, *19*, 93-114, 124, 136
ATP synthesis test, 105-112
microchemiosmotic model, 130
mosaic coupling model, 130
proton movement experiments, 103-105
summary of experiments, 125-129
Chloride, photosystem II and, 290-291
Chloride pump, 309, 311
Chlorin *d*, structure, 150
Chlorophyll
energy transfer, 256
photosynthesis and, 239
structure, 253
Chlorophyll binding proteins, labeling, 210
structure, 257-260
Chloroplast $b_6 f$ complex, 326-330
Chloroplasts
electron transport complexes, 88
genetic map, 80
membrane structure, 78, 79, 81-82, 83, 86, 135-136, 211, 212
photophosphorylation, 294
photosynthetic antenna complexes, 257-261
Chlorosomes, energy transfer, 262, 263, 264
Chromaffin granules, amine transport and, 450-452, 453
Chromatophores
formation, 105
quinone (Q) cycle and, 344
Cirrhosis, primary biliary, 121, 123
Colicins, 456
Copper binding sites, cytochrome oxidases, 314, 315-316, 317-318
Copper proteins, 169-171
Coupled reactions, rate, *22-23*

Index

Cross-correlation function, 471
Cytochrome b_5, 162
Cytochrome b_6, hydropathy plot, 333
Cytochrome bc_1 complex, 326–346
  composition, 337
  oxidant–induced reduction, 340
  properties of inhibitors, 336–337
  proton translocation, 340–345
Cytochrome $b_6 f$ complex, 326–346
  composition, 327
  properties of inhibitors, 336–337
  proton translocation, 345–346
Cytochrome c_1
  amino acid sequences, 165–169
  hydropathy plot, 167
Cytochrome c_2, photosynthetic reaction center, 278
Cytochrome c_3
  hemes, 156, 158
  models, 157
Cytochrome d, characterization, 150–151
Cytochrome f
  amino acid sequences, 165–169
  hydropathy plot, 167
Cytochrome o, characterization, 150–151
Cytochrome oxidase, 190, 311–326
  alternative, 149, 196–198
  amino acid sequence, 312, 313
  copper binding sites, 314, 315–316, 317–318
  electron diffraction, 312–314
  folding pattern, 316
  heme binding sites, 314, 315, 317–318
  mitochondrial myopathy and, 312
  oxidation-reduction reactions and, 318–319, 326
  polypeptides, 149
  as proton pump, 311–326
  proton translocation and, 318–326, 349–350
  spectral properties, 148
Cytochromes
  absorbance spectra, 141, *142*, 144, 145
  characterization, 189–190
  complementary acidic residues, 164
  redox properties, 141, 145
Cytochromes b
  amino acid sequence, 331

  characterization, 151–152
  electron transport, 152, 340, 341
  helix models, 334
  hydropathy plots, 473
  inhibitor binding, 332, 338
  polypeptides, 335
  reduction, 340, 344
  structure, 330–332
Cytochromes c
  amino acids, 153, 155
  chemical modification, 158–159, 160
  diffusion, 164–165
  lysines
    binding sites, 162–164
    chemical modification, 158–159
    labeling, 159, 161–162
  mobility, 155–156
  orientability, 164–165
  peroxidases, 162–164
  reduction by flavins, 63
  reduction by quinols, 63, 65–66
  structure, 152–155

D

Decarboxylation, sodium transport and, 433–435
Desulfovibrio desulfuricans, 156
Diffusion, 204–208, 216–217, 237–238
DNA, mitochondrial, evolution and, 82
DNA sequence, *unc* mutants, 368–372

E

E. coli, respiratory chain, 91, 150–151
Electric fields, ATP synthesis and, 109–110
Electrical work, Gibbs Free Energy and, *16–17*
Electrochemical potential
  Gibbs Free Energy and, *19–20*
  sodium and, 429
Electrochromism, 100
Electron paramagnetic resonance, 173–177
Electron transfer reactions, 57, 75, 118
  effect of temperature, 69–70
  Gibbs Free Energy and, 57–60
  long distance, 67–69

Electron transfer reactions (*cont.*)
  mechanisms, 57–71
  nuclear tunneling and, 69–70
  proteins, 63, 64, 65, 66–68
  rate, *61–62, 66–67*
Electron transport
  cytochrome *b* and, 340, 341
  effect of ADP, 116–117
  effect of uncouplers, 116–118
  free energy storage and, 92–114
  high energy bond model, 92–93
  mitochondrial complex I and, 186
  quinones and, 230
  role of tyrosine, 283–284
  thylakoids, 208, 216–217
Electron transport chains, 94, 191
  cyclic, 291–292, 294, 295, 297
  cytochromes *b*, 152
  iron-sulfur proteins and, 171–172
  mitochondria, 90
  noncyclic, proton translocation, 292, 293, 294, 295, 297
  photosynthesis and, 91
  proton translocation and, 346–348
  quinones and, 196–198
Enthalpy, defined, *4*
Entropy
  second law of thermodynamics and, *6–10*
  work and, *10–11*
Eubacterial ATPases, *see* ATPases, F type
Eubacterial proteolipids, evolution, 393
Eukaryotes, transport systems, 449–455
Evolution
  ATP and, 29
  ATP synthase and, 369
  eubacterial proteolipids, 393
  membrane structure and, 81–82
  sodium based systems, 123
Excited energy states, *239*

F
Ferredoxin:NADP$^+$ oxido-reductase, 178, 180–181
Ferredoxins
  characterization, 191

oxidation-reduction, 172–173
  in oxygenic photosynthesis, 178–181
Flagella, transport systems and, 447–449
Fluorescence, photosynthesis and, 240–241, 242–243
FRAP technique, 205–206, 208
Free energy, *see* Gibbs Free Energy

G
γ-aminobutyric acid (GABA), transport, 452–453, 454
Genetic maps
  chloroplasts, 80
  mitochondria, 81
Gibbs Free Energy
  activation energy, 57, 58, 59, 60, *61–62*
  ATP hydrolysis and, *26–27*
  ATP synthesis and, 27
  change in, *38–39*
  chemical reactions and, *14–15*
  chemical work and, *17–18*
  concentration dependence, *13*
  defined, *11–12*
  direction of reaction and, *15*
  electrical work and, *16–17*
  electrochemical potential and, *19–20*
  electron transfer rate and, 57–60
  high energy bonds and, 23–24, 26
  oxidation-reduction potential and, 35–36
  proton electrochemical potential and, *106–107*
Glyceraldehyde-3-phosphate dehydrogenase reaction, 92
Gramicidin, ionophoric capacity, 131, 133, 134, 135
Grana, structure, 213

H
Halorhodopsin
  amino acid sequence, 310
  bacteriorhodopsin and, 309–311
  helical wheel projection, 311
  photochemical cycle, 310

Index 541

Heme
  proteins, 141–142
  structure, 142, 143, 148
Heme binding sites, cytochrome oxidases, 314, 315, 317–318
Herbicides, resistance to, 232, 233, 235
High-energy bonds, defined, 23–24
Histidine, lactose symport and, 426
Hydrophobicity indices, membrane proteins, 470–474
Hydroxide antiport, proton symport and, 416–417

I
Import and secretion, proteins, 456–465
Ion gradients, thermodynamics, *18–19*
Ionophores
  membrane potential and, 130–135
  transmembrane pH gradient and, 130–135
Iron-sulfur proteins, 171–172; *see also* Ferredoxins; NADH:ubiquinone oxidoreductase; Rieske protein; Succinate:ubiquinone oxidoreductase
  high potential, 177–178

K
Krebs cycle, 187

L
lac permease
  activity, 420
  amino acids, 423, 425
  and proton symport, 426
  structure, 424, 427
Lactose transport, 414–415, 419–422, 426
Le Chatalier's Principle, 16
Leader peptides, 461
Light, absorption, 239, 240–241
Light energy transfer, *see* Photosynthesis, light energy transfer
Light harvesting complex (LHC), 257–261
Luft disease, 116, 118

M
Macromolecules, transport systems, 455–464
Manganese, oxygen evolution and, 287–291
Melibiose, sodium transport and, 431–433
Membrane potential
  ATP synthesis and, 107–109
  charge movement and, 103
  ionophores and, 130–135
  measurement, 94–95, 98, *99, 100,* 102–103
  protein translocation and, 458–461
  proton translocation and, 73–74
  surface potential and, 102–103
Membrane proteins
  folding, 470–474
  hydrophobicity indices, 470–474
Membrane structure
  bacteria, 78, 79, 81–82, 86, 135–136
  chloroplasts, 78, 79, 81–82, 83, 86, 135–136
  evolution and, 81–82
  mitochondria, 78, 79, 81–82, 84, 86, 135–136
Menaquinone, structure, 194, 195
Metallo-porphyrin, oxidation-reduction potential and, 50–51
Metalloproteins, EPR properties, 174
Mitochondria
  see also Brown fat mitochondria
  electrogenicity, 357
  electron transport chain, 90
  energy transduction, 82
  evolution and, 82
  fatty acids, 90
  genetic map, 81
  inner membranes, diffusion coefficients, 207
  lipids, 86, 89
  loose coupling, 118
  membrane structure, 78, 79, 81–82, 84, 86, 135–136
  proteins, 87, 89, 459
  respiration rates, 86
Mitochondrial complex I, properties, 184, 185, 186, 191

Mitochondrial complex II, properties, 187–189, 191
Mitochondrial complex III
 composition, 327, 328, 330
 mitochondrial myopathy and, 330
 proton translocation, 326–328, 330
Mitochondrial complex IV, cytochrome oxidase, 311–326
Mitochondrial myopathy
 cytochrome oxidases and, 312
 redox therapy, 330
Molecular biology, 283–284
Monensin, ionophoric capacity, 131, 133, 135
Motility, transport systems and, 447–449

N

NADH:ubiquinone oxidoreductase, 182–185
Nernst equation, *16–17*, *40*
Neurotransmitters, transport and, 449–455
Nigericin, ionophoric capacity, 131, 133, 135
Nuclear tunneling, electron transfer reactions and, 69–70

O

Oxidation-reduction potential
 changes in, 74–75
 concentration dependence, *39–40*
 cross-relation and, *62–63*
 effect of axial ligands, 50–51
 effect of ionic strength, 49, 50
 effect of ligand binding, 49–50
 effect of membrane potential, 51
 Gibbs Free Energy and, 35–36
 group transfer potential, 37–38
 measurement, 47–48
 metallo-porphyrin and, 50–51
 midpoint potential, *54–55*
 photosynthesis and, 55–57
 quinones, *52–54*
 reaction pathway and, *46*
 scale of, *36–37*

Oxidation-reduction reactions
 cytochrome oxidases and, 318–319, 326
 ferredoxins, 172–173
Oxygen evolution
 manganese and, 287–291
 plants, 281–283
 polypeptides and, 286
Oxygen free radicals, medical relevance of, 46–47

P

Peptides, leader, protein translocation and, 461–464
Periplasmic proteins, transport systems and, 440–443
Periplasmic space, transport systems and, 440–447
Phagocytes, oxygen free radicals and, 46–47
Phase separation, methodology, 214, 215
Phenomenological stoichiometry, defined, *23*
Phosphate compounds, High energy, transport and, 435–440
Phospho-ATPases, *see* ATPases, P type
Phosphorylation
 light energy and, 260
 uncouplers, 112–115
Phosphotransferase systems, sugar uptake and, 438–440
Photophosphorylation, 112
 chloroplasts, 294
Photosynthesis
 electron transport and, 91
 ferredoxins in, 178–181
 light energy regulation, 260
 light energy transfer, 239–241, 295–296
  phycobilisomes and, 248–251
  resonance energy, 241, *242–244*, 245–247, *248*
  spectroscopy and, *243–245*, 246–248
 water splitting, 281–283

Index 543

Photosynthetic antenna complexes
  bacteria, 252–257
  chloroplasts, 257–261
  structure, 252–264
Photosynthetic reaction centers
  amino acid sequences, 231–232
  bacteria, structure, 269–278, 296
  center-center distance, 228
  cytochrome c_2 and, 278
  EPR signals, 224
  helices, 228
  photosystem I, plants, 278, 280
  photosystem II, plants, 280–281
  plants, 278–281, 296–297
  properties, 265
  protein structure, 266–268, *269*
  quinones and, 223, 225–235
  redox properties, 55–57, 277, 286
  representation, 227
  sequence identity, 230, 232
  structure, 264–278
  subunits, hydropathy plots, 474
Photosystem I
  components, 216–217, 236–237
  photosynthetic reaction center, 277, 278, 280
  topography, 279
Photosystem II
  calcium and, 290–291
  chloride and, 290–291
  chlorophyll-protein complexes, 257–260
  components, 216–217, 236–237
  manganese and, 287–291
  particle preparations, 284–287
  photosynthetic reaction center, 280–281
  polypeptides, 285, 290, 291
  proton release, 281–283
Phycobilisomes, photosynthesis and, 248–251
Planar bilayers, 131
Plasma membrane ATPases, *see* ATPases, P type
Plastocyanin, 169–171
Plastoquinone, structure, 194, 195
Potassium transport, ATP and, 435–437
Potential energy, 5

Protein folding, in membranes, 470–474
Protein translocation
  leader peptides and, 461–464
  membrane potential and, 458–461
  mitochondria, 459
  transmembrane pH gradient and, 458–461
Protein transport, energy sources, 456–461
Proton:sugar stoichiometry, 414–415, 419
Proton electrochemical potential
  ATP and, *21*, 356–357, 388–390
  bacteria, 410–412
  Gibbs Free Energy and, *106–107*
  sugar uptake and, 412, 414
  transport and, *19–21*
Proton exchange, purple membranes, 111–112
Proton pump
  bacteriorhodopsin and, 299–311
  cytochrome oxidases and, 311–326
Proton transfer reactions
  laser techniques, 71–72
  proteins, 71–74
Proton translocation, 75, 349–350
  bacteriorhodopsin, 72–73
  cytochrome b_6f complex, 345–346
  cytochrome oxidases and, 318–326
  electron transport chains and, 346–348
  mechanisms of, 299
  membrane potential and, 73–74
  mitochondria, 326–328, 330
  movement direction, 367
  noncyclic electron transport and, 292, 293, 294, 295, 297
  number translocated, *346–347*
  quinone (Q) cycle and, 326
  stoichiometry, 347
Proton transport
  amino acids and, 427, *428*
  hydroxide antiport and, 416–417
  lac permease and, 426
  sugar transport and, 412–419
Proton-motive force, *19*
Purple membranes, proton exchange, 111–112

Q

Quinone (Q) cycle, 326, 340, *341*, 342, *343–344*, 345–346, 350
Quinones
 absorption spectra, 219, 220
 arrangement in photosynthetic reaction center, 229
 binding proteins, 217–218, 220, 221, 222, 226–235
 binding sites, 232–235, 236–237, 332, 338
 branch point, 196–198
 as carriers of electrons and protons, 193
 diffusion coefficient, *204–205*
 electron acceptors, 223–226, 273, 276
 in electron transport, 196–198, 223, 225–226, 230
 EPR properties, 222, 224
 function, 198
 lateral mobility, 204–205, 208, 216–217
 membrane connection, 199–204
 oxidation-reduction potential, *52–54*
 pools, 195
  location, 198–199
 structure, 193–194, 218

R

Redox potential, *see* Oxidation-reduction potential
Respiration, sodium pump and, 123, 431
Respiratory chain, bacteria, 91
Rhodopsin, 300
Rieske iron-sulfur protein
 amino acid sequence, 181–182
 inhibitors, 336–337
 redox reactions and, 338–339

S

Serotonin, transport, 453–454
Site-directed mutagenesis, 283–284
Sodium gradients, thermodynamics, *123*
Sodium pump, respiration-linked, 123, 431
Sodium transport, *428*, 429–435
 decarboxylation and, 433–435
Sodium/proton antiports, 429–431
Spectra, electric field, 101
Spectrophotometry
 cytochromes, 142, 145
 magnetic field and, 175, 176, 177
 methodology, 145–147, *173–175*, 176, 177
Standard potential, defined, 37
Stroma, structure, 213
Succinate:ubiquinone oxidoreductase, 187–189
Sugar transport
 kinetic characteristics, 419–422
 proton transport and, 412–419
Sugar uptake
 phosphotransferase systems and, 438–440
 proton electrochemical potential and, 412, 414
Symport reactions, 20, 21

T

Thermodynamics
 first law, *3–4*
 of ion gradients, *18–19*
 nonequilibrium, *22–23*
 proton electrochemical potential and, *19–21*
 second law
  defined, *4–6*
  entropy and, *6–10*
  Gibbs Free Energy and, *12*
Thermogenesis, brown fat mitochondria and, 115, 137
Thylakoids
 ATP synthesis, 109–110
 diffusion coefficients, 207
 electron transport components, 208, 216–217
 fatty acids, 90
 lipids, 86, 89
 structure, 83, 209
Translocation, macromolecules, 455–464
 energy sources, 457, 460

Transmembrane charge separation, reaction center proteins and, 266–268, *269*
Transmembrane pH gradient
　ATP synthesis and, 107–109
　chemiosmotic hypothesis and, 105–107
　ionophores and, 130–135
　measurement, 94–95, *96*, *97–98*
　protein translocation and, 458–461
Transport systems
　bacteria, 465
　classification, 406–407
　techniques, 408–409
　eukaryotes, 449–455
　macromolecules, 455–464
　motility and, 447–449
　neurotransmitters and, 449–455
　periplasmic space and, 440–447
　phosphate compounds and, 435–440
　protein carrier mediated, 407–408
　shockable, sequences, 444–445

Tyrosine, electron transport and, 283–284

U

Ubiquinone
　diffusion coefficient, 205
　structure, 194, 195
Uncouplers
　effect on electron transport, 117
　mechanism, 112–115
　structures, 113
Uncoupling proteins, brown fat mitochondria, 115
Uniport reactions, 20

V

Vacuolar ATPases, *see* ATPases, V type
Valinomycin, ionophoric capacity, 131, 132, 133, 135, 137
Van't Hoff equation, temperature and, *15–16*
Viral infections, bacteria, 455–456

Preface to Solutions Section

This student edition of *Energy Transduction in Biological Membranes* appears approximately one year after the original hardcover version was tested in the one semester graduate course in the Department of Biological Sciences at Purdue. Thus, there are a set of errata attached the set of problem solutions.

The graduate course covered the first eight chapters of the book. It was realized that though the answers to the problems published as Appendix I can be helpful to the student working the homework problems, it is often important for teaching purposes to ultimately have access to detailed solutions. The discussion attached to many of the problem solutions often implements that in the text and thus was useful in providing additional perspective. On the other hand, having the solutions available for the homework may detract from part of the value of the problems, because there is no substitute in the learning process for independent thought. Therefore, it is likely that the best use of the problems and solutions would involve their utilization in mix with original problems created by the course instructor.

W. A. Cramer
February 1, 1991
West Lafayette, IN

Solutions to Homework Problems

CHAPTER 1

$dE = dQ + dW$
$dE = dQ - pdV$, for p-V work;
$dQ = dE + pdV$, and
$dQ = d(E + pV)$, if p is constant, or
$dQ_p = d(E + pV) = dH$
where $H \equiv E + pV$.

$\Delta E = \Delta Q + \Delta W$
$= 20$ kJ

(a) $\Delta Q = +40$ kJ
$\Delta W = -20$ kJ
(b) $\Delta Q = +30$ kJ
$\Delta W = -10$ kJ; because work done by the system is negative in sign, more work is done in the reversible transition (a).

$G \equiv E + pV - TS$
$\partial G_T = \partial E_T + pdV + Vdp - TdS$; at constant temperature
$\partial E_T = dQ + dW$; first law
$= dQ - pdV$; only p-V work
$= TdS - pdV$; reversible, $dQ = TdS$.
Substituting for ∂E_T,
$\partial G_T = Vdp$.

From p. 8, $\Delta S = 2.3 \, R \log \left(\dfrac{V_2}{V_1}\right) = 2.3 \, R \log \left(\dfrac{V_2}{V_1}\right) = 2.3 \, R \log_{10} 10 = 2.3 \, R.$

5.(a) $\Delta S_{melt} = \Delta Q_{melt}/T_m = n\, \Delta Q^o(melt) / T_m$; n=9/18=0.5.
  =(0.5)(5980) / (273) = 11 J / °K; $T_m = 273$ ° K.

b) $\Delta S_{expan} = nR\ln\left(\frac{V_2}{V_1}\right) = 11$ J / ° K;

then, $\ln\left(\frac{V_2}{V_1}\right)$ = 11 / (0.5) (8.3) = 2.65 and V_2 / V_1 = 14.

6. $n_A A + n_B B \rightarrow n_C C + n_D D$;

$\overline{G}_A = \overline{G}_A^o + RT \ln (A) \equiv$ molar free energy of 'A' at the concentration used in the reaction.

$\overline{G}_B = \overline{G}_B^o + RT \ln (B) \equiv$ molar free energy of 'B' at the concentration used in the reaction.

$\overline{G}_C = \overline{G}_C^o + RT \ln (C) \equiv$ molar free energy of 'C' at the concentration used in the reaction.

$\overline{G}_D = \overline{G}_D^o + RT \ln (D) \equiv$ molar free energy of 'D' at the concentration used in the reaction.

If the mole ratio of A : B : C : D in the reaction is $n_A : n_B : n_C : n_D$, then

$\Delta G = (n_C \overline{G}_C + n_D \overline{G}_D) - (n_A \overline{G}_A + n_B \overline{G}_B)$; $\Delta G \equiv G_{products} - G_{reactants}$

$= n_C (\overline{G}_C^o + RT \ln C) + n_D (\overline{G}_D^o + RT \ln D)$
$\quad - \left[n_A (\overline{G}_A^o + RT \ln A) + n_B (\overline{G}_B^o + RT \ln B) \right]$
$= \Big\{ (n_C \overline{G}_C^o + n_D \overline{G}_D^o) - (n_A \overline{G}_A^o + n_B \overline{G}_B^o)$
$\quad + \left[n_C RT \ln (C) + n_D RT \ln (D) \right]$
$\quad - \left[n_A RT \ln (A) + n_B RT \ln (B) \right] \Big\}$

$= \Delta G^o + RT \ln \dfrac{(C)^{n_C} (D)^{n_D}}{(A)^{n_A} (B)^{n_B}}$

where $\Delta G^o = (n_C \overline{G}_C^o + n_D \overline{G}_D^o) - (n_A \overline{G}_A^o + n_B \overline{G}_B^o)$

= the difference in standard free energies of products and reactants.

At equilibrium, $\Delta G = 0$ and $\dfrac{(C)^{n_C} (D)^{n_D}}{(A)^{n_A} (B)^{n_B}} \equiv K_{eq}$, by definition

then, $\Delta G^o = -RT \ln K_{eq}$; Q.E.D.

(a) $\Delta G^O = -2.3RT \log K_{eq}$;
7.5 kcal / mol = -1.36 $\log K_{eq}$;
5.5 = $\log K_{eq}$;
$K_{eq} = 3.2 \times 10^5$.

(b) $-7.5 = -16.4 - 298(\Delta S^O)$;
$\Delta S^O = -8,900 / 298 = -30$ cal / °K-mol.

$\Delta G^O = \Delta H^O - T\Delta S^O \le 0 = 40,100 - T(107.4)$.
$T \ge 40,100 / 107.4 = 373$ °K.

$\Delta G^O = \Delta H^O - T\Delta S^O = -11.9 - (298)(-25.8) / 1000 = -11.9 + 7.7 = -4.2$ kcal/mol.

Chymotrypsin(native) -> chymotrypsin(denatured);
[(product)/(reactant)]$_{eq}$ = 1.
$\Delta G^O = -RT \ln K_{eq} = \Delta H^O - T_m \Delta S^O = 0$.
$T_m = \Delta H^O / \Delta S^O = 317$ °K = 44 °C.

. G-6-P -> Glucose + P$_i$;

$\Delta G = \Delta G^O + 2.3RT \log \left(\dfrac{\text{products}}{\text{reactants}}\right)$

$= -13.8 + 2.3\ (8.31)\ (310) \log \left(\dfrac{(5 \times 10^{-3})^2}{10^{-3}}\right)$, at 37°C, in kJ/mol;

$= -13.8 + 5.93 \log (25 \times 10^{-3})$
$= -13.8 + 5.93\ (-1.60)$
$= -13.8 - 9.5$
$= -23.3$ kJ/mol = -5.6 kcal/mol (37 °C).

(a) $\Delta G^O = \Delta H^O - T\Delta S^O$;
$4,640 = 0 - T\Delta S^O$
$\Delta S^O = -15.9$ cal/°K-mol, because of the "order" of the H$_2$O molecules forced to be in contact with the non-polar molecule in aqueous solvent.

(b) $\Delta G^O = \Delta H^O - T\Delta S^O$;
$-600 = +21,000 - 298 \Delta S^O$; [n. b., the signs given for ΔH^O and ΔG^O in the problem for the folding reaction and should be reversed for the denaturation and unfolding]; then,
$\dfrac{16 \times 10^4}{-298} = \Delta S^O = +72.5$ cal / °K-mol for the thermal denaturation.

13.

Amino Acid	ΔG^0(cal/mol)	K_{eq}
Glycine	+394	0.52
Alanine	-51	1.09
Leucine	-198	1.39
Valine	-247	1.52
Glutamate	-57	1.10
Lysine	-6	1.01

All but glycine will form > 50% alpha helix.

14.

Hairpin → coil	T(°C)	1-α	α	$K_{eq}=(1-\alpha)/\alpha$	$\Delta G° = -2.3 RT \log K$
α → (1-α)	10	0.04	0.96	0.041	+1.80 kcal/mol
	20	0.10	0.90	0.11	+1.29
	30	0.26	0.74	0.34	+0.65
	40	0.50	0.50	1.0	0
	50	0.70	0.30	2.3	-0.55
	60	0.85	0.15	5.7	-1.14
	70	0.90	0.10	9.0	-1.48
	80	0.95	0.05	19.0	-2.07

15. Use R = 0.082 atmosphere- liter/ ° K - mol
p = RTc = (0.082) (0.1) (298) = 2.44 atmospheres, at 25 ° C.

16. (a) ΔG^0 = -2.3RT log K_{eq} = -1.37 log K_{eq}, in kcal/mol at 27 ° C;
Initially, K_{eq} = K_o ; if the final value of K_o = 10 K_o ,
then, $(\Delta G^0)_{final}$ = -1.37 log (10K_o)
and, $(\Delta G^0)_{final} - (\Delta G^0)_{initial}$ = -1.37 { log(10K_o) - log (K_o) }
    = -1.37 log (10) = -1.37 kcal/mol.
Thus, a 10-fold change in the value of K_{eq} causes a change of 2.3 RT = -1.37 kcal/mol the value of ΔG^0.

(b) If ΔG^0 becomes more negative by 2.3 RT, then K_{eq} has increased by 10-fold, as argued in (a).
[*n.b.*, 2.3RT = 1.37 kcal/mol at 27 ° C, not 1.36 as stated in problem]

$\Delta G = \Delta G^O + -2.3 RT \log \left(\frac{products}{reactants}\right)$; $2.3RT = 1.25$ kcal/mol at $0\,°C$

(a) if $(p) = (r) = 5$ mM, then
$\Delta G = \Delta G^O = -1.25$ kcal/mol, and the direction is
$r \to p$; another way to visualize this result is that $(p)/(r) = 10$ at equilibrium because $\Delta G^O = -1.25$ kcal/mol.

(b) $\Delta G = \Delta G^O + -2.3RT \log \left(\frac{products}{reactants}\right)$
$= -1.25 + 1.25\log\left(\frac{0.91}{9.1}\right) = -2.5$ kcal/mol and the direction is $r \to p$.

(c) $\Delta G = -1.25 + 1.25\log\left(\frac{9.1}{0.91}\right) = -1.25 + 1.25 = 0$, so that the reaction is at equilibrium and neither direction is favored.

(d) $\Delta G = -1.25 + 1.25\log\left(\frac{9.11}{0.09}\right) = -1.25 + 2.50 = +1.25$, so that the reaction proceeds from $p \to r$.

(a) $\Delta\mu^o = -1934 - 771(2)$; for ethane, C_2H_6, and
$\Delta\mu^o = -3,476$ cal/mol.
(b) $\Delta\mu^o = -1934 - 771(5)$; for pentane, C_5H_{12}, and
$\Delta\mu^o = -5,789$ cal/mol.

(a) $\Delta G = -1.36\,(pK - pH) = -1.36(\,3.7 - 7) = -1.36\,(-3.3) = -4.49$ kcal/mol.
(b) $\Delta G^o = \Delta H^O - T\Delta S^O = 16\,(-5) - (25) = -105$ kcal/mol.

$\Delta G^o = \Delta H^O - T\Delta S^O$;
$-8 = -4 - T\Delta S^O$;
$T\Delta S^O = 4$ kcal/mol;
$\Delta S^O = +13.4$ cal/$°$K- mol, or $+13.4$ e. u.; ΔH^O arises from the charge repulsion of the β,γ phosphates of the ATP; ΔS^O arises from the increase in number of resonant forms of the orthophosphate product, and the decrease in the degree of order of the water bound to the less highly charged products.

Firstly, because we know that ΔH^O is negative, we know that K_{eq} is larger at $25\,°C$ than at $37\,°C$.

$\frac{d}{dt}(\ln K_{eq}) = \frac{\Delta H^O}{RT^2}$;

$d\,(\ln K_{eq}) = \frac{\Delta H^O}{RT^2}\,dT = \frac{\Delta H^O}{R}\,\frac{dT}{T^2}$;

$\int d(\ln K_{eq}) = \frac{\Delta H^o}{R} \int \frac{dT}{T^2} = \frac{\Delta H^o}{R} (\frac{1}{T_1} - \frac{1}{T_2})$; n. b., $\int \frac{1}{x^2} = -\frac{1}{x_2} + \frac{1}{x_1}$;

where states 1 and 2 refer to T = 25 ° C and 37 ° C, respectively. Integrating from T_1 to T_2,

$\ln K_{eq}(37 °C) - \ln K_{eq}(25 °C) = \ln \frac{K_{eq}(37 °C)}{K_{eq}(25 °C)} = \frac{\Delta H^o}{R}(\frac{1}{T_1} - \frac{1}{T_2}) = \frac{-4000}{1.99}(\frac{1}{310} - \frac{1}{298})$

$= -0.252$; then, $\frac{K_{eq}(37 °C)}{K_{eq}(25 °C)} = 0.78$, $K_{eq}(25 °C) = 2.56 \times 10^5$,

and $\Delta G^o = -1.36 \log_{10}(2.56 \times 10^5) = -7.36$ kcal/mol, at 25°C.

22. ATP + H_2O -> ADP + P_i, with $K_{eq} = 10^6$, so that the reaction is shifted far to the right at equilibrium.

Concentration of ATP = 10^{-3} M.

Let x ≡ the concentration of ATP at equilibrium.

Then, (10^{-3} - x) = the concentration of ADP and P_i at equilibrium.

and $\left[\frac{(ADP)(P_i)}{(ATP)}\right]_{eq} = \frac{(10^{-3} - x)^2}{x} = K_{eq} = 10^6$.

To solve this eqn., assume that x is very small (i.e., << 10^{-3} M). Therefore, if we neglect the concentration of x in the numerator,

$\frac{(10^{-3})^2}{x} = \frac{10^{-6}}{x} = 10^6$, and x = 10^{-12} M.

23. (a) From eqn. (27), Chapt. 1, $\log_{10}\left(\frac{S_i}{S_o}\right) = n \Delta pH - (n + z)\frac{\Delta\psi}{Z}$, with $Z \equiv \frac{2.3\, RT}{F}$,

$\log_{10}\left(\frac{S_i}{S_o}\right) = (2)(1) - (2 + (-2))\frac{\Delta\psi}{Z}$;

$\log_{10}\left(\frac{S_i}{S_o}\right) = 2$, and

$\left(\frac{S_i}{S_o}\right) = 10^2$.

(b) From (a), $\log_{10}\left(\frac{S_i}{S_o}\right) = 2 - (2 - (0))(\frac{-59}{59}) = 2 - (2)(-1) = 4$.

$\left(\frac{S_i}{S_o}\right) = 10^4$.

24. (a) From eqn. (28), $\log_{10}\left(\frac{S_i}{S_o}\right) = -z\frac{\Delta\psi}{Z} = -(-1)\frac{\Delta\psi}{Z} = \frac{\Delta\psi}{Z}$.

(b) If $\Delta\psi = +59$ mV, $\log_{10}\left(\frac{S_i}{S_o}\right) = (\frac{59}{59}) = 1$, and $(\frac{S_i}{S_o}) = 10$.

If $\Delta\psi = +177$ mV, $\log_{10}\left(\frac{S_i}{S_o}\right) = 3$, and $(\frac{S_i}{S_o}) = 10^3$.

increase in the accumulation ratio is $10^3 - 10 = 990$.

a) Starting with the hexokinase reaction:
glucose + ATP -> glucose-6-phosphate + ADP, $\Delta G^{o'} = -3.3$ kcal/mol,
reaction glucose-6-phosphate + H_2O -> glucose + phosphate, $\Delta G^{o'} = -4.4$ kcal/mol, must be coupled to it to yield ATP + H_2O -> ADP + P_i, $\Delta G^{o'} = -7.7$ kcal/mol.

b) Starting with the galactokinase reaction:
galactose + ATP -> galactose-1-P + H_2O, $\Delta G^{o'} = -1.9$ kcal/mol,
reaction galactose-1-P + H_2O -> galactose + P_i, $\Delta G^{o'} = -5.0$ kcal/mol,
be coupled to it to yield ATP + H_2O -> ADP + P_i, $\Delta G^{o'} = -6.9$ kcal/mol.

From p. 27, K_{eq} (pH 7, 37 °C) = 1.6×10^5.
From prob. #21, K_{eq} (pH 7, 25 °C) = 2.05×10^5,
at ΔG^o(pH 7, 25 °C) = $-1.36 \log_{10}(2.05 \times 10^5) = -1.36 (5.31) = -7.22$ kcal/mol.
From Fig. 1.6C, the $\Delta G(ATP)$ at pH 8.0 is approximately 1.4 kcal/mol more negative at 8.0 compared to 7.0, because of charge repulsion of the negative phosphates.
Therefore, ΔG^o(pH 8, 25 °C) = -8.6 kcal/mol. [n. b., the value of -8.2 kcal/mol on p. 466 assumed a difference of -1 kcal/mol between the ΔG^o values at pH 8 vs. 7].

From eqn. (39), $\dfrac{J_p}{J_e} = 3[\, 0.97 + 3\,(\dfrac{42}{-210})\,] / [\, 1 + 0.97\,(3)\,(\dfrac{42}{-210})\,]$

$= 3[\, 0.97 - 0.6] / [\, 1 - 0.582\,] = 3\,(\dfrac{0.37}{0.418}) = 2.66.$

a) $G \equiv E + pV - TS;$
$dE = dQ + dW;$ first law.
$dW = -pdV + dW_{other};$ pressure-volume plus other work.
$dG = dE + pdV + Vdp - TdS - SdT.$
$G_{P,T} = dE + pdV - TdS;$
tituting for dE,
$G_{P,T} = (\,dQ + dW) + pdV - TdS\,;$
use $dQ = TdS,$
$G_{P,T} = (\,TdS - pdV + dW_{other}) + pdV - TdS = dW_{other}.$

b) $\Delta W_{other} = $ (charge)(electrical potential change)
$= nFz\,\Delta E,$ where E in this problem is the electrical potential and the notation erwise defined as in section 1.9;
$\Delta W_{other} = -nF\,\Delta E,$ if $z = -1$ for an electron;
$= -F\,\Delta E,$ for 1 mol of electrons = 9.65×10^4 coulombs; then,
$\Delta W_{other} = -9.65 \times 10^4\,(1)$ coulomb-volt for a 1 volt change of potential;
$= -9.65 \times 10^4$ joules/Faraday, because 9.65×10^4 coulomb-volt $\equiv$ 1 joule,
$= -23.06$ kcal/Faraday, or -23.06 kcal/mol.

CHAPTER 2

29. Using R = 1.99 cal/°K-mol, and F = 23.06 kcal/mol-volt, the values of the Nernst Coefficient are:

°C	2.3 RT (kcal/mol)	2.3 RT/F (mV) [n = 1]	[n = 2]
0	1.25	54.2	27.1
25	1.36	59.1	29.6
37	1.42	61.5	30.8

30. $\Delta G^o = -RT \ln K_{eq} = -nF\Delta E^o$; [n.b., there is a mistake in the right-hand side of eqn (1) chapt. 2, which should read $-nF\Delta E$]; then,

$$\Delta E^o = \frac{RT}{nF} \ln K_{eq} = 2.3 \frac{RT}{nF} \log_{10} K_{eq} = \frac{59}{n} \log_{10} K_{eq}, \text{ at } 25\,°C, \text{ in mV}.$$

31. Half-cell reactions: $b(ox) + e^- \rightarrow b(red)$ $E_m = +0.05$ V
   $c(ox) + e^- \rightarrow c(red)$ $E_m = +0.25$ V

   $b(red) + c(ox) \rightarrow b(ox) + c(red)$ $\Delta E_m = 0.25 - 0.05 = 0.2$ V

then, $\Delta E = \Delta E_m - 0.059 \log \frac{b(o)}{b(r)} \cdot \frac{c(r)}{c(o)}$

$= 0.2 - 0.059 \log \left(\frac{1}{10} \cdot \frac{1}{10}\right) = 0.2 - 0.059 \log (10^{-2})$

$= +0.32$ V

(a) $\Delta G = -nF\Delta E = -(1)(23)(0.32) = -7.3$ kcal/mol for n = 1
(b) $= -14.6$ kcal/mol for n = 2.

32. Energy is made available for ATP synthesis through a coupled reaction involving the transfer of 2 electrons between cytochromes b and c.

2 cyt b (r) + 2 cyt c (o) + ADP + $P_i \rightarrow$ 2 cyt b (o) + 2 cyt c (r) + ATP + H_2O;
$\Delta G_{tot} = \Delta G_1 + \Delta G_2$, where
ΔG_1 is the energy made available from the electron transport reaction, and
ΔG_2 = the energy used for ATP synthesis;
$\Delta G_1 = -14.6$ kcal/mol from prob. #31;
Because $\Delta G_{tot} = 0$ if all of the energy available is used for ATP synthesis,
$\Delta G_2 = -\Delta G$ (ATP) $= +14.6$ kcal/mol;
ΔG (ATP) $= \Delta G° + 2.3\,RT \log_{10} \frac{(ATP)}{(ADP)(P_i)}$;

If $\Delta G° = +8$ kcal/mol,
$$\Delta G \text{ (ATP)} = 8 + 1.36 \log_{10} \frac{\text{(ATP)}}{(10^{-5})(10^{-3})},$$
and $14.6 = 8 + 1.36 \log_{10} (10^8) \text{(ATP)}$
$4.87 = \log_{10} (10^8) \text{(ATP)}$

Then, $\text{(ATP)} = 10^{-3.13}\text{M} = 7.5 \times 10^{-4}\text{M}$ = the highest concentration of ATP at which the reaction can be run.

$E = E_m - \dfrac{59}{n} \log_{10}\left(\dfrac{\text{(red)}}{\text{(ox)}}\right)$; $Q \equiv \dfrac{\text{(red)}}{\text{(ox)}}$, and $\dfrac{\text{(red)}}{\text{((red)+(ox))}} = \dfrac{Q}{Q+1}$, then,

$Q = \dfrac{(E_m-E)}{59/n}$;

or n = 1, when

Q =	$\dfrac{Q}{Q+1}$ =	then E(mV) =
1	0.5	E_m = +260 mV
2, $\dfrac{1}{2}$	$\dfrac{2}{3}, \dfrac{1}{3}$	E_m- 18, E_m+ 18 = 242, 278 mV
4, $\dfrac{1}{4}$	$\dfrac{4}{5}, \dfrac{1}{5}$	E_m- 36, E_m+ 36 = 224, 296 mV
10, $\dfrac{1}{10}$	$\dfrac{10}{11}, \dfrac{1}{11}$	E_m- 59, E_m+ 59 = 201, 319 mV
100, $\dfrac{1}{100}$	$\dfrac{100}{101}, \dfrac{1}{101}$	E_m- 118, E_m+ 118 = 142, 378 mV,

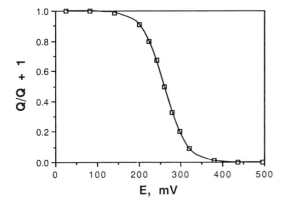

(b)

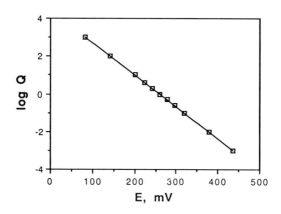

34. For n =2, when

Q =	$\dfrac{Q}{Q+1}$ =	then E(mV) =
1	0.5	E_m = +260 mV
2, $\dfrac{1}{2}$	$\dfrac{2}{3}$, $\dfrac{1}{3}$	E_m- 9.5, E_m+ 9.5 = 250.5, 269.5 mV
4, $\dfrac{1}{4}$	$\dfrac{4}{5}$, $\dfrac{1}{5}$	E_m- 18, E_m+ 18 = 242, 278 mV
10, $\dfrac{1}{10}$	$\dfrac{10}{11}$, $\dfrac{1}{11}$	E_m- 29.5, E_m+ 29,5 = 230.5, 289.5 mV
100, $\dfrac{1}{100}$	$\dfrac{100}{101}$, $\dfrac{1}{101}$	E_m- 59, E_m+ 59 = 201, 319 mV,

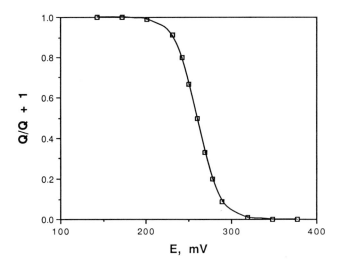

$E_{mh} = E^o - 59(pH)$ at 25 °C ;
$\Delta E_m / \Delta pH = -59$ mV/pH unit ;
fore, E_m (pH 9) = E_m (pH 7) - 118 mV,
E_m (pH 9) = 142 mV; the graph of $\frac{Q}{Q+1}$ vs E at pH 9 looks the same as that of em 34 if 118 mV is subtracted from the values on the abscissa of that graph.

(a) The redox reactions in the three pH ranges of (i) high, (ii) intermediate, and (iii) low pH, relative to the pK_{ox} and pK_{red} values as described in section 2.5, are:

(i) $C_o + e^- \to C_r$; $E_m = E_{mb}$
(ii) $C_o + e^- + H^+ \to C_rH$; $\Delta E_{mh}/\Delta pH = -59$ mV
(iii) $C_oH + e \to C_rH$; $E_m = E_{ma}$;

in addition, $K_o = \frac{(ox)(H^+)}{(oxH)}$, $K_r = \frac{(red)(H^+)}{(red\ H)}$;   (i)

$E_h = E_{mh} - 59 \log \frac{(red) + (red\ H)}{(ox) + (oxH)}$; substituting from (i),

$E_h = E_{mh} - 59 \log \frac{[(red\ H)(K_r)/H^+ + (red\ H)]}{[(oxH)(K_o)/H^+ + (oxH)]}$,   (ii)

$= E_{mh} - 59 \log \left(\frac{redH}{oxH}\right) \left(\frac{1 + K_r/H^+}{1 + K_o/H^+}\right)$,

$= E_{mh} - 59 \log \left(\frac{redH}{oxH}\right) - 59 \log \left(\frac{1 + K_r/H^+}{1 + K_o/H^+}\right)$;   (iii)

in the low pH (≤ 2) region, $E_h = E_{ma} - 59 \log \left(\frac{redH}{oxH}\right)$; (iv)

because there is a value of E_h at which expressions (iii) and (iv) are equal,

$$E_{mh} = E_{ma} + 59 \log \left(\frac{1 + K_{red}/H^+}{1 + K_{ox}/H^+}\right) \text{ for pH} > 2;$$

for this problem, $E_{mh} = 500 + 59 \log \left(\frac{1 + 10^{-9}/H^+}{1 + 10^{-3}/H^+}\right)$,

= 482 mV, at pH 3
= 439 mV, at pH 4
= 382 mV, at pH 5
= 323 mV, at pH 6
= 264.5 mV, at pH 7
= 207.5 mV, at pH 8
= 164 mV, at pH 9
= 148.4 mV, at pH 10
= 146 mV, at pH 11.

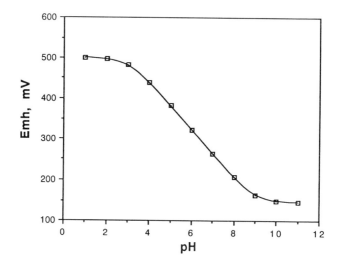

(b) E_{mb} is the asymptotic value, 146 mV, of the midpoint potential at high pH.

(c) In the acidic pH region,
$(C_o\text{-}L) + e^- \rightarrow C_r + L$;

$$E_L = E_{ma} - 59 \log_{10} \frac{(C_r)(L)}{(C_o\text{-}L)} = E_{ma} - 59 \log (L) - 59 \log \frac{(C_r)}{(C_o\text{-}L)};$$

then, $E_{mL} = E_{ma} - 59 \log (L)$ at this midpoint, and
therefore E_{mL} decreases by 118 mV when (L) increases by 100-fold.

There are 4 pathways from the initial state to the final state. Calculate the ΔG, or ΔG_m, for the transition calculated in terms of F, the Faraday constant.

(i) $O_2 + 4e^- + 4H^+ \rightarrow 2H_2O$, for a 4 e^- reaction, and
$E_{m7} = +0.82$ V;
$\Delta G = -nF\Delta E = -4(F)(-82) = -3.28$ F.

(ii) $O_2 + 2e^- + 2H^+ \rightarrow H_2O_2$, for a 2$e^-$ reaction to a peroxide intermediate, and
$E_{m7} = +0.31$V;
$\Delta G_1 = -(2)(F)(-31) = -.62$ F;
$H_2O_2 + 2e^- + 2H^+ \rightarrow 2H_2O$, for the 2$e^-$ reaction from peroxide to water, and
$E_{m7} = +1.35$ V;
$\Delta G_2 = -(2)(F)(1.35) = -2.70$ F;
$\Sigma \Delta G = \Delta G_1 + \Delta G_2 = -3.32$ F.

(iii) $O_2 + e^- \rightarrow O_2^{\cdot -}$, for a 1$e^-$ reaction to the superoxide anion radical, and
$E_m = -.33$ V;
$\Delta G_1 = +.33$ F;
$O_2^- + e^- + 2H^+ \rightarrow H_2O_2$, for a 1$e^-$ reaction to peroxide, and
$E_{m7} = +.94$ V;
$\Delta G_2 = -.94$ F;
$H_2O_2 + 2e^- + 2H^+ \rightarrow 2H_2O$, for the 2$e^-$ reaction from water to peroxide, and
$E_{m7} = +1.35$ V;
$\Delta G_3 = -(2)(F)(1.35) = -2.70$ F;
$\Sigma \Delta G = \Delta G_1 + \Delta G_2 + \Delta G_3 = -3.31$ F.

(iv) $O_2 + e^- \rightarrow O_2^-$; 1 e^- reaction
$O_2^- + 3e^- + 4H^+ \rightarrow 2H_2O$; 3 e^- reaction;
$\Delta G = -3.27$ F, calculated as above.

Thus, the ΔG's for the four different pathways are the same within the experimental errors ($\sim \pm 1\%$ of the mean ΔG).

$E_{m1} - E_{m2} = 59 \log_{10} K_s = 59 \log_{10} (4 \times 10^{-2})$
$= 59 (-1.4) = -83$ mV

$E_{m1} - E_{m2} = -83$

$\dfrac{E_{m1} + E_{m2} = +90}{2}$

$\Rightarrow 2 E_{m1} = 97$ mV

$E_{m1} = +48.5$ mV

$E_{m2} = +131.5$ mV

39. $E = h\nu = h\dfrac{c}{\lambda}$; $\lambda = 865$ nm for reaction center.

$$= \dfrac{(6.6 \times 10^{-34})(3 \times 10^8)}{8.65 \times 10^{-7}} = 2.29 \times 10^{-19} \text{ joules}$$

Since 1 electron-volt = 1.6×10^{-19} joules, the photon energy is 1.43 eV. Without correcting for vibrational relaxation, the midpoint potential of the excited state is $0.36 - 1.43 = -1.07$ V.

——— E_m, P/P*, -1.07 V

——— ground state, P+/P, $E_m = +0.36$ V

40. Azurin (red) + c_{553} (ox.) $\xrightarrow{k_{12}}$ Azurin (ox) + c_{553} (red);

For the azurin from *Ps. aeruginosa*, $E_m = +304$ mV,
$k_{11} = 9.9 \times 10^5$ M^{-1}s^{-1};

For the c_{553} from *B. filiformis*, $E_m = +335$ mV,
$k_{22} = 2.8 \times 10^8$ M^{-1} s^{-1};

Using $\Delta E_m = 59 \log_{10} K_{eq}$, at 25°C,
$(335-304) = 59 \log_{10} K_{eq}$,
$31 = 59 \log_{10} K_{eq}$,
$\log_{10} K_{eq} = 0.53$, and
$K_{eq} = 3.4$; then,

$k_{12} = \sqrt{k_{11} k_{22} k_{eq}} = \sqrt{(9.9 \times 10^5)(2.8 \times 10^8)(3.4)}$

$= \sqrt{94.2 \times 10^{13}} = \sqrt{9.42 \times 10^{14}} = 3.1 \times 10^7$ M^{-1} s^{-1}

41. $k_{12} \approx \sqrt{k_{11} k_{22} K_{eq}}$

The k_{11} for all reactions is 2.8×10^8 M^{-1} s^{-1}, and the four values of k_{22} are given. From Table 3, the E_m values are:

B. filiformis,	+335 mV
c_{551},	+286 mV
azurin,	+304 mV
plastocyanin,	+350 mV
horse heart c,	+260 mV

$\Delta E_m = \dfrac{59}{n} \log_{10} K_{eq} = 59 \log_{10} K_{eq}$ since n = 1.

	For c_{553} →	c_{551}, →	azurin, →	plastocyanin, →	horse
ΔE_m =		-49 mV	-31 mV	$+15$ mV	-75 mV
log K_{eq} =		-0.83	-0.52	$+0.255$	-1.27
K_{eq} =		0.15	0.30	1.8	0.054
k_{22} =		4.6×10^7	9.9×10^5	6.6×10^2	1.5×10
k_{11} =		2.8×10^8	2.8×10^8	2.8×10^8	2.8×10
$k_{11}k_{22}K_{eq}$ =		1.9×10^{15}	8.3×10^{13}	3.3×10^{11}	2.3×10
k_{12} =		4.4×10^7	9.1×10^6	5.7×10^5	4.8×10
		↑			↑
		(fastest)			(slowest)

When a graph is made of the pairs of $\log_{10}k_{12}$ and ΔE_m values, as shown below, the fit to the data is $\log_{10}k_{12} = 5.18 + \dfrac{\Delta E_m(mV)}{143}$; the correlation coefficient, $R^2 =$ ($R^2 = 1.0$ for perfect fit).

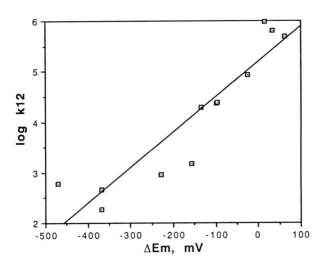

(a) $\Delta G^{\ddagger} = (\Delta G^{\circ} + \lambda_{12})^2/4\lambda_{12}$; $\lambda_{12} = (\lambda_{11} + \lambda_{22})/2 = 23$ kcal/mol
$= (-23 + 23)/4(23)$
$= 0$

(b) $k_{12} = \nu_{12}\, k(r)\, e^{-\Delta G^{\ddagger}/RT}$
$= \nu_{12}\, e^{-\alpha(r-r_0)}\, e^{-\Delta G^{\ddagger}/RT}$
$= 10^{13}\, e^{-\alpha(r-r_0)}\, e^{-\Delta G^{\ddagger}/RT}$

(c) $10^6 = 10^{13}\, e^{-\alpha(r-r_0)}$ since $-\Delta G^{\ddagger} = 0$
$= 10^{13}\, e^{-0.8(r-r_0)}$

$10^7 = e^{-0.8(r-r_0)} = 10^{-\left(\frac{0.8}{2.3}\right)(r-r_0)}$

$-7 = -[(0.8)/2.3]\, (r-r_0)$; taking the $\log_{10}$ of both sides,
$(r-r_0) = 20.1$ Å.

44. Derive eqn. (35): The energies, **E**, are really free energies, **G**.
$\Delta G^* = -2.3\ RT \log K^* = 2.3\ RT \cdot pK^*$ for dissociation from the excited state;
$\Delta G_o = -2.3\ RT \log K^\circ = 2.3\ RT \cdot pK^\circ$ for dissociation from the ground state;
then, $pK^* - pK^\circ = (\Delta G^* - \Delta G_o)/2.3\ RT$.

Adding energies, $\Delta G_{HA} + \Delta G^* = \Delta G_o + \Delta G_{A^-}$, where ΔG_{HA} and ΔG_{A^-} are the fr energies associated with the excited state of the neutral and anionic forms, assayed by t energy of the fluorescence emission; the respective emission peaks of the neutral and anionic peaks are at 445 and 510 nm; and $pK^* - pK^\circ = (\Delta G_{A^-} - \Delta G_{HA})/2.3\ RT$
$= N(h\nu_{A^-} - h\nu_{HA})/2.3\ RT$, where N is Avogadro's number; then,
since R/N = k, the Boltzmann constant,

$pK^* - pK^\circ = (h\nu_A - h\nu_{HA})/2.3\ RT$

$= \dfrac{hc}{2.3\ kT}\left(\dfrac{1}{\lambda_{A^-}} - \dfrac{1}{\lambda_{AH}}\right)$

$= \dfrac{(6.6 \times 10^{-27})(3 \times 10^{10})}{2.3 \times 1.38 \times 10^{-16} \times 298}\left\{\dfrac{1}{5.1 \times 10^{-5}} - \dfrac{1}{4.45 \times 10^{-5}}\right\}$

$= \dfrac{19.8 \times 10^{-22}}{2.3 \times 1.38 \times 2.98 \times 10^{-24}}\left[\dfrac{-0.65}{(5.1)(4.45)}\right] = -6.0$

Thus, if the pH of the ground state is 7.0, that of the excited state is 1.0.

45. $v = \dfrac{m \cdot \Delta\psi}{d} = \dfrac{(10^{-4})(8 \times 10^{-2})}{4 \times 10^{-7}} = 20$ cm/s

$t = \dfrac{d}{v} = \dfrac{4 \times 10^{-7}}{20} = 2 \times 10^{-8}$ s.

CHAPTER 3

$$\Delta pH = \log_{10}\left[\frac{Q}{(1-Q)} \cdot \frac{1}{v}\right]; \quad \text{eqn (1e) of chapt. 3;}$$

$$Q = 1 - \frac{F_e}{F_d} = 0.5; \quad v = 0.05 \frac{ml}{mg\ Chl} \cdot 0.02 \frac{mg\ Chl}{ml\ soln} = 10^{-3};$$

then, $\Delta pH = \log_{10}\left[\frac{0.5}{0.5} \cdot \frac{1}{10^{-3}}\right] = \log_{10} 10^3 = 3.$

Volume of cell = $4/3\ \pi\ r^3 = 4/3\ \pi\ (5 \times 10^{-5})^3 = 5.22 \times 10^{-13}\ cm^3$
  $= 5.22 \times 10^{-16}\ l;$
therefore the number of K^+ at 0.2 M in this space is
$(0.2)\ (6 \times 10^{23})\ (5.22 \times 10^{-16}) = 6.26 \times 10^7;$
the surface area of cell = $4\pi \cdot (5 \times 10^{-5})^2 = 3.14 \times 10^{-8}\ cm^2$
  $= 3.14 \times 10^8\ \text{Å}^2.$
It was calculated in section 3.5 that a potential of 100 mV is generated when 0.6 charges move across $10^4\ \text{Å}^2$ of membrane surface. Therefore, the number of charges that move across the surface of the *E. coli* cell to generate $\Delta\psi = -100\ mV$

$$= (0.6)\ \frac{(3.16 \times 10^8)}{10^4} = 1.9 \times 10^4.$$

The fractional change in internal K^+ concentration is then $\frac{1.9 \times 10^4}{6.26 \times 10^7} = 3.0 \times 10^{-4}.$

Thus, it is fair to say that the diffusion potential is formed without a significant change in the internal K^+ concentration. It is interesting to consider what happens to the fractional change in internal ion concentration as the vesicle becomes smaller. At what size would the fractional change become significant?

Area of mitochondrion with 0.5 µ radius = $4\pi\ (5000)^2\ \text{Å}^2$
$= 12.56 \times 2.5 \times 10^7 = 3.14 \times 10^8\ \text{Å}^2$
From section 3.5, the number of H^+ extruded = $\frac{3.14 \times 10^8}{10^4} \times 0.6$
$= 1.9 \times 10^4.$

$\left(\frac{100}{1.9 \times 10^4}\right)^{1/2} = \frac{r}{5000};\ r = 5000 \times \frac{10}{1.37 \times 10^2}\ \text{Å} = 365\ \text{Å}$

and the diameter = 730 Å.

50. From section 3.4.2, $\Delta\lambda = \dfrac{\lambda_m^2}{2hc}\left\{[\Delta\alpha\,(\vec{E}_o + \vec{E})]\cdot\vec{E}\right\}$;

because energy density (i.e., energy/volume) in esu units = $E^2/8\pi$, the expression in cu[rly] brackets has dimensions of energy. Because hc has dimensions of (energy) (length), energy cancels out and the right-hand-side reduces to length, as anticipated.

Then, $\Delta\lambda = \dfrac{(5\times 10^{-5})^2}{2\,(6.6\times 10^{-27})\,(3\times 10^{10})}\;\dfrac{(4\times 10^{-21})}{1}\;\dfrac{(2\times 10^6 + 4\times 10^5)}{(300)}\;\dfrac{(4\times}{(3(}$

$= 27\times 10^{-7}\,\text{cm} = 27\,\text{Å} = 2.7\,\text{nm}$.

51. $\Delta G_{ATP} = n_{H^+}\cdot(\Delta\tilde{\mu}_{H^+})$; *vide*, section 3.6.2. $\Delta\tilde{\mu}_{H^+} = F\cdot\Delta\psi - 2.3\,RT\cdot\Delta p$
 14.2 kcal = $n_{H^+}\,(4.56)$; $= 0 - 1.36\,(-3.35);\;\Delta pH = pH_i -$
 $n_{H^+} = 3.11$; $= 4.56\,\text{kcal/mol}$.

 therefore, 3 H$^+$ are required for the synthesis of one molecule of ATP.

52. (a) Basal rate (μmol O$_2$/mg Chl-hr) = $\dfrac{0.03}{0.02}\,(60) = 90$ μmol/mg-hr.

 (b) If the O$_2$ evolution rate is stimulated by a factor of 3 for 1 min, then the amount of evolved by 20 μg Chl in 1 min = 90 nmol. Depending on whether a correction is mad[e] for basal electron transport or not, the electron transport rate is 120 or 180 nmol O/ml. 120 μM ADP (120 nmol/ml) is converted to ATP during this time, the ADP/O ratio is, respectively, 1.0 or 0.67.

CHAPTER 4

$\Delta A_{563-575} = [\Delta\varepsilon_{mM} (c) (l)] (0.5)$, where c is the concentration (mg/ml) and l is the path length (1 cm); then,
$c = (0.02 \text{ mg/ml}) (3 \times 10^{-9} \text{ mol/mg}) = 6 \times 10^{-11} \text{ mol/ml}$
$= 6 \times 10^{-8} \text{ M} = 6 \times 10^{-5} \text{ mM};$
$\Delta A_{563-575} = [(17) (6 \times 10^{-5}) (1)] (0.5) = 5.1 \times 10^{-4}.$

(a) $h\nu = g\beta H$
$(6.63 \times 10^{-34} \text{ J} \cdot \text{s}) (9.13 \times 10^{-9} \text{ s}^{-1}) = g(9.27 \times 10^{-24} \text{ J} \cdot \text{T}^{-1}) (0.11 \text{ T})$, and
$g = 5.94.$

If the 1/2 spin is common to both spin systems, and the β values scale with the mass, then formula (9) applies:

$h\nu = g\beta H$, with $\beta = \dfrac{1}{1836} (\beta_{electron}) = \dfrac{9.27 \times 10^{-24}}{1836}$, and $g = 2.79;$

(a) then, $\beta_{proton} = \dfrac{9.27 \times 10^{-24}}{1836} = 5.0 \times 10^{-27} \text{ J} \cdot \text{T}^{-1}.$

(b) $(6.63 \times 10^{-34}) \nu = (2.79) (5.0 \times 10^{-27}) (10^5)$
$\nu = 2.1 \times 10^8 \text{ Hz}.$

56. (a) $U = -\vec{\mu} \cdot \vec{E}$, the vector dot product, $= -\mu E \cos\theta = \dfrac{-\mu E}{2} = -310 (3.3 \times 10^{-30}$
$(\text{C} \cdot \text{m}) (3 \times 10^7 \text{V});$ [n.b., 1 debye = 3.3×10^{-30} coulomb-meters]
$= -3.1 \times 10^{-20} \text{ J}.$

(b) $kT = (1.38 \times 10^{-23} \text{ J}/°\text{K}) (298°\text{K})$ at $25°\text{C}$
$= 4.1 \times 10^{-21}$ J, so that the orientation energy is 7.5 times the thermal energy that would cause disorientation.

(a) From section 2.8,
$E_m = (E_{m1} + E_{m2})/2$, for the $2e^-$ reaction,
$= [-130 + (-30)]/2 = -80 \text{ mV}$

(b) $(E_{m1} - E_{m2}) = 59 \log_{10} K_s;$
$-130 - (-30) = 59 \log_{10} K_s;$
$(-100) = 59 \log_{10} K_s,$
$-1.69 = \log_{10} K_s$, and
$K_s = 2.05 \times 10^2.$